MICROBIAL ECOLOGY
OF THE OCEANS

WILEY SERIES IN
ECOLOGICAL AND APPLIED MICROBIOLOGY

EDITED BY

Ralph Mitchell
Division of Applied Sciences
Harvard University

ADVISORY BOARD

Ilan Chet
Faculty of Agriculture
Hebrew University of Jerusalem

Madilyn Fletcher
Belle W. Baruch Institute for Marine
  Biology and Coastal Research
University of South Carolina

Peter Hirsch
Institut für Algemeine
Mikrobiologie Universität Kiel

David L. Kirchman
College of Marine Studies
University of Delaware

Kevin Marshall
School of Microbiology
University of New South Wales

James T. Staley
Department of Microbiology
University of Washington

David White
Institute for Applied Microbiology
University of Tennessee

Lily Y. Young
Center for Agricultural Molecular
  Biology
Cook College, Rutgers University

RECENT TITLES

THERMOPHILES: General,
Molecular, and Applied Microbiology
Thomas D. Brock, Editor, 1986

INNOVATIVE APPROACHES TO
PLANT DISEASE CONTROL
Ilan Chet, Editor, 1987

PHAGE ECOLOGY
Sagar M. Goyal, Charles P. Gerba,
and Gabriel Bitton, Editors, 1987

BIOLOGY OF ANAEROBIC
MICROORGANISMS
Alexander J.B. Zehnder, Editor, 1988

THE RHIZOSPHERE
J.M. Lynch, Editor, 1990

BIOFILMS
William G. Characklis and Kevin C.
Marshall, Editors, 1990

ENVIRONMENTAL MICROBIOLOGY
Ralph Mitchell, Editor, 1992

BIOTECHNOLOGY IN PLANT
DISEASE CONTROL
Ilan Chet, Editor, 1993

ANTARCTIC MICROBIOLOGY
E. Imre Friedmann, Editor, 1993

EFFECTS OF ACID RAIN ON FOREST
PROCESSES
Douglas L. Godbold and Aloys
Hütterman, Editors, 1994

MICROBIAL TRANSFORMATION
AND DEGRADATION OF TOXIC
ORGANIC CHEMICALS
Lily Y. Young and Carl E. Cerniglia,
Editors, 1995

BACTERIAL ADHESION: Molecular
and Ecological Diversity
Madilyn Fletcher, Editor, 1996

EXTREMOPHILES: Microbial Life in
Extreme Environments
Koki Horikoshi and W.D. Grant,
Editors, 1998

WASTEWATER MICROBIOLOGY
2nd Edition
Gabriel Bitton, 1999

MICROBIAL ECOLOGY OF THE
OCEANS
David L. Kirchman, Editor, 2000

BIOFILMS II: Process Analysis
and Applications
James D. Bryers, Editor, 2000

MICROBIAL ECOLOGY OF THE OCEANS

Edited by

David L. Kirchman

Graduate College of Marine Studies
University of Delaware
Lewes, Delaware

WILEY-LISS

A JOHN WILEY & SONS, INC., PUBLICATION

New York · Chichester · Weinheim · Brisbane · Singapore · Toronto

This book is printed on acid-free paper. ⊗

Copyright © 2000 by Wiley-Liss, Inc. All rights reserved.

Published simultaneously in Canada.

For information and customer service call 1-800-CALL-WILEY

Library of Congress Cataloging-in-Publication Data:
Microbial ecology of the oceans / edited by David L. Kirchman.
    p. cm.
  Includes index.
  ISBN 0-471-29993-6 (cloth : alk. paper) -- ISBN 0-471-29992-8 (pbk.)
   1. Marine microbiology. 2. Marine ecology. 3. Carbon cycle (Biogeochemistry) I.
Kirchman, David L.

QR106.M53 2000
579'.177--dc21   99-045537

Printed in the United States of America.

10 9 8 7 6 5 4 3 2

CONTENTS

PREFACE

One of the most important findings in biological oceanography and aquatic ecology is that microbes, especially heterotrophic bacteria, are large and essential components of food webs and elemental cycles in the oceans and other aquatic systems. Although studies on microbes and microbial processes are well represented in several leading journals and a few symposium volumes, I felt that no book summarized the essentials of modern marine microbial ecology. *Microbial Ecology of the Oceans* is an attempt to correct this deficit.

Several chapters review selected topics in marine microbial ecology. But I want the book to be more than a collection of reviews. A reader new to the field should be able to pick up enough basics from this book to appreciate the importance of microbes in the oceans and to understand the questions that have been and are still being addressed by marine microbial ecologists. Some chapters were specifically designed to provide this basic information. Authors of all chapters were asked to write for a general audience even if space limitations meant that they could not cite every reference or cover all topics equally thoroughly.

I am particularly interested in hearing about areas of microbial ecology that are not adequately addressed in the book. E-mail can be sent directly to me at kirchman@udel.edu.

Each chapter was reviewed by other chapter authors and colleagues not involved with the book. Although all the authors contributed to this review process, I thank especially Barry and Ev Sherr, Peter J. le B. Williams, and Hugh Ducklow for doing more than their share of the reviewing. The following colleagues also helped with looking at one or more chapters: Dave Karl, Niels Jørgensen, Steve Wilhelm, Mike Pace, and Lars Tranvik. The chapter authors and I thank them for their service. Matt Cottrell looked over several chapters and helped with mailing off nearly all of them to Wiley while I was in Germany. Ralph Mitchell provided encouragement and advice at critical times during the assembly of this book. Connie Edwards at the University of Delaware and Luna Han at Wiley patiently dealt with e-mails, snail mail, and faxes in moving manuscripts from various parts of world to New York via Germany.

I acknowledge the financial support of the National Science Foundation and Department of Energy. The final stages of my work on the book were completed while I was a fellow at the Hanse Wissenschaftskolleg in Delmenhorst, Germany; I am most grateful for that fellowship. The Hedges, the Eglintons, other Hanse fellows, and my wife Ana Dittel did their best to distract me, but I thank them nevertheless.

DAVID L. KIRCHMAN

Lewes, Delaware

CONTRIBUTORS

Douglas G. Capone, University of Southern California, Department of Biological Sciences and Wrigley Institute for Environmental Studies, Los Angeles, CA

David A. Caron, University of Southern California, Department of Biological Sciences, Los Angeles, CA

Jonathan J. Cole, Institute of Ecosystem Studies, Millbrook, NY

Paul A. del Giorgio, University of Maryland, Horn Point Laboratory, Center for Environmental Science, Cambridge, MD

Hugh Ducklow, Virginia Institute of Marine Science, College of William and Mary, Gloucester Point, VA

Jed Fuhrman, University of Southern California, Department of Biological Sciences, Los Angeles, CA

Stephen Giovannoni, Oregon State University, Department of Microbiology, Corvallis, OR

David L. Kirchman, University of Delaware, Graduate College of Marine Studies, Lewes, DE

Mary Ann Moran, University of Georgia, Department of Marine Sciences, Athens, GA

Toshi Nagata, University of Tokyo, Ocean Research Institute, Tokyo, Japan

Hans W. Paerl, University of North Carolina at Chapel Hill, Institute of Marine Sciences, Morehead City, NC

Michael Rappé, Oregon State University, Department of Microbiology, Corvallis, OR

Evelyn Sherr, Oregon State University, College of Oceanic and Atmospheric Sciences, Corvallis, OR

Barry Sherr, Oregon State University, College of Oceanic and Atmospheric Sciences, Corvallis, OR

Suzanne L. Strom, Western Washington University, Shannon Point Marine Center, Anacortes, WA

T. Frede Thingstad, University of Bergen, Department of Microbiology, Bergen, Norway

Bess B. Ward, Princeton University, Department of Geosciences, Princeton, NJ

Peter J. le B. Williams, University of Wales, Bangor, School of Ocean Sciences, Meanai Bridge, United Kingdom

Jonathan P. Zehr, University of California–Santa Cruz, Ocean Sciences Department, Santa Cruz, CA

Richard G. Zepp, U.S. Environmental Protection Agency, Ecosystems Research Division, National Exposure Research Laboratory, Athens, GA

1

INTRODUCTION

David L. Kirchman

College of Marine Studies,
University of Delaware,
Lewes, Delaware

Peter J. le B. Williams

School of Ocean Sciences,
Marine Sciences Laboratory
University of Wales, Bangor, United Kingdom

The title, *Microbial Ecology of the Oceans* could be used for several different books on marine microbes and their interactions with the marine environment. The version presented here introduces many of the organisms found in the incredibly diverse microbial world, including protists, nitrogen-fixing cyanobacteria, chemoautotrophic nitrifying bacteria, and even viruses, which would be included in only a generous definition of "microbe." Various chapters discuss these microbes (and viruses) and the processes they mediate in the water column of the oceans, although many topics are equally relevant in sediments and even freshwaters. But the organisms featured perhaps in the most chapters are the aerobic heterotrophic bacteria, that is, microbes in the *Bacteria* domain that oxidize organic material using oxygen as the terminal electron acceptor. Archaea may have a role in some of the processes discussed here (see Chapter 3), but we do not know enough of their ecological functions

Microbial Ecology of the Oceans, Edited by David L. Kirchman.
ISBN 0-471-29993-6   Copyright © 2000 by Wiley-Liss, Inc.

to say much at this date.

Heterotrophic bacteria and the processes they mediate in the water column deserve to be highlighted in a book. These microbes are the critical link in the microbial loop that starts with the production of dissolved organic matter (DOM) and ends up fueling much respiration, nutrient cycling, and the growth of bacteria and organisms grazing on bacteria. Although now well recognized to be crucial parts of marine food webs and carbon cycling, microbial loop organisms and processes have not been given their deserved center stage in earlier books. Sediment microbes and phytoplankton have been the featured organisms in other monographs (e.g., Falkowski and Raven 1997; Fenchel et al. 1998). Cyanobacteria such as *Synechococcus* and *Prochlorococus* are taxonomically within the *Bacteria* domain and study of them is an important branch of microbial ecology, but these autotrophic microbes are ecologically more appropriately considered along with the eukaryotic phytoplankton. The book needs all the space it can get to do justice to examining microbial loop organisms and processes.

The choice of microbes discussed in the various chapters says much about this book's contents, but an exegesis of the "microbial ecology" part of the title may say even more. We use a bit of history of marine microbiology to discuss different types of "microbial ecology" and to introduce the type represented by this book.

WHAT IS MICROBIAL ECOLOGY?

There are at least two types of microbial ecology that differ in research topics and questions, methodology, and not of least importance, the background of their practitioners. One type, "microbial autecology," focuses on individual bacterial "species" and follows specific reactions mediated by and interactions among specific microbes. Microbial autecologists often start by isolating a bacterium targeted for study in order to identify it and to examine its metabolic capacities. From controlled laboratory experiments the physiological and ecological roles of the microbe in nature can be deduced. The foundation of this philosophy can be traced back to Robert Koch (1841–1910), who showed the importance of isolating and identifying the causative agent of several diseases. Microbial autecologists, characteristically trained first as microbiologists and only second as ecologists or oceanographers, tended to view the oceans from a laboratory perspective. The huge advantage of this approach is that highly controlled experiments can be conducted with potentially (if not actually) well-understood microbes under well-defined environmental conditions.

Microbial autecologists dominated the early days of microbial ecology, but they soon ran into major obstacles when traditional microbiological techniques were used to examine bacteria in the oceans. Perhaps most seriously, autecologists found that few bacteria can be isolated and cultivated on solid media (agar plates), suggesting that bacteria were *not* very numerous in the oceans. However, Jannasch and Jones (1959) showed that simple light microscopy gives estimates of bacterial abundance several orders of magnitude greater than those revealed by the plate count method (or in tests based on liquid media), the so-called viable count method. But these results also suggested that bacteria observable only by microscopic techniques were dormant or maybe even dead, even if they were more numerous than previously thought (ZoBell 1946). Autecologists suspected that their culturing methods could be improved and that some bacteria may be active but just unable to grow in cultures. But the prevailing wisdom then was that many bacteria were inactive due to the low organic matter concentrations in the oceans.

As a result, bacteria were not thought to be central parts of biological processes in the ocean. At best bacteria were perceived only as decomposers of organic material and producers of inorganic nutrients to support phytoplankton growth. This perception was shared by other marine scientists, as perhaps best illustrated in the book by Steele (1974), where heterotrophic bacteria are not included in any marine food webs.

The discrepancy between plate and microscopic counts of marine bacteria created an additional problem for pioneering microbial ecologists. If many bacteria observable only by microscopic methods were in fact viable, even if not terribly active, then the inability to culture them meant that they could not be identified by traditional taxonomic methods. This general problem was shared with limnologists and soil ecologists. The lack of a taxonomic name— and the information carried by its phylogenetic position—inhibited microbiologists from using autecological approaches to study marine microbes.

Microbial ecologists have argued since ZoBell (1946) about the number of "viable" bacteria in the oceans. We now know that although defining "viability" is not easy, many, if not most, bacteria in the oceans exhibit various signs of life, many more than are indicated by the "viable count" method, or more accurately the "plate count" method. Although controlled laboratory experiments are still not possible with uncultured bacteria, we now can at least identify them, thanks to various molecular methods, as discussed in Chapter 3. The new molecular methods now make possible autecological studies of complex bacterial assemblages under near–in situ conditions. However, even if these methods had been available 30 years ago, it seems doubtful that the autecological approach alone would have established the oceanographic importance of marine bacteria and other microbes, one reason being that the taxonomic composition ("community structure") of uncultivated microbes is quite complex (see Chapter 3). This complexity is a formidable barrier to piecing together, using only autecological approaches, the general roles of microbes in various biogeochemical cycles and food web dynamics. Another approach for studying natural microbes with different methods and perspectives was necessary.

The beginnings of this other approach can be traced at least to the 1960s (e.g., Parsons and Strickland 1962; Wright and Hobbie 1965), although one could claim antecedents in the mixed culture experiments of Louis Pasteur (1822–1895). Chapter 2 gives a historical account of some work conducted during the 1970s and early 1980s. Hobbie and Williams (1984) used the term "synecology" to describe this work, but "microbial ecology" is simpler and as accurate, since today's microbial ecologist can identify specific microbial species (autecology) while also examining biogeochemical cycles (synecology). Microbiologists practicing this type of microbial ecology often know much about other aquatic organisms (e.g., phytoplankton and heterotrophic protists) and general problems in limnology and oceanography. Arguably these early microbial ecologists and their followers have provided most of the evidence indicating the importance of bacteria in various biogeochemical cycles and in marine food web dynamics.

Undeterred by the inability to cultivate microbes and thus to identify them with traditional methods, microbial ecologists have asked questions about microbial processes and about rates and bulk properties of heterotrophic bacteria, often relative to analogous phytoplankton parameters or to total community properties (e.g., total biomass or community respiration). For many processes, it was (and still is) quite satisfactory to treat various microbial components as single, taxonomically homogeneous groups, each being defined only by the process it mediates regardless of the taxonomy of the individuals making up the group. Heterotrophic bacteria, which are often discussed as if they were a single taxonomic group even though the diversity of prokaryotes capable of mineralizing organic material has been well known for decades, comprise an important example. This attitude toward microbes is what autecologists may call a "black box" approach: one may know the arrows (fluxes) going in and out of a box (bacteria and other microbes), but the internal dynamics within a box are ignored. It may be illuminating to note that many of the early practitioners of black box microbial ecology were not trained in microbiology; they were zoologists, chemists, biochemists, or food technologists. The exception may be the Kiel group, which contributed much during the 1970s (Rheinheimer 1977), including work on bacterial biomass (the first use of epifluorescence microscopy; Zimmerman and Meyer-Reil 1974) and the number of active bacteria (Hoppe 1976; Meyer-Reil 1978) in natural microbial assemblages.

Figure 1 summarizes some of the microbes and processes that microbial ecologists have been examining over the last 30 years. In part to keep it simple, the figure does not include findings past roughly 1980. It does not mention the many microbes reviewed in Chapter 2, nor the mixotrophic microbes and symbiotic relationships discussed in Chapter 16, nearly all of which were discovered in the last 20 years. Viruses are also not included (see Chapter 11). The figure has no arrows indicating uptake of plant nutrients such as ammonium and phosphate by heterotrophic bacteria, a process first examined extensively in the mid-1980s (Chapter 9). Perhaps the most important omissions are arrows from various grazers to the DOM pool; Nagata in Chapter 5 concludes that grazing, not direct phytoplankton excretion as

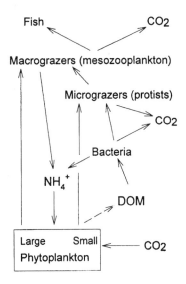

Figure 1. Schematic diagram illustrating some organisms and pathways examined by microbial ecologists. The text discusses the many omissions in this oversimplified diagram.

implied by the figure, is the most important mechanism producing DOM. Still, the figure serves to illustrate some of the seemingly simple questions asked by this type of microbial ecology: What are the standing stocks and production rates of various microbial groups? What is the flux of DOM through bacteria to higher tropic levels — that is, how much primary production is processed by the microbial loop versus herbivores grazing on phytoplankton? What is the contribution of heterotrophic bacteria to mineralization of C and N, especially as sources of nutrients like ammonium for primary production? In short, microbial ecologists examining these and other questions seek to understand the role of microbes in food web dynamics and biogeochemical cycles in the oceans. They are very much interested in problems faced by bacteria living in nutrient-poor environments like the oceans, but the justification for examining these problems is to understand how microbes impact oceanographic and ecological processes.

Many of these processes center on carbon cycling. Even when another element is examined, the underlying justification for the work is often to understand the role in carbon cycling of that element (e.g., P or Fe). Carbon is the currency of choice for examining the fate of primary production in the oceans: mineralization back to CO_2, transfer up to larger organisms (higher trophic levels), eventually to fish, or storage as refractory DOM or burial in sediments. The fate of primary production and the impact of microbes on various geochemical cycles are perennial topics in oceanography, but examining especially carbon takes on greater urgency as we consider the role of the oceans in climate change forced by increases in atmospheric CO_2. Because of the importance of carbon, Chapters 5–7 discuss sources and fates of oceanic DOM, one of the largest pools of carbon on the planet. Nitrogen fixation (Chapters 13 and 15) and nitrification (Chapter 14) are key reactions in the nitrogen cycle, but these reactions acquire additional importance because they impact new production, a useful concept for examining carbon storage and mineralization in the oceans (Chapter 15).

Examining these microbial processes characterizes the type of microbial ecology mentioned in the title of this book. Microbial ecologists examining fluxes and standing stocks of microbes have established the importance of bacteria in oceanographic processes, arguably one of the most fundamental discoveries in oceanography over the last 20 years. Analogous studies in less salty environments have also changed limnology and terrestrial ecology. This book tries to summarize this progress, while never forgetting the need for more work.

SIGNS OF PROGRESS

The last 20 years have seen major technical and conceptual advances in microbial ecology—arguable the greatest in environmental sciences. To see how far we have come in understanding especially marine microbes, it is useful to look back at the four questions listed by Hobbie and Williams (1984), who summarized results from a 1981 workshop on "heterotrophic activity" in the sea. The list that follows indicates which chapters here answer or begin to answer these questions. The questions are given in their original form, although questions 3 and 4 are reversed from the order listed by Hobbie and Williams (1984).

1. What is the identity and abundance of marine microbial heterotrophs such as bacteria and microflagellates? One of the more spectacular advances in microbial ecology has been the application of molecular tools to identify marine bacteria that have resisted cultivation and thus identification by more traditional methods. Chapter 3 by Giovannoni and Rappé discusses this work, all of which was published after Hobbie and Williams (1984). We also know more about "microflagellates," or "protists" as they are now more commonly called, as discussed in Chapter 2. More work is needed on the phylogenetic composition of protists, since these microbes also are very difficult to culture, and natural assemblages of protists have rarely been examined with molecular tools.

We also now have a fairly good general picture of microbial abundance in the oceans, as discussed in Chapter 4 by Ducklow. A more urgent requirement is for better data on microbial biomass, not just cell abundance, as discussed by Ducklow. He emphasizes the range in bacterial biomass that results from assuming different factors for carbon per bacterial cell or carbon per bio-volume, which in turn impacts whether bacterial biomass is low, equal to, or even greater than phytoplankton biomass. It is of no consolation to remember that phytoplankton biomass is also uncertain because of variation in carbon to chlorophyll ratios. Hobbie and Williams (1984) mentioned that "much work still needs to be done on... conversion factors" [to convert cell numbers to biomass], words that still apply today.

2. What are the rates of heterotrophic processes, such as growth, respiration, and mineralization of nitrogen and phosphorus? To simplify examining our progress in answering this question, let us focus only on processes mediated by heterotrophic bacteria.

The first paper on using thymidine incorporation to measure bacterial "growth," more precisely bacterial production, of natural bacterial assemblages was published just before Hobbie and Williams (1984). Since then, there has been an explosion of studies on bacterial growth rates and production in various oceanic regimes, although mostly in coastal areas. Certainly there has been substantial progress on this topic over the last 20 years, as summarized in Chapter 4 by Ducklow. Nonetheless, it is sobering to see the large uncertainties in bacterial production estimates still remaining even after years of intense efforts examining various methods.

In Chapter 6, Williams continues to plead for more measurements on respiration, a process that remains understudied in spite of the call for more data nearly 20 years ago by Hobbie and Williams (1984). Chapter 6 discusses the recent controversy of whether bacterial respiration can ever exceed primary production in oligotrophic oceanic regimes. Given that we have so few data on respiration, especially compared to production measurements, the uncertainties about even the magnitude of bacterial respiration perhaps should not be surprising.

We have made much progress in understanding the relationship between respiration and growth (i.e. the growth efficiency of bacteria) as discussed in Chapter 10 by del Giorgio and Cole. A major finding from the last 20 years of work is that bacterial growth efficiencies are lower than the 50% assumed by early studies and are perhaps as low as 15% in the open ocean. With growth efficiencies lower than 50%, heterotrophic bacteria must be a net sink, not a link in transferring carbon routed through the DOM pool back to higher trophic levels; more than 50% of the DOM flux must be mineralized to CO_2 by bacteria. Although it is well accepted that growth efficiencies are lower than 50%, the very low growth efficiencies (ca. 15%) create problems in reconciling bacterial production data with primary production, as discussed in Chapter 6.

Chapter 9 summarizes the relatively few studies on mineralization of nitrogen by heterotrophic bacteria and points out that there are even fewer studies on phosphorus mineralization. In addition to the technical difficulties in examining these processes, another reason for the paucity of studies on mineralization is that our focus shifted to the opposing process, inorganic nutrient uptake by heterotrophic bacteria, a topic examined most intensively after 1981, as reviewed in Chapter 9. The finding that heterotrophic bacteria take up ammonium, which is radically different from the old view of bacteria being mere mineralizers, resonated with other work indicating that bacteria were important prey and food for larger microbes: bacteria were not mere mineralizers of organic material, sitting on the edges of the major biological processes in the oceans, as depicted by earlier work. Rather than mainly providing inorganic nutrients to support plant growth, work in the late 1980s suggested that bacteria may compete with phytoplankton for growth-limiting nutrients. Some studies suggest that heterotrophic bacteria, at least as viewed at the community level, simultaneously excrete and take up inorganic nutrients, as discussed in Chapter 9. In short, we still do not know whether heterotrophic bacteria are net sinks or sources of compounds and elements like ammonium and iron.

3. What are the controls of various processes that make up heterotrophy in the sea? Again, we restrict our discussion to the heterotrophic bacteria, but Hobbie and Williams (1984) had the same focus. They guessed correctly that grazing by various protists is a major control of bacterial biomass and helps explain the relatively constant numbers of bacteria over time and space in the oceans. As discussed by Strom in Chapter 12, the methodology for measuring grazing that was developed after 1981 allowed for grazing rates to be matched up with bacterial production. According to Strom, grazing is equal to bacterial production in oligotrophic marine systems but is less than bacterial production in highly productive regimes. The other main cause of bacterial mortality is viral lysis; viruses in the oceans were discovered after 1981, although perhaps "rediscovered" would be more accurate, inasmuch as bacteriophages were known to ZoBell (1946). According to Fuhrman in Chapter 11, viruses account for 10–40% of bacterial mortality on average, but this percentage varies greatly. Unfortunately, very few studies have compared grazing and viral mortality directly, and we have little idea about the factors governing the relative importance of grazing versus viruses in accounting for bacterial mortality.

Hobbie and Williams (1984) could say little about what controls bacterial growth rates in the oceans because the methods for measuring growth were just being published at the time. We now have a much better idea about various controlling factors, as discussed in Chapter 8 by Thingstad. The chapter mentions that the supply of DOM is often assumed to regulate bacterial growth, and there are data supporting that hypothesis (see also Chapter 6). But Thingstad also discusses recent work showing the importance of phosphate in determining bacterial growth in both real oceans and conceptual models. Interestingly, he points out that there is little evidence that inorganic nitrogen has any direct affect on bacterial growth. In regions like the Southern Ocean, iron may exert an influence. Temperature can never be forgotten, although bacterial production relative to primary production does not appear to be any lower in perennially cold oceans than in warm ones (Chapter 4). In short, nearly 20 years of work have given us large amounts of data to discuss, but the question of what controls bacterial growth rates (and how those controls may vary) remains open.

Hobbie and Williams (1984) discuss controls on DOM uptake, one of the first examples of "heterotrophic activity" to be examined by marine microbial ecologists. We now know much more about DOM concentrations and composition, although Chapter 6 points out the paucity of studies examining variations in concentrations and fluxes with depth and over the seasons. Perhaps the biggest advance since Hobbie and Williams (1984) is the discovery of the impact of photochemistry on DOM-related processes. Most of the processes summarized by Moran and Zepp in Chapter 7 were only discovered during the last 5 years, but it is already clear that photochemical and other abiotic reactions greatly impact DOM turnover and DOM–bacteria interactions. Williams in Chapter 6 emphasizes these points as well.

4. What is the role of marine heterotrophs in the food web of the sea? We now know much about the various roles of heterotrophic protists in the ocean, including the observation that many are not just heterotrophs but are often capable of autotrophy, as discussed in Chapter 16. We also know that various protists graze on heterotrophic bacteria, cyanobacteria, and small eukaryotic phytoplankton, mineralizing that biomass to inorganic nutrients and serving as prey for larger grazers (zooplankton). The role of heterotrophic bacteria has also become clearer, but many questions remain.

Whether bacteria and the microbial loop were sinks or links in carbon cycling was an open question back in 1981. We now know that heterotrophic bacteria are mainly sinks, as mentioned before, but Williams in Chapter 6 points out that the microbial loop can still support much growth at higher trophic levels (i.e. zooplankton secondary production) even while mineralizing more than 50% of assimilated DOM. Whether heterotrophic bacteria are net mineralizers or net sinks for inorganic nutrients remains unclear, as mentioned before.

Hobbie and Williams (1984) raised questions about the role of bacteria in modifying organic matter, such as in a "microbially mediated process of humification." We should interpret "humification" to mean the transformation of labile organic material to more refractory compounds, rather than the formation of specific geopolymers such as the humics substances found in soil. Regardless of our definition, the few advances in answering these questions are discussed in Chapter 5 by Nagata, Chapter 6 by Williams, and Chapter 7 by Moran and Zepp. Microbial ecologists need to collaborate with organic geochemists in understanding the formation of refractory DOM, one of the most pressing problems in geochemistry. How this organic carbon survives more than 1000 years remains intriguing and largely unknown. Williams discusses a possible explanation in Chapter 6.

We know about many roles for heterotrophic microbes in the oceans in addition to those outlined by Hobbie and Williams (1984). Although this book tries to cover some of these (e.g., denitrification is discussed in Chapter 15), summarizing all of marine microbiology and bacterial oceanography would take multiple books. Furthermore, several microbes have been identified (see Chapters 2 and 16) that do not fit easily into a simple compartmental model such as that given in Figure 1. We do not know even what all the heterotrophic bacteria are doing in the oceans.

Perhaps the biggest question in microbial ecology is, What is the role of the various and varied bacteria found in the oceans? Although there has been much progress in describing the diversity of marine bacteria and archaea (see Chapter 3), we still know very little about the "function" of these microbes. Their location within the water column (shallow vs deep, free-living particle-associated) gives us some clues that not all bacteria are mediating the same biogeochemical processes. We already knew that pathways such as nitrification are mediated by specific bacterial groups (Chapter 14), but it is not clear whether the many other pathways, such as DOM use and mineralization of inorganic nutrients, are similarly dominated by selected groups of marine microbes. Do oceans really need bacteria of all these different types for various biogeochemical cycles and food webs to operate? Answering these questions will require modern microbial ecologists to use methods and concepts from autecological studies of individual bacterial groups and from ecological studies of microbial assemblages in the oceans, that is, the two types of microbial ecology discussed in this introductory chapter.

Once the links between function and community structure have been elucidated, microbial ecologists will then need to turn to convincing other ecologists and oceanographers that these links are important in understanding general questions in aquatic ecology, biogeochemistry, and oceanography. Scientists other than microbial ecologists already can appreciate that who is present among the phototrophic bacteria has a big impact on various aquatic processes. For example, it is well recognized that the nitrogen-fixing cyanobacteria (Chapter 13) have huge impacts on the regulation of biological production in the oligotrophic gyres, as discussed by Capone in Chapter 15. We know why it matters to recognize the difference between the nitrogen fixer *Trichodesmium* and non-nitrogen-fixing strains of *Synechococcus*. But it is less obvious why it matters to know about the phylogenetic composition and function of many other parts of the microbial world. Undoubtedly it does matter, but microbial ecologists still need to demonstrate this to other aquatic ecologists.

Perhaps one sign of the vitality of marine microbial ecology is that we end with more questions than answers. In spite of exponential growth over the last 20 years, the unknowns in microbial ecology still outnumber the established facts, leaving much room for further work. However, it probably will take much less than 20 years of work to fill another book entitled *Microbial Ecology of the Oceans*.

REFERENCES

Falkowski, P. G., and Raven, J. (1997) *Aquatic Photosynthesis.* Blackwell Science, Malden, MA.

Fenchel, T., King, G., and Blackburn, H. T. (1998) *Bacterial Biogeochemistry.* Academic Press, San Diego, CA.

Hobbie, J. E., and Williams, P. J. le B. (eds.) (1984) *Heterotrophic Activity in the Sea:* Plenum Press, New York.

Hoppe, H.-G. (1976) Determination and properties of actively metabolizing heterotrophic bacteria in the sea, investigated by means of micro-autoradiography. *Mar. Biol.* 36:291–302.

Jannasch, H. W., and Jones, G. E. (1959) Bacterial populations in sea water as determined by different methods of enumeration. *Limnol. Oceanogr.* **4**:128–139.

Meyer-Reil, L.-A. (1978) Autoradiography and epifluorescence microscopy combined for the determination of number and spectrum of actively metabolizing bacteria in natural waters. *Appl. Environ. Microbiol.* 36:506–512.

Parsons, T. R., and Strickland, J. D. H. (1962) On the production of particulate organic carbon by heterotrophic processes in sea water. *Deep-Sea Res.* 8:211–222.

Rheinheimer, G. (ed). (1977) *Microbial Ecology of a Brackish Water Environment.* Springer-Verlag, Berlin.

Steele, J. H. (1974) *The Structure of Marine Ecosystems.* Cambridge University Press, Cambridge.

Wright, R. T., and Hobbie, J. E. (1965) The uptake of organic solutes in lake water. *Limnol. Oceanogr.* 10:22–28.

Zimmermann, R., Meyer-Reil, L.-A., and Hobbie, John E. (1965) A new method for fluorescence staining of bacterial populations on membrane filters. *Kieler Meeresforsch., Sonderh.* 30:24–27.

ZoBell, Claude E. (1946) *Marine Microbiology.* Chronica Botanica, Waltham, M.A.

2

MARINE MICROBES
AN OVERVIEW

Evelyn Sherr and Barry Sherr

*College of Oceanic and Atmospheric Sciences,
Oregon State University,
Corvallis, Oregon*

This chapter is an introduction to the general groups of marine microbes. "Microbes" is a term that includes all single-celled organisms: autotrophic and heterotrophic prokaryotes (bacteria and cyanobacteria) and autotrophic and heterotrophic eukaryotes (algae and phagotrophic protists), as well as viruses (Table 1). The length difference between the smallest bacterial cells in seawater (about $0.3 \mu m$ in diameter) and the largest ciliate grazers in the microplankton (about $150 \mu m$ long) is similar in magnitude to the difference between a 4 cm krill and a 20 m whale. Microbes include all of the three deepest divisions, or domains, of life on our planet: Bacteria, Archaea, and Eukarya. These domains are identified by genetic distance in the composition of the 16S or 18S subunits of ribosomal ribonucleic acid molecules (rRNA) (Woese et al. 1990). The first two, Bacteria and Archaea, are divisions of prokaryotes, organisms with, usually, a rigid cell wall, and DNA loosely organized in a region of the cell termed a nucleoid. Humans are in the third domain, Eukarya, organisms with a membrane-bound nucleus, as are all algae (except "blue-green algae," correctly termed "cyanobacteria"), flagellated and ciliated protists, fungi, plants, and animals. Modern microorganisms have a long (on the order of 3.5 billion years) evolutionary history, which has been played out largely in marine

Microbial Ecology of the Oceans, Edited by David L. Kirchman.
ISBN 0-471-29993-6   Copyright © 2000 by Wiley-Liss, Inc.

Table 1. General groups of pelagic microbes in the sea[a]

Size Category	Microbial Group	Size Range (μm)	Discussed in Chapters
Femtoplankton	Viruses	0.01–0.2	11
Picoplankton	Prokaryotes		
	Bacteria		
	Photoautrophic	0.5–1.0	3, 13
	Prochlorophytes	0.5–2.0	
	Coccoid cyanobacteria	1.0–2.0	
	Filamentous cyanobacteria	7–10 wide × ⩽ 100s long	
	Chemoautrophic	0.3–1.0	14
	Heterotrophic	0.3–1.0	3–12, 15
	Archaea		3
	Eukaryotes		
	Picoalgae, picoheterotrophic flagellates	1.0–2.0	2, 12, 16
Nanoplankton	Nanoalgae, nanoheterotrophic protists (mainly flagellates)	2–20	2, 12, 16
Microplankton	Microalgae		
	Microheterotrophic protists (mainly ciliates and heterotrophic dinoflagellates)	20–200	2, 12, 16

[a]Microbial size categories are based on the biovolume size classes for marine plankton proposed by Sieburth et al. (1978). The focus of this book is on prokaryotic microbes (bacteria) and heterotrophic eukaryotic microbes (phagotrophic protists). Viruses are also considered. Photosynthetic eukaryotes (algae) are mentioned only briefly.

environments. The variety and distribution in marine systems of microbes, especially of prokaryotes, has been shaped by changes in environmental conditions in the oceans over geological time. Chapter 3 presents details of the evolution and phylogenetic diversity of marine prokaryotes. Other resources for the history of evolution of microbes include Madigan et al. (1997), Dyer and Obar (1994), and Fenchel and Finlay (1995).

MARINE PROKARYOTES

Prokaryotic organisms in the ocean are morphologically simple: microscopic rods, spheres, and filaments generally less than $1–2 \mu$m in size. [A dramatic exception to this size range is the largest bacterium known: *Epulopiscium fishlesoni*, a gram-positive species 200–800 μm long, which lives in the guts of a Red Sea fish (Angert et al. 1993).] However, prokaryotes are highly diverse in terms of both taxonomy and metabolism. In some cases—for example, the methanogens—metabolism is associated with phylogenetic affiliation; in other cases, genetically related organisms have diverse metabolic modes. Different species in the group of nonpurple sulfur bacteria, for instance, may be either heterotrophic or autotrophic. Two types of evidence support the antiquity of microbes in the earth's history: (1) paleobiological evidence in the sedimentary record (microfossils, organic matter, stable isotope composition of bioactive elements), and (2) molecular genetic evidence that prokaryotic microbes are at the base of the family tree of all living organisms (see Chapter 3). It should be emphasized that all living prokaryotes are adapted to modern environmental conditions, and no group should be considered to be simply a remnant "living fossil" from the Precambrian era.

Fundamental requirements of all living organisms include (1) a source of energy to regenerate ATP from ADP + phosphate (i.e., phosphorylation; (2) a source of elements for biosynthesis (carbon, nitrogen, phosphorus, sulfur, and trace elements such as iron and manganese), and (3) a source of reducing equivalents (electrons) to produce organic molecules and organic polymers from more oxidized compounds, either inorganic compounds or small molecular weight organic compounds. The basic metabolic divisions are between *autotrophy* (self-feeding: i.e., obtaining all requirements for life from inorganic compounds and chemical or light energy) and *heterotrophy* (other-feeding: i.e., obtaining virtually all requirements for life from organic compounds).

Distinctions between metabolic modes are also made in terms of sources of energy for production of ATP. *Phototrophic* organisms obtain energy from light. *Chemotrophic* organisms gain energy from oxidation of reduced chemicals, including inorganic and organic compounds (Table 2). A further distinction may be made on the basis of source of elements and reducing equivalents (electrons) for biosynthesis: *lithotrophs* can exist solely on inorganic compounds, while *organotrophs* require organic compounds for biosynthetic pathways. Combinations of these terms are sometimes used to denote more specific

Table 2. Comparison of energy yields (kilocalories produced per mole of reductant oxidized) from various redox reactions

Process	Redox Pair: Reductant + Oxidant	Energy Yield (kcal/mol)
Aerobic respiration	glucose + O_2	686
Nitrate respiration (denitrification)	glucose + NO_3^-	649
Sulfide oxidation	$HS^- + O_2$	190
Fermentation	glucose $\rightarrow$ lactate	58
NH_4^+ oxidation (nitrification)	$NH_4^+ + O_2$	66
NO_2^- oxidation (nitrification)	$NO_2^- + O_2$	18
Sulfate respiration	Lactate + SO_4^{2-}	9.7
Methanogenesis	$H + CO_2$	8.3

Source: Fenchel and Blackburn (1979).

functional groups; for example, *chemoautotroph* and *chemolithotroph* are terms used to describe bacterial autotrophs such as nitrifying bacteria or sulfur-oxidizing bacteria, which do not require light or organic compounds for growth. The terms "photosynthetic" and "chemosynthetic" refer to autotrophic organisms that obtain the energy needed for biosynthesis either from light or from oxidation of inorganic chemicals. A brief description of major types of prokaryotic metabolism follows. More information can be found throughout this book. Also, excellent treatments of this subject are presented in Madigan et al. (1997) and Fenchel et al. (1998).

Photoautotrophs

Anoxygenic, anaerobic phototrophs occur in marine systems, although they are largely confined to shallow sediments or found in association with decomposing organic matter. Anaerobic phototrophic bacteria are relatively unimportant in pelagic marine systems. In the Black Sea, the world's largest body of anoxic marine water, however, oxygen-depleted water masses extend up to the base of the euphotic zone. In this system, blooms of the anaerobic phototroph *Chlorobium* spp. can be significant (Repeta et al. 1989). In contrast, two groups of oxygenic photosynthetic bacteria have recently been found to be ubiquitous in marine pelagic systems, and at times to comprise a significant fraction of total phytoplankton biomass and productivity. These are coccoid cyanobacteria and prochlorophytes. These phototrophs are closely related to each other based on 16S rRNA sequences (Palenik and Haselkorn 1992; Urbach et al. 1992).

Most of the coccoid cyanobacteria are in the genus *Synechococcus*, around 1 μm diameter cells that fluoresce bright orange when excited with blue light (Waterbury et al. 1986). Their pigments are chlorophyll a and phycobili-proteins; cyanobacteria lack chlorophyll b. *Synechococcus* occurs abundantly (10^2–10^5 cells mL^{-1}) in the euphotic zone of both coastal and open ocean waters; the only part of the ocean in which these cyanobacteria do not appear to be important are polar seas (Joint 1986; Waterbury et al. 1986). In oligotrophic open ocean gyres, phototrophs smaller than 1–3 μm have been estimated to contribute 60–80% of water column primary productivity (Platt et al. 1983; Takahashi and Bienfang 1983).

Two other types of cyanobacteria are at times significant in pelagic marine ecosystems: *Trichodesmium* spp. and *Richelia intracellularis*. Both are filamentous forms, capable of nitrogen fixation. (The process of nitrogen fixation in marine systems is treated in detail in Chapter 13.) *Richelia intracellularis* is, as its species name suggests, an intracellular symbiont of large oceanic diatoms such as *Rhizosolenia* sp. and *Hemiaulus* sp. (Venrick 1974; Villareal 1991). *Richelia* obtains energy for nitrogen fixation in the form of organic carbon compounds produced by the host diatom, thus providing the diatom with an extra source of nitrogen in a nitrogen-limited environment.

Pelagic prochlorophytes, in the genus *Prochlorococcus*, are phototrophic cells slightly smaller than *Synechococcus* (0.7 μm vs 1.0 μm) and have a different suite of photopigments compared to cyanobacteria (Chisholm 1992). Prochlorophytes have divinyl chlorophyll a, a precursor molecule of the monovinyl chlorophyll a that is present in cyanobacteria and chloroplasts; and they contain divinyl chlorophyll b as a major accessory pigment (Chisholm 1992). Since prochlorophytes fluoresce very dimly, they are difficult to enumerate via epifluorescence microscopy. Flow cytometric enumeration of prochlorophytes in seawater indicates that this group of prokaryotic autotrophs is usually more abundant than *Synechococcus* (on the order of 10^4 to $>10^5$ *Prochlorococcus* cells mL^{-1}) (Chisholm 1992; Campbell and Vaulot 1993). Recent studies have shown that there are genetically different populations adapted for growth at either high-light or low-light intensities in the open ocean (Moore et al. 1998).

To date, most work on marine pelagic prochlorophytes has been done in open ocean systems: the Sargasso Sea (Chisholm 1992) and the equatorial Pacific (Campbell and Vaulot 1993). Prochlorophytes appear to be an important part of the biomass and productivity of phytoplankton in such regions. Li et al. (1992) found that in the Sargasso Sea during September, prochlorophyte carbon biomass was twice that of cyanobacteria, 50% that of photosynthetic eukaryotes, and 25% that of heterotrophic bacteria. Campbell and Vaulot (1993) reported that *Prochlorococcus* contributed, on average, 45% of phytoplankton carbon biomass in the euphotic zone at Station ALOHA off Hawaii, and prochlorophytes comprised a significant fraction ($\leqslant 30\%$) of the "heterotrophic" bacteria enumerated via epifluorescence microscopy in these tropical waters.

Chemoautotrophs

The general types of prokaryotic chemoautotroph are methanogens, sulfuroxidizing bacteria, and nitrifying bacteria. All these microbes are either anaerobes (methanogens) or aero tolerant organisms that require both oxygen and reduced chemical substrates characteristic of suboxic environments (e.g., sulfide, ammonium, nitrite).

Methanogens   Methanogens (methane producers), one of the three major groups of Archaea, are strict anaerobes that live by gaining chemical energy in the oxidation of hydrogen. Some methanogens can convert a few other substrates (acetate, formate, and methyl compounds) to methane. Methanogens can use carbon dioxide both as the oxidant for energy generation and as the source of carbon for biosynthesis. The energy-generating reaction of methanogens is:

$$4H_2 + H^+ + HCO_3^- \rightarrow CH_4 + 3H_2O$$

The surface waters of the ocean can be supersaturated with methane, which is presumptive evidence that methanogenic bacteria are ubiquitous even in well-oxygenated water columns. Sieburth (1987, 1993) proposed that methanogens exist in microniches such as the interior of suspended particles in which oxygen has been used up via respiration. Methanogenesis occurs in regions of intense upwelling in which sub–euphotic zone waters have reduced oxygen concentrations owing to oxidation of sinking organic matter. Methane production in the Arabian Sea was associated with degradation of high phytoplankton biomass resulting from monsoon-driven upwelling of nutrient-rich water along the Arabian coast (Owens et al. 1991).

Sulfur-Oxidizing Bacteria The sulfur-oxidizing bacteria are phylogenetically diverse chemoautotrophs, including thermophilic members of the Archaea as well as thermophilic and mesophilic Bacteria in the purple sulfur bacterial group. Both elemental sulfur (S°) and hydrogen sulfide (H_2S) are oxidized to yield energy for phosphorylation. The oxidation reactions are:

$$H_2S + 2O_2 \rightarrow SO_4^{2-} + 2H^+$$
$$S^\circ + H_2O + \tfrac{3}{2}O_2 \rightarrow SO_4^{2-} + 2H^+$$

The final product of microbial sulfur oxidation is sulfate; oxidation of sulfur and sulfide by bacteria over geological time is the dominant source of sulfate in the biosphere (Fenchel et al. 1998). In marine systems, chemoautotrophic sulfur bacteria form the base of the food web at hydrothermal vents, where volcanic processes spew sulfides into oxygenated seawater. Rift worms and clams have tissues packed with symbiotic sulfur-oxidizing bacteria; white mats of filamentous *Beggiatoa* grow at the expense of reduced sulfur from the vents. In the pelagic zone sulfur-oxidizing bacteria are important mainly in hypoxic water masses.

Nitrifying Bacteria Obligate lithotrophs, nitrifying bacteria are a consortium of two genetically distinct groups of Bacteria: the *ammonium-oxidizing* bacteria, and the *nitrite-oxidizing* bacteria. These bacteria and the processes they mediate are described in more detail in Chapter 14. These two groups of bacteria have an important role in closing the nitrogen cycle in the sea by removing ammonium, which at high levels is toxic to eukaryotes, and by forming nitrate. Because of the small number of electrons and the low amount of free energy liberated from oxidation of ammonium and nitrite, coupled with the energy requirements of carbon fixation, nitrifying bacteria operate at the practical limit for a chemoautotrophic mode of existence. Based on cell yields of marine nitrifying bacteria in laboratory cultures, it appears that ammonium-oxidizing and nitrite-oxidizing bacteria generate only 1–10% of the cell yield expected from a heterotrophic bacterium with the same initial amount of energy-generating substrate (Fenchel et al. 1998).

Ammonium oxidation is the sole energy source of one group of nitrifying bacteria, ammonium-oxidizing bacteria. Approximately 35 moles of ammonium must be oxidized to gain sufficient energy to fix one mole of carbon dioxide. The reaction occurs in two steps, with hydroxylamine (NH_2OH) formed as an intermediary compound:

$$NH_3 + O_2 + H^+ \rightarrow NH_2OH + H_2O$$

$$NH_2OH + O_2 \rightarrow NO_2^- + H_2O + H^+$$

Nitrite oxidation is carried out via a single-step reaction by nitrite-oxidizing bacteria:

$$NO_2^- + \tfrac{1}{2}O_2 \rightarrow NO_3^-$$

Each reaction generates only 1 ATP; for each mole of carbon dioxide fixed, a minimum of 15 moles of nitrite must be oxidized. The greenhouse gas nitrous oxide, N_2O, is a by-product of nitrification, particularly when oxygen concentrations are low. In the world ocean, major zones of intense nitrification occur beneath regions of upwelling (e.g., off Peru and in the Arabian Sea), as a result of the rapid rate of ammonium regeneration from the large flux of sinking organic detritus at the base of the euphotic zone in these regions (Wada and Hattori 1991).

Both photoautotrophic and chemoautotrophic bacteria are characterized by layers of energy-generating membranes which take up a significant part of the cell volume. As a result of the extra space required for these membranes, autotrophic bacterial cells typically have a larger average cell size than heterotrophic bacteria in the sea. For example, cells of the ammonium-oxidizing bacterium *Nitrosococcus oceanus* have a diameter of 1.8–2.2 μm (Ward 1986), while most open ocean heterotrophic bacteria are on the order of 0.3–0.5 μm in diameter. It has been suggested that the larger cell size of nitrifying bacteria makes them more susceptible to grazing mortality by bacterivorous protists (Lavrentyev et al. 1997).

Heterotrophic Prokaryotes

The bacterioplankton assemblage is numerically dominated by strains of gram-negative, heterotrophic Bacteria, which live via oxidation of organic substrates. In oxygenated water, microbes grow by catabolizing organic molecules via aerobic respiration, in which oxygen is the terminal electron acceptor. Under conditions of low oxygen availability, either in a water mass or in microsites of intense microbial activity, nitrate and sulfate may be used as alternate terminal electron acceptors via anaerobic respiration. Fermentation may also become important when oxygen is low or completely absent.

Aerobic Respiration   The optimal trophic mode for organisms is aerobic respiration. Extra energy is not required for the reduction of carbon dioxide to organic precursors for biosynthesis, as is the case for autotrophs. The energy yield from each mole of organic substrate oxidized is large (Table 2). Heterotrophic growth is not dependent on either light or on availability of reduced inorganic chemicals. Because of the high energy yield of aerobic respiration, most organic compounds can eventually be degraded by heterotrophic bacteria growing in an oxygenated environment.

Heterotrophs assimilate only low molecular weight organic compounds, which can be transported directly across the cell membrane. Such substrates include C_2 compounds (e.g., acetate, ethanol), C_3 compounds (e.g., pyruvate), C_4 compounds (e.g., succinate), and C_5 and C_6 compounds (e.g., sugars, amino acids). Small molecular weight organic compounds are easily transported across bacterial cell walls; these are termed labile substrates and are generally present at very low concentrations (ng L^{-1} to μg L^{-1}) in seawater. In contrast, high molecular weight organic compounds such as polysaccharides and proteins can be utilized by bacteria only after extracellular cleavage, or hydrolysis, of monomer (one-subunit) or dimer (two-subunit) molecules from the polymer, which can then be transported into the bacterial cell. Such hydrolysis is carried out by extracellular enzymes associated with the bacterial cell surface (Chrost 1991). Chapters 5–10 provide more information on the interactions of heterotrophic bacteria with dissolved and particulate organic matter in the sea.

Some aerobic heterotrophs, C_1 bacteria, can grow on reduced carbon compounds containing only one carbon atom. These bacteria generate energy by oxidizing C_1 carbon substrates, and in addition obtain carbon and reducing equivalents from such compounds. C_1 bacteria, which have been identified in both coastal and oceanic habitats, appear to be a widespread and diverse component of bacterioplankton assemblages (Sieburth 1993; Sieburth et al. 1993; Kelley et al. 1998). Their quantitative significance with respect to carbon flow in pelagic microbial food webs is unknown. There are two recognized groups of C_1 bacteria:

1. Methylotrophs. A variety of heterotrophic bacteria, including species in the common genera *Pseudomonas* and *Vibrio*, can grow on diverse C_1 compounds such as methanol (CH_3OH), methylamine (CH_3NH_2), and formate (HCOO). Some methylotrophs can also utilize methane (CH_4). Methylotrophs that cannot use methane tend to be metabolic generalists capable of growing on larger organic molecules as well as on C_1 compounds.

2. Methanotrophs. A subset of methylotrophs can grow only on methane and a few other C_1 compounds. Methanotrophs are similar in some ways to chemoautotrophic bacteria; they have a complex internal membrane system that is involved in methane oxidation. Methanotrophs are unique among marine prokaryotes in having relatively large amounts of sterols in their membranes. Sterols are rigid lipid molecules that serve to stabilize the structure of eukaryotic cell membranes. As a rule, prokaryotic membranes do not have sterols, since the cell walls of bacteria are sufficiently strong to keep the cell shape intact. The exceptions are the methanotrophs, as well as some myco-plasmic bacteria which lack cell walls (Madigan et al. 1997).

Anaerobic Respiration When oxygen is unavailable for aerobic respiration, some bacteria are able to carry out respiration (i.e., generation of ATP via an electron transport system of coenzymes), using oxidized substrates instead of oxygen as the terminal electron acceptor. The two major compounds used as electron acceptors in anaerobic respiration are nitrate and sulfate.

Nitrate Respiration When reduced all the way to dinitrogen gas via denitri-fication, nitrate is nearly equivalent to oxygen as a terminal electron acceptor in terms of energy yield (Table 2). Many aerobic bacteria are facultative nitrate respirers; they are able to substitute nitrate as the terminal electron acceptor when no oxygen is present. The capacity for nitrate respiration is widespread among genera of heterotrophic bacteria. The end products of nitrate respir-ation may be nitrite, ammonium, or in the case of denitrification, nitrous oxide or dinitrogen gas.

Nitrate Respiration — Denitrification A subset of heterotrophic, nitrate-respiring bacteria are able to reduce nitrate and nitrite through sequential steps to nitrogen gas, a process termed denitrification:

$$NO_3^- \rightarrow NO_2^- \rightarrow NO \rightarrow N_2O \rightarrow N_2$$

Utilization of nitrate as a terminal electron acceptor is dissimilative nitrate *reduction,* as opposed to assimilative nitrate reduction, in which nitrate is utilized as a nitrogen source in anabolic pathways. Denitrification in marine systems is discussed in more detail in Chapter 15.

Sulfate Respiration Sulfur-reducing bacteria are obligate anaerobes that can use sulfate, thiosulfate, and even elemental sulfur as terminal electron acceptors in respiration. Sulfide, or hydrogen sulfide, is the major by-product. The "rotten egg" stench of hydrogen sulfide gas is a powerful diagnostic of anoxic water. Reduction of oxidized sulfur compounds produces much less free energy than reduction of oxygen or nitrate. Oxidation of a mole of lactate using nitrate as the terminal electron acceptor yields 298 kcal, whereas if sulfate is the electron acceptor, only 9.7 kcal of free energy is liberated. ATP can be generated using sulfate as electron acceptor only if the electron donor (the substrate being oxidized) can yield reduced coenzymes along the electron transport chain. Such substrates include hydrogen and organic acids (e.g., acetate). Since hydrogen and acetate are also the preferred substrates of methanogens, substrate competition by sulfate-reducing bacteria inhibits meth-anogenesis in marine anaerobic environments.

Fermentation The least efficient trophic mode for heterotrophic microbes is the process of fermentation, a strictly anaerobic process. During fermentation, electrons are directly transferred from a more reduced to a more oxidized organic compound within the cytoplasm of the cell. In the process, ATP is directly formed via substrate level phosphorylation. The classic fermentative pathway is glycolysis, in which one molecule of glucose (C_6) is split into two molecules of pyruvate (C_3), which are subsequently reduced to ethanol or lactate (C_2) plus carbon dioxide. This catabolic pathway requires an input of two molecules of ATP, and it yields four molecules of ATP for each molecule of glucose oxidized. Although sugars are the preferred substrates for fermenting bacteria, a wide variety of other low molecular weight organic compounds, as well as polymeric compounds which are cleaved into subunit molecules by extracellular enzymes, are also used.

The by-products of fermentation are hydrogen, organic acids, and alcohols. Buildup of these metabolites in the immediate environment decreases the efficiency of fermentation. Other groups of anaerobes, methanogens, and sulfur-reducing bacteria grow at the expense of the by-products of fermentation and act as sinks for hydrogen and organic acids. Thus the presence of mixed species assemblages of fermenting bacteria, methanogens, and anaerobic respiring bacteria results in improved growth rates for all these organisms. Fermentation can proceed more rapidly at low levels of metabolites, and the respirers have a constant supply of substrates required for their metabolism (Fenchel and Findlay 1995).

Anaerobic processes are not usually considered to be of significance in pelagic microbial food webs in the sea. However, anoxic conditions can occur inside organic particles such as fecal pellets (Alldredge and Cohen 1987). Also, large areas of the oceans often exhibit suboxic conditions below the euphotic zone; most of these are regions of upwelling-induced phytoplankton blooms. The Indian Ocean, for example, has a large region of low-oxygen water at intermediate depths (Owens et al. 1991). In addition, increased eutrophication of coastal regions has led to frequent formation of suboxic/anoxic water masses in which anaerobic metabolism replaces aerobic metabolism in microbial communities. Anoxia in subsurface waters has become a growing problem in European coastal waters, as well as in the Chesapeake Bay and at the mouth of the Mississippi River (Turner and Rabalais 1994). Understanding anaerobic processes is likely to assume increasing importance in the study of microbes in marine pelagic systems.

MARINE PROTISTS

Unicellular eukaryotic organisms of microscopic size are termed *protists* (Margulis et al. 1990). Older terminology for marine protists divided these organisms into *algae* to designate phototrophic cells, and *protozoa* to designate heterotrophic cells. In light of what is now known about phylogenetic affiliations and mixing of trophic modes among protists, such terminology has become inadequate (Patterson and Larsen 1992, Corliss 1995). Several examples can be cited of the problems with the use of "algae" versus "protozoa": numerous cases in which strictly heterotrophic protists are related taxonomically to phototrophic protists; phytoflagellates that also ingest bacteria (see Chapter 16); and ciliates that are either completely phototrophic or partially phototrophic as a result of chloroplasts sequestered from their phytoplankton prey. We suggest abandoning the term "protozoa," which means "first animal life," since protists are really neither plant nor animal. "Algae" remains a useful way to describe eukaryotic phototrophs as opposed to prokaryotic phototrophs. Since many volumes have already been published on marine phytoplankton, here we focus on heterotrophic protists in marine pelagic food webs (see also Chapters 12 and 16).

Phylogenetic Diversity of Marine Heterotrophic Protists

The timing of the origin of eukaryotic cells in the early history of the biosphere has not yet been well established. Besides the development of a nucleus, the major innovation of eukaryotic cells was the production of an internal cytoskeleton and an internal system of membranes (Cavalier-Smith 1991a). This resulted in cells that exceeded prokaryotes in cytoplasmic complexity and enabled a new trophic mode based on phagocytosis and internal digestion of particulate food. The original eukaryotes were undoubtedly anaerobic. There are modern anaerobic protists that do not have mitochondria, for example, microsporidians, flagellated parasites of animals (Cavalier-Smith 1991b; Dyer and Obar 1994, Figure 1). However, there is recent evidence that even these "primitive" eukaryotic microbes once harbored mitochondria (Katz 1998).

The origin of mitochondria and chloroplasts is thought to be via establishment of symbiotic associations of Bacteria with primitive, anaerobic hosts in the eukaryotic line of descent (Margulis 1981; Cavalier-Smith 1991a). This concept of the origins of eukaryotic organelles is known as the endosymbiotic theory. Presumably such symbiotic relationships developed via phagocytotic ingestion of prey cells (Cavalier-Smith 1991a; McFadden and Gilson 1995). In the modern world, examples are known of incomplete symbioses between photosynthetic nonsulfur bacteria capable of aerobic respiration and anaerobic benthic ciliates (Fenchel and Bernard 1993), and between algae and phagotrophic protists (Taylor 1990, Chapter 16).

Phylogenetic relationships among protists have been worked out via nucleotide sequencing of small subunit (18S) rRNA, in analogy to studies of bacterial phylogeny via sequencing 16S rRNA (Schlegel 1991, 1994; Knoll 1992; Wainright et al. 1993; Leipe et al. 1994; Wright and Lynn 1997). An interpretive figure of relationships based on molecular genetic information (Figure 1) illustrates the rich diversity of protist groups that occur in marine systems. The euglenoids and kinetoplastids form a separate group. Dinoflagellates are most closely affiliated with ciliates. Heterotrophic flagellates are spread throughout

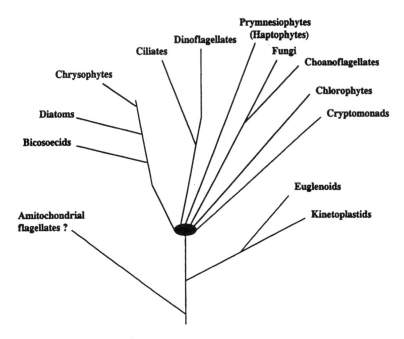

Figure 1. Interpretive diagram of general phylogenetic relationships among eukaryotic microbes that, except for the amitochondrial groups, are common in marine systems. The relationships are based on sequence comparisons of small subunit rRNA presented in Wainright et al. (1993), Leipe et al. (1994), Schlegel (1994), and Katz (1998). Genetic distances depicted here are not drawn to absolute scale. The status of the "primitive" amitochondrial flagellates is currently in question (Katz 1998). Details of phylogenetic relationships among eukaryotes are likely to change as new information becomes available. Phagotrophy (i.e., ability to ingest particles) has been demonstrated for heterotrophic and/or photosynthetic species in all the mitochondrial-bearing groups shown here except for the diatoms and fungi.

the protistan phylogenetic tree. Choanoflagellates, which have no close tax-onomic affinity to any chloroplast-bearing group of organisms, are the protists most closely related to fungi and to animals (Wainright et al. 1993). Other heterotrophic flagellates are within eukaryotic groups in which photosynthetic species are common (Schlegel 1991; Wainright et al. 1993). The chrysophyte lineage in particular includes a variety of both colorless and pigmented flagellates that are important bacterivores. The idea that these protist lineages diverged fairly early in earth history is supported by a molecular clock analysis of ciliate rRNA sequences that suggested an origin of the phylum Ciliophora about 2 billion years ago (Wright and Lynn 1997).

The spectrum of symbiotic relationships between phototrophic and phago-trophic protists that exists among modern marine protists provides insights into how ancestral symbioses may have occurred. For instance, many pelagic ciliates "enslave" the chloroplasts of their algal prey, sequestering them just under the cell membrane and obtaining fixed carbon from the still active plasmids as a supplemental food source (Stoecker et al. 1987). The ciliate *Mesodinium rubrum* is completely autotrophic via an obligate symbiosis with a cryptomonad that has lost most of its nuclear DNA (Lindholm 1985). On the other hand, many groups of phytoflagellates are primarily autotrophic but also may ingest particulate prey (Sanders and Porter 1988, Chapter 16), revealing the phagotrophic nature of the plastid-bearing eukaryotic host.

Appreciation of the phylogenetic diversity among marine protists is helpful in understanding their trophic roles in pelagic food webs. Individual species of these organisms cannot easily be segregated into traditional trophic levels of "primary producer" and "heterotrophic consumer." One might also expect a great diversity in phenotypic expression of protistan genomes separated by such great evolutionary distance: for example, diversity in behavior, nutritional modes, and adaptation to extreme environments.

Major Types of Marine Heterotrophic Protist

Although broad phylogenetic affiliations among groups of protists are now being elucidated via molecular genetics, identification of individual species, particularly among the marine flagellates, is still in a state of flux. We will discuss the general groups of marine heterotrophic protists that share morphological characteristics that can be identified by microscopy. In contrast to prokaryotic organisms, which are diverse in metabolism but not in morphology, protists are highly diverse morphologically but limited in metabolic mode to phototrophy, phagotrophy, and mixtures of the two, a combined trophic mode termed "mixotrophy" (see Chapter 16). The various taxa of marine planktonic protists are discussed under the headings of nanoflagellates (heterotrophic and mixotrophic flagellates 2–20 μm in size) and microzooplankton (heterotrophic and mixotrophic dinoflagellates and ciliates of approximately 10–200 μm). Marine sarcodines (amoebae, radiolarians, foraminiferans, acantharians, and heliozoans) are not considered here because at present there is insufficient information about how, and to what extent, these protists contribute to energy and mineral flows within marine pelagic food webs (Caron and Swanberg 1990).

Nanoflagellates (2–20 μm)   Microscopic inspection of organisms in seawater, first with inverted light microscopy, later with epifluorescence methods, yielded the routine observation that a large number of planktonic flagellated cells less than 20 μm in size did not have chloroplasts. These organisms have been referred to as "colorless, or nonpigmented, microflagellates," "heterotrophic nanoplankton (HNAN)," or as Sieburth and Estep (1985) suggested, the tongue-twisting "aplastidic nanomastigotes." Another term, "zooflagellates," is clearly inappropriate, inasmuch as many of these organisms are closely related to phytoflagellates. Such "lumping" terminology tends to reinforce the notion that all small, nonpigmented flagellates form one cohesive group. However, in view of the great phylogenetic diversity among flagellated protists, this assemblage of marine protists should not, a priori, be considered to be a coherent entity in terms of either systematics or trophic mode. Figure 2 depicts examples of heterotrophic flagellates (excluding dinoflagellates) in three size classes: small, medium, and larger flagellates. The following taxa of nonpigmented, nanoplankton-sized flagellates are frequently observed in, and isolated from, seawater.

Heterokont (Having Two Different Flagella) Flagellates   Among the most numerically abundant group of heterotrophic flagellates are 2–10 μm, spherical to oval cells with two flagella, one or both of which has flagellar hairs (mastigonemes). Individual species are difficult to distinguish by means of ordinary microscopic methods; morphological characteristics such as flagellar hairs, or distinctive scales covering the body of the cell, where present, must be visualized via scanning electron microscopy. Taxonomically identifiable groups of heterokont flagellates (Preisig et al. 1991) include the following:

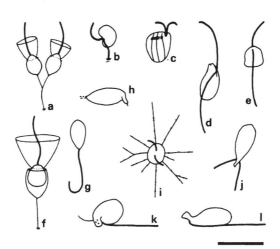

Figure 2. Examples of small (I), medium (II), and large (III) marine heterotrophic flagellates. (Ia) *Codosiga botrytis* (Ehr.) Kent, 1880; (Ib) *Cafeteria roenbergensis* Fenchel and Patterson 1988); (Ic) *Goniomonas pacifica* (Larsen and Patterson 1988) (Id) *Bordnamonas tropicana* (Larsen and Patterson 1990); (Ie) *Caecitellus parvulus* (Griessmann) (Patterson et al. 1993); (If) *Salpingoeca infusorium* Kent 1880; (Ig) *Metromonas simplex* (Griessmann) (Larsen and Patterson 1990); (Ih) *Amastigomonas debrunei* De Saedeleer 1931; (Ii) *Massisteria marina* (Patterson and Fenchel 1990); (Ij) *Telonema subtile* Griessmann 1913; (Ik) *Ancyromonas sigmoides* Kent 1880; (Il) *Rhynchomonas nasuta* (Stokes) Klebs 1892. (IIa) *Pteridomonas danica* (Patterson and Fenchel 1985); (IIb) *Bicosoeca conica* Lemmermann 1914; (IIc) *Pseudobodo tremulans* Griessmann 1913; (IId) *Bodo designis* Skuja 1948; (IIe) *Paraphysomonas imperforata* Lucas 1967; (IIf) *Ciliophrys infusionum* Cienkowski 1876. (IIIa) *Diplonema ambulator* (Larsen and Patterson 1990); (IIIb) *Ploeotia costata* (Farmer and Triemer 1988); (IIIc) *Ebria tripartita* (Schumann) Lemmermann 1899; (IIId) *Leucocryptos marina* (Braarud) (Butcher 1967); (IIIe) *Diaphanoeca grandis* Ellis 1930. Bar scale, 10 μm. Drawings kindly provided by Naja Vors.

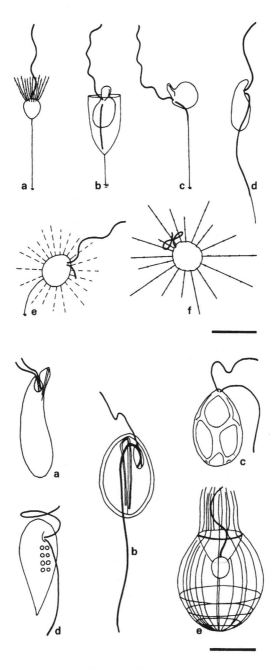

Figure 2. (*Continued*).

1. Chrysomonads. Cells with two unequal flagella, one long with hairs, and one shorter, smooth flagellum. The longer flagellum is directed forward and beats in a wavelike pattern. The most commonly isolated marine heterotrophic flagellates of this group are *Paraphysomonas* spp. particularly *Paraphysomonas imperforata* (Figure 2, IIe), which has been used by several investigators in studies of feeding and growth of marine flagellates under controlled conditions (e.g., Goldman and Caron 1985; Choi and Peters 1992). *Paraphysomonas* spp. are covered with siliceous scales diagnostic for individual species. The chrysomonad genus *Ochromonas*, noted for mixotrophy in freshwater environments (Sanders and Porter 1988), includes marine species.

2. Bicosoecids: Flagellation is similar to the chrysomonads, however, all species in this group are heterotrophic. Three genera are found in marine waters: *Bicosoeca* spp. (e.g., *Bicosoeca conica*, Figure 2, IIb), which are attached by the shorter, smooth flagellum to a cup-shaped lorica; *Pseudobodo* spp. (e.g., *Pseudobodo tremulans*, Figure 2, IIc), small spherical cells that attach to surfaces by the shorter flagellum and superficially resemble bodonids; and *Cafeteria* spp. (e.g., *Cafeteria roenbergensis*, Figure 2, Ib), similar to bicosoecids but without a lorica.

Cafeteria is a recently described genus (Fenchel and Patterson 1988), species of which have been found in coastal seawater in many locations. These flagellates grow very well under conditions of high food density. How *Cafeteria* got its name is a story in itself. D. J. Patterson and Tom Fenchel isolated the type species from Danish coastal waters. As they sat with beers at a sidewalk cafe, mulling over what to call this previously unidentified protist, they looked across the water to the site at which they had collected the sample containing the flagellate. A neon sign advertising a restaurant gleamed over the spot; it said "Cafeteria." So that became the genus name of the new species (Patterson, personal communication). The species name was derived from the small Danish village near the sampling site, thus the name of this flagellate literally means "Roenbergen's cafeteria." [Patterson and Fenchel (1990) also whimsically labeled another newly identified genus of marine flagellate *Massisteria*, Figure 2, Ii].

3. Pedinellids. These flagellates are in the crysophyte group and have distinctive tentacles in rings around the flagellum. Only one flagellum bearing hairs is present, the other being reduced to a basal body. Some species have anterior stalks with which they attach to surfaces. Common coastal pedinellids are *Pteridomonas danica* and *Ciliophys infusionum* (Larson and Souria 1991; Preisig et al. 1991) (Figure 2, IIa and IIf).

4. Choanoflagellates. Choanoflagellates are a distinctive, routinely observed, and often significant component of nanoflagellate assemblages in the sea (Thomson and Buck 1991). These flagellates are strictly heterotrophic and not related to any pigmented forms (Figure 1). Choanoflagellates are spherical to oval cells, 3–10 μm long, with a single flagellum and a collar formed of pseudopodial filaments at the oral end of the cell (Figure 2, Ia and If). One or two food vacuoles are generally found at the other end of the cell. Species of choanoflagellates in the family Acanthoecidae form elaborate siliceous tests, or loricae, around themselves (Figure 2, IIIe). Choanoflagellates may be free-swimming or attached to particles in the water. Attached choanoflagellates can form dense colonies on particles such as the frustules of senescent diatoms.

5. Bodonids. Bodonids are kinetoplastids, so named for the presence of the kinetoplast, an unusual concentration of DNA in the single mitochondrion (Zhukov 1991). Visualization of the kinetoplast after treating specimens with a DNA-specific stain is diagnostic. Species in the common marine genus *Bodo* are easily isolated from eutrophic coastal seawater (Figure 2, IIId). Small oval flagellates, usually not more than 4–10 μm long, bodonids tend to be surface-associated and are capable of gliding along a surface using one of the two flagella.

6. Nanodinoflagellates. Nonarmored dinoflagellates are often the most abundant cells in the group of nanoflagellates exceeding 5 μm in size (Verity et al. 1993). One dinoflagellate genus, *Gymnodinium* spp. contains 8–15 μm species that have been found to be abundant (10^4–10^5 cells L^{-1}), in the Northwest Pacific (Strom 1991), in estuarine waters of the Southeastern U.S. Atlantic coast (Sherr et al. 1991), and in the Antarctic Ocean (Bjørnsen and Kuparinen 1991). Nanodinoflagellates are spherical, oval, or conical cells that usually can be identified by a large nucleus, and/or by the sulcus, a groove in the middle of the cell in which one of the two flagella is inserted (Larsen and Souria 1991).

The flagellate taxa above contain species that are not difficult to isolate and culture in the laboratory. Most of the easily culturable species are isolated from food-rich coastal or estuarine waters and tend to be forms that prefer attaching to, or gliding along, surfaces. Other morphological types are often observed in seawater but have not yet been brought into culture; their taxonomic affinity to other flagellates is uncertain. One of these is *Leucocryptos* spp. (Patterson and Zolffel 1991; Vors 1992), 5–20 μm long oval cells with two equilength flagella, which superficially resemble autotrophic cryptomonads (Figure 2, IIId). When these organisms are stained with a fluorochrome such as DAPI, two rows of extrusomes may be seen. Vors (1992) speculated that the primary food of this common marine flagellate was phytoplankton. An effort to isolate previously uncultured species of marine flagellates such as *Leucocryptos* would be useful to balance our understanding of the behavior and trophic capabilities of marine nanoflagellates.

Mixotrophic Nanoflagellates   A variety of taxa of phytoflagellates have been shown to be mixotrophic: that is, capable of both photosynthesis and ingesting particulate prey, for the most part bacteria (Sanders and Porter 1988; Sanders 1991). These organisms are described in detail in Chapter 16. Marine phytoflagellates capable of ingesting prey include the prymnesiophyte genera

Prymnesium, Coccolithus, and *Chrysochromulina* (Green 1991), the prasino-phyte (a class of chlorophytes) *Micromonas pusilla* (Gonzalez et al. 1993), and several dinoflagellate genera (Bockstahler and Coats 1993; Jacobson and Anderson 1996).

Microzooplanktonic Protists (20–200 μm) The first intensively studied marine protists in the microplankton size range were tintinnids, pelagic choreotrichous ciliates that elaborate houses, or loricae, around themselves (Heinbokel 1978; Rassoulzadegan and Etienne 1981; Stoecker et al. 1983; Verity 1985). The loricae were an advantage for researchers working with this group: the shape and composition of the loricae are species-specific, and they protect the delicate ciliates inside them when samples are collected with fine-mesh plankton nets. However, it was recognized early on that tintinnids represented only a small fraction of the entire assemblage of planktonic ciliates.

Species composition of microplankton has been routinely evaluated using the Utermohl technique: whole water samples are preserved with iodine-based acid Lugol's solution, and then microbial cells are allowed to settle to the bottom of a chamber for inspection by inverted light microscopy, in which the objective lens of the microscope is beneath the specimen rather than above it. This method, as well as other approaches such as live counting (Sorokin 1980), revealed that more than 90% of all pelagic ciliates enumerated did not have loricae; rather, they were "aloricate" or "naked" (Beers and Stewart 1969; Rassoulzadegan 1977; Sorokin 1980; Smetacek 1981; Burkill 1982). Later studies showed that the assemblages of ciliates in the plankton included abundant cells smaller than 20 μm in size (Sherr et al. 1986) and that ciliates typical of benthic environments were at times present in the plankton, in association with large organic aggregates (Silver et al. 1984).

Initial research on marine protists in the "microzooplankton" focused almost exclusively on ciliates, ignoring the presence of phagotrophic dinoflagel-lates. In part this resulted from the conventional wisdom that marine dino-flagellates were phytoplankton, and in part from the use of Lugol's solution to preserve microzooplankton for enumeration. Since the iodine in Lugol's solution stains cells dark brown, autotrophic and heterotrophic dinoflagellates cannot be easily distinguished. However, protozoologists had long recognized that about half of described species of dinoflagellates did not contain chloro-plasts and thus must be heterotrophic (Gaines and Elbrächter 1987). Interest in the roles of phagotrophic dinoflagellates in marine food webs was initially sparked by visibly dense aggregations of large, nonarmored, bioluminescent *Noctiluca* spp. in coastal waters (Uhlig and Sahling 1982). Lessard and Swift (1985, 1986) later demonstrated that heterotrophic dinoflagellates were a diverse and quantitatively important component of the microzooplankton. Armored nonpigmented dinoflagellates were shown to be capable of feeding extracellularly on diatom chains larger than themselves (Gaines and Elbrächter 1987). More recent studies have revealed that assemblages of marine hetero-trophic dinoflagellates are often numerically dominated by cells smaller than

20 μm (Verity et al. 1993). Most heterotrophic marine dinoflagellates appear to be primarily herbivorous, feeding on phytoplankton cells over the entire size range of picoplankton to microplankton (Smetacek 1981; Bjørnsen and Kuparinen 1991; Hansen 1991; Lessard 1991). Dinoflagellates can consume a variety of prey particles of other types, including copepod eggs and nauplii (Jeong 1994) and either bacteria or bacterivorous protists (Lessard and Swift 1985; Lessard and Rivkin 1986). The original view that the assemblage of microzooplanktonic protists consisted mainly of ciliates larger than 20 μm has been amended to recognize that phagotrophic dinoflagellates are equally important in this size range, and also that herbivorous ciliates and dinoflagellates smaller than 20 μm may at times be an abundant component of the "microzooplankton."

Ciliates   The most common morphological types of ciliate found in the microplankton are in the subclass Choreotricha (Small and Lynn 1985). These ciliates are spherical, oval, or conical filter feeders with a crown of cilia at the oral end, and multinucleate with at minimum one macronucleus and one micronucleus (a feature diagnostic of ciliates in general). Older terms used for these planktonic ciliates include "oligotrichs" and "spirotrichs" (Maeda and Carey 1985; Maeda 1986). Broad groups of choreotrichous ciliates include subclass *Choreotricha*, order *Choreotrichia*: oral cilia form a complete circle at the anterior end of the cell. Ciliates in this order include the loricate tintinnids and aloricate species in the genera *Strombidinopsis*, *Strobilidium*, and *Lohmanniella*. Subclass *Choreotricha*, order *Oligotrichida*: oral cilia form an incomplete circle at the anterior end of the cell. These include common pelagic ciliates in the genera *Halteria*, *Laboea*, *Tontonia*, and *Strombidium*. *Laboea spiralis* is a large, distinctive ciliate with a line of cilia spiraling up the cell. This species is a well-known mixotroph that harbors chloroplasts from ingested algal prey (McManus and Fuhrman 1986). *Tontonia* spp. are unusual ciliates having at the posterior end a thick appendage, or tail, that contracts and expands during swimming. Other ciliates that occur in the plankton are haptorians, didinids, and urotrichs. *Mesodinium rubrum* is a strictly autotrophic didinid ciliate with cryptomonad-like chloroplasts; occasionally these ciliates form blooms and can be major contributors to primary production in the microplankton size range (Lindholm 1985).

Dinoflagellates Larger than 20 μm   Microzooplanktonic dinoflagellates are strong swimmers, often large. Many species are quite bizarrely shaped, and the predatory behavior of certain of these protists is quite terrible for their victims. Some dinoflagellates produce deadly neurotoxins, suck out the cytoplasm of prey with feeding tubes, or envelop their prey with a membranous feeding veil. One of these dinoflagellates, *Pfiesteria piscicida*, is responsible for massive fish kills in estuaries (Burkholder et al. 1992). Most of the 2000 dinoflagellate species so far described are not so dangerous. On the order of 40–60% of these are nonpigmented and therefore presumed to be heterotrophic (Taylor 1987).

Recent investigations in marine systems indicate that the job of assessing the diversity and trophic roles of dinoflagellates is only beginning. As mentioned earlier, there are abundant small ($<20 \, \mu$m) naked forms that resemble species in the genera *Gymnodinium*, *Gyrodinium*, and *Katodinium*. The heterotrophic dinoflagellate most often cultured for experimental work, *Oxyrrhis marina*, is a benthic species that is likely not a good model for pelagic dinoflagellates.

A "typical" dinoflagellate is oval to pear-shaped, with one groove, the girdle, encircling the middle of the cell, and a second groove, the sulcus, reaching from the girdle down one side of the cell. Motility is accomplished by two flagella, one wrapped around the girdle, and a second, posterior flagellum beating in the groove of the sulcus. Dinoflagellates are strong swimmers. Larger cells can attain speeds of 1 meter per hour and thus can undergo fairly extensive migrations in the upper water column. Dinoflagellate cells have a complex outer membrane system that contains vesicles. In some species, cellulosic plates in the vesicles fit together over the cell "like a suit-of-armor" (Taylor 1987). Dinoflagellates that have thecal plates are said to be "armored"; these include about 280 marine species of *Protoperidinium* and some 200 species of *Dinophysis*. Species without cellulosic plates are termed "unarmored" or "naked." Dinoflagellates also possess a unique nucleus, the dinokaryon, whose chromosomes never completely unwind in the nondividing stages of the life cycle. When viewed via electron microscopy or epifluorescence microscopy, the dinoflagellate nucleus has a fibrillar or grainy texture owing to the condensed chromosomes.

Variations in dinoflagellate morphology seem endless. Armored dinoflagellates assume spectacularly bizarre shapes, with flanges, horns, and spines of various shapes extending from the cell. The flagella may emerge from the middle, anterior, or posterior part of the dinoflagellate cell. Dinoflagellates may be flattened like a plate, form chains of cells, or creep around like amoebae (Taylor 1987). The $200-2000 \, \mu$m *Noctiluca* sp. consists of a large internal vacuole with a thin, peripheral layer of cytoplasm just beneath the cell membrane. Some species of dinoflagellates, including *Noctiluca* (whose generic name means "night light") are bioluminescent, producing blue-green light when disturbed. Other heterotrophic species possess an autofluorescent compound that emits blue-green light when stimulated by blue light (Shapiro et al. 1989). Blue-green fluorescence occurs among both larger-sized armored cells and nanoplanktonic naked dinoflagellates; this property is easily visualized via epifluorescence microscopy (Shapiro et al. 1989). Under unfavorable conditions, dinoflagellates may form resting stages, or cysts, which generally sink to the bottom (Taylor 1987).

Marine Fungi

Other heterotrophic eukaryotes occur in the sea: species of filamentous fungi and yeasts adapted to saline enviroments (Sieburth 1979; Newell 1993, 1994). These organisms are osmotrophic or saprotrophic; that is, they live on dissolved organic matter, or as decomposers of particulate organic matter. Marine fungi are mainly important in coastal systems, as decomposers of vascular plants such as *Spartina* (Newell 1993). Yeasts occur in the open ocean as parasites of copepods (Sieburth 1979). Since marine fungi do not play a major role in pelagic ecosystems (Newell 1994), they are not considered further here.

MICROBIAL FOOD WEBS IN THE SEA

Our conceptualization of the roles of these diverse groups of heterotrophic microbes in marine ecosystems has been evolving ever since the landmark paper of Pomeroy (1974). Prior to the mid-1970s, virtually the only microbial component of pelagic food webs given serious attention was the "net" phytoplankton (i.e., algal cells in the plankton that could be captured using the finest mesh plankton nets). The "net" phytoplankton was taxonomically dominated by centric diatoms and dinoflagellates. The original concept of marine food webs, as presented in Steele's classic book *The Structure of Marine Ecosystems* (1974), was thus based on the primary production of what are now recognized as the largest-sized phytoplankton cells in the sea. In the "classic" marine food chain model, the production of diatoms and dinoflagellates was consumed by copepods, which in turn were eaten by larger consumers; at the end of the food chain were commercially important fish. In Steele's words: "The phytoplankton of the open sea is eaten nearly as fast as it is produced, so that effectively all plankton production goes through the herbivores," by which he meant macrozooplankton. Heterotrophic bacteria were relegated to a "decomposer" role, analogous to the role of bacteria and fungi in terrestrial systems; their only formal part in Steele's food chain model was as a source of nutrition for the benthos.

Such was the status of the concept of marine food webs in the 1960s, when L. R. Pomeroy and his colleague R. E. Johannes began working together in the estuaries of the southeastern United States, at the University of Georgia Marine Institute on Sapelo Island. Early research by John Teal, Eugene Odum, and their colleagues in the salt marsh estuaries surrounding Sapelo Island was instrumental in the development of the concept of detritus-based marine ecosystems, in which food webs were largely based on nonliving organic matter. By the time Pomeroy and Johannes arrived at the Marine Institute, the University of Georgia had become a center for ecosystem ecology. Both Pomeroy and Johannes were interested in elemental cycling. Based in part on Johannes's research showing that marine heterotrophic flagellates had very high biomass-specific rates of excretion of phosphorus (Johannes 1964), the two colleagues set out to demonstrate that dark-bottle respiration, measured as decrease in oxygen concentration, was mainly due to microbes, not to larger-sized plankton such as copepods. Pomeroy and Johannes compared the relative respiration in a volume of seawater due to planktonic organisms in fractions greater and less than the mesh size of a No. 2 plankton net, or 366 µm. Their studies, as well as subsequent work by others, indicated that on a per-unit-volume basis, the smaller size class, mainly microbes, had a respiration rate on the order of 10 times greater than that of larger plankton, mainly metazoans.

By the early 1970s, Pomeroy had deduced that John Steele's food chain model was an inadequate representation of the structure of marine food webs. Drawing on his own work with Johannes, the detrital ecosystem concept, and new information about the potential importance of phytoplankton smaller in size than "net" diatoms and dinoflagellates, Pomeroy drafted a manuscript entitled "The Ocean's Food Web: A Changing Paradigm." After a difficult struggle with reviewers entrenched in the old food chain concept, the paper finally appeared in *BioScience* in 1974. It is now widely regarded as the single most influential paper spurring the subsequent interest in marine microbes and is a classic example of a revolution in scientific theory.

The central points of Pomeroy's "changing paradigm" paper were as follows:

1. The main primary producers in the sea were not "net" phytoplankton, but rather "nanoplankton," defined as phototrophic cells less than 60 µm in size.
2. Microbes, or at least organisms smaller than 366 µm, were the plankton component responsible for the bulk of respiration, and thus metabolism, in seawater.
3. Nonliving organic matter, in both dissolved and particulate forms, is an important source of food in marine food webs and is primarily consumed by heterotrophic microbes.

This last point represents an important insight: the universality of the detritus food web. Pomeroy applied the detrital ecosystem concept of salt marsh estuaries to marine food webs in general. Pomeroy theorized "that the primary consumers [of phytoplankton production in the sea] are active bacteria, and that they are converting a substantial fraction of primary photosynthate, and secondarily produced dissolved organic materials as well, into microbial protoplasm. This could channel into higher trophic levels at least 30% more energy than we now estimate." A main fate of the bacterial biomass was consumption by small protists, and both bacteria and bacterivores in turn could be directly consumed by "other microorganisms, by mucus-net feeders such as salps, or certain pteropods." In a concluding statement, Pomeroy wrote "The new paradigm of the ocean's food web that is developing, as a result of recent studies of protistan activities and alternative pathways of organic matter, may contain many unseen strands." Microbial ecologists have been hard at work studing these "unseen strands" ever since.

Other researchers were also contemplating the role of microbes in the sea during the 1970s. A high efficiency of assimilation of sugars and amino acids by marine bacteria was demonstrated using radiolabeled substrates, and high rates of respiration by the smallest size fractions of marine plankton were found in additional marine systems. A classic review paper by P. le B. Williams (1981) confirmed and extended the view of microbial processes presented by Pomeroy (1974). Williams reviewed his own work on the relative amount of biomass, surface area, and respiration for organisms in various plankton size fractions made during a set of "big bag" experimental enclosure experiments in a British Columbia estuary. The results indicated that planktonic organisms smaller than 30 μm accounted for almost all of the respiratory activity and that the surface area of bacteria overwhelmingly dominated the total living surface area (Table 3). This second point highlights a central reason why marine bacteria play such a major role in energy and material fluxes in the sea: by virtue of their large combined surface area, these small and abundant cells have the highest probability of encountering, and interacting with, dissolved substances.

The concept of a "microbial loop" as an integral part of pelagic food webs (Azam et al. 1983) drew on ideas presented by Pomeroy (1974), Sieburth et al. (1978), and Williams (1981), as well as others. A representation of the "microbial loop" component of marine food webs is shown in Figure 3. This idea solidified the notion that heterotrophic microbes — bacteria, bacterivorous flagellates, and ciliate consumers of the flagellates — served primarily as a pathway for regeneration of organic nitrogen, phosphorus, and other bioactive elements, and represented a shunt of carbon and energy from the main phytoplankton-based food web.

During the 1980s, mounting evidence from laboratory cultures and from in situ observations of herbivory by phagotrophic protists suggested that protists might be dominant grazers of phytoplankton as well as of bacteria in marine systems (Sherr and Sherr 1994). Landry and Hassett (1982) proposed a technically simple method to assess the grazing of "microzooplankton" (most

Table 3. Comparison of biomass and living surface area of various groups of plankton in CEPEX experimental enclosure CEE-2

Planktonic Group	Biomass [μg dry wt L^{-1} (% total)]	Surface Area [cm^2 L^{-1} (% total)][a]
Bacterioplankton	26 (4.6)	24.6 (69)
Protozoa	9.2 (1.7)	0.3 (0.7)
Phytoplankton	310 (56)	10.7 (30)
Macrozooplankton	206 (37)	0.3 (0.9)
Total	551.2	35.9

[a]Surface area calculations were made assuming spherical geometry.

Source: Adapted from Williams (1981), Table 5. See Chapter 4, this volume, for more details on bacterial biomass in the sea.

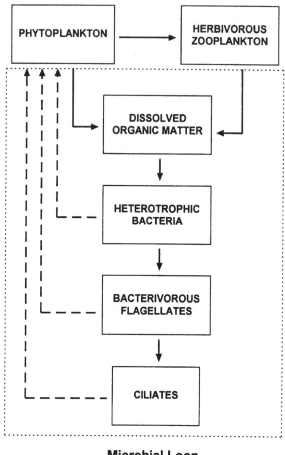

Microbial Loop

Figure 3. Conceptualization of the microbial loop, redrawn from Figure 2 of Ducklow (1983). Solid arrows represent trophic transfers of organic matter. In this model, heterotrophic bacteria consume dissolved organic matter obtained directly from phytoplankton or indirectly from herbivores feeding on phytoplankton; bacteria are consumed by flagellates, which in turn are eaten by ciliates. At each step in the microbial loop food chain, inorganic nutrients (ammonium, phosphate) are regenerated for phytoplankton uptake (dashed arrows), and carbon is lost via respiration. This conceptualization depicts the microbial loop both as a sink for organic matter and as a major source of regenerated nutrients.

of which are protists); the technique involved determining phytoplankton growth rates in water samples that had been diluted to reduce grazing mortality. By the early 1990s, the Landry–Hassett dilution method had been used to estimate microzooplankton herbivory in various regions of the sea. The results indicated that protists routinely consumed between 25 and 100% of daily phytoplankton production (Sherr and Sherr 1994).

Appreciation of the significance of other trophic links has also changed our views of the structure and functioning of marine food webs. Viral lysis may be a significant source of mortality for bacteria and can represent a "short-circuit" of the food web (Proctor and Fuhrman 1990; Bratbak et al. 1992; see Chapter 11). Some species of pigmented flagellates can consume bacteria (see Chapter 16). Heterotrophic flagellates have been found to consume high molecular weight dissolved organic matter (e.g. Sherr 1988; Tranvik et al. 1993). Heterotrophic protists have been recognized as an important food resource for copepods, fish larvae, and small gelatinous zooplankton (Stoecker and Capuzzo 1990).

A revised view of the overall pelagic microbial food web, which incorporates post-1983 information, is shown in Figure 4. This "spaghetti diagram" captures

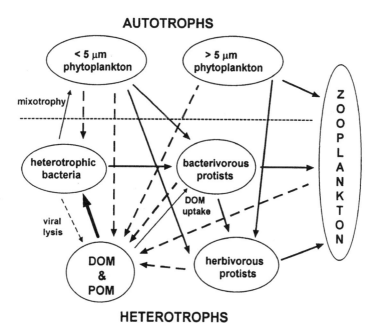

Figure 4. Conceptualization of the overall microbial food web, redrawn from the diagram of Sherr and Sherr (1988), based on information gained since the early 1980s. Solid arrows show pathways of consumption of organic matter; dashed arrows show pathways by which organic matter, both dissolved (DOM) and particulate (POM) is released from living organisms. The microbial food web is divided between autotrophic and heterotrophic microbes. Autotrophs are further separated into cells greater than 5 μm in size, which are large enough to be grazed by copepods and other multicellular zooplankton, and cells smaller than 5 μm, which are mainly consumed by protists. Zooplankton feed not only on the larger phytoplankton but also on bacterivorous protists (nanoflagellates and smaller ciliates) and on herbivorous protists (larger nanoflagellates, heterotrophic dinoflagellates, and ciliates). Three new pathways — mixotrophic consumption of bacteria by phytoflagellates, viral lysis of bacteria, and uptake of high molecular weight DOM by bacterivorous protists — are shown as well.

some of the complexity of relationships among microbes in the sea. Phytoplankton are divided into a larger size fraction (cells > about 5 μm) which can be directly consumed by zooplankton, and a smaller size fraction (cells < about 5 μm) which is consumed mainly by protists, which in turn can be food for larger protists or zooplankton. Bacteria derive nutrients from multiple pathways in the food web and are consumed by protists and by mixotrophic algae. Bacteria and protists, along with viral lysis, are responsible for most of the regeneration of nitrogen, phosphorus, and other elements from organic matter. Legendre, Rassoulzadegan, and their colleagues (Legendre and Le Fevre 1995; Legendre and Rassoulzadegan 1996; Rivkin et al. 1996; Thingstad et al. 1997) have further extended conceptualization of microbial food webs based on the newer information. These papers propose a theory for how an active versus inactive microbial loop within the overall pelagic food web can affect utilization of dissolved and particulate organic matter, and the sinking flux of biogenic carbon. Details of the roles of bacteria and heterotrophic protists as consumers of organic matter and as agents of major biogeochemical processes are explored in chapters that follow.

SUMMARY

1. Microbes, including Bacteria, Archaea, and protistan Eukarya, are ubiquitous, diverse, and exceedingly important components of marine ecosystems.
2. The metabolic diversity among marine prokaryotes is central to mediating reactions in various biochemical cycles in the sea. Microbial metabolic activities include photosynthesis, organic carbon degradation, sulfur oxidation and reduction, nitrogen fixation, nitrification, and denitrification.
3. In terms of carbon cycling, the two most important activities of prokaryotes in most pelagic marine systems are primary production by cyanobacteria and prochlorophytes and aerobic heterotrophy (i.e., the oxidatation of organic material by bacteria using oxygen as the terminal electron acceptor).
4. Phagotrophic protists are phylogenetically diverse and have vital roles in marine food webs as consumers of bacteria and of phytoplankton, as regenerators of inorganic nutrients, and as prey for copepods and other zooplankton.

REFERENCES

Alldredge, A. L., and Cohen, Y. (1987) Can microscale chemical patches persist in the sea? Microelectrode study of marine snow, fecal pellets. *Science* 235:689–691.

Angert, E. R., Clements, K. D., and Pace, N. R. (1993) The largest bacterium. *Nature* 362:239–241.

Azam, F., and Ammerman, J. W. (1984) The cycling of organic matter by bacterio-plankton in pelagic marine systems: Microenvironmental considerations. In M. J. R. Fasham ed., *Flows of Energy and Materials in Marine Ecosystems: Theory and Practice.* NATO Conference Series 4, *Marine Science.* Plenum Press, New York, pp. 345–360.

Azam, F., Fenchel, T., Field, J. G., Meyer-Reil, R. A., and Thingstad, F. (1983) The ecological role of water column microbes in the sea. *Mar. Ecol. Prog. Ser.* 10:257–263.

Beers, J. R., and Stewart, G. L. (1969) Microplankton and its abundance relative to the large zooplankton and other seston components. *Mar Biol.* 4:182–189.

Bjørnsen, P. K., and Kuparinen, J. (1991) Growth and herbivory by heterotrophic dinoflagellates in the Southern Ocean, studied by microcosm experiments. *Mar. Biol.* 109:397–405.

Bockstahler, K. R., and Coats, W. D. (1993) Spatial and temporal aspects of mixotrophy in Chesapeake Bay dinoflagellates. *J. Protozool.* 40:49–60.

Bratbak, G., Heldal, M., Thingstad, T. F., Riemann, B., and Haslund, O. H. (1992) Incorporation of viruses into the budget of microbial C-transfer. A first approach. *Mar. Ecol. Prog. Ser.* 83:273–280.

Burkholder, J. M., Noga, E. J., Hobbs, C. H., and Gusgow, H. B. (1992) New 'phantom' dinoflagellate is the causative agent of major estuarine fish kills. *Nature* 385:407–410.

Burkill, P. H. (1982) Ciliates and other microplankton components of a nearshore foodweb: Standing stocks and production processes. *Ann. Inst. Oceanogr.* Paris 58:335–350.

Butcher, R. W. (1967) An introductory account of the smaller algae of British coastal waters. IV. Cryptophyceae. *Fisheries Investigations London Series* 4:1–54.

Campbell, L., and Vaulot, D. (1993) Photosynthetic picoplankton community structure in the subtropical North Pacific near Hawaii (Station ALOHA). *Deep-Sea Res.* 40:2043-2060.

Capriulo, G. M., and Carpenter, E. J. (1983) Abundance, species composition and feeding impact of tintinnid micro-zooplankton in Central Long Island Sound. *Mar. Ecol. Prog. Ser.* 10:277–288.

Caron, D. A. (1983) Technique for the enumeration of heterotrophic and phototrophic nanoplankton, using epifluorescent microscopy, and comparison with other procedures. *Appl. Environ. Microbiol.* 46:491–498.

Caron, D. A., and Goldman, J. C. (1988) Dynamics of protistan carbon and nutrient cycling. *J Protozool.* 35:247–249.

Caron, D. A., and Swanberg, N. R. (1990) The ecology of planktonic sarcodines. *Rev Aquat. Sci.* 3:147–180.

Caron, D. A., Davies, P. G., Madin, L. P., and Sieburth, J. McN. (1982) Heterotrophic bacteria and bacterivorous protozoa in oceanic micro-aggregates. *Science* 218:795–797.

Caron, D. A., Goldman, J. C., and Dennett, M. R. (1986) Effect of temperature on growth, respiration and nutrient regeneration by an omnivorous microflagellate. *Mar Ecol Prog Ser.* 24:243–254.

Cavalier-Smith, T. (1991a) The evolution of cells. In S. Osawa, and T. Honjo, eds., *Evolution of Life: Fossils, Molecules, and Culture.* Springer-Verlag, Tokyo, pp. 271–304.

Cavalier-Smith, T. (1991b) Cell diversification in heterotrophic flagellates. In D. J. Patterson and J. Larsen, eds., *The Biology of Free-Living Heterotrophic Flagellates.* Clarendon Press, Oxford, pp. 113–131.

Chisholm, S. W. (1992) Phytoplankton size. In: P. G. Falkowski and A. D. Woodhead, eds., *Primary Productivity and Biogeochemical Cycles in the Sea.* Plenum Press, New York, pp. 213–231.

Choi, J. W., and Peters, J. (1992) Effects of temperature on two psychrophilic ecotypes of a heterotrophic nanoflagellate, *Paraphysomonas imperforata. Appl. Environ. Microbiol.* 58:593–599.

Chrost, R. J. (1991) Environmental control of the synthesis and activity of aquatic microbial ectoenzymes. In R. J. Chrost, ed., *Microbial Enzymes in Aquatic Environments.* Springer-Verlag, New York, pp. 29–59.

Cole, J. J., Findlay, S, and Pace, M. L. (1988) Bacterial production in fresh and salt water ecosystems: A cross-system overview. *Mar. Ecol. Prog. Ser.* 43:1–10.

Corliss, J. O. (1995) The ambiregnal protists and the codes of nomenclature: A brief review of the problem and of proposed solutions. *Bull Zool. Nomencl.* 52:11–17.

Ducklow, H. W. (1983) Production and the fate of bacteria in the oceans. *BioScience* 33:494–501.

Dyer, B. D., and Obar, R. A., eds. (1994) *Tracing the History of Eukaryotic Cells.* Columbia University Press, New York.

Farmer, M. A., and Triemer, R. E. (1988) Flagellar systems in the euglenoid flagellates. *Biosystems* 21:283–291.

Fenchel, T., and Bernard, C. (1993) Endosymbiotic purple non-sulfur bacteria in an anaerobic ciliated protozoon. *FEMS Microbiol. Lett.* 110:21–25.

Fenchel, T., and Blackburn, T. H. (1979) *Bacteria and Mineral Cycling.* Academic Press, London.

Fenchel, T., and Finlay, B. J. (1995) *Ecology and Evolution in Anoxic Worlds.* Oxford University Press, New York.

Fenchel, T., and Patterson, D. J. (1988) *Cafeteria roenbergensis* nov. gen., nov. sp., a heterotrophic microflagellate from marine plankton. *Mar. Microb. Food Webs* 3:9–19.

Fenchel, T., King, G. M., and Blackburn, T. H. (1998) *Bacterial Biogeochemistry.* Academic Press, San Diego, CA.

Gaines, G., and Elbächter, M. (1987) Heterotrophic nutrition. In F. J. R. Taylor, ed., *The Biology of Dinoflagellates.* Blackwell Scientific, Oxford, pp. 224–268.

Goldman, J. C., and Caron, D. A. (1985) Experimental studies on an omnivorous microflagellate: Implications for grazing and nutrient regeneration in the marine microbial food chain. *Deep-Sea Res.* 32:899–915.

Gonzalez, J. M., Sherr, E. B., and Sherr, B. F. (1993) Digestive enzyme activity as a quantitative measure of protistan grazing: The acid lysozyme assay for bacterivory. *Mar. Ecol. Prog. Ser.* 100:197–206.

Green, J. C. (1991) Phagotrophy in prymnesiophyte flagellates. In D. J. Patterson and J. Larsen, eds., *The Biology of Free-Living Heterotrophic Flagellates.* Clarendon Press, Oxford, pp. 401–414.

Hansen, P. J. (1991) Quantitative importance and trophic role of heterotrophic dinoflagellates in a coastal pelagic food web. *Mar. Ecol Prog. Ser.* 73:253–261.

Heinbokel, J. F. (1978) Studies on the functional role of tintinnids in the southern California bight. I. Grazing and growth rates in laboratory cultures. *Mar. Biol.* 52:23–32.

Jacobson, D. M., and Anderson, D. M. (1996) Widespread phagocytosis of ciliates and other protists by marine mixotrophic and heterotrophic thecate dinoflagellates. *J. Phycol.* 32:279–285.

Jeong, H. J. (1994) Predation by the heterotrophic dinoflagellate *Protoperidinium* cf. *divergens* on copepod eggs and early naupliar stages. *Mar. Ecol. Prog. Ser.* 114:203–208.

Johannes, R. E. (1964) Phosphorus excretion and body size in marine animals: Microzooplankton and nutrient regeneration. *Science* 146:923–924.

Joint, I. R. (1986) Physiological ecology of picoplankton in various oceanographic provinces. In T. R. Platt and W. K. W. Li, eds., *Photosynthetic picoplankton*, Can. Bull. Fish. Aquat. Sci. 214:287–309.

Katz, L. A. (1998) Changing perspectives on the origin of eukaryotes. *Trends Ecol. Evol.* 13:493–497.

Kelley, C. A., Coffin R. B., and Cifuentes, L. A. (1998) Stable isotope evidence for alternative bacterial carbon sources in the Gulf of Mexico. *Limnol. Oceanogr.* 43:1962–1969.

Knoll, A. H. (1992) The early evolution of eukaryotes: A geological perspective. *Science* 256:622–627.

Landry, M. R., and Hassett, R. P. (1982) Estimating the grazing impact of marine microzooplankton. *Mar. Biol.* 67:283–288.

Larsen, J., and Patterson, D. J. (1990) Some flagellates (Protista) from tropical marine sediments. *J. Nat. Hist.* 24:801–937.

Larsen, J., and Souria, A. (1991) The diversity of heterotrophic dinoflagellates. In D. J. Patterson and J. Larsen, eds., *The Biology of Free-Living Heterotrophic Flagellates.* Clarendon Press, Oxford, pp. 313–332.

Lavrentyev, P. J., Gardner, W. S., and Johnson, J. R. (1997) Cascading trophic effects on aquatic nitrification: Experimental evidence and potential implications. *Aquat. Microb. Ecol.* 13:161–175.

Legendre, L., and Rassoulzadegan, F. (1996) Food-web mediated export of biogenic carbon in oceans: hydrodynamic control. *Mar. Ecol. Prog. Ser.* 145:179–193.

Legendre, L., and Le Fevre, J. (1995) Microbial food webs and the export of biogenic carbon in oceans. *Aquat. Microb. Ecol.* 9:69–77.

Leipe, D. D., Wainright, P., and Gunderson, J. H. (1994) The stramenopiles from a molecular perspective: 16S-like rRNA sequences from *Labyrinthuloides minuta* and *Cafeteria roenbergensis. Phycologia* 33:369–377.

Lessard, E. (1991) The trophic role of heterotrophic dinoflagellates in diverse marine environments. *Mar. Microb. Food Webs* 5:49–58.

Lessard, E. J. (1993) Culturing free-living marine phagotrophic dinoflagellates. In P. F. Kemp, B. F. Sherr, E. B. Sherr, and J. J. Cole, eds., *Handbook of Methods in Aquatic Microbial Ecology.* Lewis Publishers, Boca Raton, FL, pp. 67–76.

Lessard, E. J., and Swift, E. (1985) Species-specific grazing rates of heterotrophic dinoflagellates in oceanic waters, measured with a dual-label radioisotope technique. *Mar. Biol.* 87:289–296.

Lessard, E. J., and Swift, E. (1986) Dinoflagellates from the North Atlantic classified as phototrophic or heterotrophic with epifluorescence microscopy. *J. Plankton Res.* 6:1209–1215.

Lessard, E. J., and Rivkin, R. B. (1986) Nutrition of microzooplankton and macrozooplankton from McMurdo Sound. *Antarct. J. U.S.* 21:187–188.

Li, W. K. W., Dickie, P.M., Irwin, B. D., and Wood, A. M. (1992) Biomass of bacteria, cyanobacteria, prochlorophytes, and photosynthetic eukaryotes in the Sargasso Sea. *Deep-Sea Res.* 39:501–519.

Lindholm, T. (1985) *Mesodinium rubrum* — A unique photosynthetic ciliate. *Adv. Aquat. Microbiol* 3:1–48.

Lynn, D. H., and Montagnes, D. J. S. (1991) Global production of heterotrophic marine planktonic ciliates, In P. C. Reid, C. M. Turley, and P. H. Burkill, eds., *Protozoa and Their Role in Marine Processes.* NATO ASI Series G, *Ecological Sciences,* Vol. 25. Springer-Verlag, Heidelberg and Berlin, pp. 281–308.

Madigan, M. T., Martinko, J. M., and Parker, J. eds. (1997) *Brock — Biology of Microorganisms,* 8th ed. Prentice Hall, Englewood Cliffs, NJ.

Maeda, M. (1986) An illustrated guide to the species of the Families Halteriidae and Strobilidiidae (Oligotrichia, Ciliophora), free swimming protozoa common in the aquatic environment. *Bulletin of the Ocean Research Institute of the University of Tokyo,* No. 21, 67 pp.

Maeda, M., and Carey, P. G. (1985) An illustrated guide to the species of the Family Strombidiiae (Oligotrichida, Ciliophora), free swimming protozoa common in the aquatic environment. *Bulletin of the Ocean Research Institute of the University of Tokyo,* No. 19, 68 pp.

Margulis, L. (1981) *Symbiosis in Cell Evolution: Life and Its Environment on the Early Earth.* W. H. Freeman, San Francisco.

Margulis, L., Corliss, J. O., Melkonian, M, and Chapman, D. J. eds. (1990) *Handbook of Protoctista.* Jones and Barlett, Boston.

McFadden, G., and Gilson, P. (1995) Something borrowed, something green: Lateral transfer of chloroplasts by secondary endosymbiosis. *Trends Ecol. Evol.* 10:12–17.

McManus, G. B., Fuhrman, J. A. (1986): Photosynthetic pigments in the ciliate *Laboea strobila* from Long Island Sound, USA. *J. Plankton Res.* 8:317–327.

Moestrup, O., and Andersen, R. A. (1991) Organization of heterotrophic heterokonts. In D. J. Patterson and J. Larsen, eds., *The Biology of Free-Living Heterotrophic Flagellates.* Clarendon Press, Oxford, pp. 333–360.

Moore, L. R., Rocap, G., and Chisholm, S. W. (1998) Physiology and molecular phylogeny of co-existing *Prochlorococcus* ecotypes. *Nature* 393:464–467.

Newell, S. Y. (1993) Decomposition of shoots of a salt-marsh cordgrass: Methodology and dynamics of microbial assemblages. *Adv. Microb. Ecol.* 13:301–326.

Newell, S. Y. (1994) Ecomethodology for organoosmotrophs: Prokaryotic unicellular versus eukaryotic mycelial. *Microb. Ecol.* 28:151–157.

Owens, N. J. P., Law, C. S., Mantoura, R. F. C., Burkill, P. H., and Llewellyn, C. A. (1991) Methane flux to the atmosphere from the Arabian Sea. *Nature* 354:293–296.

Palenik, B., and Haselkorn, R. (1992) Multiple evolutionary origins of prochlorophytes, the chlorophyll b–containing prokaryotes. *Nature* 355:265–267.

Patterson, D. J. (1993) The current status of the free-living heterotrophic flagellates. *J. Eukaryote Microbiol.* 40:606–609.

Patterson, D. J., and Fenchel, T. (1985) Insights into the evolution of heliozoa (Protozoa, Sarcodinia) as provided by ultrastructural studies on a new species of flagellate from the genus *Pteridomonas*. *Biol. J. Limn. Soc.* 34:381–403.

Patterson, D. J., and Fenchel, T. (1990) *Massisteria marina* Larsen and Patterson 1990, a widespread and abundant bacterivorous protist associated with marine detritus. *Mar. Ecol. Prog. Ser.* 62:11–19.

Patterson, D. J., and Larsen, J. (1992) A perspective on protistan nomenclature. *J. Protozool.* 39:125–131.

Patterson, D. J., and Zolffel, M. (1991) Heterotrophic flagellates of uncertain taxonomic position. In D. J. Patterson and J. Larsen, eds., *The Biology of Free-Living Heterotrophic Flagellates*, Clarendon Press, Oxford, pp. 427–476.

Patterson, D. J., Nygaard, K., Steinberg, G., and Turley, C. M. (1993) Heterotrophic flagellates and other protists associated with oceanic detritus throughout the water columns in the mid North Atlantic. *J. Mar. Biol. Assoc. U.K.* 73:67–95.

Pfennig, N. (1989) Ecology of phototrophic purple and green sulfur bacteria. In H. G. Schlegel and B. Bowien, eds., *Autotrophic Bacteria*, Science Tech Publications, Madison, WI, pp. 97–116.

Platt, T. R., and Li, W. K. W., eds. (1986) Photosynthetic picoplankton. *Can. Bull. Fish. Aquat. Sci.* 214:287–309.

Platt, T., Subba Rao, D. V., and Irwin, B. (1983) Photosynthesis of picoplankton in the oligotrophic ocean. *Nature* 316:747–749.

Pomeroy, L. R. (1974) The ocean's food web: A changing paradigm. *BioScience* 24:409–504.

Preisig, H. R., Vors, N., and Hallfors, G. (1991) Diversity of heterotrophic heterokont flagellates. In D. J. Patterson and J. Larsen, eds., *The Biology of Free-Living Heterotrophic Flagellates*. Clarendon Press, Oxford, pp. 361–399.

Proctor, L. M., and Fuhrman, J. A. (1990) Viral mortality of marine bacteria and cyanobacteria. *Nature* 343:60–62.

Rassoulzadegan, F. (1977) Evolution annuelle des ciliés pelagiques en Mediterranée nord-occidentale. Ciliés Oligotriches "non-tinninides" (*Oligotrichina*). *Ann. Inst. Oceanogr. Paris* 53:125–134.

Rassoulzadegan, F., and Etienne, M. (1981) Grazing rate of the tintinnid *Stenosemella ventricosa* (Clap. and Lachm.) Jorg. on the spectrum of the naturally occurring particulate matter from a Mediterranean neritic area. *Limnol Oceanogr.* 26:258–270.

Repeta, D. J., Simpson, D. J., Jorgensen, B. B., and Jannasch, H. W. (1989) Evidence for anoxygenic photosynthesis from the distribution of bacteriochlorophylls in the Black Sea. *Nature* 342:69–72.

Rivkin, R. B. et al. (1996) Vertical flux of biogenic carbon in the ocean: Is there food web control? *Science* 272:1163–1166.

Sanders, R. W. (1991) Trophic strategies among heterotrophic flagellates. In D. J. Patterson and J. Larsen, eds., *The Biology of Free-Living Heterotrophic Flagellates*, Clarendon Press, Oxford, pp. 21–38.

Sanders, R. W., and Porter, K. G. (1988) Phagotrophic flagellates. *Adv. Microb. Ecol.* 10:167–192.

Schlegel, M. (1991) Protist evolution and phylogeny as discerned from small subunit ribosomal RNA sequence comparisons. *Eur. J. Protistol.* 27:207–219.

Schlegel, M. (1994) Molecular phylogeny of eukaryotes. *Trends Ecol. Evol.* 9:330–335.

Shapiro, L., Haugen, E. M., and Carpenter, E. J. (1989) Occurrence and abundance of green-fluorescing dinoflagellates in surface waters of the Northwest Atlantic and Northeast Pacific Oceans. *J. Phycol.* 25:187–191.

Sherr, E. B. (1988) Direct use of high molecular weight polysaccharide by heterotrophic flagellates. *Nature* 335:348–351.

Sherr, E. B., and Sherr, B. F. (1988) Role of microbes in pelagic food webs: A revised concept. *Limnol. Oceanogr.* 33:1225–1227.

Sherr, E. B., and Sherr, B. F. (1994) Bacterivory and herbivory: Key roles of phago-trophic protists in pelagic food webs. *Microb. Ecol.* 28:233-235.

Sherr, E. B., Sherr, B. F., Fallon, R. D., and Newell, S. Y. (1986) Small, aloricate ciliates as a major component of the marine heterotrophic nanoplankton. *Limnol. Oceanogr.* 31:177–183.

Sherr, E. B., Sherr, B. F., and McDaniel, J. (1991) Clearance rates of <6 μm fluorescent-ly labeled algae (FLA) by estuarine protozoa: potential grazing impact of flagellates and ciliates. *Mar. Ecol. Prog. Ser.* 69:81–92.

Sieburth, J. McN. (1979) *Sea Microbes.* Oxford University Press, New York.

Sieburth, J. McN. (1987) Contrary habitats for redox-specific processes: Methanogen-esis in oxic waters and oxidation in anoxic waters. In M. A. Sleigh, ed., *Microbes in the Sea.* Ellis Horwood, Chichester, pp. 11–38.

Sieburth, J. McN. (1993) C_1 bacteria in the water column of Chesapeake Bay, USA. I. Distribution of sub-populations of O_2 tolerant, obligately anaerobic, methylotrophic methanogens that occur in microniches reduced by their bacterial consorts. *Mar. Ecol. Prog. Ser.* 95:67–80.

Sieburth, J. McN., and Estep, K. W. (1985) Precise and meaningful terminology in marine microbial ecology. *Mar. Microb. Food Webs* 1:1–16.

Sieburth, J. McN., Smetacek, V., and Lenz, J. (1978) Pelagic ecosystem structure: Heterotrophic compartments of plankton and their relationship to plankton size fractions. *Limnol. Oceanogr.* 33:1225–1227.

Sieburth, J. McN., Johnson, P. W., Church, V. M., and Laux, D. C. (1993) C_1 bacteria in the water column of Chesapeake Bay, USA. III. Immunologic relationships of the type species of marine monomethylamine- and methane-oxidizing bacteria to wild estuarine and oceanic cultures. *Mar. Ecol. Prog. Ser.* 95:91–102.

Silver, M. W., Gowing, M. M., Brownlee, D. C., and Corliss, J. O. (1984) Ciliated protozoa associated with oceanic sinking detritus. *Nature* 309:246–248.

Small, E. B., and Lynn, D. H. (1985) Phylum Ciliophora. In J. J. Lee, S. H. Hutner, and

E. C. Bovee, eds., *An Illustrated Guide to the Protozoa*. Society of Protozoologists, Lawrence, KS.

Smetacek, V. (1981) The annual cycle of protozooplankton in Kiel Bight. *Mar. Biol.* 63:1–11.

Sorokin, Yu. I. (1980) Microheterotrophic organisms in marine ecosystems. In A. R. Longhurst, ed., *Analysis of Marine Ecosystems*, Academic Press, New York, pp. 293–342.

Steele, J. H. (1974) *The Structure of Marine Ecosystems*. Harvard University of Cambridge, MA.

Stoecker, D. K., and Capuzzo, J. McD. (1990) Predation on protozoa: Its importance to zooplankton. *J. Plankton Res.* 12:891–988.

Stoecker, D. K., Davis, L. H., and Provan, A. (1983) Growth of *Favella* sp. (Ciliata: Tintinnina) and other microzooplankters in cages incubated in situ and comparison to growth in vitro. *Mar. Biol.* 75:293–302.

Stoecker, D. K., Michaels, A. E., and Davis, L. H. (1987) Large proportion of marine planktonic ciliates found to contain functional chloroplasts. *Nature* 326:415–423.

Strom, S. L. (1991) Growth and grazing rates of the herbivorous dinoflagellate *Gymnodinium* sp. from the open subarctic Pacific Ocean. *Mar. Ecol. Prog. Ser.* 78:103–113.

Takahashi, M., and Bienfang, P. K. (1983) Size structure of phytoplankton biomass and photosynthesis in subtropical Hawaiian waters. *Mar. Biol.* 76:203-211.

Taylor, F. J. R. (1987) *The Biology of Dinoflagellates*. Blackwell Scientific, Oxford.

Taylor, F. J. R. (1990) Symbiosis in marine protozoa. In G. M. Capriulo, ed., *Ecology of Marine Protozoa*. Oxford University Press, New York, pp. 323–340.

Thingstad, T. F., Hagstrom, A., and Rassoulzadegan, F. (1997) Accumulation of degradable DOC in surface waters: Is it caused by a malfunctioning microbial loop? *Limnol. Oceanogr.* 42:398–404.

Thomsen, H. A., and Buck, K. R. (1991) Choanoflagellate diversity with particular emphasis on the Acanthoecidae. In D. J. Patterson and J. Larsen, eds., *The Biology of Free-Living Heterotrophic Flagellates*. Clarendon Press, Oxford, pp. 259–284.

Turner, R, E., and Rabalais, N. N. (1994) Coastal eutrophication near the Mississippi River delta. *Nature* 368:619–621.

Tranvik, L. J., Sherr, E. B., and Sherr, B. F. (1993) Uptake and utilization of "colloidal DOM" by heterotrophic flagellates in seawater. *Mar. Ecol. Prog. Ser.* 92:301–309.

Uhlig, G., and Sahling, G. (1982) Rhythms and distributional phenomena in *Noctiluca miliaris*. *Ann. Inst. Oceanogr. Paris* 58:277–284.

Urbach, E., Robertson, D. L., and Chisholm, S. W. (1992) Multiple evolutionary origins of prochlorophytes within the cyanobacterial radiation. *Nature* 355:267–269.

Venrick, E. (1974) The distribution and significance of *Richelia intracellularis* Schmidt in the North Pacific Central Gyre. *Limnol. Oceanogr.* 19:437–445.

Verity, P. G. (1985) Grazing, respiration, excretion, and growth rates of tintinnids. *Limnol. Oceanogr.* 30:1268–1281.

Verity, P. G., Stoecker, D. K., Sieracki, M. E., Burkill, P. H., Edwards, E. S., and Tronzo, C. R. (1993) Abundance, biomass, and distribution of heterotrophic dinoflagellates during the North Atlantic Spring Bloom. *Deep-Sea Res.* 40:227–244.

Vors, N. (1992) Ultrastructure and autecology of the marine, heterotrophic flagellate *Leucocryptos marina* (Braarud) Butcher 1967 (Katablepharidaceae/Kathabelpharidae) with a discussion of the genera *Leucocryptos* and *Katablepharis/Kathablepharis*. *Eur. J. Protistol.* 28:369–389.

Villareal, T. A. (1991) Nitrogen-fixation by the cyanobacterial symbiont of the diatom genus *Hemiaulus*. *Mar. Ecol. Prog. Ser.* 76:201–204.

Wada, E., and Hattori, A. (1991) *Nitrogen in the Sea: Forms, Abundances, and Rate Processes.* CRC Press, Boca Raton, FL.

Wainright, P. O., Hinkel, G., Sogin, M. L., and Stickel, S. K. (1993) Monophyletic origins of the metazoa: An evolutionary link with the fungi. *Science* 260:340–342.

Ward, B. B. (1986) Nitrification in marine environments. In J. I. Prosser, ed., *Nitrification.* IRL Press, Oxford, pp. 157–184.

Waterbury, J., Watson, S. W., Valois, F. W., and Franks, D. G. (1986) Biological and ecological characterization of the marine unicellular cyanobacterium *Synechococcus*. In T. R. Platt and W. K. W. Li, eds., Photosynthetic picoplankton. *Can. Bull. Fish. Aquat. Sci.* 214:71–120.

Williams, P. J. le B. (1981) Incorporation of microheterotrophic processes into the classical paradigm of the planktonic food web. *Kieler Meeresforsch. Suppl.* 5:1–28.

Woese, C. R., Kandler, O., and Wheelis, M. L. (1990) Towards a natural system of organisms: Proposal for the domains Archaea, Bacteria, and Eukarya. *Proc. Natl. Acad. Sci. USA* 87:4576–4579.

Wright, A.-D. G., and Lynn, D. H. (1997) Maximum ages of ciliate lineages estimated using a small subunit rRNA molecular clock: Crown eukaryotes date back to the Paleoproterozoic. *Arch. Protistenkd.* 148:329–341.

Zhukov, B. F. (1991) The diversity of bodonids. In D. J. Patterson and J. Larsen, eds., *The Biology of Free-Living Heterotrophic Flagellates.* Clarendon Press, Oxford, pp. 177–185.

3

EVOLUTION, DIVERSITY, AND MOLECULAR ECOLOGY OF MARINE PROKARYOTES

Stephen Giovannoni and Michael Rappé

Oregon State University,
Department of Microbiology,
Corvallis, Oregon

For all organisms, the history of evolution is a story about the challenges and opportunities that have arisen for living forms as the biosphere has developed. It is also a story of adaptation and competition, and of new innovations in the structure and function of cells that have permitted cells to claim a part of the world for their species. The realm of marine bacteria, the oceans, is probably the oldest and largest ecosystem on the planet. Modern bacterioplankton are the victors in an intense competition that has spanned eons. Their success can be measured by their vast numbers, or by the tremendous resources of carbon, nitrogen, and phosphorus that they sequester. This chapter recounts what we know of bacterioplankton diversity, which is considerable, and explains how molecular biology solved an important riddle. It also examines some of the current mysteries surrounding the evolution of bacterioplankton diversity and the function of microbial species in modern systems.

Microbial Ecology of the Oceans, Edited by David L. Kirchman.
ISBN 0-471-29993-6   Copyright © 2000 by Wiley-Liss, Inc.

RIBOSOMES: DECIPHERING THE EVOLUTION OF LIFE ON EARTH

Phylogenetics is the science of reconstructing how organisms are related. From the simplest perspective, phylogenetics assumes that new species arise from existing species by a process that resembles the pattern of a phylogenetic tree, which is a network of bifurcating nodes. We also assume that organisms that possess shared common features are related; in other words, they are alike because they are both derived from the same ancestor. The real picture is probably more complicated. Organisms may inherit features from more than one ancestor by a process in which genes are transferred laterally from one species to another. Nonetheless, the basic model holds that cells are related to other cells by a thread that goes back into distant time.

Modern phylogenetics is largely based on comparisons of the nucleotide bases and amino acids that make up macromolecules. Ribosomal RNA (rRNA) gene trees are by far the most common for phylogenetic comparisons. Prior to the arrival of genome sequencing, studies of small subunit (SSU) rRNA genes provided a detailed image of relationships among organisms, and today they remain the most widely applied and useful molecule for phylogenetic studies of marine bacteria. However, the arrival of complete genome sequences has provided molecular phylogeneticists with a windfall of information that is being used to reconstruct the evolution of cells in far greater detail. This field has been dubbed "phylogenomics."

The ribosomal RNA gene is unusual in that it does not contain the code for a protein product. Instead, the ribonucleic acid molecules transcribed from rRNA genes fold to form a structural component of ribosomes. rRNAs and ribosomes together function as the protein-synthesizing machines of the cell. Therefore, it is not surprising that rRNA molecules are ancient and highly conserved structures that are found in all cells. Also, because cells need many ribosomes, the rRNAs are often present in thousands of copies per cell. The fact that the rRNA gene products are highly conserved structural RNAs is important for microbial ecologists because of the pattern in which evolution has distributed variation in the gene. Compare many copies of rRNAs side by side and you will find regions of base sequence that never change, other regions that are found only in members of a kingdom, division, or genus, and still other regions that vary from species to species. This patchwork pattern of conservation makes it possible to design probes with varying specificity that can be applied to study microbial diversity in the context of ecology.

Diverse members of two domains of life, the *Bacteria* and the *Archaea*, make up the prokaryotic component of marine plankton. An SSU rRNA phylogenetic tree that broadly depicts relationships within the domain *Bacteria* is shown in Figure 1. Technically, it is a "gene" tree of orthologs; that is, it portrays the evolution of a single gene among different species. The figure shows 18 major microbial groups, along with the subdivisions within one of these groups, the *Proteobacteria*. The *Proteobacteria* are the classical gram-negative bacteria; they are by far the most diverse and successful microbial group in the oceans. The shape of the tree shows that the bacterial divisions

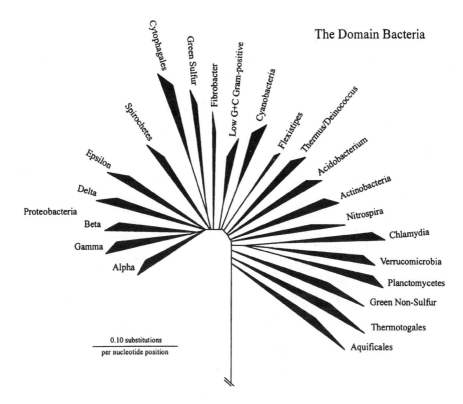

Figure 1. Consensus phylogenetic tree illustrating the major lineages of the domain *Bacteria*.

arose in an ancient radiation. A few divisions, like the *Aquificales*, the *Thermotogales*, and Green Non-Sulfur group, seem to branch early; but most appear later, apparently in rapid succession. Not shown in this tree but discussed later is the domain *Archaea*, the members of which are also prokaryotic cells that are mainly free-living.

In phylogenies such as that in Figure 1, the ribosomal RNA genes were used to examine relationships among cells that had been cultured. To place the cultured species in a phylogenetic tree, the gene was extracted and sequenced, and then suitable computer algorithms were used to produce the gene tree showing the relationship of the unknown gene to genes from other species (Hillis et al. 1996). As described next, this was just a beginning.

MOLECULAR SLEUTHS: SOLVING THE RIDDLE OF MARINE BACTERIOPLANKTON DIVERSITY

In the late 1970s and 1980s marine microbiologists discovered two groups of marine cyanobacteria, the prochlorophytes and the marine *Synechococcus* group, which were ubiquitous and abundant in seawater (Waterbury et al.

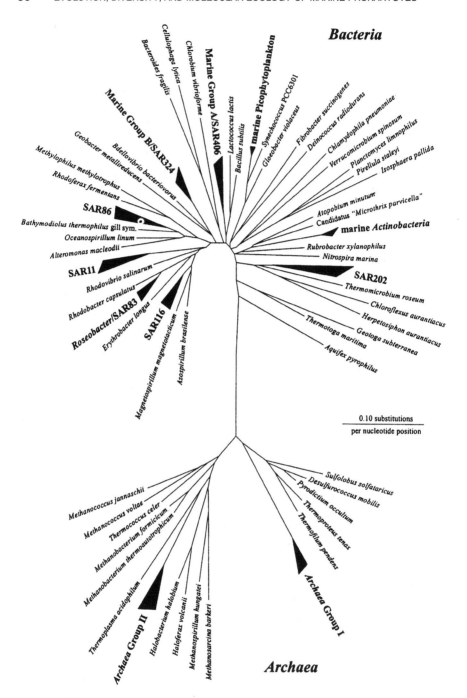

Figure 2. Composite phylogenetic tree displaying relationships among the most widespread SSU rRNA gene clusters from marine prokaryotic plankton.

1979; Chisholm et al. 1988). How did such abundant organisms defy discovery for so long? The answer is that they were difficult, but not impossible, to grow; and when viewed through a microscope without special methods, they looked much like all the other small cells in the oceans. This story illustrates a basic problem that pervaded microbiology before the 1990s. In a nutshell, if you did not know how to grow an organism, you would probably overlook it entirely. How important was this problem? As it turned out, developing new ways of studying microbial diversity was crucial for the advancement of microbial ecology.

Microbiologists have long noted a discrepancy of several orders of magnitude between the number of cells that can be seen in natural systems by direct observation (e.g., by epifluorescence microscopy) and the number of colonies that appear on agar plates. Jannasch and Jones (1959) reported such observations from seawater in 1959. Staley and Konopka (1985) named this discrepancy "the great plate count anomaly." Fundamentally, there are two explanations for this discrepancy: (1) bacterial communities are composed of known species that are capable of forming colonies on agar plates, but do so with low efficiency, or (2) bacterial communities are composed of unknown species that do not grow on common microbiological media. The actual explanations for the failure of cells to form colonies may be as diverse as the species in consideration, but these two alternative hypotheses capture the essence of the debate. Various observations, including the recognition that cells of some species may enter a viable but nonculturable state (VBNC), and measurements that indicate that a proportion of direct counts are in fact dead cells, support the first hypothesis (Morita 1985). However, the story of the discovery of marine cyanobacteria was clear evidence that the second hypothesis was not only credible but might lead to major discoveries.

The significance of the "great plate count anomaly" was considerable. At stake was the relevance of existing knowledge about cultured marine bacteria. If the first hypothesis was correct, then existing culture collections could be thought of as good representatives of marine bacterioplankton. On the other hand, if the second hypothesis were true, then an understanding of the physiology and ecology of bacterioplankton could not be based on experiments performed with the commonly known marine bacteria. The range of compounds they can metabolize, their association constants for substrates, the extracellular enzymes they possess, and their ability to attach to substrates or chemotax to a nutrient source can be determined from cells in culture. These attributes of cells do not tell us exactly what cells are doing in nature; for that, ecological experiments are required. However, knowledge of a cell's potential sets the boundary conditions for microbial ecology, and without this information their role in ecosystem dynamics can never be resolved. For example, the unusual phycourobilins of marine *Synechococcus* sp., which provide these cells with a unique capacity to harvest blue light, led directly to experiments linking the ecology of these organisms to the optical qualities of the water column (Wood 1985).

The explanation for the "great plate count anomaly" eventually came from a laboratory of molecular biologists headed by Norman Pace (Olsen et al. 1986). This group, which at that time included David Stahl, Gary Olsen, and David Lane, formulated ideas for identifying microbes from natural systems by means of gene cloning. Their approach did not involve growing the bacteria. Instead, cells were collected from the ecosystem directly and lysed to produce extracts of mixed genomic DNA from all the members of the community. In the next step, variants of a single gene, the SSU rRNA gene, were cloned into vectors and sequenced. It is important to remember that the process of gene cloning literally involves separating and replicating a single DNA molecule so that it can be studied apart from the cell. Ocean DNA became like a grab bag of marbles — each gene was studied by comparison of its nucleotide sequence to those of known genes from cultured marine bacteria.

The results of the first study of this type from marine samples were reported in 1990 (Giovannoni et al. 1990). The genes recovered included one in particular that bore only a distant relationship to its nearest neighbors in databases. It was called SAR11. In addition, the genes of two cyanobacteria groups were recovered. One of these groups was identified as the marine *Synechococcus* spp. Unknown to the authors, the other group of genes corresponded to the as-yet-uncultured marine prochlorophytes. SAR11 later turned out to be the most ubiquitous bacterial gene recovered from seawater.

Nearly a decade later, when over 616 microbial plankton genes cloned from seawater had been identified, the following conclusions were reached:

1. The most abundant rDNA genes recovered do not correspond to cultured species.
2. Marine *Archaea* are abundant and almost invariably fall within two phylogenetic groups.
3. Most (80%) marine *Bacteria* clones fall among nine phylogenetic groups.
4. These phylogenetic groups form clusters of related genes rather than single lineages.
5. In some cases the sublineages of gene clusters have different depth-specific distributions, and these may represent different species.
6. The major marine prokaryotic groups appear to have cosmopolitan distributions.
7. Particle-associated and freely suspended marine prokaryotes are different.
8. Stratification of bacterioplankton populations is typical of the ocean surface layer.

The 11 major marine prokaryotic picoplankton groups referred to earlier are shown in boldface type in Figure 2. This phylogenetic tree includes the *Archaea*, but it leaves out the eukaryotes. This tree also leaves out the very diverse marine bacteria that are not among the 11 most abundant rRNA groups. However, these organisms are described in the sections that follow.

It should be pointed out that SSU rRNA genes only rarely reveal the physiology of cells. For the most part, the functional significance of organisms such as SAR11 remains unknown, except that it is safe to assume from their abundance that they are important players in the carbon cycle. In ecological studies, SSU rRNA genes serve the role of providing markers for species that otherwise might not be differentiated from the cells of other species occupying the same habitat. As markers, SSU rRNA genes have the added asset of providing phylogenetic information on the organisms. While these SSU rRNA gene "markers" leave much unknown, they are very effective at providing a phylogenetic framework for viewing microbial diversity.

WHY CULTURABLE AND NONCULTURABLE?

Of the nine most abundant bacterioplankton groups detected by gene sequencing, only two (the marine oxygenic phototroph clade and the *Roseobacter* clade) contain members that have been cultured. Clearly, the cells of many of the major bacterioplankton groups must have unusual growth requirements, and therefore the second hypothesis presented above explains a major part of the discrepancy between direct counts and plate counts. Presently, it is uncertain what aspects of the physiology of these cells makes them so difficult to culture.

A general microbiological theory that distinguishes between cells that are adapted to low and high nutrient environments may be a source of insight into the problems of culturing oceanic bacteria (Fry 1990; Morita 1997). Heterotrophic bacteria that can reproduce at very low dissolved organic carbon concentrations are known as oligotrophs; their counterparts, which thrive at higher organic carbon concentrations, are known as copiotrophs. Species capable of growth at both low and high concentrations of organic carbon are known as facultative oligotrophs. There is no accepted definition of these categories, but typically the cutoff value to distinguish between these two groups is 1–10 mg of C per liter (mg C L^{-1}). By this definition, almost all pelagic marine bacteria are oligotrophs.

One method for obtaining pure cultures of oligotrophs (a culture containing the cells of a single species) is to dilute live cells from seawater into sterile seawater media that contain only ambient dissolved organic carbon (DOC), the so-called dilution to extinction method (Button et al. 1993). The cell cultures obtained by this method often fail to grow when transferred to high-nutrient media, indicating that these cells are physiologically unable to replicate at high DOC concentrations. These results would seem promising, but thus far, the marine oligotrophs cultured by this approach have all turned out to be gamma and alpha *Proteobacteria* unrelated to the major bacterioplankton groups recovered by gene cloning.

Despite the failures of seawater cultures to produce either true oligotrophs or the organisms corresponding to the abundant bacterioplankton rRNA groups, the theory behind the experiments may be on the correct track. The most eloquent development of this theory has come from Don Button, of the University of Alaska, who proposed that the surface-to-volume ratios of the cells of true oligotrophs are set by evolutionary selection at values that optimize the match between the cell's transport capacity and its capacity to metabolize incoming substrates (Button 1998).

It seems likely that eventually all the major bacterioplankton groups will be cultured in laboratories. Perhaps at that time answers will emerge to questions about why these organisms are so sensitive to cultivation conditions currently being used.

THE MAJOR BACTERIOPLANKTON GROUPS

Results obtained by means of modern, molecular biology techniques centered on the analysis of SSU rRNA gene sequences have made it increasingly apparent that under normal conditions in pelagic marine systems, marine bacterioplankton can be generally separated into three groups: the culturable heterotrophic bacterioplankton, the uncultured (and probably predominantly heterotrophic) bacterioplankton, and, in areas with sufficient illumination, the oxygenic photoautotrophs. Under particular conditions or in distinct regions of the water column, specialized groups such as nitrifying or particle-associated bacteria may be present, although in terms of cell numbers and biomass, they usually represent only a small fraction of the total community.

In this section, we summarize the current status of the systematics of culturable bacterioplankton based on SSU rRNA gene sequence comparisons, dealing mainly with microorganisms that have been repeatedly cultured from seawater and have traditionally been considered the major heterotrophic bacteria in the oceans. SSU rRNA gene sequence analyses have revealed that these microorganisms predominantly fall within the gamma subclass of the *Proteobacteria* and less so in the *Cytophaga–Flavobacterium–Bacteroides* group, though more recent molecular data indicate that some previously unrecognized groups in the alpha *Proteobacteria* are easily cultured from seawater as well. Other groups that are encountered infrequently in cultivation-based studies but nonetheless have cultivated representatives — methylotrophic bacteria and the *Planctomycetales* — are also considered here. The final portion of this section addresses the phylogenetic distribution of uncultured bacterioplankton revealed by analyses of SSU rRNA genes, and available information on the biogeographic distribution and ecology of these uncultured microorganisms. The utility of SSU rRNA gene sequence comparisons is highlighted by comparing environmental gene clones recovered from coastal margin areas to their open-ocean and cultivated counterparts.

Systematics and the Culturable Heterotrophic Bacterioplankton

For decades, bacteria in seawater were enumerated and studied based on their ability to form colonies on agar plates. As we discussed previously, this practice has undoubtedly led to the selection of only a particular subset of the bacteria living or residing in seawater. For example, consider the common case of the bacteria isolated from surface seawater using the spread plate technique and a standard microbiological medium (i.e., high organic nutrients) such as ZoBell's medium (Difco 2216 marine agar). What types of bacteria would this study yield? The vast majority of isolates would be gram-negative chemoorganotrophs. Most would be straight or curved rods that are motile by means of polar or peritrichous flagella, and some would probably be able to use nitrate as a terminal electron acceptor for facultative anaerobic growth. It is a safe bet that the SSU rRNA gene sequence analyses would reveal, interestingly, that most of the isolates belonged to a single phylogenetic clade, the gamma subclass of the *Proteobacteria*. It might be assumed from this observation that this group must dominate in the body of water the sample originated from. This example is a generality that amazingly holds true for a geographically wide range of oceanic environments. A microbiological description of these organisms follows.

Culturable Gamma Proteobacteria   Phylogenetically, a majority of the named species of culturable marine bacteria fall within two related clades within the gamma subclass of the *Proteobacteria*. The genus *Alteromonas* was traditionally used as a refuge for a diverse array of gram-negative aerobic, chemoorganotrophic marine bacteria that possess a single polar flagellum. This genus has since been divided into several loosely related, predominantly marine genera including the original *Alteromonas*, *Pseudoalteromonas*, *Marinomonas*, *Shewanella*, and the newly formed *Glaciecola* (Figure 3). The genera *Oceanospirillum* and *Marinobacter* form a separate clade. Many marine isolates related to these genera do not match the SSU rRNA gene sequences of characterized strains exactly, indicating that there is much diversity yet to be discovered in these groups; yet most isolates match closely enough to allow us to conclude that they belong to these branches of the evolutionary tree. Most studies assessing the ecological importance of these genera have relied on cultivation-based methodologies, which has undoubtedly led to an overestimation of their abundance in the environment due to the apparent ease at which these cells form colonies on agar plates. However, it does appear as though some gamma *Proteobacteria* may occasionally be abundant in seawater, and many can be found in association with surfaces in the marine environment.

Members of the aerobic genus *Oceanospirillum* have traditionally been distinguished from other easily cultivated marine bacteria by their unique morphology: cells are helical, rather than rod-shaped. However, helical shape has proved to be a polyphyletic trait among marine bacteria; species assigned to the genus *Oceanospirillum* form several separate lineages in the gamma

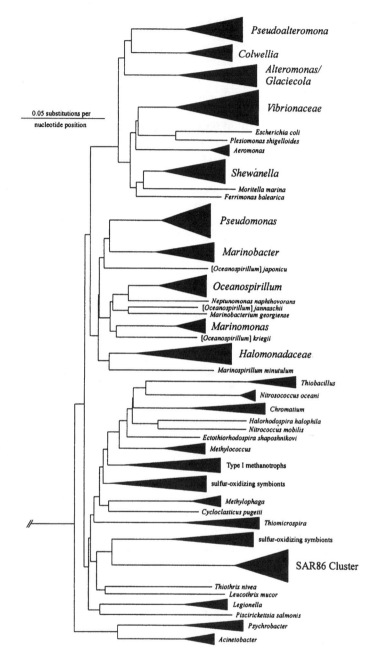

Figure 3. Phylogenetic dendrogram of the gamma subclass of the *Proteobacteria*, constructed by the neighbor-joining method from evolutionary distances calculated with the Kimura two-parameter model for nucleotide change. A data set of approximately 200 full-length sequences was used in the analysis. Members of the beta subclass of the *Proteobacteria* were used to root the tree. Clades shown in black are exclusively marine, or consist of a significant number of marine isolates. Taxa in brackets are generically misclassified.

subclass of the *Proteobacteria* (Figure 3), as well as a lineage in the alpha subclass (Figure 4). *Oceanospirillum* spp. have traditionally been more difficult to isolate than many of the other marine gamma *Proteobacteria*, and their occurrence in the oceans is still relatively unexplored.

The family *Vibrionaceae* encompasses a phylogenetically related group of bacteria that includes many characterized marine bacteria of the genera *Vibrio* and *Photobacterium* (Baumann et al. 1984). These organisms are able to grow both aerobically and in the absence of oxygen using fermentations for energy-yielding metabolism. This family has traditionally been considered to be one of the most important groups of bacteria in the marine environment, and its members were sometimes thought to dominate the heterotrophic bacterioplankton and to interact with other marine organisms in relationships ranging from symbiotic to pathogenic. Free-living members of this group are easy to culture from seawater but infrequently occur in culture-independent studies of free-living marine bacterioplankton, perhaps indicating that because of their ability to form colonies on agar plates, their importance as free-living heterotrophs in marine seawater has been historically overestimated. Phylogenetic analyses have shown the family *Vibrionaceae* is a sister clade to the enteric bacteria of the gamma *Proteobacteria*, which share many properties, particularly the capacity for facultative anaerobic growth that would seem to adapt them to a life cycle that included periods of growth in animal guts.

Members of the genus *Shewanella* also appear in culture collections of marine isolates, though usually not as often as some of the other marine genera in the gamma subclass of the *Proteobacteria* (Höfle and Brettar 1996; Bowman et al. 1997; Pinhassi et al. 1997). The gram-negative, facultatively anaerobic, motile, and rodlike *Shewanella* strains have often been isolated from surfaces in seawater such as oysters, macroalgae, sea ice, and sediments. They seem to particularly predominate among barophilic isolates (Kato and Horokoshi 1998). They are considered to be metabolically versatile, and some isolates have been targeted for study as a result of their ability to use a wide range of compounds, including ferrous iron, as electron acceptors for anaerobic respiration. Originally, members of the genus *Shewanella* were sequentially lumped in the genera *Achromobacter*, *Pseudomonas*, and *Alteromonas*, though they finally found a phylogenetically based taxonomic home with the application of nucleic acid sequencing. Gene sequence analyses indicate that *Shewanella* spp. are allied with the large clade that contains the genera *Alteromonas*, *Pseudoalteromonas*, *Colwellia*, the family *Vibrionaceae*, enteric bacteria such as *Escherichia coli*, and others (Figure 3).

Culturable Alpha Proteobacteria: The Roseobacter and Sphingomonas Clades   In comparison to the gamma subclass of the *Proteobacteria*, the alpha subclass is not represented nearly as often in culture collections from seawater, although culture-independent methods reveal that they are much more common than gamma *Proteobacteria*. There are no accepted explanations for this observation, although it suggests that perhaps some of the major

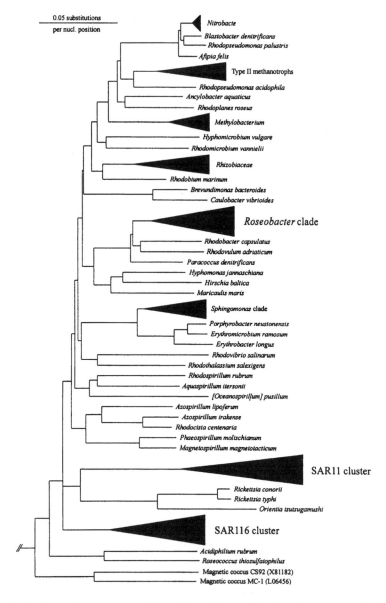

Figure 4. Phylogenetic dendrogram of the alpha subclass of the *Proteobacteria*, constructed as in Figure 3 from a set of approximately 150 full-length SSU rRNA gene sequences. Members of the gamma and beta subclasses of the *Proteobacteria* were used to root the tree.

marine alpha *Proteobacteria* groups are highly specialized species that are not very adaptable to laboratories. Whatever the case, there are two different phylogenetically coherent groups of alpha *Proteobacteria*, the *Roseobacter* and *Sphingomonas* clades, which include many culturable marine members. We describe them next.

The *Roseobacter* clade is a group of phylogenetically related, phenotypically heterogeneous, exclusively marine (or saltwater-requiring) chemoorgano-trophic bacteria (Figure 4). They are named after the first genus described from this group. *Roseobacter denitrificans* and *R. litoralis* were isolated from surfaces of marine macrophytic algae and were initially characterized as a new species of the genus *Erythrobacter* because they synthesized bacteriochlorophyll *a* aerobically (Shiba et al. 1979). The vast majority of phototrophic alpha *Proteobacteria* produce bacteriochlorophyll only under anaerobic conditions. Although they do not use sunlight for autotrophic growth, photosynthetic reaction center biochemistry appears to permit *R. denitrificans*, *R. litoralis*, and the other aerobic photoheterotrophic bacteria scattered throughout the alpha subclass of the *Proteobacteria* to synthesize ATP from sunlight under aerobic conditions. Starved *Roseobacter* cells survive much longer in the light than in the dark, which offers a potential selective advantage to *Roseobacter* spp. in oligotrophic systems. In addition to the two *Roseobacter* species described above, other genera of obligately aerobic photoheterotrophic bacteria shown in Figure 4 include *Erythrobacter* (marine), *Erythromicrobium* (freshwater), *Porphyrobacter* (freshwater), *Acidiphilum* (freshwater), and *Roseococcus* (freshwater). Based on analysis of SSU rRNA gene sequences, the aerobic photo-heterotrophic genus *Erythromonas* (freshwater) has recently been renamed *Sphingomonas*, while *Sandaracinobacter* (freshwater) has been included in the *Sphingomonas* clade in Figure 4 because it phylogenetically branches between different subgroups of this genus.

Two additional species had been included in the genus *Roseobacter*: *R. algicola* was isolated from a culture of the marine dinoflagellate *Prorocentrum lima*, and *R. gallaeciensis* was isolated from larval rearings and collections of the scallop *Pecten maximus*. Interestingly, these two species do not synthesize bacteriochlorophyll *a* under aerobic or anaerobic conditions, but SSU rRNA gene sequence analyses showed them to be clearly related to the original *Roseobacter* species. *Roseobacter algicola* has since been renamed *Ruegeria algicola*. In the last few years, isolates belonging to the *Roseobacter* clade have been cultivated from a wide range of environments including the northwest and southeast coasts of the United States, the South California Bight, the Baltic Sea, the Sargasso Sea, the Black Sea, a geothermal lake in Iceland, Arctic and Antarctic Sea ice, the Mediterranean Sea, marine phytoplankton cultures, and surfaces of marine macrophytic algae, as well as from associations with a diverse array of marine organisms. Members of the *Roseobacter* clade form a phylogenetically coherent group within the alpha *Proteobacteria* (Figure 4), and all require a saline environment for growth.

Roseobacter isolates exhibit diverse modes of metabolism, but many are tied together by their ability to utilize organic and inorganic sulfur compounds (Gonzalez et al. 1999). Strain LFR, a nonphotosynthetic, carotenoid-contain-ing isolate, degrades dimethysulfoniopropionate (Ledyard et al. 1993). Coastal isolates within this clade have been obtained using lignin as a sole carbon source for enrichment (Gonzalez et al. 1996). One strain of *Ruegeria algicola* synthesizes the toxin okadeic acid, raising the possibility that members of the

Roseobacter clade may produce allelopathic compounds in association with eukaryotic hosts (Lafay et al. 1995).

R. litoralis and *R. dentrificans*, which produce aerobic bacteriochlorophyll *a* form a tight monophyletic cluster within the larger *Roseobacter* clade, while *Ruegeria algicola* and *R. gallaeciensis* specifically associate with each other but not the original *Roseobacter* species. The two original *Roseobacter* species are the only aerobic bacteriochlorophyll *a* producers that have been described in the *Roseobacter* clade, which now includes over ten closely related genera.

At the same time as the first descriptions of the genus *Roseobacter* emerged, SSU rRNA gene clones retrieved directly from the surface of the Sargasso Sea were characterized and found to be specifically related to this group (Britschgi and Giovannoni 1991). Since that time, environmental gene clones related to the *Roseobacter* clade have been recovered from both coastal and oceanic prokaryotic plankton samples, and in diverse associations with marine organisms (González et al. 1996).

The *Roseobacter* clade is currently the second most abundant SSU rRNA gene clone type recovered from marine plankton clone libraries, and it is one of only two examples in which cultured representatives from a dominant clade of environmental gene clones have been obtained, the other example being the oxygenic photoautotrophs described later. Phenotypic heterogeneity within the *Roseobacter* clade has made it difficult to assess the ecological niche occupied by related environmental gene clones in natural samples. It is clear that members of the *Roseobacter* clade are ubiquitous in seawater environments, though the open-ocean members of this group appear to be relatively more abundant at and above the deep-chlorophyll maximum than at greater depth in the water column. Although members of the *Roseobacter* clade appear to be common in the ocean, sensitive measurements have failed to reveal bacteriochlorophyll in seawater, and it seems unlikely that photoheterotrophy is an important mode of metabolism for *Roseobacter* clade strains in the open-ocean environment (Mullins et al. 1995).

Relatives of the genus *Sphingomonas* also occasionally appear in marine culture collections though, to our knowledge, only a single marine isolate exists that has been thoroughly characterized. This genus was originally described from isolates that were considered to have clinical importance because they had been found in disease samples and infections. They are aerobic, motile, yellow-pigmented, gram-negative rods possessing a single polar flagellum (some species are not motile); they were originally lumped in various genera such as *Pseudomonas* and *Flavobacterium*. Comparisons of SSU rRNA gene sequences show that the genus *Sphingomonas* is specifically related to *Porphyrobacter*, *Erythrobacter*, and *Erythromicrobium*, three genera in the alpha subclass of the *Proteobacteria* that produce bacteriochlorophyll *a* aerobically (Figure 4). As mentioned earlier, the genus *Erythromonas* has recently been renamed *Sphingomonas*, which gives it some members that produce bacteriochlorophyll *a* as well. The genus *Sphingomonas* is also phylogenetically heterogeneous: several members of different genera branch between or close to major

Sphingomonas subgroups. One well-characterized marine strain is *Sphingomonas* str. RB2256, isolated by dilution culture from Resurrection Bay, Alaska (Schut et al. 1993). This strain has been labeled an "ultramicrobium" owing to its small cell size and small genome, and it is of interest because it is a marine oligotroph that when grown under nutrient-limiting conditions, possesses a high-affinity nutrient uptake system.

Marine Methylotrophs   Microorganisms known as methylotrophs are capable of using reduced carbon substrates that contain no carbon–carbon bonds (as in, e.g. methane, methanol, methylated amines, and methylated sulfur compounds) for cellular carbon and energy. In general, methylotrophs are considered to have global importance because of their ability to moderate methane flux by acting as a sink for methane. In addition to methane, the marine environment may harbor at least 15 different single-carbon compounds including some, such as dimethylsulfoniopropionate, having global significance. Since a phylogenetically and taxonomically diverse array of marine microorganisms may be facultative methylotrophs, we limit the current discussion to microorganisms that make their living by obligate methanotrophy or methylotrophy.

Aerobic, obligately methylotrophic bacteria can generally be divided into two groups: one with the ability to grow on methane (methanotrophs) and one that is not capable of growth on methane but can utilize methanol and other methylated compounds (Hanson and Hanson 1996). Based on the intracytoplasmic membranes they contain and the pathway used for carbon assimilation, methanotrophs have further been divided into two groups. Type I methanotrophs, including members of the genera *Methylomonas*, *Methylobacter*, *Methylomicrobium*, *Methylococcus*, and *Methylosphaera*, contain intracytoplasmic membranes in bundles and utilize the ribulose monophosphate (RuMP) pathway for carbon assimilation. Type II methanotrophs, including members of the genera *Methylosinus* and *Methylocystis*, contain intracytoplasmic membranes arranged around the periphery of the cell and utilize the serine pathway for carbon assimilation. SSU rRNA gene sequence analyses of methanotrophs have supported the distinction between the two types: Type I methanotrophs fall within a related set of clades in the gamma subclass of the *Proteobacteria* (Figure 3), while Type II methanotrophs form a monophyletic clade within the alpha subclass (Figure 4). Methanotrophs isolated from the marine environment have exhibited a narrow range of phenotypic and genotypic diversity, owing in part to the difficulty in isolating them, which in turn is due partially to a lack of adequate techniques for their isolation. *Methylomicrobium pelagicum*, isolated from surface waters of the Sargasso Sea, has been well characterized, though only a few other strains have been isolated from seawater (Sieburth et al. 1987). In fact, very little is known regarding aerobic methane oxidation in seawater.

The criteria used to distinguish methanotrophs have also been used to divide into two groups the aerobic, obligate methylotrophs that are incapable of assimilating methane (Lidstrom 1992). Type I methylotrophs, including the genera *Methylobacillus* and *Methylophilus* as well as some misclassified *Methylomonas* species, also utilize the RuMP pathway but are not phylogenetically related to the RuMP Type I methanotrophs of the gamma subclass of the *Proteobacteria*. They instead form a monophyletic clade within the beta subclass. In the alpha *Proteobacteria*, methylotrophs of the genus *Methylobacterium* use the serine pathway for carbon assimilation but are in fact not obligate methylotrophs and can use multicarbon compounds for heterotrophic growth.

The Cytophaga–Flavobacterium–Bacteroides Group
The *Cytophaga–Flavobacterium–Bacteroides* (CFB) group is one of the main phylogenetic branches of the domain *Bacteria*, whose members were not fully recognized as forming a coherent clade until after the characterization of SSU rRNAs by Woese and coworkers (Figure 1) (Paster et al. 1985). Overall, the CFB group is morphologically and phenotypically quite diverse, though most of the marine bacteria in this group fall into a fairly homogenous phylogenetic clade referred to as the *Flavobacterium–Cytophaga* complex, originally called the "*Cytophaga* subgroup" by Woese and coworkers. Most marine CFB isolates are strictly aerobic or facultatively anaerobic chemoorganotrophs, many display gliding motility, and they are potentially important in marine (and other) environments for their ability to degrade biomacromolecules such as chitin, agar, DNA, and cellulose (Reichenbach 1992). Phenotypic characteristics such as the presence or absence of gliding motility have traditionally been used to delineate genera in this group, though recent genetic data have revealed many of the delineating phenotypic traits to be poor predictors of true phylogenetic relationships. Since the genera *Cytophaga*, *Flavobacterium*, and *Flexibacter* are polyphyletic, the CFB group is in the midst of major taxonomic revision. As the taxonomy now stands, most marine members of the CFB group appear to have been excluded from membership in both the genera *Flavobacterium* and *Cytophaga*. The majority of marine members of the CFB group are clearly unrelated to *Cytophaga hutchinsonii*, a soil bacterium that is the type species of the genus. Instead, most marine isolates phylogenetically fall within the recently described family *Flavobacteriaceae* (Bernardet et al. 1996). *Cytophaga lytica*, a misclassified marine cytophaga, was recently placed in the newly formed genus *Cellulophaga* along with two new marine species, *Cellulophaga baltica* and *Cellulophaga fucicola* (Johansen et al. 1999).

In recent years, both cultivation-based and culture-independent approaches have indicated that members of the CFB group are quite widespread in the marine environment and are commonly associated with surfaces such as marine snow, macro- and microalgae, and mollusk shells, though they are also found free-living in seawater. As mentioned earlier, the ecological significance of marine CFB bacteria is assumed to lie in their ability to degrade biomacromolecules via the production of hydrolytic exoenzymes, and in their surface-associated gliding motility. However, many of the marine isolates (and all environmental gene clones) recently reported from the CFB group have not yet been investigated for these traits. In fact, many of the SSU rRNA gene sequences obtained from recently isolated bacteria and gene clones recovered directly from marine environments form unique lineages within the family *Flavobacteriaceae*, indicating that many currently undescribed clades are in culture, waiting for proper taxonomic characterization or, in the case of the environmental gene clones, for cultivation in the laboratory.

Planctomycetales   The planctomycetes, also referred to as the order *Planctomycetales*, form one of the main phylogenetic branches of the domain *Bacteria* (Figure 1). This unique group contains many phenotypic and molecular features that set it apart from the rest of the bacteria, including the lack of a peptidoglycan layer in the cell wall, the presence of craters or pits on the surfaces of cells, budding division, an unusually short 5S rRNA, unlinked 16S and 23S rRNA genes (in some members), and unique 16S (SSU) rRNA gene sequences (Fuerst 1995).

Although they were originally regarded as freshwater microorganisms, in the last few years there have been more reports of *Planctomycetales* from various marine environments (DeLong et al. 1993; Schlesner 1994; Rappé et al. 1997). To date, two of four *Planctomycetales* genera, *Planctomyces* and *Pirellula*, have been cultured from marine environments. These genera correspond to distinct phylogenetic clades within the *Planctomycetales* based on SSU rRNA gene sequence comparisons. *Planctomyces* are stalked, budding bacteria that form rosettes of aggregated cells, while *Pirellula* are stalkless, though they divide by budding and form cell aggregates as well. The importance of *Planctomycetales* in the marine environment may be underappreciated in cultivation-based studies because of the characteristic slow growth of these organisms and their requirement for dilute (low-DOC) media for cultivation. Gene cloning and sequencing studies of environmental SSU rRNA have revealed more diversity in the *Planctomycetales* than had been recognized from culture-based studies alone, with several new, apparently uncultivated phylotypes present in the marine environment. Among these is a clade of as-yet uncultured marine bacteria that, based on phylogenetic analyses of SSU rRNA genes, appears to form the deepest branch of the *Planctomycetales* lineage thus far discovered (Rappé et al. 1997). Interestingly, *Planctomycetales* may also be underrepresented in many SSU rRNA gene cloning and sequencing studies because their SSU rRNA genes contain mismatches with many of the most common "bacterial" PCR primers and probes currently in use (Vergin et al. 1998).

Little is currently known about the ecology or physiology of marine *Planctomycetales*. Most characterized strains are chemoorganotrophic, and for the most part they are obligate aerobes, although some facultative anaerobes have been isolated. Data from rRNA gene cloning and sequencing indicates that *Planctomycetales* cells are mainly associated with marine snow in the marine environment, suggesting that they are found attached to particles

(DeLong et al. 1993). Further genotypic and phenotypic characterization of this group of marine bacteria via laboratory cultivation is of direct interest to microbiologists interested in the origin and diversification of this unique bacterial phylum.

Oxygenic Phototrophs: The Cyanobacteria   The oxygenic (oxygen-producing) photoautotrophic prokaryotes are one of the main lineages of the domain *Bacteria* (Figure 1). In seawater environments small, unicellular cyanobacteria of the provisional genus *Synechococcus* and even smaller, unicellular prochlorophytes of the genus *Prochlorococcus* often dominate the picophytoplanton, particularly in the open oceans. Most cyanobacteria, including the marine *Synechococcus* group, possess chlorophyll *a* and phycobilisomes, while prochlorophytes possess divinyl chlorophylls *a* and *b*.

Interestingly, these two phenotypically different types of prokaryotic primary producer are quite closely related. Phylogenetic analyses have shown that marine members of the genus *Synechococcus* and members of the genus *Prochlorococcus* belong to a single clade referred to as the marine picophytoplankton (Figure 2; Urbach et al. 1998). Originally, the term "picophytoplankton" referred to a size class, but it has since come to have a phylogenetic meaning as well. *Prochlorococcus* is not specifically related to the other prochlorophytes, indicating that the ability to produce divinyl chlorophylls *a* and *b* originated more than once during evolutionary history.

It is not surprising that cyanobacterial SSU rRNA genes were recovered in clone libraries of culture-independent studies of pelagic marine samples, since it was known that unicellular oxygenic photoautotrophs often constitute a significant proportion of the total prokaryotes in these environments. Environmental gene clones related to the marine picophytoplankton clade are for the most part closely related to cultured *Prochlorococcus* and *Synechococcus* strains, and all cyanobacterial environmental gene clones obtained from natural marine plankton communities so far are affiliated with the marine picophytoplankton clade.

From an evolutionary perspective, it is noteworthy that the marine picophytoplankton clade appears to have a relatively recent origin as deduced from phylogenetic trees. Undoubtedly cyanobacteria have occupied the oceans from a time close to the origin of oxygenic photosynthesis, if not since the origin of the process. Therefore, the data suggest that some biological adaptation led to the emergence of the picophytoplankton clade and eliminated many older lineages of pelagic marine cyanobacteria. Evolutionary logic dictates that the key event in the evolution of the clade must have predated the origin of divinyl chlorophylls *a* and *b* within the group, since otherwise it would be necessary to postulate an unparsimonious reversion to phycobilin photosynthesis from a chlorophyll-based light-harvesting system. The dating of this event is not presently possible because of uncertainties regarding the rate of the picophytoplankton "molecular clock," but a conservative interpretation would place this event sometime in the Cambrian period.

The Dominant Uncultured Bacterioplankton Groups

The first marine prokaryotic plankton communities analyzed by SSU rRNA gene cloning and sequencing came from surface samples of oligotrophic, subtropical regions of the Atlantic and Pacific Oceans. From the Atlantic Ocean, planktonic biomass was collected from the surface of the Sargasso Sea, and the polymerase chain reaction (PCR) was used in conjunction with general bacterial primers to amplify SSU rRNA genes from genomic DNA isolated from the sample (Giovannoni et al. 1990). The resulting PCR amplification product was cloned, and selected clones were subsequently grouped into similar types and sequenced. For the Pacific Ocean sample, a different approach was used: genomic DNA fragments retrieved from the collected biomass were cloned directly into lambda phage, and SSU rRNA gene-containing clones were subsequently identified and sequenced (Schmidt et al. 1991). Initial results from these two studies were astounding: though the samples originated from two different oceans and employed different methods for obtaining SSU rRNA genes, several close phylogenetic relatives were recovered among the analyzed clones. In addition, the vast majority of the SSU rRNA genes characterized in these initial samples were unrelated to previously recognized groups of cultivated heterotrophic marine bacteria.

Following the two initial studies, the range of depths and locales sampled for molecular-based analyses has steadily increased (Table 1). These analyses showed the prokaryotic fraction of marine planktonic communities to be much more phylogenetically diverse than had thus far been revealed by culturing, with up to seven different bacterial phyla and both major divisions of the *Archaea* represented (Figure 2). However, these studies have generally confirmed the results of the early sequencing studies. Different investigators, using different methods and sampling from different depths and locations in the world's oceans, repeatedly uncovered the same phylogenetic lineages (Tables 1 and 2, Figure 2).

An examination of the approximately 660 gene sequences retrieved by the studies listed in Table 1 shows that about 80% of all bacterial SSU rDNAs obtained from marine plankton using culture-independent techniques fall into one of nine distinct phylogenetic groups (Figure 5). Furthermore, nearly all the SSU rDNA clones recovered from marine planktonic *Archaea* fall into one of two clades. Of the major bacterial clades shown in Figure 2, only the *Roseobacter* clade of the alpha *Proteobacteria* and the marine picophytoplankton clade of the cyanobacteria/prochlorophyte radiation, both discussed earlier, contain members that have been cultured. Together, these two clades account for only about one-fourth of the total bacterial clones recovered; the remaining groups are completely composed of uncultured (or uncharacterized) bacterial phylotypes.

The uniqueness of the uncultured marine bacteria is underscored by the observation that the median similarity between marine bacterial genes retrieved from nature and genes from cultured marine microbial species is only

Table 1. Summary of SSU rRNA gene clone libraries from marine prokaryotic plankton

Clone Library Prefix	Seawater Source (depth, location)	Method of Construction[a]	Clones Analyzed[b]	Refs.
SAR(1-199)	Surface water, Sargasso Sea, Atlantic Ocean	PCR-bacterial	44	Giovannoni et al. (1990), Britschgi & Giovannoni (1991), Mullins et al. (1995)
ALO	Surface water, north central Pacific Ocean	Shotgun clone library	15	Schmidt et al. (1991)
SBAR	Surface water, Santa Barbara Channel, Pacific Ocean	PCR-archaeal	20	DeLong (1992)
WHAR	Surface water, Woods Hole, east coast of U.S., Atlantic Ocean	PCR-archaeal	20	DeLong (1992)
NH25	100 m, northeastern Pacific Ocean	PCR-universal	13	Fuhrman et al. (1992), Fuhrman et al. (1993)
NH49	500 m, northeastern Pacific Ocean	PCR-universal	10	Fuhrman et al. (1992), Fuhrman et al. (1993)
NH16	100 m, northeastern Pacific Ocean	PCR-universal	15	Fuhrman et al. (1993)
NH29	100 m, northeastern Pacific Ocean	PCR-universal	9	Fuhrman et al. (1993)
BDA1	10 m, Sargasso Sea, Atlantic Ocean	PCR-universal	14	Fuhrman et al. (1993)
FL	Surface water, Santa Barbara Channel, Pacific Ocean	PCR-bacterial	20	DeLong et al. (1993)
ANTARCTIC	Surface water, Arthur Harbor, Antarctica	PCR-archaeal	14	DeLong et al. (1994)
OAR	Surface water, Oregon coast, Pacific Ocean	PCR-archaeal	—[c]	DeLong et al. (1994)

Clone	Location	Method	No.	Reference
SAR(400–599)	80 m, Sargasso Sea, Atlantic Ocean	PCR-bacterial	92	Gordon & Giovannoni (1996), unpubl. data
SAR(200–399)	250 m, Sargasso Sea, Atlantic Ocean	PCR-bacterial	113	Giovannoni et al. (1996), Wright et al. (1997), unpubl. data
OCS	10 m, Oregon coast, Pacific Ocean	PCR-bacterial	114	Suzuki et al. (1997), unpubl. data
SB95	Surface and 200 m, Santa Barbara Channel, Pacific Ocean	PCR-archaeal	575	Massana et al. (1997)
OM	10 m, eastern continental shelf of U. S., Atlantic Ocean	PCR-bacterial	169	Rappé et al. (1997)
p712	500 m, Pacific Ocean basin	PCR-universal	18	Fuhrman & Davis (1997)
pN1	3000 m, Pacific Ocean basin	PCR-universal	23	Fuhrman & Davis (1997)
pB1	1000 m, Atlantic Ocean basin	PCR-universal	21	Fuhrman & Davis (1997)
C	10 m, west Irish coast, Atlantic Ocean	PCR-archaeal	7	McInerney et al. (1997)
PM	0, 100, and 500 m, northeast Atlantic Ocean	PCR-archaeal	3	McInerney et al. (1997)
pC2	Surface water, Long Island Sound, New York, Atlantic Ocean	PCR-universal	17	Fuhrman & Ouverney 1998
pC8	Surface water, Long Island Sound, New York, Atlantic Ocean	PCR-universal	7	Fuhrman & Ouverney 1998
pM	Surface water, Malibu Pier, California, Pacific Ocean	PCR-universal	16	Fuhrman & Ouverney 1998
pM9	Surface water, Monterey Bay, California, Pacific Ocean	PCR-universal	5	Fuhrman & Ouverney 1998

[a] PCR-bacterial, PCR amplification employing SSU rDNA oligonucleotide primers specific for the domain *Bacteria*; PCR-universal, primers targeting *Eukarya*, *Archaea*, and *Bacteria*; PCR-archaeal, primers specific for *Archaea*.

[b] Includes clones characterized by sequencing, restriction fragment length polymorphism, oligonucleotide probe hybridization, and other screening procedures. In cases where the total number of characterized clones was not clear, a conservative value is reported.

[c] Not reported.

Table 2. Distribution of the major bacterial SSU rRNA gene clone clusters in libraries from marine prokaryotic plankton[a]

Clone Library Prefix	SAR11	SAR83	SAR86	SAR116	Marine Pico	Marine Actino	SAR202	SAR324	SAR406
SAR (0 m)	•	•	•	•	•	•			
ALO	•	•	•	•	•				
NH25	•				•	•			
NH49	•		•						•
NH16	•	•	•		•	•			•
NH29	•		•		•	•			
BDA1	•		•		•	•			
FL	•								
SAR (80 m)	•	•	•	•	•	•			•
SAR (250 m)	•	•					•	•	
OCS	•	•	•	•		•			•
OM	•	•	•	•		•			
p712	•							•	•
pN1	•		•				•	•	•
pB1	•						•	•	•
pC2	•	•							
pC8	•	•		•					
pM	•	•	•						
pM9	•				•				

[a] Marine Pico., marine picophytoplankton clade; marine Actino, marine Actinobacteria cluster.

about 87%. To put this value in perspective, consider that bacterial SSU rDNAs diverge at an average rate of 1% per 50 million years. Members of the same bacterial species typically have sequence similarities greater than 97%, while similarity values between species within a genus are often greater than 93% (Stackebrandt and Göebel 1994). Even allowing for broad variation in rates of evolution, the data still support the conclusion that most marine bacteria do not share recent (i.e., Cambrian or later) common ancestors with cultured species.

What follows is a short summary of each of the major uncultivated marine planktonic bacterial and archaeal clades, including a description of their phylogenetic position and what limited information exists on their biogeographic distribution.

The Ubiquitous SAR11 Cluster SSU rRNA gene clones of the SAR11 cluster were some of the first environmental gene clones recovered from surface seawater samples of the Sargasso Sea and provided the first indication that novel microbial groups unrelated to the major groups of cultured heterotrophic marine bacteria made up a significant fraction of this and other marine bacterioplankton communities. Interestingly, it is a rapidly evolving, deeply branching clade of the alpha subclass of *Proteobacteria*, rather than the gamma

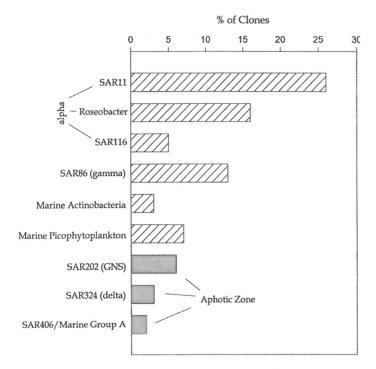

Figure 5. Frequency of the most common bacterioplankton SSU rRNA gene clusters. Frequencies were determined by dividing the number of clones from a particular gene cluster by the total number of clones in the data set (578). Clones from marine snow and sediments, and *Archaea*, were not included in the data set.

subclass prevalent in cultivation-based studies (Figure 2). Phylogenetic relationships between deeply branching members of the alpha *Proteobacteria* are difficult to resolve with current data and methods, so the precise order of branching between many of these lineages is unclear. Members of the SAR11 cluster show less than 82% sequence similarity to cultivated members of the alpha *Proteobacteria*; a glance at Figure 4 shows that it is phylogenetically very different from all characterized alpha *Proteobacteria*. Since its original discovery in the oligotrophic Sargasso Sea, members of the SAR11 cluster have been recovered in every SSU rRNA gene clone library constructed with universal or bacterial PCR primers from marine prokaryotic plankton samples, including such diverse locales as shallow coastal lagoons and depths up to 3000 m in the northeast Pacific Ocean (Table 2). It is the most abundant marine prokaryotic plankton SSU rRNA gene clone type recovered to date, and quite possibly the most abundant microorganism in seawater. It now appears that phylogenetic relatives of SAR11 are not limited to pelagic seawater environments, for SSU rRNA gene clone lineages related to the SAR11 cluster have been recovered from permanently anaerobic marine sediments and from several freshwater lakes.

The SAR11 cluster appears to encompass multiple subgroups that are specialized. These are described in greater detail later when we discuss the phenomenon of gene clusters.

The SAR116 Cluster   Members of the SAR116 cluster were also uncovered in the original Sargasso Sea plankton SSU rRNA gene clone library and, as with the SAR11 cluster, they also form a novel lineage within the alpha subclass of the *Proteobacteria* (Figure 4). This clade branches from within the alpha *Proteobacteria* where, as mentioned earlier, the precise order of branching is not resolvable. Phylogenetic analyses of the SAR116 gene clade often place it as a distant sister clade to the genera *Azospirillum, Magnetospirillum,* and *Rhodospirillum,* though this affiliation is not very robust. Much that was said previously concerning phylogenetic relationships within the SAR11 cluster applies to the SAR116 cluster as well; like the SAR11 cluster, the SAR116 cluster contains several well-defined subgroups that differ from each other by up to 10% sequence divergence over nearly complete SSU rRNA genes. Marine bacterioplankton with SAR116 cluster SSU rRNA genes are ubiquitous in seawater (though not as prevalent as SAR11) and appear to reside in both open-ocean and coastal environments. Preliminary data suggest that this group of bacteria generally prefers the ocean surface layer and may be relatively more abundant in coastal margin environments. No members of this broad phylogenetic clade are in culture, and thus their physiological characteristics are completely unknown.

The Uncultivated Gamma Proteobacteria: SAR86   Though the majority of easily cultivatable heterotrophic bacteria in seawater belong to the gamma subclass of the *Proteobacteria,* the alpha subclass has so far been much more prevalent in culture-independent studies. One clade of gamma *Proteobacteria* SSU rRNA genes has been repeatedly recovered in culture-independent studies of marine bacterioplankton, however. This gene clone cluster, named the SAR86 cluster after some of the first clones recovered from the Pacific and Atlantic Oceans and the Sargasso Sea, is a phylogenetically unique clade within the gamma *Proteobacteria* (Figure 3; see Fuhrman et al. 1993; Mullins et al. 1995). As we discussed previously, the cultivatable heterotrophic marine bacteria of the gamma *Proteobacteria* occupy two major subclusters of this subclass, and many are specifically related to one another. Figure 3 demonstrates that based on SSU rRNA gene sequence comparisons, the SAR86 cluster is clearly unrelated to these genera and instead has evolved from a separate clade in the gamma *Proteobacteria.* A closely related, cultivated relative of the SAR86 cluster has not yet been identified, and all available SSU rRNA gene sequences from cultivated or characterized microorganisms are less than 90% similar to full-length gene sequences from members of this cluster.

Although it is a relatively isolated gene clade, the SAR86 cluster does appear to have some phylogenetic affinities. The SAR86 gene clone lineage branches from within an interesting clade of bacteria and bacterial endosymbionts that includes the Type I methanotrophs, *Methylophaga*, autotrophic and facultative sulfur-oxidizing chemolithotrophs, the photosynthetic, sulfur-oxidizing genus *Chromatium*, and chemolithotrophic sulfur-oxidizing symbionts of various marine eukaryotes (Figure 3). A loose phylogenetic relationship between the clade of chemolithotrophic bacterial endosymbionts which includes the sulfur-oxidizing symbiont of *Bathymodiolus thermophilus*, a deep-sea hydrothermal vent-associated bivalve, and members of the SAR86 cluster has been recovered in multiple analyses, though this affiliation is not well supported statistically. At least two distinct lineages, which are about 95% similar, are evident within the SAR86 cluster. Though the SAR86 cluster is the only numerically abundant gamma *Proteobacteria* SSU rRNA gene type identified in marine plankton clone libraries to date, others may be identified as more plankton community samples are analyzed and full-length SSU rRNA gene sequences are character-ized. Several environmental gene clones that are represented by only short gene sequences, predominantly from the deep ocean, potentially form deep-branching lineages affiliated with the SAR86 cluster and related microorgan-isms and sulfur-oxidizing endosymbionts, but the data currently available are not conclusive (Fuhrman and Davis 1997).

Gram-Positive Bacterioplankton: the Marine Actinobacteria Clade

Though they are usually not abundant, it has been known for some time that gram-positive bacteria are relatively easy to enrich for and cultivate from seawater. For the most part, the marine isolates have been shown to be closely related to well-characterized genera in the low G + C and high G + C (class *Actinobacteria*) clades of the domain *Bacteria*. All gram-positive environmental gene clones recovered from marine prokaryotic plankton so far form a phylogenetically coherent clade within the class *Actinobacteria* (Figure 2; Fuhrman et al. 1993). These clones form a very closely related cluster that branches very early within the high G + C gram-positive bacteria and are distantly related to their nearest cultivated relatives (Figure 6; Rappé et al. 1999). Based on their unique phylogenetic position, the marine bacteria with SSU rRNA genes in the marine *Actinobacteria* clade probably represent a previously unidentified subclass or order within the class *Actinobacteria*. Members of the marine *Actinobacteria* clade do not appear to be as abundant as some of the other major clades of uncultivated bacterioplankton, though they are fairly ubiquitous in seawater environments (Table 2). The limited data collected so far suggest that members of the marine *Actinobacteria* clade generally reside in the photic zone: no gene clones within this cluster have been recovered from the limited number of studies from below this region of the ocean.

SAR202 and the Mesopelagic Green Non-Sulfur Species

Initial culture-independent investigations uncovered closely related groups of prokaryotic plankton residing in surface waters of both the Atlantic and Pacific Oceans, but early hybridization experiments with taxon-specific oligonucleotide probes

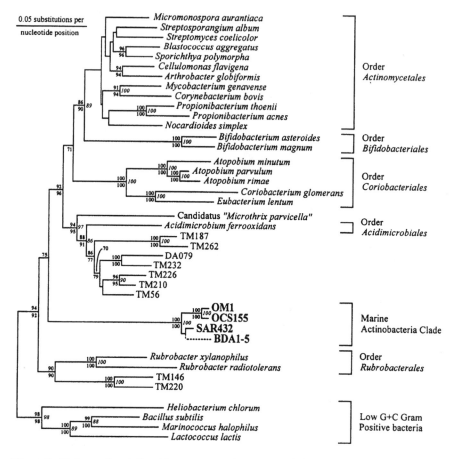

Figure 6. Phylogenetic relationships among environmental SSU rRNA gene clones recovered from seawater to other members of the class *Actinobacteria*, inferred by the neighbor-joining method from a matrix of LogDet distances. A total of 1061 nucleotide positions were included in the analysis. The percentage of bootstrap replicates supporting the proposed branching order are shown above (LogDet distance calculations) below (transversion distance calculations) and to the right of (in italics, GG95 distance calculations) the relevant nodes, from 1000 bootstrapped resamplings. (Reproduced from Rappé et al. 1999).

showed that many of the surface phylotypes were dominant only in the upper surface layer. Armed with the knowledge that unique microbial communities may inhabit deeper regions of the oceans, the breadth of prokaryotic plankton communities analyzed with molecular methods has been slowly expanding to include communities from the lower surface layer and the deep sea (Table 1). The SAR202 gene clone lineage was not initially discovered in prokaryotic plankton communities obtained from surface seawater samples, but instead was recovered in a plankton sample taken from a depth of 250 m in the Sargasso Sea (Giovannoni et al. 1996). This clade of prokaryotic plankton SSU rRNA genes occupies a unique phylogenetic position as it is potentially the deepest branch of the Green Non-Sulfur (GNS) phylum, one of the main branches of the domain *Bacteria* (Figures 1, 2). This phylum appears to have diverged before the radiation of most other main bacterial lineages, and it contains only four genera of cultivated microbes: *Chloroflexus, Herpetosiphon, Heliothrix,* and *Thermomicrobium,* all of which have some thermophilic members. Hybridization studies with taxon-specific oligonucleotide probes targeting members of the SAR202 cluster in the upper 250 m of both the Atlantic and Pacific Oceans indicated that this group of bacterioplankton reached a distinct maximum at the lower boundary of the chlorophyll maximum. Recently, SSU rRNA gene clones of the SAR202 cluster have also been recovered in samples from 3000 m in the northeastern Pacific and 1000 m in the subtropic Atlantic, extending the range of this unique group (Table 2; Fuhrman and Davis 1997). The genera currently included in the GNS phylum are phenotypically quite diverse, and it has not been possible to make reliable inferences regarding the physiology of the SAR202 cluster except to say that they are adapted to life in the deep ocean.

The Marine Group A Clade   Originally discovered in environmental gene clone libraries from 100 and 500 m in the northeastern Pacific Ocean, the Marine Group A clade has since been recovered in samples from the Sargasso Sea and the Pacific coast off Oregon (Fuhrman et al. 1993; Gordon and Giovannoni 1996). Though the first fragmentary sequences did not offer clear conclusions, subsequent analyses of nearly complete SSU rRNA gene sequences have indicated that the SAR406 lineage may be a distant phylogenetic relative of the genus *Fibrobacter* and the green sulfur bacteria, which includes members of the genus *Chlorobium* (Figure 2). It is sufficiently divergent from these two lineages, however, to conclude that members of the SAR406 cluster probably represent a unique, previously undetected major branch of the bacterial radiation. Hybridization data obtained with probes targeting the SAR406 cluster have indicated that members of this gene cluster are vertically stratified in the water column of both the Atlantic and Pacific Oceans, with peak abundance occurring below the deep chlorophyll maximum (Gordon and Giovannoni 1996). Additionally, there are regular seasonal oscillations in the relative abundance of this group in surface waters of the western Sargasso Sea, with a strong positive correlation to surface chorophyll *a* values (Gordon and Giovannoni 1996). Like many of the other major bacterioplankton gene clone clades, the SAR406 cluster appears to be a cosmopolitan group of bacterioplankton; it has recently been recovered in clone libraries from the deep sea (Table 2).

The Marine Group B/SAR324 Clade   The SAR324 cluster, also referred to as Marine Group B, was discovered in simultaneous studies from 250 m in the Sargasso Sea, 500 and 3000 m in the northeastern Pacific Ocean, and from 1000 m in the subtropical Atlantic Ocean (Fuhrman and Davis 1997; Wright et al. 1997). Nearly complete SSU rRNA gene clone sequence analyses have shown that this gene clone lineage represents a unique clade within the delta *Proteobacteria,* but it does not have a cultivated close relative in this group

(Figure 2). Thus, as with most of the other major clades of prokaryotic plankton identified in culture-independent studies, characteristic physiological traits of this major group of bacterioplankton are not currently known. Consistent with clone library findings, hybridization analyses using taxon-specific oligonucleotide probes specific for the SAR324/Marine Group B cluster have demonstrated that this group of bacterioplankton is proportionally more abundant in the aphotic zone of both the Atlantic and Pacific Oceans, indicating that this group is functionally specialized for life in the deep sea.

The Marine *Archaea*

The recovery of *Archaea* is one of the more exciting results from culture-independent studies of marine prokaryotic plankton communities (DeLong 1992; Fuhrman et al. 1992). Using oligonucleotide primers specific for archaeal SSU rRNA genes as well as universal primers targeting nearly all *Bacteria*, *Archaea*, and *Eukarya*, archaeal SSU rRNA gene clones have been recovered from a diverse array of plankton communities, including the eastern and western continental shelves of North America, Antarctic coastal waters, and several depths and locations in the Pacific and Atlantic Oceans. The archaeal marine plankton clades form novel lineages within the two major divisions of the *Archaea*: *Archaea* Group I are peripherally related to the Crenarchaeota, while *Archaea* Group II share a common ancestry with the Euryarchaeote *Thermoplasma acidophilum* (Figure 2). Phylogenetic analyses indicate that within the thermophilic Crenarchaeota, *Archaea* Group I belongs to a broad clade of low-temperature-derived phylotypes (Massana et al. 1997). Within the Euryarchaeota, which primarily consists of methanogens, extreme halophiles, and some thermophiles, *Archaea* Group II may also share a common ancestry with mesophilic phylotypes obtained from benthic marine microbial communities.

Using molecular methods such as oligonucleotide probe–based assays and denaturing gradient gel electrophoresis (DGGE), recent studies have focused on the distribution and dynamics of planktonic marine *Archaea* (Massana et al. 1997; Murray et al. 1998). In general, the relative abundance of archaeal rRNA appears to be lower in surface water samples than at depths greater than 100 m. In addition, *Archaea* Group II appears to dominate the *Archaea* in surface water samples, while *Archaea* Group I dominates at depth. *Archaea* can make up a significant fraction of marine prokaryotic plankton communities (>20% of the total rRNA, 14% of total DAPI counts in in situ studies), and may also follow seasonal patterns that are negatively correlated with chlorophyll *a*.

GENE CLUSTERS AND BACTERIOPLANKTON POPULATION GENETICS

Even though a relatively small number of microbial groups seem to dominate bacterioplankton, genetic variability within these populations is very high.

From a phylogenetic perspective, gene clusters from the environment resemble gene clades from cultivated organisms, but because gene cluster sequences come from uncultivated organisms, it is not known how many species are represented. Differences among the multiple rRNA genes that can be found in a single cell may explain the shallowest branches in gene clusters, but this explanation is not sufficient to explain the observed diversity. In the SAR11 cluster it has been shown that some of this 16S rRNA variability is the result of speciation related to adaptation to different depths within the water column (Field et al. 1997). Very high variability has also been shown for the *rpo*C1 genes, and *pet*B/D genes of cultivated and uncultivated marine cyanobacteria (Ferris and Palenik 1998; Urbach and Chisolm 1998). As with SAR11, some of this variability within the marine cyanobacteria correlates with vertical stratification.

COASTAL VERSUS OPEN-OCEAN BACTERIOPLANKTON SPECIES

Recent culture-independent studies have started to provide a picture of the types of bacteria present in coastal bacterioplankton communities. Though coastal sites can differ markedly from oligotrophic open oceans in that the former have quite different trophic structures and higher productivity and are influenced by freshwater inputs, bacterioplankton communities in coastal regions and open oceans appear quite similar. These similarities are evident in environmental gene clones affiliated with the SAR86 cluster, the SAR116 cluster, the SAR11 cluster, the marine *Actinobacteria* clade, Marine Group A, and the *Roseobacter* clade. Despite recent emphasis on the remarkable diversity of microbes, the common elements of natural microbial communities are in some cases more striking than the differences.

However, recent culture-independent studies of coastal prokaryotic plankton communities have uncovered members of the beta subclass of the *Proteobacteria* that had not been observed in SSU rRNA gene clone libraries constructed from open-ocean environments (Rappé et al. 1997; Fuhrman and Ouverney 1998). Several unique lineages within the beta *Proteobacteria* have been identified, including one lineage closely allied to a clade of obligate methylotrophs that includes *Methylophilus methylotrophus*. No cultured microorganisms related to this clade have been described from seawater, and it is unknown whether these organisms possess the metabolism to oxidize C_1 compounds. It seems plausible that they might, given the distribution of methane hydrate deposits as well as the production of C_1 compounds as breakdown products of dimethylsulfoniopropionate metabolism.

Though it is plausible that the bacterioplankton represented by these gene lineages are adapted to continental shelf or coastal seawater environments, recent studies have revealed a potential freshwater origin for these beta *Proteobacteria*. Both clone library and oligonucleotide probe hybridization data have uncovered large, active populations of beta *Proteobacteria* in freshwater bacterioplankton communities, including relatives of the coastal beta *Proteobacteria* SSU rRNA gene lineages (Methé et al. 1998). Thus, it will be interesting to determine whether the beta *Proteobacteria* environmental clones recovered from coastal environments represent active marine bacteria or simply are transported freshwater microorganisms.

BACTERIOPLANKTON POPULATION DYNAMICS

The Stratification of Bacterioplankton Populations

One of the earliest observations to emerge from rRNA probe hybridization studies was that microbial populations in the ocean surface layer are highly stratified. Vertical stratification of communities first observed in the western Sargasso Sea were confirmed by measurements off the Oregon coast near the continental shelf break point, and similar observations have been reported from the Santa Barbara Channel and Antarctica. The biggest distinction in microbial communities occurs at the boundary of the photic and aphotic zone. Microbial groups that appear to be much more abundant in the aphotic zones include the SAR202 (GNS) clade (Giovannoni et al. 1996), the SAR324 (Marine Group B) clade of delta *Proteobacteria* (Wright et al. 1997), Marine Group A (Gordon and Giovannoni 1996), and the Group I *Archaea* (Massana et al. 1997).

Figure 7 illustrates stratification of microbial communities, showing prochlorophyte and SAR202 rRNAs in the western Sargasso Sea. The ocean surface layer extends from the surface to the lower extent of the deepest winter mixing. At BATS (the Bermuda Atlantic time-series study site), where the data for Figure 7 were collected, winter mixing extends to 250 or sometimes 300 m, and the transition between the photic and aphotic zones occurs at around 120 m.

LINKS BETWEEN COMMUNITY STRUCTURE AND BIOGEOCHEMICAL CYCLES

At present the factors that control the diversity and distributions of the major bacterioplankton species are unknown. The ideas discussed here amount to reasonable conjecture based on scant facts. Their main usefulness is that they provide a conceptual framework for considering the types of environmental factor that might control microbial community structure. Many of these comments are organized around explaining the vertical stratification of bacterioplankton, and the clear partitioning of species between those found on particles and those suspended freely in the water. The factors controlling total biomass levels and biomass production of the entire bacterioplankton, which are discussed in Chapter 8, will probably have different impacts on the different clades within the bacterial community.

Organic particles offer a niche that invites specialization. All bacteria essentially obtain nutrients by transporting small molecules (e.g., monomers or

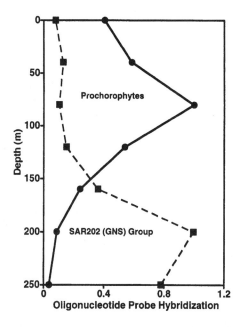

Figure 7. The average distributions of prochlorophytes and the uncultured SAR202 group in the surface layer in the Sargasso Sea, from 10 depth profiles. The data were obtained by hybridizing radioactive oligonucleotide probes to ribosomal RNA extracted from seawater. (Reproduced from Giovannoni et al. 1996.)

oligomers) into the cell, unlike eukaryotes, which can assimilate particles by pinocytosis. However, particle-associated microflora may obtain nutrients by degrading solid organic matter using extracellular hydrolytic enzymes—exoenzymes (Chróst 1991). These enzymes are not typically released by the cell, which would be an "expensive" strategy from a standpoint of cellular energetics. Instead, they appear to be tethered to the cell surface, where they are active when the cell comes into contact with macromolecular organic matter (e.g., organic particles). Cells employing this strategy may still rely on organic carbon uptake systems—typically permeases—to obtain nutrients. However, it is presumed that by hydrolysis and diffusion they create a zone around them that is enriched in DOC (Martinez et al. 1996). Although it is easiest to imagine these adaptive strategies at work on particle surfaces, marine DOC is a heterogeneous mixture of molecules that includes a macromolecular (high molecular weight) fraction that probably could not be used by prokaryotic cells without extracellular processing by hydrolytic enzymes.

DeLong et al. (1993) showed that bacteria belonging to the *Cytophaga–Flavobacterium-Bacteroides* group and the order *Planctomycetales* were associated with particles. A majority of *Cytophaga* spp., *Flavobacterium* spp., and species from the order *Planctomycetales* have adaptations to allow cells to attach to surfaces. In addition, most *Cytophaga* spp. are motile by means of gliding, and a majority of *Planctomycetales* produce small, flagellated swarmer cells that probably are involved in the colonization of new surfaces (Fuerst 1995).

In contrast to bacteria that may be attached to surfaces, freely suspended bacterioplankton probably compete for ambient dissolved organic matter (DOM) that is released by a variety of processes, including the lysis and leakage of phytoplankton (see Chapter 5). For cells that specialize in the uptake of ambient DOM, efficient transport and metabolism of DOM are likely to be important adaptive strategies. Motility and chemotaxis may also be modes of specialization that would allow some of bacterioplankton to exploit microscale patchiness in the environment.

Both DOC and particles can be found throughout the water column, but at different depths the amount and composition of DOC and particles can vary greatly (Chapter 6). There is a strong correlation between microbial diversity and depth in the oceans (Table 2). Vertical clines in the composition of DOC, which are as yet poorly resolved, are likely to influence microbial community composition. It is not unusual for temperature and nitrogen and phosphorus availability to vary dramatically on the vertical axis, so these factors must be considered together to understand the vertical stratification of microbial communities.

The upper surface layer harbors a variety of species but seems to be dominated by members of the alpha *Proteobacteria*, particularly the *Roseobacter* clade, SAR116, and SAR11. If we apply Occam's razor and assume that these organisms subsist at the expense of DOM produced by phytoplankton, then the major selective force that remains, in terms of nutrients, is competition for nitrogen and phosphorus, which is a severe constraint. Characteristics of the alpha *Proteobacteria* that adapt them to life in the photic zone have not been identified, but one possibility is that the ocean surface alpha *Proteobacteria* are more competitive when nitrogen and phosphorus are limiting.

From a purely phylogenetic perspective the microbial populations that occur in the aphotic zone are far more interesting and diverse that their counterparts in the photic zone. In particular, the abundance of *Archaea* and the presence of the diverse marine GNS group typified by SAR202, suggest that the resources of the aphotic zone are partitioned between organisms that have come from very different evolutionary origins.

The resources available to sustain the microbial population of the lower surface layer are far different from those of the upper surface layer. Nitrogen and phosphorus are present in sufficient amounts and are almost certainly not limiting. By contrast, organic matter originating from primary production is in short supply — the lower surface layer is physically remote from the source of energy that drives it but well situated to intercept organic matter that rains from above just as it enters the region where competition for inorganic nutrients is no longer a factor.

Biotic interactions may influence the distributions of some planktonic microorganisms. For example, *Vibrio* and *Photobacterium* species are capable of anaerobic growth by means of fermentative metabolism. This ability of

Vibrio and *Photobacterium* spp. suggests that these organisms spend part of their life cycle in anoxic habitats — perhaps the guts of zooplankton, or else on organic-rich particles where oxygen diffusion is insufficient to keep up with the demands of microbial metabolism.

The *Pseudoalteromonas* genus of gamma *Proteobacteria* have relatively large genomes (Lanoil et al. 1996) but are obligate aerobes, unlike *Vibrio* spp., which are facultative anaerobes. As with the other marine groups, the ecological role of *Pseudoalteromonas* spp. is unknown; however, in some cases interesting interactions between *Pseudoalteromonas* spp. and algae have been observed. Several *Pseudoalteromonas* spp. have been shown to secrete factors that are toxic to algal species (Lovejoy et al. 1998). Typically, we think of interactions between bacterioplankton and primary producers as indirect, but there is no a priori reason to exclude from consideration the possibility that direct interactions occur.

RESEARCH HORIZONS

In the future, research on marine bacterioplankton diversity will almost certainly focus on the role of these species in global biogeochemical cycles. There are compelling reasons to think that the growth of specific microbial populations in planktonic ecosystems is linked to the amount and composition of DOM. Nevertheless, microbial ecologists have rarely been able to establish links between ecosystem scale processes and the activities of specific microbial populations. In part, this outcome is a natural consequence of the evolution of methods that tended to focus on the activities of bacteria as if they were a single element in the system, without considering the ecology of individual "species." Thus, the partitioning of resources by sympatric heterotrophic bacterioplankton species and the role of seasonal and spatial distributions of microbial populations in DOM flux were left as horizons open for exploration.

There are numerous approaches that might be useful for establishing a functional role for bacterioplankton species. One of the most obvious approaches is to culture the organisms in question, and to determine their physiology in a laboratory setting. This approach is not straightforward for two reasons: (1) the organisms may be very difficult to cultivate, and (2) physiological experiments may reveal little about the in situ activity of a microorganism unless the experimentalist is particularly insightful. Indeed, the sequencing of complete microbial genomes now routinely reveals that organisms have capacities for types of metabolism that were never suspected because the appropriate tests were not performed.

A number of approaches are emerging for determining the activity of microorganisms that are uncultured. The uptake of the DNA precursor bromodeoxyuridine has been used to identify growing bacterial cells (Urbach et al. 1999), and ^{13}C incorporation into lipids has been used to identify organisms that assimilate acetate (Boscher et al. 1998). A combination of autoradiography and in situ hybridization techniques also holds promise for identifying the cells able to assimilate specific organic compounds (Ouverney and Fuhrman 1999).

The term "environmental genomics" has been coined to encompass scientific endeavors that utilize nucleic acids from natural microbial ecosystems as the starting point for investigative research. The excitement in this field stems from recognition that microbial diversity provides a vast and unexplored resource of biological potential. It is likely that much will be learned about bacterio-plankton by studying genomic DNA recovered from seawater, but a combination of these genomics methods and in situ chemical measurements seems to hold the greatest promise for yielding insight into bacterioplankton ecology.

SUMMARY

1. The most abundant bacterioplankton have never been cultured.

2. The major marine prokaryotic groups appear to have cosmopolitan distributions.

3. A relatively small number of uncultured marine bacterioplankton clades (nine) account for 80% of marine *Bacteria* 16S rRNA gene clones recovered from seawater.

4. Marine *Archaea* are abundant, and almost invariably fall within two phylogenetic groups.

5. There is much genetic diversity within the major prokaryotic plankton groups. It is not yet known how much ecological specialization occurs among the species that make up groups, but in some cases members of these groups are distributed differently with depth.

6. Particle-associated and freely suspended marine prokaryotes are different.

7. Stratification of bacterioplankton populations is typical of the ocean surface layer.

REFERENCES

Baumann, P., Furniss, A. L., and Lee, J. V. (1984) *Vibrio.* In N. R. Krieg and J. G. Holt, eds., *Bergey's Manual of Systematic Bacteriology.* Williams & Wilkins, Baltimore, pp. 518–544.

Bernardet, J.-F., Segers, P., Vancanneyt, M., Berthe, F., Kersters, K., and Vandamme, P. (1996) Cutting a Gordian knot: Emended classification and description of the genus *Flavobacterium,* emended description of the family *Flavobacteriaceae,* and proposal of *Flavobacterium hydatis* nom. nov. (basonym, *Cytophaga aquatilis* Strohl and Tait 1978). *Int. J. Syst. Bacteriol.* 46:128–148.

Boscher, H. T. S., Nold, S. C., Wellsbury, P., Bos, D., de Graaf, W., Pel, R., Parkes, R. J., and Cappenberg, T. E. (1998) Direct linking of microbial populations to specific biogeochemical processes by ^{13}C-labelling of biomarkers. *Nature* 392:801–805.

Bowman, J. P., McCammon, S. A., Nichols, D. S., Skerratt, J. H., Rea, S. M., Nichols, P. D., and McMeekin, T. A. (1997) *Shewanella gelidimarina* sp. nov. and *Shewanella frigidimarina* sp. nov., novel Antarctic species with the ability to produce eicosapentaenoic acid and grow anaerobically by dissimilatory Fe(III) reduction. *Int. J. Syst. Bacteriol.* 47:1040–1047.

Britschgi, T. B., and Giovannoni, S. J. (1991) Phylogenetic analysis of a natural marine bacterioplankton population by rRNA gene cloning and sequencing. *Appl. Environ. Microbiol.* 57:1313–1318.

Button, D. K. (1998) Nutrient uptake by microorganisms according to kinetic parameters from theory as related to cytoarchitecture. *Microbiol. Mol. Biol. Rev.* 62:636–645.

Button, D. K., Schut, F., Quang, P., Martin, R., and Robertson, B. R. (1993) Viability and isolation of marine bacteria by dilution culture: Theory, procedures, and initial results. *Appl. Environ. Microbiol.* 59:1707–1713.

Chisholm, S. W., Olsen, R. J., Zehler, E. R., Goericke, R., Waterbury, J. B., and Welschmeyer, N. A. (1988) A novel free-living prochorophyte abundant in oceanic euphotic zone. *Nature* 334:340–343.

Chróst, R. J. (1991) Environmental control of the synthesis and activity of aquatic microbial ectoenzymes. In R. J. Chróst, ed., *Microbial Enzymes in Aquatic Environments*. Springer-Verlag, New York, pp. 22–59.

DeLong, E. F. (1992) Archaea in coastal marine bacterioplankton. *Proc. Natl. Acad. Sci. USA* 89:5685–5689.

DeLong, E. F., Franks, D. G., and Alldredge, A. L. (1993) Phylogenetic diversity of aggregate-attached vs. free-living marine bacterial assemblages. *Limnol. Oceanogr.* 38:924–934.

Ferris, M. J., and Palenik, B. (1998) Niche adaptation in ocean cyanobacteria. *Nature* 396:226–228.

Field, K. G., Gordon, D., Wright, T., Rappé, M., Urbach, E., Vergin, K., and Giovannoni, S. J. (1997) Diversity and depth-specific distribution of SAR11 cluster rRNA genes from marine planktonic bacteria. *Appl. Environ. Microbiol.* 61:63–70.

Fry, J. C. (1990) Oligotrophs. In C. Edwards, ed., *Microbiology of Extreme Environments*. McGraw-Hill, New York, pp. 93–116.

Fuerst, J. A. (1995) The Planctomycetes: Emerging models for microbial ecology, evolution and cell biology. *Microbiology* 141:1493–1506.

Fuhrman, J. A., and Ouverney, C. C. (1998) Marine microbial diversity studied via 16S rRNA sequences: cloning results from coastal waters and counting of native archaea with fluorescent single cell probes. *Aquat. Ecol.* 32:3–15.

Fuhrman, J. A., McCallum, K., and Davis, A. A. (1992) Novel major archaebacterial group from marine plankton. *Nature* 356:148–149.

Fuhrman, J. A., McCallum, K., and Davis, A. A. (1993) Phylogenetic diversity of subsurface marine microbial communities from the Atlantic and Pacific Oceans. *Appl. Environ. Microbiol.* 59:1294–1302.

Fuhrman, J. A., and Davis, A. A. (1997) Widespread Archaea and novel Bacteria from the deep sea as shown by 16S rRNA gene sequences. *Mar. Ecol. Prog. Ser.* 150:275–285.

Giovannoni, S. J., Britschgi, T. B., Moyer, C. L., and Field, K. G. (1990) Genetic diversity in Sargasso Sea bacterioplankton. *Nature* 345:60–63.

Giovannoni, S. J., Rappé, M. S., Vergin, K. L., and Adair, N. L. (1996) 16S rRNA genes reveal stratified open ocean bacterioplankton populations related to the Green Non-Sulfur bacteria. *Proc. Natl. Acad. Sci. USA* 93:7979–7984.

González, J. M., Whitman, W. B., Hodson, R. E., and Moran, M. A. (1996) Identifying numerically abundant culturable bacteria from complex communities: An example from a lignin enrichment culture. *Appl. Environ. Microbiol.* 62:4433–4440.

Gonzalez, J. M., Kiene, R. P., and Moran, M. A. (1999) Transformation of sulfur compounds by an abundant lineage of marine bacteria in the α-subclass of the class *Proteobacteria*. *Appl. Environ. Microbiol.* 65:3810–3819.

Gordon, D. A., and Giovannoni, S. J. (1996) Stratified microbial populations related to *Chlorobium* and *Fibrobacter* detected in the Atlantic and Pacific Oceans. *Appl. Environ. Microbiol.* 62:1171–1177.

Hanson, R. S., and Hanson, T. E. (1996) Methanotrophic bacteria. *Microbiol. Rev.* 60:439–471.

Hillis, D. M., Moritz, C., and Mable, B. K. (1996) *Molecular Systematics*. Sinauer, Sunderland, MA.

Höfle, M. G., and Brettar, I. (1996) Genotyping of heterotrophic bacteria from the central Baltic Sea by use of low-molecular-weight RNA profiles. *Appl. Environ. Microbiol.* 62:1225–1228.

Jannasch, H. W., and Jones, G. E. (1959) Bacterial populations in seawater as determined by different methods of enumeration. *Limnol. Oceanogr.* 4:128–139.

Johansen, J. E., Nielsen, P., and Sjoholm, C. (1999) Description of *Cellulophaga baltica* gen. nov., sp. nov. and *Cellulophaga fucicola* gen. nov., sp. nov. and reclassification of *Cytophaga lytica* to *Cellulophaga lytica* gen. nov., comb. nov. *Int. J. Syst. Bacteriol.* 49:1231–1240.

Kato, N. Y., and Horokoshi, K. (1998) Taxonomic studies of the deep-sea barophilic *Shewanella* strains and description of *Shewanella violacea* sp. nov. *Arch. Microbiol.* 170:331–338.

Lafay, B., Ruimy, R., Rauch de Traubernberg, C., Breittmayer, V., Gauthier, M. J., and Christen, R. (1995) *Roseobacter algicola* sp. nov., a new marine bacterium isolated from the phycosphere of the toxin-producing dinoflagellate *Prorocentrum lima*. *Int. J. Syst. Bacteriol.* 45:290–296.

Lanoil, B. D., Ciufettii, L. M., and Giovannoni, S. J. (1996) *Pseudalteromonas haloplanktis* has a complex genome structure composed of two separate genetic units. *Genome Res.* 6:1160–1169.

Ledyard, K. M., DeLong, E. F., and Dacy, J. (1993) Characterization of a DMSP-degrading bacterial isolate from the Sargasso Sea. *Arch. Microbiol.* 160:312–318.

Lidstrom, M. E. (1992) The aerobic methylotrophic bacteria. In A. Barlows et al., eds., *The Prokaryotes*. Springer-Verlag, New York, pp. 431–445.

Lovejoy, C., Bowman, J. P., and Hallegraeff, G. M. (1998) Algicidal effects of a novel marine *Pseudoalteromonas* isolate (class *Proteobacteria*, gamma subdivision) on harmful algal bloom species of the genera *Chattonella*, *Gymnodinium* and *Heterosigma*. *Appl. Environ. Microbiol.* 64:2806–2813.

Martinez, J., Smith, D. C., Steward, G. F., and Azam, F. (1996) Variability in ectohydrolytic enzyme activities of pelagic marine bacteria and its significance for substrate processing in the sea. *Aquat. Microb. Ecol.* 10:223–230.

Massana, R. A., Murray, A. E., Preston, C. M., and DeLong, E. F. (1997) Vertical distribution and phylogenetic characterization of marine planktonic *Archaea* in the Santa Barbara Channel. *Appl. Environ. Microbiol.* 63:50–56.

McInerney, J. O., Mullarkey, M., Wernecke, M. E., and Powell, R. (1997) Phylogenetic analysis of Group I marine archaeal rRNA sequences emphasizes the hidden diversity within the primary group *Archaea. Proc. R. Soc. London* 264:1663–1669.

Methé, B. A., Hiorns, W. D., and Zehr, J. P. (1998) Comparison of marine and freshwater bacterial community structure: Analyses of communities in Lake George, NY, and six other Adirondack lakes. *Limnol. Oceanogr.* 43:368–374.

Morita, R. Y. (1997) *Bacteria in Oligotrophic Environments.* Chapman & Hall, New York.

Morita, R. Y. (1985) Starvation and miniaturisation of heterotrophs with special emphasis on maintenance of the starved viable state. In M. Fletcher and G. Floodgate, eds., *Bacteria in Natural Environments: The Effect of Nutrient Conditions.* Academic Press, London, pp. 111–130.

Mullins, T. D., Britschgi, T. B., Krest, R. L., and Giovannoni, S. J. (1995) Genetic comparisons reveal the same unknown bacterial lineages in Atlantic and Pacific bacterioplankton communities. *Limnol. Oceanogr.* 40:148–158.

Murray, A. E., Preston, C. M., Massana, R., Taylor, L. T., Blakis, A., Wu, K., and DeLong, E. F. (1998) Seasonal and spatial variability of bacterial and archaeal assemblages in the coastal waters near Anvers Island, Antarctica. *Appl. Environ. Microbiol.* 64:2585–2595.

Olsen, G. J., Lane, D. L., Giovannoni, S. J., Pace, N. R., and Stahl, D. A. (1986) Microbial ecology and evolution: A ribosomal RNA approach. *Annu. Rev. Microbiol.* 40:337–366.

Ouverney, C. C., and Fuhrman, J. A. (1999) Combined autoradiography–16S rRNA probe technique for the determination of radioisotope uptake by specific microbial cells types in situ. *Appl. Environ. Microbiol.* 65:1746–1752.

Paster, B. J., Ludwig, W., Weisburg, W. G., Stackebrandt, E., Hespell, R. B., Hahn, C. M., Reichenbach, H., Stetter, K. O., and Woese, C. R. (1985) A phylogenetic grouping of the bacteroides, cytophagas, and certain flavobacteria. *System. Appl. Microbiol.* 6:34–42.

Pinhassi, J., Zweifel, U. L., and Hagström, Å. (1997) Dominant marine bacterioplankton species found among colony-forming bacteria. *Appl. Environ. Microbiol.* 63:3359–3366.

Rappé, M. S., Kemp, P. F., and Giovannoni, S. J. (1997) Phylogenetic diversity of marine coastal picoplankton 16S rRNA genes cloned from the continental shelf off Cape Hatteras, North Carolina. *Limnol. Oceanogr.* 42:811–826.

Rappé, M. S., Gordon, D. A., Vergin, K. L., and Giovannoni, S. J. (1999) Phylogeny of Actinobacteria small subunit (SSU) rRNA gene clones recovered from marine bacterioplankton. *Syst. Appl. Microbiol.* 22:106–112.

Reichenbach, H. (1992) The order *Cytophagales*. In A. Barlows et al., eds., *The Prokaryotes*. Springer-Verlag, New York, pp. 3631–3675.

Schlesner, H. (1994) The development of media suitable for the microorganisms morphologically resembling *Planctomyces* spp., *Pirellula* spp., and other *Planctomycetales* from various aquatic habitats using dilute media. *Syst. Appl. Microbiol.* 17:135–145.

Schmidt, T. M., DeLong, E. F., and Pace, N. R. (1991) Analysis of a marine picoplankton community by 16S rRNA gene cloning and sequencing. *J. Bacteriol.* 173:4371–4378.

Schut, F., de Vries, E. J., Gottschal, J. C., Robertson, B. R., Harder, W., Prins, R. A., and Button, D. K. (1993) Isolation of typical marine bacteria by dilution culture: Growth, maintenance, and characteristics of isolates under laboratory conditions. *Appl. Environ. Microbiol.* 59:2150–2161.

Shiba, T., Simidu, U., and Taga, N. (1979) Distribution of aerobic bacteria which contain bacteriochlorophyll *a*. *Appl. Environ. Microbiol.* 38:43–45.

Sieburth, J. M., Johnson, P. W., Eberhardt, M. A., Sieracki, M. E., Lidstrom, M., and Laux, D. (1987) The first methane-oxidizing bacterium from the upper mixing layer of the deep ocean: *Methylomonas pelagica* sp. nov. *Curr. Microbiol.* 14:285–293.

Stackebrandt, E., and Göebel, B. M. (1994) Taxonomic note: A place for DNA–DNA reassociation and 16S sequence rRNA analysis in the present species definition in bacteriology. *Int. J. Syst. Bacteriol.* 44:846–849.

Staley, J. T., and Konopka, A. (1985) Measurement of in situ activities of nonphotosynthetic microorganisms in aquatic and terrestrial habitats. *Annu. Rev. Microbiol.* 39:321–346.

Suzuki, M. T., Rappé, M. S., Haimberger, Z. W., Winfield, H., Adair, N., Ströbel, J., and Giovannoni, S. J. (1997) Bacterial diversity among small-subunit rRNA gene clones and cellular isolates from the same seawater sample. *Appl. Environ. Microbiol.* 63:983–989.

Urbach, E., and Chisholm, S. W. (1998) Genetic diversity in *Prochlorococcus* populations flow cytometrically sorted from the Sargasso Sea and Gulf Stream. *Limnol. Oceanogr.* 43:1615–1630.

Urbach, E., Scanlan, D. J., Distel, D. L., Waterbury, J. B., and Chisholm, S. W. (1998) Rapid diversification of marine picophytoplankton with dissimilar light-harvesting structures inferred from sequences of *Prochlorococcus* and *Synechococcus*. *J. Mol. Evol.* 46:188–201.

Urbach, E., Vergin, K. L., and Giovannoni, S. J. (1999) Immunochemical detection and isolation of DNA from metabolically active bacteria. *Appl. Environ. Microbiol.* 65:1207–1213.

Vergin, K. L., Urbach, E., Stein, J. L., DeLong, E. F., Lanoil, B. D. and Giovannoni, S. J. (1998) Screening of a fosmid library of marine environmental genomic DNA fragments reveals four clones related to members of the order *Planctomycetales*. *Appl. Environ. Microbiol.* 64:3075–3078.

Waterbury, J. B., Watson, S. W., Guillard, R. R., and Brane, L. E. (1979) Widespread occurrence of a unicellular, marine, planktonic cyanobacterium. *Nature* 277:293–294.

Wood, A. M. (1985) Adaptation of photosynthetic apparatus of marine ultraphytoplankton to natural light fields. *Nature* 316: 253–255.

Wright, T. D., Vergin, K., Boyd, P., and Giovannoni, S. J. (1997) A novel deltaproteobacterial lineage from the lower ocean surface layer. *Appl. Environ. Microbiol.* 63:1441–1448.

4

BACTERIAL PRODUCTION AND BIOMASS IN THE OCEANS

Hugh Ducklow

Virginia Institute of Marine Science,
College of William and Mary,
Gloucester Point, Virginia

The study of bacterial growth dynamics in the sea is a relatively new field of investigation. The subject of bacterial growth is not treated by ZoBell in his classic treatise *Marine Microbiology* (ZoBell 1946), even though Henrici (1938) provided a surprisingly familiar look at bacterial dynamics in freshwater lakes. Brock (1971) reviewed other early attempts to investigate bacterial growth processes in nature, but few of his references are to the marine realm. Yet following the introduction of new methods for assessing bacterial abundance and production rates, bacterial production studies became commonplace on oceanographic cruises and in the literature. Nearly all approaches until quite recently have been limited to addressing bacterioplankton as a homogeneous assemblage, which may explain the success of the field; the new measurements were directly amenable to compartmental modeling just when that activity began its own renaissance, aided by the rapid evolution of the personal computer. In this chapter, I survey recent developments in technique and provide a synthesis of current understanding of bacterioplankton productivity and biomass levels in the sea.

Microbial Ecology of the Oceans, Edited by David L. Kirchman.
ISBN 0-471-29993-6   Copyright © 2000 by Wiley-Liss, Inc.

WHAT IS BACTERIAL PRODUCTION?

Bacterial production is *secondary production*: the synthesis of bacterial biomass, primarily from organic precursors with some inorganic nutrients. The net effect is to move organic matter from one pool to another. Bacterial production can be expressed as the rate of synthesis of cells (N) or cell mass $B*$:

$$P = \mu B \tag{1}$$

where μ is the specific growth rate of the population expressed in units of inverse time t^{-1},

$$\mu = \frac{1}{B}\frac{dB}{dt} \tag{2}$$

and B (or N) is the mass (number) of cells, expressed per unit of volume. As we'll see later on, the definition, while exact mathematically, contains in practice an element of ambiguity, or circularity. This is because we do not always derive estimates of P through *a priori* measurements of μ and B. In fact, one message of this chapter is that it is easiest in a practical sense to measure P, but the key to understanding the meaning and regulation of bacterial production is still precise and unambiguous determination of in situ values of μ (see, e.g., Ducklow et al., 1992, 1999). Kemp et al. (1993) provide a practical guide to measurements of the three terms in Equation (1). There are many treatments of bacterial growth in the laboratory, but few in nature. Cooper (1991) is the standard text on the physiology and biochemistry of bacterial growth at the cellular level. Appendix 6 in Cohen (1995) provides a lucid, insightful, and entertaining discussion of the mathematics of *human* population growth, which can be applied to bacteria with few if any modifications. I will try to provide a guide for understanding the biological processes and practical pitfalls associated with measuring and understanding bacterial production in the sea. I will consider in turn measurement of μ and P. I discuss N and B in less detail, mostly insofar as they pertain to understanding and measuring μ and P. Finally I review the magnitude of bacterial production in various marine systems.

Is Bacterial Production Net or Gross?

The development of practical and reliable (but see below) approaches for measuring bacterial production allowed a meaningful dialog between marine bacteriologists and biological oceanographers for the first time. Only after bacterial processes could be expressed in the same units used by other oceanographers could bacteria be fitted into current paradigms of marine

---

*In this chapter, I address production mostly in terms of the biomass as carbon produced, but cell numbers are also discussed. Most readers can consider mass and numbers to be more or less interchangeable for general understanding.

trophodynamics (Williams 1981, 1984; Azam et al. 1983). However there is still some misunderstanding concerning the meaning of bacterial production, especially when bacteriologists talk to phytoplankton ecologists. Scientists working on phytoplankton have the luxury of being able to specify primary production directly in terms of measured fluxes of mass or energy. They measure carbon fixation rates using $^{14}CO_2$ (Steeman-Nielsen 1952) or ^{18}O-labeled water (Bender and Grande 1987), determine changes in total CO_2 or O_2 dissolved in seawater (Williams 1993, Emerson et al. 1993), or quantify light absorption with optical sensors (Marra et al. 1999). From such measurements two quantities, *gross* and *net* primary production, can be estimated. Gross primary production is the total fixation of carbon during photosynthesis in the light, whereas net primary production (NPP) is the gross production in the light less the amount of carbon respired by phytoplankton over a 24-hour period (Falkowski and Raven 1997). Unambiguous determination of gross and net primary production rates is complicated by the presence in most water samples of microheterotrophs, which carry out respiration in addition to that accomplished by the phytoplankton (Williams 1998). But the point here is that phytoplankton ecologists are alert to the distinctions between net and gross production, and they expect bacteriologists to be also. What is it that bacteriologists measure?

The short answer is that they measure net bacterial production (BP_{net}) but usually just call it bacterial production, as I will do in the remainder of this chapter (similarly, when I say "primary production," I am referring to NPP). *All* approaches for determining bacterial production provide estimates in some fashion or other of the *net* rate of biomass synthesis, without including bacterial respiration in the estimates (Jahnke and Craven 1995). Bacterial biomass synthesis plus respiration (R) can be termed gross production (BP_g) in loose analogy to phytoplankton production. Further misunderstanding arises from the use of the term *bacterial carbon demand* (BCD) in place of gross production. In physiological terms, BCD is determined from the gross growth efficiency Y (Lancelot and Billen 1986), or GGE (Goldman et al. 1987):

$$BCD = BP_g = \frac{BP_{net}}{Y} = BP_{net} + R \qquad (3)$$

$$GGE = Y = \frac{BP_{net}}{BP_g} \qquad (4)$$

where Y is a traditional term from bacterial physiology for *fractional growth yield*, defined as the biomass synthesized per unit total limiting nutrient utilized (Stanier et al. 1976). Bacterial growth yields are addressed in detail in Chapter 10 and in del Giorgio and Cole (1998). Here I simply clarify the relationships among bacterial growth efficiency, carbon demand, and net and gross bacterial production. I should also note that while the GGE can be defined physiologically, it is harder to specify its meaning, or value ecologically, especially over longer time scales (Jahnke and Craven 1995).

One further comment about bacterial production is in order. It is common to try to assess the significance of particular bacterial production estimates by comparing the bacterial production to a simultaneous determination of primary production, usually using carbon-14. Since many routine primary production measurements do not include estimates of the production of DOC (and many estimates do include it: see Cole et al. 1988), one should be careful to avoid using these ratios to claim "bacterial production is $X\%$ of primary production," as I occasionally say in this chapter. This form of statement implies that bacteria are directly using some share of the contemporaneous primary production, when in fact what we really wish to convey is that they are using an amount of carbon equivalent to $X\%$ of particulate primary production. Only if a total estimate of primary production (particulate plus dissolved) is available can one properly claim that bacterial production was some fraction of the primary production, in that time and place.

WHY MEASURE BACTERIAL PRODUCTION?

The advantages of measuring bacterial production should be obvious to bacterial ecologists, but the rationale for focusing on this measurement may still not be immediately apparent to other biological oceanographers and biogeochemists. In fact, there are clear advantages for bacteriologists and biogeochemists, as implied by the relationships shown in Eqations (1)–(4). This rationale for bacterial production measurements is articulated nicely in Cole and Pace (1995).

Importance of the Microbial Loop

Establishing the existence, functioning, and magnitude of the microbial loop and microbial food webs in the sea has been a major theme of biological oceanography over the past two decades, since the introduction of more easily used methods for determining bacterial production (Fuhrman and Azam 1980; Ducklow 1983; Azam 1998). The term "microbial loop" per se refers to the bacterial recovery through uptake and metabolism of dissolved organic matter (DOM) otherwise "lost" from the trophic system via excretion, exudation, and diffusion (Azam et al. 1983; Jumars et al. 1989). Bacterial production is the key process originating the flux of DOM through the loop, and so estimates of bacterial production establish the importance of the microbial loop and of microbial food webs initiated by bacterivory, in marine ecosystems.

Quantifying Biogeochemical Fluxes of Carbon and Other Elements

Marine bacterioplankton are usually free-living and are sustained by the flux of low molecular weight DOM (LMW-DOM) into the cell. Furthermore for all practical purposes, they dominate DOM incorporation (Azam and Hodson 1977). Only molecules below 500–1000 Da are recognized and transported through cell membranes by bacterial permeases. Some variable and perhaps large fraction of the LMW-DOM is derived from the breakdown of high molecular weight DOM (HMW-DOM) by extracellular enzymes (Somville 1983; Hoppe 1983; D. C. Smith et al. 1992). Both the LMW and HMW DOM pools consist potentially of hundreds or thousands of individual compounds. These pools cannot yet be fully characterized chemically, and we cannot measure the aggregate fluxes directly (see Chapters 5–7 for discussion of DOM composition and dynamics). In other words it is not yet possible to measure directly the total flux of DOM into bacteria, BP_g. The most practical approach is to determine bacterial production and the GGE and then use equations (3) and (4) to derive the BCD, even though because of interbacterial and viral carbon cycling, the BCD is not a unique function of bacterial production and GGE, (Jahnke and Craven 1995). Even without direct measurements of GGE, we can make first-order estimates showing that DOM fluxes are large terms in the budgets of organic carbon in marine ecosystems, just by knowing that bacterial production is an appreciable fraction of primary production (see below).

Estimating Growth Rates

It is fiendishly difficult to measure bacterial growth rates in nature (Brock 1971). Although in principle it is straightforward to determine the right-hand terms in Equation (2), in practice it is seldom possible to obtain unambiguous estimates of dB/dt. Rates of change of cell populations in nature are usually underestimates of the actual growth rate because there is simultaneous removal of prey cells by predators (Landry and Hassett 1982; Ducklow and Hill 1985a; Chapter 12) and viruses (Chapter 11). Strategies that have been employed for minimizing or independently accounting for the removal terms include dilution, size fractionation, and specific metabolic inhibitors; these are reviewed elsewhere in this book (Chapters 12). Growth rates are of course intrinsically interesting to know, and they are required to parameterize models. Further, if we could measure growth rates unambiguously, and relate them to other, more easily measured variables (e.g., chlorophyll, temperature), then the derived growth rates could be used to estimate bacterial production from equation (1) for large-scale system comparison. But in practice, it is easier to measure bacterial production and biomass, and calculate μ from equation (1), instead. Thus, measuring bacterial production remains our best approach to obtaining large data sets on growth rates. The approach is flawed, however, unless we can specify the fraction of bacterial production (or N) that is actually growing (Zweifel and Hagström 1995; Sherr et al. 1998).

METHODS: A SURVEY AND UPDATE

Marine bacteriology has always been challenged by methodological difficulty, imposed in large part by the exceptionally small size and dilute concentrations of cells in a complex mix of contaminating organisms and dead particles (Kemp et al. 1993). Any treatment of bacterial biomass dynamics has to address methods to place critical understanding of the data in proper context. It is important to recognize that very few direct determinations of bacterial biomass or production have ever been made in unmanipulated or minimally manipulated samples. There is no carbon-14 assay for bacterial production. Instead, both biomass and production are derived from measurements of related quantities through application of conversion factors. Both the choice of property analyzed and the conversion factor values influence the conclusion of the measurements. The following review is meant to guide the reader toward both deeper and more comprehensive treatments of each subject (e.g., Karl 1986).

Bacterial Biomass

The overwhelming majority of published studies are based on microscopic determinations of bacterial abundance. There are other methods for estimating bacterial numbers or biomass, notably detection of gram-negative cell walls using *Limulus* amebocyte lysate (LAL; Watson et al. 1977). I limit this review to a discussion of direct detection of cells by microscopy or flow cytometry, but it is worthwhile noting that new, automated, and sensitive colorimetric assays of LAL may make this technique more attractive. I am not aware of any published observations using this modernized approach in field study.

Epifluorescence Microscopy Direct microscopy had long been understood to yield substantially higher numbers of bacterial cells in lake waters (Henrici 1933, 1938) and seawater (Jannasch and Jones 1959) than plate count and other cultural techniques. But the difficulty in resolving cells by light or phase contrast optics limited the application of the method, and so cultural estimates remained in favor, despite the lower estimates. As ZoBell (1946, p. 52) concluded, "At best direct counts give data which only supplement and aid in the interpretation of results obtained by cultural procedures."

In spite of the shortcomings of direct microscopy, the approach was followed by Soviet oceanographers, who obtained bacterial biomass estimates of the same order of magnitude as other plankton groups. They formed a modern dynamic viewpoint about oceanic bacterioplankton considerably in advance of Western bacteriologists (e.g., Sorokin 1964). Direct microscopy was finally adopted widely following the introduction of a practical method for concentrating bacteria on optically flat polycarbonate filters for direct counting with acridine orange epifluorescent microscopy (AODC) (Hobbie et al. 1977; Watson et al. 1977). Other, brighter and/or more specific DNA fluorochromes have been introduced (e.g., DAPI; Porter and Feig 1980; SYBR Green: Noble and Fuhrman 1998), but the original AODC protocol remains largely unmodified, irrespective of the dye employed. Direct microscopy remains the most widely used approach to measuring bacterioplankton abundance and is irreplaceable as a ground truth baseline on which microbiological interpretation can be based. Experienced microscopists are remarkable image processors and data reducers, but it is hard to document properly the information obtained visually during direct counting.

Flow Cytometry With the development of sensitive optics, practical laser systems, and higher fluorescence yield fluorochrome dyes, the flow cytometric detection and enumeration of marine heterotrophic bacteria is becoming an attractive alternative to microscopy. The method is preferable because many more cells are counted in each sample than is possible by epifluorescence microscopy and because the heterotrophic bacteria can be distinguished from prochlorophytes and coccoid cyanobacteria of similar size and fluorescence characteristics (Campbell et al. 1994); these phototrophs are usually counted as heterotrophic bacteria with microscopy. Sample preparation is much less labor intensive, requiring only an adequate supply of liquid nitrogen at sea. Because of the ease of sample throughput and the greater yield of information per unit effort, flow cytometry should replace epifluorescence microscopy for most routine applications over the next few years. The two approaches appear to detect essentially the same population of cells (Figure 1), although region- or perhaps cruise-specific variability remains to be resolved (Ducklow et al. in press).

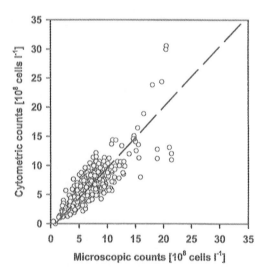

Figure 1. Comparison of epifluorescence microscope versus flow cytometric counts of heterotrophic bacteria in the upper 200 m of northwestern Arabian Sea in January, March, and December 1995. The model II regression line is $Y = 1.04X - 0.87$; $r^2 = 0.69$, $n = 421$. (Data from Ducklow et al. in press; Campbell et al. 1999.)

Cell Volume and Mass   Bacterial biomass cannot be measured directly. Rather, biomass estimates are derived from abundance or biovolume measurements multiplied some factor for carbon per cell. Several papers (Cho and Azam 1990; Ducklow and Carlson 1992; Fukuda et al. 1998) show graphs of bacterial biomass versus chlorophyll *a*, indicating that bacterial biomass tends to equal or exceed phytoplankton mass at low chlorophyll concentrations. Table 1 indicates why this might be true, using typical carbon conversion factors (CCF) for deriving biomass. The critical dependence of relative biomass level on assumptions about conversion factor values is obvious.

Estimation of cell volume remains technically difficult by microscopy and cytometry. Cell volume is now routinely measured with epifluorescence microscopy using digital image analysis of video images (Bjørnsen 1986; Ducklow et al. 1995; Carlson et al. 1996; Pomroy and Joint 1999). Especially when large numbers of samples must be analyzed for oceanographic surveys, the analysis is limited for practical considerations to about 300–400 cells per sample, usually without replication. A serious problem with epifluorescent determina-

Table 1. Phytoplankton and bacterial biomass in the ocean, estimated using different C:Chl ratios for calculating phytoplankton carbon (Phyto-C) from chlorophyll (Chl) and carbon conversion factors (CCF) for estimating bacterial carbon (Bact-C) from cell counts

Regime	Chl (μg L^{-1})	Number of Bacteria (10^9 cells L^{-1})	C:Chl (μg μg^{-1})	CCF (fg C cell^{-1})	Phyto-C (μg C L^{-1})	Bact-C (μg C L^{-1})
Open sea	0.1	0.5	50	10	5	5
Coastal	1	1	50	10	50	10
Estuary	10	5	50	10	500	50
Open sea	0.1	0.5	100	10	10	5
Coastal	1	1	100	10	100	10
Estuary	10	5	100	10	1000	50
Open sea	0.1	0.5	50	20	5	10
Coastal	1	1	50	20	50	20
Estuary	10	5	50	20	500	100
Open sea	0.1	0.5	100	20	10	10
Coastal	1	1	100	20	100	20
Estuary	10	5	100	20	1000	100
Open sea	0.1	0.5	50	30	5	15
Coastal	1	1	50	30	50	30
Estuary	10	5	50	30	500	150
Open sea	0.1	0.5	100	30	10	15
Coastal	1	1	100	30	100	30
Estuary	10	5	100	30	1000	150

tion of cell volume is lack of authentic standards. Fluorescent microspheres are commonly used to calibrate measuring algorithms, but their emission wavelengths and fluorescence yields differ from those of native bacterioplankton, (i.e., they are different colored and brighter) and this complicates edge detection by image processing. Thus it is not possible to compare cell volumes objectively. For example, Wiebinga et al. (1997) reported relatively large mean cell volumes of 0.11 μm^3 cell^{-1} for the northwestern Indian Ocean during the Southwest Monsoon in 1992, while Pomroy and Joint found cells averaging 0.03 μm^3 cell^{-1} slightly further northeast in 1994. The latter value is more characteristic of oceanic regimes, yet while regional and interannual differences cannot be discounted, the standardization problem renders the debate somewhat futile. Newer, brighter fluorochromes might alleviate this problem.

Deciding which objects seen or detected under the microscope should be counted and measured as bacterial cells presents another difficulty. Most image analysts rely on experience and subjective criteria such as shape, size, and brightness to edit noncellular objects prior to analysis. This approach is generally reliable for open-ocean samples, which contain smaller numbers of detrital particles and other contaminants of bacterial image fields; but even experienced microscopists report difficulty when analyzing inshore and estuarine waters with high and diverse populations of noncellular objects. Recently, Blackburn et al. (1998) reported an image discrimination technique using neural network–based algorithms whereby operators can "teach" image analyzers to reject certain classes of particles. With new stains and more sophisticated numerical approaches, microscope-based determination of cell number and size should remain a benchmark for some time to come.

Flow cytometry appears to offer some hope of improvement. Mie theory suggests that light-scattering characteristics should be a function of cell volume or mass, making it possible to derive mass from cytometric determination of mean forward angle light scatter (FALS) per sample. Robertson et al. (1998) report a good relationship between FALS and dry mass for cultured cells and a natural population. A current drawback is that marine bacterioplankton still lie right near the lower limit of resolution for most flow cytometers, rendering extrapolation of the relationship between FALS and cell volume down to the sizes characteristic of native populations uncertain (P. del Giorgio, personal communication). Another cytometric approach is to relate the mean fluorescence per sample to cell volume, determined on parallel samples with image analysis. This approach is intuitively reasonable, inasmuch as larger cells containing more DNA should absorb more stain and fluoresce more brightly (Sherr et al. 1999). However calibration relies on microscopy, with the difficulties already noted. Further, the presence of inactive, nondividing cells with low DNA content, and of small, rapidly growing cells with multiple genomes (Wiebinga et al. 1999) would also confound straightforward interpretation of cell-specific fluorescence information. Nonetheless the large sampling rates and multiparameter data collecting capability of flow cytometers make approaches toward cytometric cell sizing highly attractive.

Reliable translation to cell mass is required to convert bacteriological measurements of abundance and cell volume into biogeochemically useful mass units. Initially, Fuhrman and Azam (1980), followed by others, used a value expressed in femtograms of carbon per micrometer cubed, namely, $120\,\mathrm{fg\,C\,\mu m^{-3}}$ cited in Watson et al. (1977) to derive relevant biomass estimates. That value was based on measurements of cultured *E. coli* which are 100-fold larger than native bacterioplankton. Bratbak and Dundas (1984) and Bratbak (1985) triggered a small revolution in bacterial appreciation with new estimates ranging $160–930\ \mathrm{fg\,C\ \mu m^{-3}}$, which overnight increased bacterial standing stocks in the sea by a factor of 3 or more. These estimates were based on pure cultures of marine bacteria grown on lab media and natural samples grown in enriched seawater. The first estimates of carbon per cell for native bacterioplankton grown on naturally occurring substrates were given by Lee and Fuhrman (1987), who grew natural assemblages from small ($<0.8\ \mu m$) filtrates in particle-free seawater and related microscopic volume estimates to C and N masses measured with a CHN analyzer. Interestingly, they determined that carbon per unit volume was itself inversely proportional to cell volume, such that carbon per cell was relatively constant at 20 fg over the observed size range of $0.036–0.077\ \mu m^3$. Twenty femtograms of carbon per cell has assumed something of a canonical status in marine bacteriology (Cho and Azam 1990; Ducklow and Carlson 1992). Using similar cellular mass estimates, many investigators have concluded that bacterial biomass must equal or even exceed phytoplankton biomass in many oceanic regions (Fuhrman et al. 1989; Cho and Azam 1990; Li et al. 1992).

The $20\,\mathrm{fg\,C\,cell^{-1}}$ value represents a high cellular carbon density, especially for the small ($<0.05\ \mu m^3$) cells characteristic of many oceanic regimes ($\geqslant 400\,\mathrm{fg\,C}\ \mu m^{-1}$). This is toward the upper end of the higher estimates reported by Bratbak and Dundas and others. Independent estimates of bacterial biomass indicate that bacterial carbon content might be somewhat lower. Several investigators with access to synoptic data on biomass of a comprehensive range of plankton groups ranging from phytoplankton and bacteria through protozoans and zooplankton attempted to constrain the bacterial carbon content using plankton biomass and total particulate carbon and/or living carbon. Christian and Karl (1994) and Caron et al. (1995) obtained estimates of $10–15\ \mathrm{fg\,C\ cell^{-1}}$ for these "constrained" conversion factors. In another novel approach Carlson et al. (1999) used high precision analyses of DOC and TCO_2 to recover the bacterial carbon by difference in a mass balance approach, which also yielded GGE estimates. Their bacterial carbon content ranged from 7 to $13\ \mathrm{fg\,C\ cell^{-1}}$ for cells of $0.06–0.09\ \mu m^3$.

The cellular carbon conversion factors reported by Lee and Fuhrman (1987) and others address cultured material derived from natural bacterial assemblages, but they are not direct estimates of actual in situ bacterial populations. As pointed out by Fukuda et al. (1998), these estimates might be biased by species succession during culture and by growth on substrates supplied as artifacts of the filtrations used to prepare the seawater culture media.

Table 2. Carbon content and carbon density of bacterial cells

Region	Density (fg C μm^{-3})	Content (fg C cell^{-1})	Method	References
Pure cultures	160–930		CHN analysis	Bratbak (1985)
Estuarine: coastal shelf				
Norwegian fjord		7–12	X-ray diffraction	Fagerbakke et al. (1996)
Long Island Sound	210–600	15–24	CHN analysis	Lee and Fuhrman (1987)
Otsuchi Bay, Japan		17–53	CHN analysis	Kogure and Koike (1987)
Ross Sea Antartica	77–165	7–13	C mass balance	Carlson et al. (1999)
Oceanic				
Hawaii		10	Biomass constraints	Christian and Karl (1994)
Bermuda		15	Biomass constraints	Caron et al. (1995)
Southern Ocean		12	Direct measurement	Fukuda et al. (1998)

Fukuda et al (1998) reported the first direct measurements of the carbon and nitrogen content of marine bacterial assemblages. After preparing filtrates of surface waters from a wide range of coastal and oceanic sites, with minimal phytoplankton contamination, they analyzed carbon using the high temperature catalytic oxidation (HTCO) methodology. The analytical approach avoided the need to concentrate samples on GF/F filters for CHN analysis, with the attendant loss of small (possibly carbon-dense) cells through the filters. Fukuda et al. (1998) report mean carbon per cell of 12.4 ± 6.3 and 30.2 ± 12.3 fg C cell^{-1} for oceanic and coastal locations, respectively, noting that if their estimates are representative of most marine areas, use of a uniform factor like 20 fg C cell^{-1} would overestimate bacterial biomass in oceanic habitats and underestimate it in coastal regions. Bacterial carbon content is summarized in Table 2, and some estimates of bacterial standing stocks are given in the sections that follow.

BACTERIAL PRODUCTION

Bacterial production, defined in equation (1), is commonly measured indirectly using radioisotope-labeled precursors of DNA and/or protein synthesis to yield synthesis rates, which must be converted to production rates using empirical factors. The two most common approaches use [^{3}H]-thymidine (Fuhrman and Azam 1980) and [^{3}H]-leucine (Kirchman et al. 1985), and these approaches (especially thymidine) have been reviewed and debated extensively (Moriarty 1985; Robarts and Zohary 1993; Karl 1986; Karl and Winn 1984; Ducklow and Carlson 1992; Kemp et al. 1993). The reader is directed to these reviews for more detailed discussion of bacterial production methodology.

Earlier Approaches

Karl (1979) proposed the first modern method for estimating bacterial production in the sea. He initially measured incorporation of [^{3}H]-adenine into RNA in upper- and midwater samples from the Caribbean Sea, and later extended studies to pure cultures of phytoplankton and bacteria as well as various oceanic and other aquatic environments (Karl et al. 1981; Karl and Winn 1984). In many (but possibly not all: see Fuhrman et al. 1986) marine environments, [^{3}H]-adenine is incorporated by bacteria, phytoplankton, and perhaps other microorganisms, which prompted Karl to apply the adenine technique to estimating total microbial production rates. The ambiguity or nonspecificity of this approach is the main reason for the decision of many investigators not to adopt it in studies calling for specific information on heterotrophic bacterial production. Nonetheless the adenine approach had several advantages that have not yet been approached by other methods. With adenine it is possible to measure the intracellular specific activity of the labeled tracer precisely by extracting [^{3}H]- and/or [^{32}P]-labeled ATP, which quickly achieves isotopic equilibrium with the other macromolecular constituents of the cell (Karl et al. 1981). If one knows the turnover rates of macromolecular pools, it is possible to estimate the specific turnover rates (= growth rates) of the population directly from isotopic data. Finally, [^{3}H]-adenine is taken up very rapidly by microbial assemblages, providing a high sensitivity method. Christian et al. (1982), Hanson and Lowry (1983), and Ducklow et al. (1985) applied this approach to various marine environments.

At about the same time the adenine method was introduced, Hagström et al. (1979) proposed that the frequency of dividing cells (FDC) in bacterial assemblages could be used to derive growth rates, and thus, production rates. This approach is based on both theoretical and empirically established relationships between the frequency of dividing cells (cells that have formed invaginations in the cell wall and a division plate but have not separated) in a population and the population division rate. A major attraction of this approach is that once it is calibrated, no incubations are required to obtain growth rate information—a simple collection of preserved samples can be examined following cruises or experiments to recover the rate data. However as Hagström et al. (1979) showed, the relationship between FDC and growth rates is temperature dependent and nonlinear, so calibration requires incubations, much as in the thymidine method and other approaches (see below).

Christian et al. (1982) showed that in experimentally manipulated samples, FDC was related to adenine and changes in cell numbers. I am not aware of a systematic comparison of FDC with more widely used bacterial production methods. More significantly, the difficulty of resolving dividing cells precisely, especially small oceanic cells, renders the method impractical for many marine environments, even given the no-incubation advantage. The FDC method probably should be reevaluated, given the recent wider application of flow cytometry and image analysis techniques for investigating cell morphology and DNA content.

Thymidine and Leucine Incorporation

The introduction of $[^3H]$-thymidine (TdR) incorporation as a measurement of heterotrophic bacterial production (Fuhrman and Azam 1980, 1982; Fuhrman et al. 1980) really did usher in a new era in the study of bacterial dynamics in marine and freshwater. The TdR approach offered a relatively simple protocol that was specific for estimating the production rate of actively growing bacterial cells. The pros and cons of TdR have been debated at great length (Karl and Winn 1984; Moriarty 1985), and there are several drawbacks (Hollibaugh 1988; Robarts 1993), but it remains the most widely used of all bacterial production methods (Kemp et al. 1993).

Two recent developments extend the utility of the TdR method. An adaptation allowing processing of 1–2 mL samples in a microcentrifuge has vastly reduced the volumes of radioisotope and reagents required, decreasing cost and waste production (D. C. Smith and Azam 1992). Steward and Azam (1999) propose the application of bromodeoxyuridine as an alternative, non-radioactive precursor of DNA synthesis for estimating bacterial production. These two approaches will help investigators who are limited by money or by radioactive materials prohibitions on vessels. Both approaches may even turn out to be more sensitive than the original approach.

The measurement of $[^3H]$-leucine incorporation (Leu) into bacterial protein was proposed as an alternative to TdR by Kirchman et al. (1985). As Kirchman (1992) and several others have shown, TdR and Leu incorporation rates do covary over a variety of time and space scales, suggesting that both methods address bacterial production–related processes. Comparison of bacterial production estimates from the two methods provides some idea of the uncertainty in determining bacterial production. As originally described by Kirchman, the Leu approach required an empirical conversion factor (CF) for estimating bacterial production, in a manner analogous to TdR. Kirchman (1992) used empirically determined conversion factors to estimate bacterial production by the two methods in the subarctic Pacific Ocean. He obtained slopes of 0.92 ± 0.09 and 0.76 ± 0.08 for (model I) regressions of Leu-BP on TdR-BP in the 0–40 and 40–80 m layers, respectively. This data set points to the complications involved in comparing the two methods, which measure separate though related physiological processes.

Agreement between Leu- and TdR-based bacterial production estimates will be exact if unbiased conversion factors can be obtained and if the ratio of the conversion factors is equal to the inverse ratio of the incorporation rates. Kirchman (1992) found that the mean ratio of conversion factors for TdR and Leu (Leu:TdR) was 16.1 ($n = 17$ and 14 for TdR and Leu), and also reported that the mean ratio of Leu:TdR incorporation rates for the entire water column data set was 16.8 ($n = 481$). This is very good agreement, accounting for the good match between the bacterial production estimates. But since the ratio of the conversion factors was slightly less than the incorporation ratio, the TdR-based estimates of bacterial production were slightly greater than the Leu-based estimates. If Leu:TdR incorporation rates vary substantially through a water column, reliance on a single set of conversion factors (usually derived from just one or two depths), will probably yield differing estimates of bacterial production.

Variations in Leu:TdR incorporation ratios can be substantial (Table 3), but there have been just a few investigations of the meaning of this variability. Chin-Leo and Kirchman (1990) showed how Leu:TdR can change with changes in growth rates and/or physiological state of bacterial assemblages. Departures from balanced growth should change the incorporation ratio, since cellular composition changes when cells shift up or down to a new growth rate (Cooper 1991). The meaning of balanced versus unbalanced growth is not straightforward in mixed natural populations probably growing at a range of rates (see below). Tibbles (1996) showed that Leu:TdR is temperature dependent, with ratios increasing with temperature. Since Leu and TdR incorporation appear to differ in their temperature dependence, conversion factors should not be constant with respect to temperature. Shiah and Ducklow (1997), following

Table 3. Ratio of leucine to thymidine incorporation rates (pmol:pmol) in the upper 200 m at selected oceanic sites

	Region		
	North Atlantic (47°N, 20°W)	Equatorial Pacific (0°N, 140°W)	
	May 18–31, 1989	March 23–April 1992	October 2–21, 1992
Mean Leu: TdR ratio	23.0	19.7	12.8
Standard deviation	18.6 (81%)	6.1 (31%)	2.9 (22%)
Range	2.6–116.3	9.1–52.5	4.6–23.0
N	156	206	223

Source: Ducklow et al. (1993, 1995).

Brunschede et al. (1977), suggested that "unfavorable" conditions (e.g., excessively low or high temperatures in temperate estuaries) led bacterial populations to invest more cellular resources in biomass synthesis (measured by Leu) than cell division (TdR). These studies all address relative changes in Leu:TdR ratios, but no one to date has been able to explain quantitatively the significance of the value of the ratio in any given sample, or regional differences in the ratio (Table 3). Further insight into relationships will probably require improved models of bacterioplankton physiology, explicitly addressing protein, RNA, and DNA synthesis, and cell division. Numerical simulation models of bacterial biosynthetic pathways exist (Stephanopoulos and Vallino 1991; Vallino et al. 1996) but have not been applied to bacterial production measurement scenarios.

Simon and Azam (1989) introduced another approach to estimating bacterial production employing Leu. They showed that the ratio of protein to carbon was highly invariable in bacterioplankton cells. From this observation it can be deduced that carbon-based bacterial production can be derived from Leu incorporation without recourse to a cell-based conversion factor or knowledge of carbon per cell (see above). If the intracellular isotope dilution is known or can be measured, a single conversion factor can be established for converting Leu to bacterial production. Simon and Azam (1989) claimed intracellular dilution of $[^3H]$-leucine varied by a factor of about 2 and suggested that this conversion factor would have a range of just 1.5–3 kg C per mol of Leu incorporated. The proliferation of converstion factors for TdR and Leu complicates cross-system comparison of bacterial production (Ducklow and Carlson 1992). Simon and Azam's (1989) approach provides a somewhat universal factor for bacterial production estimation.

Moreover the ratio of biomass production from Leu to cellular production from TdR:

$$\frac{\mu g \ C^{-1} l^{-1} h^{-1}}{cells \ L^{-1} h^{-1}} \tag{5}$$

yields the biomass of the average newly produced cell ($\mu g \ C \ cell^{-1}$). For example, Ducklow et al. (1993) derived a TdR CF of 2.65×10^{18} cells mol^{-1} to estimate bacterial production in the North Atlantic in 1989. The mean Leu:TdR ratio given in Table 3 and Simon and Azam's (1989) Leu CF indicate new cells had 13–26 fg C $cell^{-1}$ [i.e., $(23 \times 1500 \ to \ 3000)/(2.65 \times 10^{18})$], a value within the range of cell masses reviewed above.

If ambient cell mass were known reliably, it could be compared to the estimated mass of daughter cells to make inferences about removal rates and size-selective grazing. If, for example, newly produced cells are small, but the mean observed cell sizes were significantly greater, one might assume that bacterivory was not intense, or at least that grazers did not seem to be selecting larger cells to ingest. Unfortunately, as suggested earlier, we cannot easily or routinely specify cell mass reliably.

GROWTH RATES AND VIABILITY

Cell Kinetics; Bacterial Abundance and Biomass

In most natural water samples, simultaneous removal of bacterial cells by bacteriovores and viruses can reduce or balance the specific growth rate of the prey bacterial population, giving the appearance that the bacteria are not growing, as shown for phytoplankton preyed upon by microzooplankton (Landry and Hassett 1982). This is a very common case for natural bacterial assemblages in the ocean. In experiments, however, grazing and viral lysis can be minimized if not eliminated completely by dilution with filtered water; grazers can be eliminated by filtration. In these experiments, bacterial abundance increases over time, which provides an estimate of bacterial production and specific growth rates.

Observation of changes in cell numbers or mass over time is the most direct but perhaps not the easiest way to measure bacteria growth. Nonetheless the approach is worth the effort: it affords us the most fundamental access to cell growth per se, since it is based on direct visualization of the cell assemblage. Further, by providing an estimate of production independent of precursor incorporation assays, this approach is commonly used to derive empirical values for conversion factors (Kirchman et al. 1982). The slope of a plot of the natural log of cell numbers versus time yields the specific growth rate μ (equation (1), Figure 2A). Figure 2A illustrates a uniform population of cells (e.g., a pure culture or a seawater culture of a natural assemblage dominated by a single population) growing exponentially with $\mu = 1.0 \ d^{-1}$. This population is incorporating thymidine at a constant cell-specific rate of 2.8×10^{-20} mol $cell^{-1} \ h^{-1}$. As Kirchman et al. (1982) pointed out, the slope of the plot of the incorporation rate versus time is identical to the abundance plot. When cells are in balanced, exponential growth, specific growth rates can be determined from the increase in incorporation rate. This direct approach utilizing either the cell or incorporation plots can be used to estimate growth rates and conversion factors (Christian et al. 1982; Cuhel et al. 1982; Li, 1984; Ducklow and Hill 1985a, 1985b; LaRock et al. 1988; Ducklow et al. 1992; Chrzanowski et al. 1993). Kirchman et al. (1982) suggested that it was easier to measure incorporation rates than to count cells, a view I suggest is a matter of individual choice.

One problem is the presence of nongrowing cells in the culture assemblage, or more generally, heterogeneity of population growth rates. This problem has not been systematically treated even though new techniques are beginning to show how it may be solved. Kirchman et al. (1982), Cuhel et al. (1982), and Li (1984) all suggested that plots of incorporation rates would reveal the mean growth rate of the cell population or fraction thereof actually incorporating the isotopes employed. If TdR were used, this by definition would address the dividing cells. If only part of the total cell population were growing or actively incorporating isotope, the incorporation rates would increase faster than the

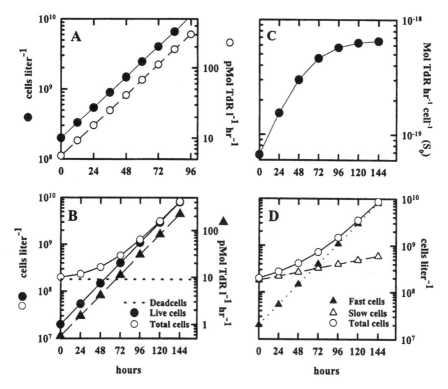

Figure 2. Exponential growth in idealized bacterial assemblages. (A) Solid circles, 100% active cells, growing at $1 \, d^{-1}$; open circles, thymidine incorporation by the growing cells. (B) 90% nongrowing "dead" cells and 10% "live" cells growing at $1 \, d^{-1}$; triangles, thymidine incorporation by growing cells. (C) Apparent cell-specific rate of Tdr incorporation by the total (live + dead) cell population in (B). (D) A small "fast" and a 10-times-larger, "slow" population growing at 1 and $0.2 \, d^{-1}$, respectively. Note the nonexponential growth of the total population.

total cell count because the cells are "diluted" by the inactive or nongrowing fraction. Figure 2B shows a bacterial assemblage with initially 10% of the cells growing at $1 \, d^{-1}$ as in Figure 2A, and the remainder of the assemblage nongrowing and constant. TdR incorporation by the growing fraction again parallels the growing cell plot. Note though, that the total population appears to have a lag period, caused by the gradual overgrowth of the nongrowing cells by the active fraction (Zweifel and Hagström, 1995). Extrapolation from the later stages of growth in the total counts back toward time zero provides an estimate of the original size of growing fraction, provided population structure has not changed. Torreton and Dufour (1996) used a similar "nongrowing fraction" model to estimate that just 0.1–5% of the total cells were active in the coral atoll lagoons of the Tuamoto Archipelago.

This example shows how the problem of nongrowing cells can be addressed simply by counting the total cell population long enough to permit extrapolation back to time zero. The scenario requires that nongrowing cells remain nongrowing in culture. In most cases in which bacterial growth in seawater cultures has been studied, this scenario has not been addressed directly — for example, by means of autoradiography or vital stains like CTC to monitor the growth of the active fraction. Choi et al. (1996: their Figure 3A) show results from a seawater culture incubation with total (DAPI) and "active" (CTC-stained) cell counts. The total counts have an apparent lag period similar to the one shown in Figure 2B, while the CTC counts grew without lag and converged with the total counts, consistent with the "inactive subpopulation" model.

Figure 2D shows a more general case with two growing populations. It is easy to see from this plot that increasing contrasts in growth rate and/or size of the two populations will enhance the curvature of the "lag" period. Even with greatly contrasting populations, the lag period may not be detected if

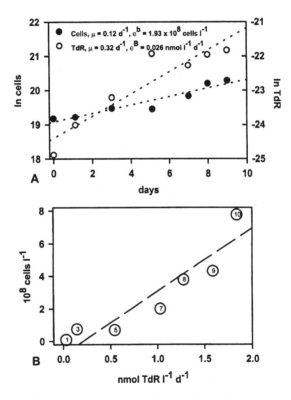

Figure 3. Seawater culture growth experiment from the Ross Sea, Antarctica. (A) Growth of cells and increase in TdR incorporation rates during a 10-day incubation. (B) Cumulative cells produced plotted against integral TeR incorporation. The slope of this plot is the thymidine conversion factor (Bjørnsen and Kuparinen 1991). Days sampled are given by numbers in circles.

sampling is infrequent, as the example shows: even daily sampling over 4–6 days would not yield a significant departure from linearity to confirm existence of the telltale "lag" phase. Using flow cytometry, Wiebinga et al. (1999) detected growing (at 2.4 d^{-1}) and nongrowing bacterial populations in seawater cultures from the Arabian Sea, showing a plot almost precisely like Figure 2B. But with samples taken at approximately 3, 6, 12, and 18 hours, curvature in the plot of total cells cannot be demonstrated, and the experiment did not last long enough for the plots to converge. However because the investigators detected both subpopulations directly (using cytofluorimetrically detected DNA content), the essential structure of the assemblage was apparent.

When a total cell count is made up of growing and nongrowing, or fast and slow populations, the incorporation rates will increase faster than cells, as noted earlier. This effect was observed by Ducklow and Hill (1985b), Ducklow et al. (1992), and Pomeroy et al. (1994). Figure 2C shows how incorporation rates which increase faster than cells result in increasing cell-specific incorporation rates. The leveling off of the curve is a reflection of the convergence of the active and total populations (cf. Figure 2B). From the simple models shown here, and indirectly suggested by Choi et al. (1996) and Wiebinga et al. (1999), it now appears that observations of differential specific growth rates determined from cell kinetics and incorporation rates can be explained by postulating two or more populations with different growth rates. I am not aware of a direct test of this hypothesis using autoradiography or cytometric sorting of labeled cells to follow the active subpopulation responsible for the growth of the incorporation plot.

Application to Determination of Conversion Factors

Fuhrman and Azam (1980, 1982) originally proposed a "theoretical" value for the TdR conversion factor of $0.2-2.4 \times 10^{18}$ cells produced per mole of TdR incorporated and showed that the TdR-derived bacterial production matched that independently estimated by counting cells in filtrates smaller than 3 μm. With an independent estimate of the specific growth rate or production rate, the conversion factor (CF) required to derive the bacterial production from the TdR incorporation, T, is just

$$CF_{der} = \tau \frac{N_0}{T_0} \tag{6a}$$

where τ is the slope of ln T versus time, and with all three parameters derived from the plot shown in Figure 3A. This approach was termed the "derivative approach" (Kirchman et al. 1982) because it was calculated from the slope of the regressions shown in Figure 3A. A difficulty arises when there is heterogeneity among the active and total populations. To address this problem, Ducklow et al. (1992) and Kirchman and Ducklow (1993) proposed a modified form of equation (6a):

$$CF_{mod} = \mu \frac{e^B}{e^b} \tag{6b}$$

and used regressions of cells and TdR incorporation versus time to estimate the parameters B and b, the y-intercepts of the regressions. This is just the same as equation (6a), except that observed N_0 and T_0 are replaced by the derived time zero estimates, e^B and e^b. It is easy to see that if there is significant curvilinearity in either cell or incorporation plots as shown in Figure 2, equation (6a) will probably result in high CF values as a result of dividing a high N_0 by a low T_0 value. Equation (6B) is an attempt moderate this effect, but most comparisons still indicate that this approach tends to yield high CF values.

The so-called cumulative approach (Bjørnsen and Kuparinen (1991) is an alternative empirical approach, that also employs time-course data on cell growth and incorporation rates. In this approach the cumulative cells produced to each time point are regressed on the integrated thymidine incorporation, and the calculated slope is the CF. This method tends to give lower values than those of Kirchman et al. (1982): either the "derivative" method [equation (6a)] or its modified form [(equation (6b)]. This is because more rapidly increasing TdR incorporation relative to cumulative cell production lowers the slope of the plot (Figure 3B).

Figure 3 shows results of an experiment conducted in the Ross Sea in November 1994 (Ducklow et al. 1999: temperature $\approx -1°C$) to illustrate cell dynamics and calculation of conversion factors. The upper plot shows time courses of ln cells and ln TdR incorporation. The slope of the TdR data is clearly greater than the slope of the cells plot, suggesting that not all cells were incorporating TdR. Although it is difficult to discern clearly, there was an apparent lag period in the cells plot, also indicative of slower growing or nongrowing cells, consistent with the larger TdR slope. Extrapolation of the cell counts between days 5 and 9 yields an estimate of 5×10^8 cells L^{-1} for the growing cell population, which is 20% of the observed initial cell abundance. Conversion factor values calculated from these data using equations (6a) and (6b) give values of 4.4 and 1.6×10^{18} cells mol^{-1}, respectively. Cumulative cell production is plotted against integral TdR incorporation for the same data set in the lower graph. The model II regression slope of the plot yields a conservative CF estimate of 3.9×10^{17} cells mol^{-1}.

Thus three methods of deriving conversion factors from the same data set result in estimates that range over an order of magnitude. Which value is correct? The original derivative approach is correct if there are indeed nongrowing or slowly growing cells in the assemblage. The high value of this estimate makes sense. A smaller population incorporating TdR will probably have low incorporation rates requiring a large CF to obtain the correct bacterial production. Without independent observations to verify that some

Table 4. Bacterial production and conversion factors in the Ross Sea[a]

Cell mass (fg C cell^{-1})	Bacterial Production (μg C L^{-1} d^{-1})		
	Derivative	Modified Derivative	Cumulative
10	1.14	0.43	0.10
20	2.27	0.85	0.20
30	3.41	1.28	0.30

[a]Rates based on TdR incorporation, using conversion factors calculated by the indicated procedures [equation (5) and the different C per cell factors. Rates based on Leu incorporation were 0.12–0.42 μg C L^{-1}d^{-1}.

percentage of the cells was nongrowing, there is no a priori way to decide whether this is the valid formula to use. The cumulative approach makes the fewest assumptions about cell growth and composition of the populations and their growth, but it clearly yields an underestimate if the growing population is small.

At present there is no very satisfying alternative to decide on which approach to take if only total cell counts are available. One approach would be to compare the bacterial production estimates to those obtained from leucine incorporation using the conversion factor suggested by Simon and Azam (1989). A difficulty is that these two estimates cannot be compared directly without also assuming a carbon-per-cell factor to convert the TdR-based estimates of bacterial production into carbon production. For the experiment shown in Figure 3, L_0, the initial rate of leucine incorporation, was 0.14 nM d^{-1}, which using Simon and Azam's (1989) conversion factor of 1.5–3 kg C mol^{-1}, gives a bacterial production of 0.21–0.42 μg C L^{-1} d^{-1}. The T_0 of 0.03 nM d^{-1} gives bacterial production of 0.1–3 μg C L^{-1} d^{-1}, using the three conversion factor estimates and three choices of C per cell (Table 4). The derivative approach seems to yield high values of bacterial production, whereas the other conversion factors give values closer to the leucine value, depending on the cell mass.

THE ECOLOGY OF GROWING AND NONGROWING CELLS

All the models discussed thus far are based on the assumption that there can be substantial populations of inactive or nongrowing cells in natural assemblages. There is no question that at least some bacterial species are exquisitely well adapted for long-term survival (days to centuries) in media with no energy sources. There is an enormous literature on the physiology and biochemistry of bacterial starvation–survival in cultures, and in various natural media and aquatic environments (Poindexter 1981; Kjelleberg et al. 1993; Morita 1997). Morita (1997) put forward the view that most of the biosphere is highly oligotrophic with respect to bacterial nutrition, and most bacteria in most habitats are in the starvation–survival state. There is substantial evidence that sometimes sizable fractions (from < 10 to $> 75\%$) of marine bacterial assemblages are not active, as indicated by autoradiography (Hoppe 1976; Douglas et al. 1987) or vital respiratory stains (e.g., CTC: del Giorgio et al. 1997; Sherr et al. 1998). Because of the lack of good operational definitions for these terms, and methods to address them, it is difficult and perhaps impractical to establish whether all the inactive cells are truly dormant, inactive, or nongrowing.

It seems unlikely that, as posited in some of the foregoing models, a static cell population could be maintained at a fixed size in nature for an extended period of time. Such cells would be cropped from the population by grazers unless they were nutritionally inferior to growing cells, and grazers strongly preferred growing cells. Nonetheless this view was reenergized by Zweifel and Hagström (1995), who used a modified DAPI staining/destaining technique to suggest that many marine bacterial cells lacked nucleoids, hence by definition were nongrowing "ghosts." Later Choi et al. (1996) showed that cells initially observed to be ghosts grew actively in seawater culture and later had nucleoids. Morita (1997) showed that starving cells undergo a loss of DNA as part of the starvation–survival adaptation. Thus Zweifel and Hagström's "ghosts" are probably similar to cells showing up as inactive in autoradiographic assays or CTC stained samples: they contain nucleoids too small to show up on the microscope, but they are viable cells.

The importance of these observations is as Zweifel and Hagström (1995), del Giorgio and Cole (1998), and others have suggested. Bacterial assemblages in nature appear to be dominated by small, highly active subpopulations coexisting with larger groups of less active, or perhaps temporarily inactive cells. The ubiquity of removal processes (e.g., bacterivory, viral lysis, adsorption, sinking) demands exchange between the active and inactive fractions of the bacterial assemblage. Perhaps some fraction of the growing population is intermittently or continuously "turned off" while parts of the inactive populations are reactivated, as observed by Choi et al. (1996). Blackburn et al. (1996) attempted to account for the presence of ghost cells in a numerical simulation model by including processes by which cells could be inactivated, but they did not include reactivation and recruitment of cells back into the growing fraction.

It is important to point out that production rates and fluxes measured by current methods are not affected by these considerations, but specific rates of growth and activity must be higher if fewer cells are actively engaged in carrying out measured activities. This argument opens up the possibility that bacteria in nature might be growing at rates substantially greater than estimated from bulk considerations, to maintain measured production rates. Mean growth rates are commonly reported to lie in the range $0.1–1$ d^{-1} for habitats reaching from the equator (Kirchman et al. 1995) to the poles

(Ducklow et al. 1999; Rich et al. 1998). If the actively growing fractions of these assemblages are as small as 10% of the total assemblage, growth rates must be scaled up accordingly. But if the active and inactive fractions exchange substantially on the time scales of growth, then growth rates integrated over the exchange timescale may lie somewhere in between these extremes. Clearly, better insight into the biology and ecology of bacterioplankton requires reliable ways to penetrate and resolve the demographic structure of natural assemblages.

BACTERIOPLANKTON STANDING STOCKS AND PRODUCTION RATES

Bacterioplankton biomass and production estimates are uncertain at least by a factor of 2, owing to unexplained variability and imprecisely specified conversion factors, discussed earlier. Further, variability in conversion factors and independent behavior of TdR and Leu incorporation (Table 4) complicate comparisons of stock and production estimates in the literature. Ducklow and Carlson (1992) approached this problem by using mean values for cell and thymidine conversion factors to back-calculate a consistent set of estimates. One possible objection to this approach is that it ignores empirical conversion factors calibrated for a particular study (Rivkin et al. 1996). Another complication is the basis of comparison. Many estimates use the euphotic zone (depth of 1% surface irradiance) as a common depth for integration. This is a logical basis for comparison to phytoplankton stocks and photosynthetic rates, but it leaves open the question of what processes support bacterial stocks below the illuminated layer.

Wiebinga et al. (1997) and others have pointed out that carbon produced in the euphotic zone must ultimately support bacterial carbon demand throughout the water column. Choice of some greater depth is arbitrary unless one could specify the time and space scales over which the products of local photosynthesis are dispersed prior to bacterial utilization. This exercise could be accomplished using three-dimensional numerical models, but only if we also knew how to parameterize the carbon flux relevant to bacterial metabolism! Mixed layer comparisons have the added complication that mixed layers vary over diel to interannual time scales (Gardner et al. 1995; Michaels and Knap 1996). There is no satisfying answer to this minor dilemma, except perhaps to provide estimates integrated over several characteristic scales. Here I compare euphotic zone integrals, simply because I wish to scale the resulting estimates to phytoplankton properties.

Cole et al. (1988) synthesized data extending over seasonal to annual scales to conclude that bacterial production was equivalent to about 20–30% of the local primary production. They reported the data as originally presented, using the conversions applied in each respective study. Their exercise should be repeated. In the ensuing decade a great many additional studies have been published, including many for the open sea, which was not well represented in the original summary. It is not clear that the ratio of Cole et al. (1988) holds for the open sea. Bacterial production was 5–15% of ^{14}C primary production in the subarctic Pacific in 1987–1988 (Kirchman et al. 1993). The researchers used an empirical conversion factor of 1.7×10^{18} cells mol^{-1} TdR and 20 fg C cell^{-1} to derive their estimates. Ducklow et al. (1993) concluded that bacterial production was 15–80% of the ^{14}C primary production during the spring phytoplankton bloom in the subarctic NE Atlantic, using a mean TCF of 2.7×10^{18} cells mol^{-1}, and 20 fg C cell^{-1}. Li et al. (1993) used TCF of 1–2.3×10^{18} cells mol^{-1} and 20 fg C cell^{-1} to conclude that bacterial production was 8–18% of the ^{14}C primary production in the northwestern Atlantic at the same time. Later, Kirchman et al. (1995) and Ducklow et al. (1995) again used 20 fg C cell^{-1} to estimate bacterial production in the central equatorial Pacific during the 1992 El Niño. They did not determine empirical conversion factor but used a mean value from the literature of 2.2×10^{18} cells mol^{-1} to convert TdR to bacterial production, obtaining bacterial production to primary production ratios of 12–20%. In the Sargasso Sea off Bermuda, bacterial production was 15% of primary production, using 1.6×10^{18} cells mol^{-1} TdR and a volumetric conversion factor of 120 fg C μm^{-3}. Wiebinga et al. (1997) and Pomroy and Joint (1999) estimated bacterial production in the Arabian Sea using a variety of TCF and CCF, obtaining a larger range of 3–50% of primary production (average 10–18%). In general, bacterial production averages about 15% of primary production, excluding a few nonequilibrium situations like decaying blooms (Ducklow et al. 1993). These estimates all used conversion factors within a factor of about 2 of each other, lending at least ease of comparability, if not absolute reliability to the estimates.

Recent observations of heterotrophic bacterial stocks and production rates are summarized in Table 5 and compared to corresponding, synoptic phytoplankton data. I tried to minimize some of the concerns cited above by reporting observations for which the raw data were accessible, and by using a consistent set of conversion factors to derive the carbon-based estimates. These data are discussed in more detail in Ducklow (1999). This summary provides data on bacteria in open-ocean regimes including the subarctic North Atlantic and Pacific, oligotrophic gyres in both basins, the Arabian Sea region influenced by monsoonal upwelling, the Antarctic shelf seas, and the central equatorial Pacific. The subarctic and equatorial Pacific are "high-nutrient, low-chlorophyll" (HNLC) regimes (Longhurst 1998). As a simple generalization, bacterial biomass in the euphotic zone averages about 1–2 g C m^{-2}, except in the Antarctic, where it is usually much lower. Production rates, and thus mean turnover rates, vary more widely, as do relationships with phytoplankton properties. However it is notable that with a few exceptions, bacterial production is usually about 10–20% of the corresponding primary production.

It is generally accepted that bacterial populations decline in size from estuaries and inshore areas of greater organic and inorganic enrichment toward the more oligotrophic open sea (Sieburth 1979). Volumetric estimates of

Table 5. Bacterioplankton and phytoplankton properties in the open sea[a]

Property	N Atlantic[b]	Eq Pac–Spr[c]	Eq Pac–Fall[d]	Sub N Pac[e]	Arabian[f]	Hawaii[g]	Bermuda[h]	Ross Sea[i]
Euphotic zone m	50	120	120	80	74	175	140	45
Biomass, mg C m^{-2}								
Bacteria	1000	1200	1467	1142	1448	1500	1317	217
Phytoplankton	4500	1700	1940	1274	1248	447	573	11450
B:P	0.2	0.7	0.75	0.9	1.2	3.6	2.7	0.02
Production, mg C m^{-2} d^{-1}								
Bacteria	275	285	176	56	257	nd	70	5.5
Phytoplankton	1083	1083	1548	629	1165	486[j]	465	1248
B:P	0.25	0.26	0.11	0.09	0.22	nd	0.18	0.04
Growth Rates, d^{-1}								
Bacteria	0.3	0.13	0.12	0.05	0.18	nd	0.05	0.25
Phytoplankton	0.3	0.64	0.8	0.50	0.93	1.1	0.81	0.11
B:P	1	0.2	0.15	0.1	0.19	nd	0.06	2.3

[a]All stock estimates based on 20 fg C cell^{-1}. Data may overestimate actual heterotrophic eubacterial biomass as a consequence of lower C contents and/or interference by *Prochlorococcus* and *Archaea*. Production estimated from 3000 g C mol^{-1} leucine incorporation.
[b]Eastern North Atlantic spring phytoplankton bloom, 47°N, 20°W; May, 1989, $n = 13$ (Ducklow et al. 1993).
[c]Equatorial Pacific, 0°N, 140°W; March–April 1992, $n = 18$ (Ducklow et al. 1995).
[d]Equatorial Pacific, 0°N, 140°W; September–October 1992, $n = 19$ (Ducklow et al. 1995).
[e]Subarctic North Pacific, 45°N, Kirchman et al. (1993).
[f]Northwest Arabian Sea, 10–20°N, 165°E, January–December 1995, $n = 21$ (Ducklow et al. in press).
[g]Hawaiian Ocean Time Series (HOT); 1995–1997; $n = 21$; (http://hahana.soest.hawaii.edu/hot/methods/pprod.html).
[h]Bermuda Atlantic Time Series (BATS); 1991–1998, $n = 106$ paired comparisons; for BP and phytoplankton biomass calculations, see Carlson et al. (1996). The ratios are means of the ratios, not ratios of the means. BP calculated from TdR (1.6 × 10^{18} cells mol^{-1}).
[i]Ross Sea, Antarctica; 76°S, 180°W; 1994–1997; Carlson et al. (1998); Ducklow, unpublished data.
[j]1989–1996; $n = 64$. Data source as for note g.

109

bacterial abundance (cells per liter) range from about $1-5 \times 10^8$ cells L^{-1} in the most oligotrophic regions studied (Cho and Azam 1990), to over 2×10^{10} cells L^{-1} in rich estuaries (Ducklow and Shiah 1993). Therefore it is surprising to compare integrated bacterial stocks in euphotic zones from different oceanic regimes (Figure 4A). Estimates from four areas of the North Atlantic region [Chesapeake Bay, the Sargasso Sea and nearby Gulf Stream, and the northeastern subarctic (*sensu* Longhurst 1998)] range from 0.5 to 120×10^{12} cells m^{-2}. There is a significant but not strong positive relationship with euphotic zone depth ($Y = 0.22X + 31.74$, $r^2 = 0.26$, $n = 320$). Chesapeake Bay is rich in nutrients, chlorophyll a, organic matter and light-absorbing material, with correspondingly shallow euphotic zones (mean depth, 5 m: Malone et al. 1988). The shallow euphotic depth obviously counteracts the effect of enrichment on bacterial accumulation in the estuarine water column (Figure 4B). In spite of high abundance, the standing stock integrated over the euphotic zone is about 50% lower than in the northwestern Sargasso Sea off Bermuda (Figure 4A and cf. Carlson et al. 1996). There are exceptions to the general trend. The euphotic zone in the northeastern Atlantic subarctic became highly enriched with bacteria following the peak of the 1989 spring bloom (Ducklow et al. 1993). In coastal Antarctic waters of the Ross Sea (Figure 4A) and other polar regions

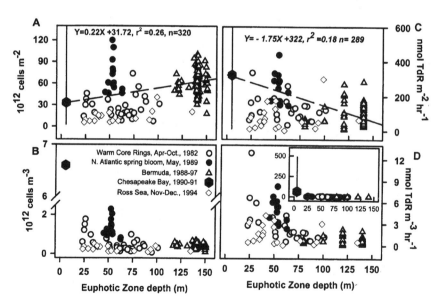

Figure 4. Bacterial standing stock and production in the euphotic zone. (A) Euphotic zone (EZ) standing stocks. (B) EZ mean abundance (stock/depth). (C) EZ integrated TdR incorporation. (D) EZ mean TdR incorporation rates (integrated rate/depth). *Inset:* Same plot including Chesapeake Bay data, presented as mean and range (heavy line) of 162 monthly measurements made at 10 stations up and down the bay in 1990–1991 (Shiah and Ducklow 1994). Other data sources cited in text.

(Karl 1993), bacterial accumulation seems to be strongly suppressed by temperature or factors related to plankton community structure (Carlson et al. 1998). Generally, however, even though the mean primary production per unit volume is clearly higher in inshore habitats, including the Ross Sea (W. O. Smith and Gordon 1997), oligotrophic regimes with deep euphotic zones have greater integrated bacterial biomass than estuaries and coastal oceans.

This pattern becomes more surprising when we analyze bacterial production (as thymidine incorporation). There is a significant *negative* relationship between euphotic zone thymidine incorporation, in nanomoles per square meter per hour, and depth of the photosynthetic layer (Figure 4C; $Y = -1.75X + 322$; $r^2 = 0.18$, $n = 289$). In this case bacterial activity is six times greater in the estuary (Figure 4D, inset), but there is no difference among integrated production rate in the other sites. Bacterial utilization of organic matter results in greater euphotic zone production levels in estuaries, but also seems to result in greater integrated biomass accumulation in the open sea. It follows from this that turnover rates (specific growth rates) are about tenfold greater in the estuaries.

We do not have a satisfactory theory of the regulation of bacterial stocks and production in the sea. There is indirect evidence for bottom-up control by resource availability (Billen et al. 1990; Ducklow 1992). Bacterial abundance and production are significantly correlated with phytoplankton biomass (chlorophyll *a*) and primary production, respectively (Cole et al. 1988). The latter relationships are reflections of the strong and ultimate dependence of bacterial metabolism on local primary production. This dependence may be obscured or nonexistent in estuaries where bacterial productivity can exceed phytoplankton production (Ducklow and Shiah 1993) and is supported by exogenous inputs of terrestrial organic matter. Dominant bottom-up effects suggest that bacterial stocks should increase with increasing organic matter supply. Removal processes (top-down controls) counterbalance the bottom-up effect. Sanders et al. (1992) compared the relationships between bacteria and heterotrophic nanoplankton bacteriovores in high and low productivity regimes of both marine and freshwater habitats. They concluded that bottom-up effects were predominant in oligotrophic systems, while top-down control was stronger in eutrophic systems. The data presented here indicate that top-down effects exert a strong effect across the trophic spectrum. Removal processes maintain relatively uniform bacterial abundance throughout the ocean outside the land–sea margin (Figure 4B). Suppression of estuarine water column stocks seems especially intense (Figure 4A), in spite of high local inputs of organic matter and high bacterial production (Figure 4D).

Thus it appears that in the open sea over a wide range of geographic and trophic habitats, integrated bacterial stocks are high (about twice as high as in estuaries) but possibly not as productive (averaged over the active and inactive assemblage) as once believed. This is apparently because bacterial GGE is low, also averaging about 20% (Chapter 10). If these mean figures are truly accurate and representative, bacterial carbon demand is about the same magnitude as the local particulate net primary production estimated from ^{14}C measurements. Better estimates of total and bacterial respiration rates would provide a rigorous constraint on this conclusion (Jahnke and Craven 1995). Bacterioplankton are forced to scavenge DOM from diverse sources (Pomeroy 1974; Williams 1981; Azam 1998; Chapter 6). The effective flux of DOM into bacterial cells must approach the magnitude of the daily primary production. Top-down controls (including viral lysis) appear to be weaker in the open sea than in estuaries. Bacteria may constitute a stronger link to higher trophic levels inshore than offshore, where large amounts of carbon are sequestered in the bacterial biomass. But methods for specifying DOM flux, removal rates, bacterial production, and respiration must be improved by at least an order of magnitude in precision before these conclusions can be verified.

There is one important caveat to the foregoing speculation. Bacterial abundance is not biomass and thymidine is not carbon production. The patterns shown in Figure 4 will not reflect actual stocks and production in nature if, for example, cell volumes and carbon content are larger in estuaries than in the open sea (Fukuda et al. 1998), or if TdR conversion factors vary significantly and systematically (e.g., Rivkin et al. 1996). Patterns of carbon flux are valuable for biogeochemical studies and modeling, but actual patterns of abundance and biovolume are also important: bacteriovores select and ingest cells, not carbon units. Both approaches are needed for complete understanding.

SUMMARY

1. Bacterial standing stocks in the euphotic zone average about 0.5–2 g C m^{-2} across a range of oceanic systems. The ratio of bacterial to phytoplankton stocks varies widely, from less than 0.1 in polar coastal seas to over 2.0 in the oligotrophic gyres.

2. Bacterial production is maintained in a remarkably constant ratio to primary production, averaging about 0.15–0.2 across oligotrophic and oceanic HNLC and upwelling and blooming systems. Bacterial production is generally much lower during polar coastal blooms, but it can be high following the peak phase of blooms in temperate and subpolar regimes.

3. Bacterial stocks seem to be limited principally by resource limitation in lower productivity systems; but removal processes are more intense in coastal and estuarine systems, suppressing integrated standing stocks to below oceanic levels.

4. Estimating bacterial biomass and production in geochemical mass units (C- or N-based estimates) is still technically difficult and uncertain. Order-of-magnitude increases in precision and perhaps accuracy are needed to gain deeper understanding of bacterial ecology in the sea.

5. Better recognition, detection, and understanding of inactive cells are needed to specify rates and mechanisms of bacterial growth.

REFERENCES

Azam, F. (1998) Microbial control of oceanic carbon flux: The plot thickens. *Science* 280:694–696.

Azam, F., and Hodson, R. E. (1977) Size distribution and activity of marine microheterotrophs. *Limnol. Oceanogr.* 22:492–501.

Azam, F., Fenchel, T., Field, J. G., Gray, J. S., Meyer-Reil, L. A., and Thingstad, F. (1983) The ecological role of water-column microbes in the sea. *Mar. Ecol. Prog. Ser.* 10:257–263.

Bender, M. L., and Grande, K. D. (1987) Production, respiration and the isotope geochemistry of O_2 in the upper water column. *Global Biogeochem Cycles* 1:49–60.

Billen, G., Servais, P., and Becquevort, S. (1990) Dynamics of bacterioplankton in oligotrophic and eutrophic aquatic environments: Bottom-up or top-down control? *Hydrobiologia* 207:37–42.

Bjørnsen, P. K. (1986) Automatic determination of bacterioplankton biomass by image analysis. *Appl. Environ. Microbiol.* 51:1199–1204.

Bjørnsen, P. K., and Kuparinen, J. (1991) Determination of bacterioplankton biomass, net production and growth efficiency in the Southern Ocean. *Mar. Ecol. Prog. Ser.* 71:185–194.

Blackburn, N., Hagström, Å., Wikner, J., Cuadros-Hansson, R., and Bjørnsen, P. K. (1998) Rapid determination of bacterial abundance, biovolume, morphology, and growth by neural network–based image analysis. *Appl. Environ. Microbiol.* 64:3246–3255.

Blackburn, N., Zweifel, U. L., and Hagström, Å. (1996) Cycling of marine dissolved organic matter. II. A model analysis. *Aquat. Microb. Ecol.* 11:79–90.

Bratbak, G. (1985) Bacterial biovolume and biomass estimates. *Appl. Environ. Microbiol.* 49:1488–1493.

Bratbak, G., and Dundas, I. (1984) Bacterial dry matter content and biomass estimations. *Appl. Environ. Microbiol.* 48:755–757.

Brock, T. D. (1971) Microbial growth rates in nature. *Bacteriol. Rev.* 35:39–58.

Brunschede, H., Dove, T. L., and Bremer, H. (1977) Establishment of exponential growth after a nutritional shift-up in *Escherichia coli* B/r: Accumulation of deoxyribonucleic acid, ribonucleic acid, and protein. *J. Bacteriol.* 129:1020–1033.

Campbell, L., Nolla, H. A., and Vaulot, D. (1994) The importance of *Prochlorococcus* to community structure in the central North Pacific Ocean. *Limnol. Oceanogr.* 39:954–961.

Campbell, L., Landry, M. R., Constantinou, J., Nolla, H. A., Brown, S. L., Liu, H., and Caron, D. A. (1999) Response of microbial community structure to environmental forcing in the Arabian Sea. *Deep-Sea Res.* II 45:301–325.

Carlson, C., Ducklow, H. W., and Sleeter, T. D. (1996) Stocks and dynamics of bacterioplankton in the northwestern Sargasso Sea. *Deep-Sea Res. II* 43:491–516.

Carlson, C. A., Ducklow, H. W., Hansell, D. A., and Smith, W. O. J. (1998) Organic carbon partitioning during spring phytoplankton blooms in the Ross Sea polynya and the Surgasso Sea. *Limnol. Oceanogr.* 43:375–386.

Carlson, C. A., Bates, N. R., Ducklow, H. W., and Hansell, D. A. (1999) Estimation of bacterial respiration and growth efficiency in the Ross Sea, Antarctica. *Aquat. Microb. Ecol.* 19:229–244.

Caron, D.A. Dam, H. G., Kremer, P., Lessard, E. J., Madin, L. P., Malone, T. C., Napp, J. M., Peele, E. R., Roman, R. M., and Youngbluth, M. J. (1995) The contribution of microorganisms to particulate carbon and nitrogen in surface waters of the Sargasso Sea near Bermuda. *Deep-Sea Res.* 42:943–972.

Chin-Leo, G., and Kirchman, D. L. (1990) Unbalanced growth in natural assemblages of marine bacterioplankton. *Mar. Ecol. Prog. Ser.* 63:1–8.

Cho, B. C., and Azam, F. (1990) Biogeochemical significance of bacterial biomass in the ocean's euphotic zone. *Mar. Ecol. Prog. Ser.* 63:253–259

Choi, J. W., Sherr, E. B., and Sherr, B. F. (1996) Relation between presence–absence of a visible nucleoid and metabolic activity in bacterioplankton cells. *Limnol. Oceanogr.* 41:1161–1168.

Christian, J. R., and Karl, D. M. (1994) Microbial community structure at the US-JGOFS Station ALOHA: Inverse methods for estimating biochemical indicator ratios. *J. Geophys. Res.* 99:14269–14276.

Christian, R. R., Hanson, R. B., and Newell, S. Y. (1982) Comparison of methods for measurement of bacterial growth rates in mixed batch cultures. *Appl. Environ. Microbiol.* 43:1160–1165.

Chrzanowski, T. H., Simek, K, Sada, R. H., and Williams, S. (1993) Estimates of bacterial growth rate constants from thymidine incorporation and variable conversion factors. *Microb. Ecol.* 25:121–130.

Cohen, J. E. (1995) *How Many People Can the Earth Support?* Norton, New York.

Cole, J. J., and Pace, M. L. (1995) Why measure bacterial production? A reply to the comment by Jahnke and Craven. *Limnol. Oceanogr.* 40:441–444.

Cole, J. J., Pace, M. L., and Findlay, S. (1988) Bacterial production in fresh and saltwater ecosystems: A cross-system overview. *Mar. Ecol. Prog. Ser.* 43:1–10.

Cooper, S. (1991) *Bacterial Growth and Division. Biochemistry and Regulation of Prokaryotic and Eukaryotic Division Cycles.* Academic Press, New York.

Cuhel, R. L., Taylor, C. D., and Jannasch, H. W. (1982) Assimilatory sulfur metabolism in marine microorganisms: Considerations for the application of sulfate incorporation into protein as a measurement of natural population protein synthesis. *Appl. Environ. Microbiol.* 43:160–168.

Del Giorgio, P. A., and Cole, J. J. (1998) Bacterial growth efficiency in natural aquatic systems. *Annu. Rev. Ecol. Syst.* 29:503–541.

Del Giorgio, P. A., Prairie, Y. T., and Bird, D. F. (1997) Coupling between rates of bacterial production and the number of metabolically active cells in lake bacterioplankton, measured by CTC reduction and flow cytometry. *Microb. Ecol.* 34:144–154.

Douglas, D. J., Novitsky, J. A., and Fournier, R. O. (1987) Microautoradiography-based enumeration of bacteria with estimates of thymidine-specific growth and production rates. *Mar. Ecol. Prog. Ser.* 36: 91–99.

Ducklow, H. (1983) Production and fate of bacteria in the oceans. *BioScience* 33: 494–499.

Ducklow, H. W. (1992) Factors regulating bottom-up control of bacterial biomass in open ocean plankton communities. *Arch. Hydrobiol. Beih. Ergebn. Limnol.* 37:207–217.

Ducklow, H. W. (1999) The bacterial component of the oceanic euphotic zone. *FEMS Microbiol. Ecol.* 30:1–10.

Ducklow, H. W., and Carlson, C. A. (1992) Oceanic bacterial productivity. *Adv. Microb. Ecol.* 12:113–181.

Ducklow, H. W. and Hill, S. (1985a) The growth of heterotrophic bacteria in the surface waters of warm core rings. *Limnol. Oceanogr.* 30:241–262.

Ducklow, H. W., and Hill, S. (1985b) Tritiated thymidine incorporation and the growth of bacteria in warm core rings. *Limnol. Oceanogr.* 30:263–274.

Ducklow, H. W., and Shiah, F.-K. (1993) Estuarine bacterial production. In Ford, T., ed., *Aquatic Microbiology: An Ecological Approach.* Blackwell, London, pp. 261–284.

Ducklow, H. W., Gardner, W. and Hill, S. 1985. Bacterial growth and the decomposition of particulate organic carbon in sediment traps. *Cont. Shelf Res.* 4:445–464.

Ducklow, H. W., Kirchman, D. L., and Quinby, H. L. (1992) Bacterioplankton cell growth and macromolecular synthesis in seawater cultures during the North Atlantic spring phytoplankton bloom, May 1989. *Microb. Ecol.* 24:125–144.

Ducklow, H. W., Kirchman, D. L., Quinby, H. L., Carlson, C. A., and Dam, H. G. (1993) Stocks and dynamics of bacterioplankton carbon during the spring phytoplankton bloom in the eastern North Atlantic Ocean. *Deep-Sea Res.* 40:245–263.

Ducklow, H. W., Quinby, H. L., and Carlson, C. A. (1995) Bacterioplankton dynamics in the equatorial Pacific during the 1992 El Niño. *Deep-Sea Res. II* 42:621–638.

Ducklow, H. W., Carlson, C. A., and Smith, W. O. (1999) Bacterial growth in experimental plankton assemblages and seawater cultures from the *Phaeocystis antarctica* bloom in the Ross Sea, Antarctica. *Aquat. Microb. Ecol.* 19:215–227.

Ducklow, H. W., Campbell, L., Landry, M. R., Quinby, H. L., Smith, D. C., Steward, G., and Azam, F. Heterotrophic bacterioplankton distributions in the Arabian Sea: Basinwide response to high primary productivity. *Deep-Sea Res. II* In press.

Emerson, S., Quay, P., Stump, C., Wilbur, D., and Schudlich, R. (1993) Determining primary production from the mesoscale oxygen field. *ICES Mar. Sci. Symp.* 197:196–206.

Fagerbakke, K. M., Heldal, M., and Norland, S. (1996) Content of carbon, nitrogen, oxygen, sulfur and phosphorus in native aquatic and cultured bacteria. *Aquat. Microb. Ecol.* 10:15–27.

Falkowski, P. G., and Raven, J. A. (1997) *Aquatic Photosynthesis.* Blackwell Scientific, Malden. MA. 375 pp.

Fuhrman, J. A., and Azam, F. (1980) Bacterioplankton secondary production estimates for coastal waters of British Columbia, Antarctica, and California. *Appl. Environ. Microbiol.* 39:1085–1095.

Fuhrman, J. A., and Azam, F. (1982) Thymidine incorporation as a measure of heterotrophic bacterioplankton production in marine surface waters: Evaluation and field results. *Mar. Biol.* 66:109–120.

Fuhrman, J. A., Ammerman, J. W., and Azam, F. (1980) Bacterioplankton in the coastal euphotic zone: Distribution, activity, and possible relationships with phytoplankton. *Mar. Biol.* 60: 201–207

Fuhrman, J., Ducklow, H, Kirchman, D., Hudak, J., Bell, T., and McManus, G. (1986) Does adenine incorporation into nucleic acids measure total microbial production? *Limnol. Oceanogr.* 31:627–636.

Fuhrman, J. A., Sleeter, T. D., Carlson, C. A., and Proctor, L. M. (1989) Dominance of bacterial biomass in the Sargasso Sea and its ecological implications. *Mar. Ecol. Prog. Ser.* 57:207–217.

Fukuda, R., Ogawa, H., Nagata, T., and Koike, I. (1998) Direct determination of carbon and nitrogen contents of natural bacterial assemblages in marine environments. *Appl. Environ. Microbiol.* 64:3352–3358.

Gardner, W. D., Chung, S. P., Richardson, M. J., and Walsh, I. D. (1995) The oceanic mixed layer pump. *Deep-Sea Res. II* 42:757–776.

Goldman, J., Caron, D. A., and Dennett, M. R. (1987) Regulation of gross growth efficiency and ammonium regeneration in bacteria by substrate C:N ratio. *Limnol. Oceanogr.* 32: 1239–1252.

Hagström, Å., Larrson, U., Horstedt, P., and Normark, S. (1979) Frequency of dividing cells, a new approach to the determination of bacterial growth rates in aquatic environments. *Appl. Environ. Microbiol.* 37:805–812.

Hanson, R. B., and Lowery, H. K. (1983) Nucleic acid synthesis in oceanic microplankton from the Drake Passage, Antarctica: Evaluation of steady-state growth. *Mar. Biol.* 73:79–89.

Henrici, A. T. (1933) Studies of freshwater bacteria. I. A direct microscopic technique. *J. Bacteriol.* 25:277–287.

Henrici, A. T. (1938) Studies of freshwater bacteria. IV. Seasonal fluctuations of lake bacteria in relation to plankton production. *J. Bacteriol.* 35:129–139.

Hobbie, J. E., Daley, R. J., and Jasper, S. (1977) Use of Nuclepore filters for counting bacteria by fluorescence microscopy. *Appl. Environ. Microbiol.* 33:1225–1228.

Hollibaugh, J. T. (1988) Limitation of the [^{3}H]thymidine method for estimating bacterial productivity due to thymidine metabolism. *Mar. Ecol. Prog. Ser.* 43:19–30.

Hoppe, H.-G. (1976) Determination and properties of actively metabolizing heterotrophic bacteria in the sea, investigated by means of micro-autoradiography. *Mar. Biol.* 36:291–302.

Hoppe, H.-G. (1983) Significance of exoenzymatic activities in the ecology of brackishwater: Measurements by means of methyl–umbelliferyl substrates. *Mar. Ecol. Prog. Ser.* 11:299–308.

Jahnke, R. A., and Craven, D. B. (1995) Quantifying the role of heterotrophic bacteria in the carbon cycle: A need for respiration rate measurements. *Limnol. Oceanogr.* 40:436–441.

Jannasch, H. W., and Jones, G. E. (1959) Bacterial populations in sea water as determined by different methods of enumeration. *Limnol. Oceanogr.* 4:128–139.

Jumars, P. A., Penry, D. L., Baross, J. A., Perry, M. J., and Frost, B. W. (1989) Closing the microbial loop: Dissolved organic carbon pathway to heterotrophic bacteria from incomplete ingestion, digestion and absorption in animals. *Deep-Sea Res.* 36:483–495.

Karl, D. M. (1979) Measurement of microbial activity and growth in the ocean by rates of stable ribonucleic acid synthesis. *Appl. Environ. Microbiol.* 38:850–860.

Karl, D. M. (1993) Microbial processes in the southern oceans. In E. I. Friedmann, ed., *Antarctic Microbiology.* Wiley, New York, pp. 1–63.

Karl, D. M., and Winn, C. D. (1984) Adenine metabolism and nucleic acid synthesis: Applications to microbiological oceanography. In J. E. Hobbie and P. J. le B. Williams, eds., *Heterotrophic Activity in the Sea.* Plenum, New York, pp. 197–216.

Karl, D. M., Winn, C. D., and Wong, C. L. (1981) RNA synthesis as a measure of microbial growth in aquatic environments. I. Evaluation verification and optimization of methods. *Mar. Biol.* 64:1–12.

Karl, D. M. (1993) Microbiol processes in the Southern Oceans. In E. I. Friedmann, ed., *Antarctic Microbiology.* Wiley, New York, pp. 1–63.

Karl, D. M., (1986) Determination of in situ microbial biomass, viability, metabolism, and growth. In J. S. Poindexter and E. R. Leadbetter, eds., *Bacteria in Nature,* volume 2, Plenum Press, New York, pp. 85–176.

Kemp, P. F., Sherr, B., Sherr, E., and Cole, J. J. (1993) *Handbook of Methods in Aquatic Microbial Ecology,* Lewis Publishers, Boca Raton, FL.

Kirchman, D. L. (1992) Incorporation of thymidine and leucine in the subarctic Pacific: application to estimating bacterial production. *Mar. Ecol. Prog. Ser.* 82:301–309.

Kirchman, D. L., and Ducklow, H. W. (1993) Estimating conversion factors for the thymidine and leucine methods for measuring bacterial production. In P. Kemp, B. Sherr, E. Sherr, and J. J. Cole, eds. *Handbook of Methods in Microbial Ecology.* Lewis Publishers, Boca Raton, FL, pp. 513–518.

Kirchman, D., Ducklow, H., and Mitchell, R. (1982) Estimates of bacterial growth from changes in uptake rates and biomass. *Appl. Environ. Microbiol.* 44:1296–1307.

Kirchman, D., K'nees, E., and Hodson, R. (1985) Leucine incorporation and its potential as a measure of protein synthesis by bacteria in natural waters. *Appl. Environ. Microbiol.* 49: 599–607.

Kirchman, D. L., Keil, R. G., Simon, M., and Welschmeyer, N. A. (1991) Biomass and production of heterotrophic bacterioplankton in the oceanic subarctic Pacific. *Deep-Sea Res. I* 40:967–988.

Kirchman, D. L., Rich, J. H., and Barber, R. T. (1995) Biomass and biomass production of heterotrophic bacteria along 140°W in the equatorial Pacific: Effect of temperature on the microbial loop. *Deep-Sea Res. II* 42:621–639.

Kjelleberg, S., Flardh, K. B. G., Nysstrom, T., and Moriarty, D. J. W. (1993) Growth limitation and starvation of bacteria. In T. Ford, ed., *Aquatic Microbiology: An Ecological Approach.* Blackwell, London, pp. 289–321.

Kogure, K., and Koike, I. (1987) Particle counter determination of bacterial biomass in seawater. *Appl. Environ. Microbiol.* 53:274–277.

Lancelot, C., and Billen, G. (1986) Carbon–nitrogen relationships in nutrient metabolism of coastal marine ecosystems. *Adv. Aquat. Microbiol.* 3:263–321.

Landry, M. R., and Hassett, R. P. (1982) Estimating the grazing impact of marine microzooplankton. *Mar. Biol.* 67:283–288.

LaRock, P. A., Schwartz, J. R., and Hofer, K. G. (1988) Pulse labeling: A method for measuring microbial growth rates in the ocean. *J. Microb. Methods* 8:281–297.

Lee, S., and Fuhrman, J. A. (1987) Relationships between biovolume and biomass of naturally-derived marine bacterioplankton. *Appl. Environ. Microbiol.* 52:1298–1303.

Li, W. K. W. (1984) Microbial uptake of radiolabelled substrates: Estimates of growth rates from time course measurements. *Appl. Environ. Microbiol* 47:184–192.

Li, W. K. W., Dickie, P. M., Irwin, B. D., and Wood, A. M. (1992) Biomass of bacteria, cyanobacteria, prochlorophytes and photosynthetic eukaryotes in the Sargasso Sea. *Deep-Sea Res.* 39:501–519.

Li W. K. W., Dickie, P. M., Harrison, W. G., and Irwin, B. D. (1993) Biomass and production of bacteria and phytoplankton during the spring bloom in the western North Atlantic Ocean. *Deep-Sea Res. II* 40:307–329.

Longhurst, A. (1998) *Ecological Geography of the Sea.* Academic, San Diego, CA.

Malone, T. C., Crocker, L. H. Pike, S. E., and Wendler, B, W. (1988) Influences of river flow on the dynamics of phytoplankton production in a partially stratified estuary. *Mar. Ecol. Prog. Ser.* 48:235–249.

Marra, J., Dickey, T. D., Ho, C., Kinkade, C. S., Sigurdson, D. E., Weller, R. A., and Barber, R. T. (1999) Variability in primary production as observed from moored sensors in the central Arabian Sea in 1995. *Deep-Sea Res. II* 45:2253–2267.

Michaels, A. F., and Knap, A. H. (1996) Overview of the US JGOFS Bermuda Atlantic Time Series Study and the Hydrostation S Program. *Deep-Sea Res. II* 43:157–198.

Moriarty, D. J. W. (1985) Measurement of bacterial growth rates in aquatic systems using rates of nucleic acid synthesis. *Adv. Microb. Ecol.* 9:245–292.

Morita, R. Y., ed. (1997) *Bacteria in Oligotrophic Environments. Starvation-Survival Lifestyle.* Chapman & Hall, New York.

Noble, R. T., and Fuhrman, J. A. (1998) Use of SYBR Green I for rapid epifluorescence counts of marine viruses and bacteria. *Aquat. Microb. Ecol.* 14:113–118.

Poindexter, J. S. (1981) Oligotrophy: Fast and famine existence. *Adv. Microb. Ecol.* 5:63–91.

Pomeroy, L. R. (1974) The ocean's food web, a changing paradigm. *BioScience* 24:499–504.

Pomeroy, L. R. Sheldon, J. E., Sheldon, W. M., Jr. (1994) Changes in bacterial numbers and leucine assimilation during estimations of microbial respiratory rates in seawater by the precision Winkler method. *Appl. Environ. Microbiol.* 60:328–332.

Pomroy, A., and Joint, I. (1999) Bacterioplankton activity in the surface waters of the Arabian Sea during and after the 1994 SW monsoon. *Deep-Sea Res. II* 46:767–794.

Porter, K. G., and Feig, Y. S. (1980) The use of DAPI for identifying and counting aquatic microflora. *Limnol. Oceanogr.* 25: 943–948.

Rich, J., Gosselin, M., Sherr, E., Sherr, B., and Kirchman, D. L. (1998) High bacterial production, uptake and concentrations of dissolved organic matter in the Central Arctic Ocean. *Deep-Sea Res. II* 44:1645–1663.

Rivkin, R. B., Anderson, M. R., and Lajzerowicz, C. (1996) Microbial processes in cold oceans. 1. Relationship between temperature and bacterial growth rate. *Aquat. Microb. Ecol.* 10: 243–254.

Robarts, R. D., and Zohary, T. (1993) Fact or fiction–bacterial growth rates and production as determined by methyl-^{3}H-thymidine? *Adv. Microb. Ecol.* 13:371–425.

Robertson, B. R., Button, D. K., and Koch, A. L. (1998) Determination of the biomasses of small bacteria at low concentrations in a mixture of species with forward light scatter measurements by flow cytometry. *Appl. Environ. Microbiol.* 64:3900–3909.

Sanders, R. W. Caron, D. A., and Berninger, U. G. (1992) Relationships between bacteria and heterotrophic nanoplankton in marine and fresh waters: An inter-ecosystem comparison. *Mar. Ecol. Prog. Ser.* 86:1–14.

Sherr, B. F., del Giorgio, P., and Sherr, E. B. (1999) Estimating abundance and single-cell characteristics of actively respiring bacteria via the redox dye CTC. *Aquat Microb. Ecol.* 18:117–131.

Shiah, F.-K., and Ducklow, H. W. (1994) Temperature and substrate regulation of bacterial abundance, production and specific growth rate in Chesapeake Bay, USA. *Mar. Ecol. Prog. Ser.* 103:297–308.

Shiah, F.-K., and Ducklow, H. W. (1997) Biochemical adaptations of bacterioplankton to changing environmental conditions: responses of leucine and thymidine incorporation to temperature and chlorophyll variations. *Aquat. Microb. Ecol.* 13:151–159.

Sieburth, J. McN. (1979) *Sea Microbes.* Oxford University Press, New York.

Simon, M., and Azam, F. (1989) Protein content and protein synthesis rates of planktonic marine bacteria. *Mar. Ecol. Prog. Ser.* 51:201–213

Smith, D. C., and Azam, F. (1992). A simple, economical method for measuring bacterial protein synthesis rates in seawater using ^{3}H-leucine. *Mar. Microb. Food Webs* 6:107–114.

Smith, D. C., Simon, M., Alldredge, A. L., and Azam, F. (1992) Intense hydrolytic enzyme activity on marine aggregates and implications for rapid particle dissolution. *Nature* 359:139–142.

Smith, W. O., Jr., and Gordon, L. I. (1997) Hyperproductivity of the Ross Sea (Antarctica) polynya during austral spring. *Geophys. Res. Lett.* 24:233–236.

Somville, M. (1983) Measurement and study of substrate specificity of exoglucosidase in natural water. *Appl. Environ. Microbiol.* 48:1181–1185.

Sorokin, Y. (1964) A quantitative study of the microflora in the central Pacific Ocean. *J. Cons. Int. Explor. Mar.* 29:25–35.

Stanier, R. Y., Adelberg, E. A., and Ingraham, J. (1976) *The Microbial World,* 4th ed. Prentice-Hall, Englewood Cliffs, NJ.

Steeman-Nielsen, E. (1952) The use of radioactive carbon (^{14}C) for measuring organic production in the sea. *J. Cons. Int. Explor. Mer.* 18:117–140.

Stephanopoulos, G., and Vallino, J. J. (1991) Network rigidity and metabolic engineering in metabolite overproduction. *Science* 252:1675–1681.

Steward, G. F., and Azam, F. (1999) Bromodeoxydeuridine as an alternative to ^{3}H-thymidine for measuring bacterial productivity in aquatic samples. *Aquat. Microb. Ecol.* 19:57–66.

Tibbles, B. J. (1996) Effects of temperature on the incorporation of leucine and thymidine by bacterioplankton and bacterial isolates. *Aquat. Microb. Ecol.* 11:239–250.

Torreton, J.-P., and Dufour, P. (1996) Bacterioplankton production determined by DNA synthesis, protein synthesis and frequency of dividing cells in Tuamoto Atoll lagoons and surrounding oceans. *Microb. Ecol.* 32:185–203.

Vallino, J. J., Hopkinson, C. S., Hobbie, J. E. (1996) Modeling bacterial utilization of dissolved organic matter: Optimization replaces Monod growth kinetics. *Limnol. Oceanogr.* 8:1591–1609.

Watson, S. W., Novitsky, T. J., Quinby, H. L., and Valois, F. W. (1977) Determination of bacterial number and biomass in the marine environment, *Appl. Environ. Microb.* 33:940–946.

Wiebinga, C. J., Veldhuis, M. J. W., and De Baar, H. J. W. (1997) Abundance and productivity of bacterioplankton in relation to seasonal upwelling in the northwest Indian Ocean. *Deep-Sea Res. I* 44:451–476.

Wiebinga, C. J., de Baar, H. J. W., and Veldhuis, M. J. W. On the assessment of rates of bacterial growth in the sea: Nucleic acid and protein synthesis in incubation experiments of marine bacterial assemblages. *Aquat. Microb. Ecol.* (in press).

Williams, P. J. le B. (1981) Incorporation of microheterotrophic processes into the classical paradigm of the planktonic food web. *Kieler Meeresforsch.* 5:1–28.

Williams, P. J. le B. (1984) Bacterial production in the marine food chain: The emperor's new suit of clothes? In M. Fasham ed., *Flows of Energy and Materials in Marine Ecosystems: Theory and Practice.* Plenum Press, New York, pp. 271–299.

Williams, P. J. le B. (1993a) On the definition of plankton production terms. In *Measurement of Primary Production from the Molecular to the Global Scale.* ICES Science Symposia, Vol. 197, pp. 9–19 .

Williams, P. J. le B. (1993b) Chemical and tracer methods of measuring plankton production. In *Measurement of Primary Production from the Molecular to the Global Scale.* ICES Marine Science Symposia, Vol. 197, pp. 20–36.

Williams, P. J. le B. (1998) The balance of plankton respiration and photosynthesis in the open oceans. *Nature* 394:55–57.

ZoBell, C. E. (1946) *Marine Microbiology.* Cronica Botanica, Waltham, MA.

Zweifel, U. L., and Hagström, Å. (1995) Total counts of marine bacteria include a large fraction of non-nucleoid-containing bacteria (ghosts). *Appl. Environ. Microbiol.* 61:2180–2185.

5

PRODUCTION MECHANISMS OF DISSOLVED ORGANIC MATTER

Toshi Nagata

Ocean Research Institute,
The University of Tokyo,
Tokyo, Japan

INTRODUCTION

A considerable fraction of the oceanic food web is driven by the energy that flows through dissolved organic matter (DOM). About 50% of daily photosynthetic production is released as DOM, which supports the production of heterotrophic bacteria (Ducklow and Carlson 1992; Chapter 4). Bacteria, in turn, are consumed by viruses, protozoa, and metazoan grazers (Chapters 2, 11, 12), connecting DOM production to diverse planktonic organisms. The DOM–microbial food chains also influence primary producers by the regeneration of or competition for inorganic nutrients (Chapter 9). Therefore, the inputs of DOM to seawater have far-reaching implications for community and ecosystem processes in oceanic environments.

The production of DOM is also important in oceanic biogeochemistry. Although marine bacteria are highly efficient in utilizing DOM, selected components of DOM resist degradation for reasons that are still not entirely clear (Lee and Wakeham 1992; Benner 1998; Chapter 6). These refractory components accumulate in seawater and form a large pool of organic carbon in the oceans. In fact, oceanic DOM represents one of the largest active

Microbial Ecology of the Oceans, Edited by David L. Kirchman.
ISBN 0-471-29993-6   Copyright © 2000 by Wiley-Liss, Inc.

reservoirs of organic carbon on earth (Hedges 1992) and is important for understanding global carbon cycles and changes in the concentration of atmospheric carbon dioxide, the most critical greenhouse gas on our planet (Siegenthaler and Sarmiento 1993).

In this chapter, I discuss how DOM is produced in marine environments. The ultimate source of the organic carbon in oceanic DOM is phytoplankton. Nonetheless, the processes that lead to the release of DOM cannot be fully understood from phytoplankton physiology alone. Apparently, several food web processes are closely related to the production of DOM. Complexity also stems from the diversity of DOM molecules with variable biological reactivity—different DOM molecules may be produced by quite different mechanisms. Our discussion first deals with specific biological processes involved in the DOM production: the release by phytoplankton; the egestion, excretion, and "sloppy feeding" by grazers; and cell lysis induced by viruses. Then the individual mechanisms will be integrated into the framework of DOM–food web dynamics in open waters, which should help to evaluate the dominant controls of DOM production in oligotrophic oceanic environments. Finally, I will introduce emerging ideas on production pathways of refractory DOM in seawater, with an emphasis on microbial food web processes.

RELEASE OF DOM BY PHYTOPLANKTON

The extracellular release of DOM by phytoplankton has been extensively studied for more than three decades. This section focuses on variability in the percent extracellular release (PER) of dissolved organic carbon (DOC) relative to total (dissolved plus particulate) primary production. PER is an important variable for assessing the quantitative role of DOM release by phytoplankton. A major goal is to address the following questions: How variable is PER? What factors can influence PER? What kinds of models have been proposed to explain algal extacellular release?

Culture experiments have demonstrated that isolates of marine phytoplankton release as DOM photosynthetic products including carbohydrates (mono-, oligo-, and polysaccharides), nitrogenous compounds (amino acids, peptides, and proteins), organic acids (glycollate), and lipids (Hellebust 1965; Fogg 1983). In cultures, PER is typically in the range of 2–10% (average 5%) for exponentially growing cells (Figure 1). The PER tends to be higher when nutrients are depleted (Obernosterer and Herndl 1995), growth conditions (irradiance, temperatures) are suboptimal (Ignatiades and Fogg 1973; Verity 1981; Zlotnik and Dubinsky 1989), or cells are in stationary or senescent growth phase (Berman and Holm-Hansen 1974; Obernosterer and Herndl 1995). Also, abrupt changes in light intensity appear to cause a transient increase in PER (Hellebust 1965; Fogg 1983).

In natural environments, the extracellular release of DOM by phytoplankton has been determined by measuring production rate of [^{14}C]DOC in

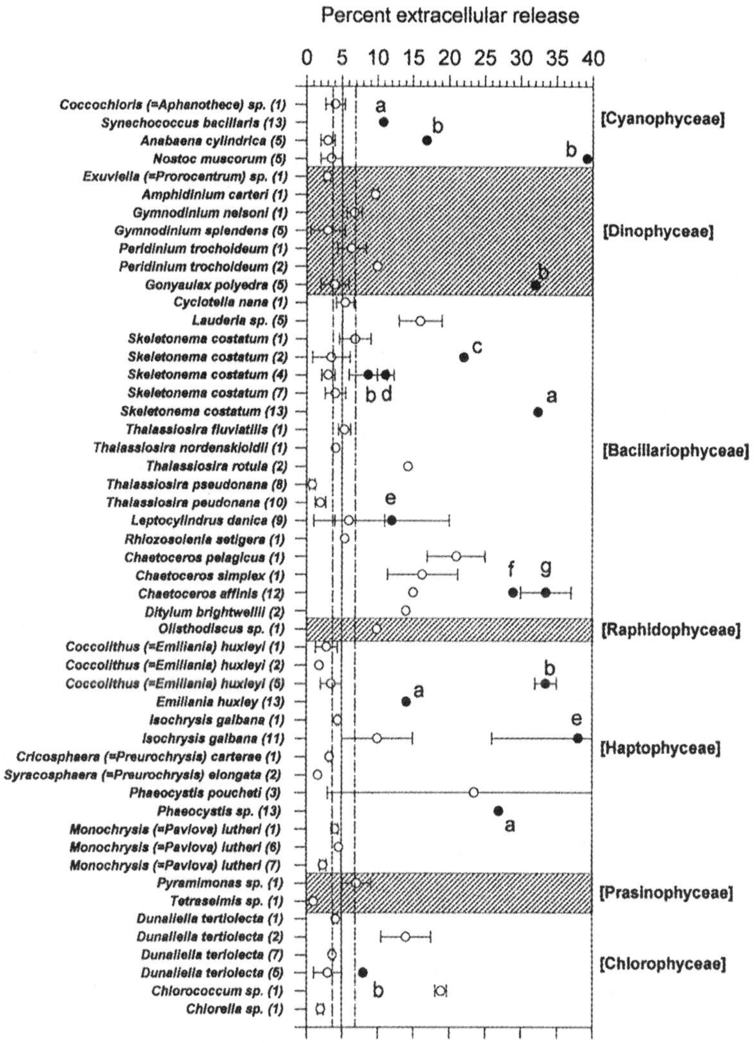

Figure 1. Percent extracellular release (PER) of DOC relative to total primary production by cultured phytoplankton. Mid-range values are plotted with error bars that indicate ranges except when only single values are reported in the source literature. Open symbols are results for exponentially growing cells, whereas closed symbols are for cells that have been exposed to suboptimum conditions indicated as follows: a, long-term culture; b, senescent culture; c, low growth rate; d, high light intensity; e, low or high temperature; f, nitrogen-limited; g, stationary growth phase. Vertical lines indicate the geometric mean (solid line, 5.1%) and 99% confidence interval (broken lines, 3.7–6.9%) of PER for exponentially growing cells. The number in parentheses after a Latin name indicates the source of the data: 1, Hellebust (1965), 2, Eppley and Sloan (1965); 3, Guillard and Hellebust (1971); 4, Ignatiades and Fogg (1973); 5, Berman and Holm-Hansen (1974); 6, Laws and Caperon (1976); 7, P. J. le B. Williams and Yentsch (1976); 8, Sharp (1977); 9, Verity (1981); 10, R. E. H. Smith and Platt (1984); 11, Zlotnik and Dubinsky (1989); 12, Obernosterer and Herndl (1995); 13, Biddanda and Benner (1997). Shaded areas help identify the algal class attached on the right Y-axis.

seawater samples incubated with [^{14}C]bicarbonate (Lancelot 1979; R. E. H. Smith and Platt 1984; P. J. le B. Williams 1990). Table 1 summarizes PER values reported in various marine environments. The values vary greatly but the average PER is usually within the range of 10–20%, higher than those for actively growing cells in cultures (average = 5%, Figure 1). It should be noted that the ^{14}C tracer technique may overestimate PER, particularly when applied to oligotrophic waters, because of potential errors due to ^{14}C blanks, rupture of fragile cells by filtration, or changes of physiological state of phytoplankton during sampling (Sharp 1977). Furthermore, heterotrophic processes (grazing, viral infection) may contribute to [^{14}C]DOC production during the incubation (Jumars et al. 1989).

High PER (up to 80%) has been observed during the declining phase of phytoplankton blooms accompanied with nutrient depletion (Larsson and Hagström 1982; Lancelot 1983). This observation is consistent with results in culture studies (Figure 1; Berman and Holm-Hansen 1974), although viral infection may also contribute to the increase in DOC release (Gobler et al. 1997; see below). High PER during blooms has been attributed to the release of large amounts of carbon-rich DOM compounds such as carbohydrates (Guillard and Wangersky 1958; Myklestad et al. 1989; Ittekkot et al. 1981; Biddanda and Benner 1997). Other studies have suggested that dissolved carbohydrates may support a significant portion of bacterial carbon and energy demand (Amon and Benner 1994; Norrman et al. 1995). Also, these compounds may facilitate the formation of large organic aggregates in seawater (Passow et al. 1994).

PER tends to be higher near the surface with high irradiance (Thomas 1971; Berman and Holm-Hansen 1974), possibly as a result of cell damage (Hellebust 1965; Watanabe 1980) or photorespiration (see below). The light-induced increase of PER may enhance bacterial production in surface waters (Jones et al. 1996), but it probably has little effect on overall production in the euphotic zone. The variability of PER may be also related to the species composition of phytoplankton (Wolter 1982; Lancelot 1983), although taxon-specific patterns of PER are not apparent in the available database (Figure 1); the variability associated with growth conditions and methodology is overwhelming.

Although PER varies greatly depending on local environmental conditions and species compositions, the average PER does not vary systematically over a wide range of productivity regimes (Table 1). In fact, upon compiling many literature values from various environments (mostly coastal marine, estuarine, and freshwater systems), Baines and Pace (1991) concluded that the extracellular release is linearly related to total primary production, indicating that PER is constant (average = 13%) across diverse aquatic systems. Given the scarcity of data on PER in open oceans, it is premature to conclude that this value can be used as a "global average" of PER. However, empirical evidence indicates that the extracellular release by phytoplankton, with average PER = 13%, is not sufficient to fulfill bacterial carbon demand in many marine environments; bacteria consume on average 40–50% of primary production

Table 1. Percent extracellular release (PER) of DOC in marine environments

Locations	Mean PER (range)	Remarks	Ref.[a]
Northeast Pacific (coastal and offshore)	14.5 (7–26)		1
Southern California Bight (coastal and offshore)	15.2 (5.9–27.3)	High PER at near the surface and at the bottom of the euphotic zone	2
Western Sargasso Sea (oceanic)	20 (1–44)	High PER at near the surface	3
Western North Atlantic (shelf water)	2.3 (0–10)		3
Mediterranean (oligotrophic)	11		4
Arctic (oceanic, ice-covered)	25 (0–70)	High PER under the condition of possible light- and N-limitation	5
Eastern North Atlantic (coastal upwelling)	11.3 (2.2–29)	No significant correlation of PER with ambient nutrient concentrations and light intensities	6
Tropical Atlantic (coastal)	6.9 (<1–23)		7
North Sea and North Atlantic (coastal)	0–80	High PER during the declining phase of flagellate blooms	8
Caribbean and Peruvian shelf waters	3.8 (1.5–16)		9
Gulf of Maine (coastal)	11.3 (5–30)		10
Baltic Sea (coastal)	10 (1–23)	High PER during the declining phase of a spring bloom	11
Baltic sea (coastal)	10.8 (2.3–16.7)	High PER during summer when nutrient concentration was low	12
Kiel Fjord, Germany (estuary)	18.7 (<1–45)	PER varied depending on dominant phytoplankton taxa	13
Randers Fjord, Denmark (estuary)	15 (13–17)		14

[a]1, Anderson and Zeutschel (1970); 2, Berman and Holm-Hansen (1974); 3, Thomas (1971); 4, Hagström et al. (1988); 5, Gosselin et al. (1998); 6, W. O. Smith et al. (1977); 7, P. J. le B. Williams and Yentsch (1976); 8, Lancelot (1983); 9, Sellnar (1981); 10, Mague et al. (1980); 11, Larsson and Hagström (1982); 12, Lignell (1990); 13, Wolter (1982); 14, Jensen (1983).

across systems (Cole et al. 1988; Ducklow and Carlson 1992). Thus, over two-thirds of bacterial carbon demand must be met by DOM sources other than direct release by phytoplankton.

MECHANICAL MODELS OF DOM RELEASE

The physiological mechanism of DOM release by phytoplankton is not fully understood. Here I examine two simple mechanistic models, overflow and leakage. The release of DOM compounds that have specific roles for phytoplankton (e.g., vitamins, toxins, organic metal chelators such as siderophores) are not considered because the contribution of these compounds to the bulk DOM flux is probably minor.

Overflow Model

Under sufficient light and low nutrient concentration, carbon fixation may exceed incorporation into cell material, resulting in extracellular release (overflow) of photosynthate (Fogg 1983; Wood and Van Valen 1990). In support of this hypothesis, relatively high PER has been observed under conditions of high irradiance and nutrient depletion (Berman and Holm-Hansen 1974; Watanabe 1980; Lancelot 1983; Obernosterer and Herndl 1995). This model has been used to explain the release of glycollate (Fogg 1983; Al-Hasan and Fogg 1987), a compound produced by photorespiration (Beardall 1989). However, controversy remains over the role of glycollate in marine DOC dynamics: Fogg and coworkers (Fogg 1983; Al-Hasan and Fogg 1987) suggest that glycollate is an important component of DOC released by phytoplankton (see also Leboulanger et al. 1997), whereas others suggest that the production of glycollate by photorespiration is negligible in marine phytoplankton (Beardall 1989; P. J. le B. Williams 1990). To what extent the release of other important DOC compounds such as carbohydrates can be explained by the overflow model is unclear.

Leakage Model

The leakage model suggests that low molecular weight DOM is continuously released from the cells because of passive permeation through the membrane (Bjørnsen 1988). In seawater, the leakage flux is driven by a steep concentration gradient (G_m) of low molecular weight DOM across membranes: G_m = (intracellular concentrations of 10^{-3} M) − (extracellular concentrations of 10^{-9} M) (Fuhrman 1987). Bjørnsen (1988) estimate that the daily loss by leakage of the intracellular pool of a 10 μm phytoplankton cell is 50% (equivalent to 5% of total cell mass), a loss that is likened to a "property tax." This model is consistent with observations that the DOM release by some phytoplankton is constant during light and dark cycles (Mague et al. 1980).

Also, the release by phytoplankton of nitrogen-rich, low molecular weight DOM such as dissolved free amino acids (Hellebust 1965; Myklestad et al. 1989; Bronk and Glibert 1993) could be explained by the leakage model.

PRODUCTION OF DOM BY GRAZERS

A substantial portion of primary and bacterial production is consumed by various grazers that differ in their food selectivity and feeding strategies. These grazers act not only as trophic links and nutrient regenerators, but also as active transformers of prey material into DOM, an important role that is often overlooked in traditional food web models. Here I examine data on release rates of DOM by grazers, especially relative to their ingestion rates. Mechanisms and controls of DOM production by grazers also are discussed, referring to the question of why grazers should waste reduced carbon and nutrients as DOM. Two classes of grazer communities are considered separately: protozoa and metazoa (hereafter, zooplankton). The former mainly eat small phytoplankton and bacteria, while the latter generally eat large phytoplankton.

Release of DOM by Protozoa

Data on release rates of DOM by protozoan grazers are summarized in Tables 2 and 3. Although some methodological problems remain to be solved for accurately measuring the DOM release by protozoa (Nagata and Kirchman 1990, 1992a; Nagata in press), the growing list of studies demonstrates that protozoa can release a significant fraction (10–30%) of ingested prey organic matter as DOM (Nagata and Kirchman 1992a; Chase and Price 1997; Strom et al. 1997; Ferrier-Pagès et al. 1998; Pelegrí et al. 1998). These values suggest that protozoa are potentially a dominant source of DOC in many regions of oceans where primary production is dominated by small phytoplankton that is mostly consumed by protozoa (Sherr and Sherr 1994). Assuming that phytoplankton growth is balanced by protozoan grazing, as often is the case in open oceans, the data suggest that 10–30% of particulate primary production can be transformed to DOC during protozoan herbivory. In addition, DOC is produced during bacterivory. The magnitude of potential DOC release by protozoa is equivalent to or even exceeds that by phytoplankton (average PER = 13%; see above).

Besides their role in DOC flux, protozoa may contribute to the release of DOM compounds that are rich in nutritional elements including N, P, and Fe. One example is dissolved free amino acids (Andersson et al. 1985; Nagata and Kirchman 1991; Ferrier-Pagès et al. 1998), which are highly labile and may support a large fraction of bacterial C and N demand (Keil and Kirchman 1991). Also, flagellates release dissolved organic phosphorus (Andersen et al. 1986) such as dissolved DNA (Turk et al. 1992), a potentially important source of P for bacteria (Ammerman and Azam 1985). Furthermore, high amounts of

Table 2. Release of DOM by flagellates grazing on bacteria and protists (flagellates are marine isolates unless noted.)

Flagellates	Prey	DOM[a]	Quantity	Remarks	Ref.[b]
Paraphysomonas imperforata	Diatom (*Pheodactyrum tricornum*)	DOC	10% of C ingestion		1
Oxyrrhis marina	Dinoflagellates (*Prorocentrum minimum*)	DOC	29% of C ingestion	Corrected for bacterial uptake of DOC	2
Oxyrrhis marina	Flagellates (*Cryptomonas* sp.)	DOC	28% of C ingestion	Corrected for bacterial uptake of DOC	2
Pteridomonas danica	Bacteria (*Escherichia coli*)	DOC	23–34% of C ingestion		3
Paraphysomonas imperforata	Bacteria	DOC	8–27% of C ingestion	High release when prey cells are Fe-stressed	4
Paraphysomonas imperforata	Bacteria (*Vibrio splendidus*)	DFAA	4% of N ingestion	Corrected for DFAA turnover	5
Paraphysomonas imperforata	Mixed bacteria	DFAA	22% of N ingestion	Corrected for DFAA turnover	5
Ochromonas sp.	Bacteria (*Escherichia coli*)	DFAA	7% of N ingestion	Corrected for DFAA turnover	6
Pseudobodo sp.	Bacteria	DFAA	10% of N ingestion	Heat-killed bacteria	7
Paraphysomonas imperforata	Bacteria (*Vibrio splendidus*)	DCAA	1% of N ingestion		5
Paraphysomonas imperforata	Diatom (*Phaeodactylum tricornum*)	DOP	15–20% of total P release		8
Paraphysomonas imperforata	Mixed bacteria	DOP	70% of total P release		8
Pterioochromonas malhaensis[c]	Bacteria (*Pasteurella* sp.)	DOP	15–20% of P ingestion	Heat-killed bacteria	9

[a]DFAA, dissolved free amino acids; DCAA, dissolved combined amino acids; DOP, dissolved phosphorus.
[b]1, Caron et al. (1985); 2, Strom et al. (1997); 3, Pelegri et al. (1998); 4, Chase and Price (1997); 5, Nagata and Kirchman (1991); 6, Andersson et al. (1985); 7, Ferrier-Pagès et al. (1998); 8, Andersen et al. (1986); 9, Caron et al. (1990)
[c]Freshwater isolate.

Table 3. Release of DOM by ciliates grazing on bacteria and protists

Ciliates	Prey	DOM[a]	Quantity	Remarks	Ref.[b]
Uronema sp.	Mixed bacteria	DOC	9% of C ingestion		1
Euplotes sp.	Mixed bacteria	DOC	3% of C ingestion		1
Mixed assemblage	Mixed bacteria	DOC	20–88% of C ingestion		1
Strombidinopsis acuminatum	Flagellates (*Cryptomonas* sp.)	DOC	30% of C ingestion		2
Strombidinopsis acumination	Dinoflagellates (*Prorocentrum minimum*)	DOC	37% of C ingestion		2
Strombidium sulcatum	Bacteria	DFAA	16% of N ingestion	Heat-killed bacteria	3
Euplotes vannus	Mixed bacteria and flagellates	DOP	26% of total P release	Measured without prey	4

[a]See Table 2, note *a*.
[b]1, Taylor et al. (1985); 2, Strom et al. (1997); 3, Ferrier-Pagès et al. (1998); 4, Johannes (1965).

Fe and trace metals in bacterial and cyanobacterial cells (Tortell et al. 1996) are efficiently regenerated by protozoan grazers (Hutchins and Bruland 1994; Chase and Price 1997), and some regenerated metals may be associated with DOM compounds released by grazers (Twiss and Campbell 1995). DOM–Fe complexation may affect greatly the availability of Fe for phytoplankton and bacteria (Hutchins 1995), which has implications for oceanic ecosystems limited by Fe (Martin et al. 1991). In short, the release of DOM by protozoa could have a direct influence on N, P, and Fe cycling and nutrient stoichiometry in oceanic waters.

Protozoa appear to release DOM during egestion. To illustrate this, Figure 2 schematically depicts a general sequence of feeding in planktonic protozoa, which includes phagocytotic engulfment of prey cells, enzymatic digestion within acidic food vacuoles, assimilation of digested molecules, and finally egestion of incompletely digested or unassimilated materials. The egestion event is characterized by the fusion of vacuoles with cytoplasmic membranes, leading to evacuation of vacuole contents consisting of prey remains, unassimilated DOM, and even digestive enzymes (Cole and Wynne 1974; Fenchel 1982; Nagata and Kirchman 1992b). The mass balance can be described as follows:

$$I = G + R + E$$

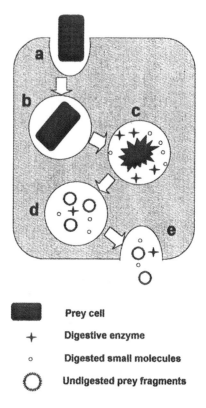

■	**Prey cell**
+	**Digestive enzyme**
o	**Digested small molecules**
○	**Undigested prey fragments**

Figure 2. Schematic representation of the feeding sequence for protozoan grazers. Prey cells engulfed by phagocytosis (a) are hydrolyzed by digestive enzymes secreted into acidic food vacuoles (b). Digestion products (small molecules) are absorbed through vacuolar membranes to become available for metabolic processes of protozoa (c, d). Incompletely digested prey materials, unabsorbed small molecules, and a part of digestive enzymes are released to seawater along with egestion (e). Structural components of prey cells (membranes and walls) form liposome-like vesicles (Cole and Wynne 1974; Fenchel 1982, 1986; Nagata and Kirchman 1992b), which accumulate in food vacuoles at later stages of digestion. These colloidal vesicles are released to seawater as "picopellets" (Nagata in press).

where I = ingested organic carbon, G = organic carbon used for growth, R = loss of organic carbon to respiration; and E = egested organic carbon in both particulate and dissolved forms. In practice, E is difficult to measure separately (Nagata in press), but it can be calculated as a deficit of the mass balance: that is, the assimilation efficiency $[(G + R)/I \times 100)]$ of 60–70% (Fenchel 1982, 1987; Nagata in press) implies egestion of 30–40% of ingested material. Note that a large fraction of protozoan egesta, particularly that of flagellates, consists of colloids ("picopellets": Nagata in press; see also Pelegrí et al. 1998), which are operationally classified in the "DOM" pool in oceanography (Nagata and Kirchman 1997). Ciliates may produce larger pellets

(Stoecker 1984; Gowning and Silver 1985), but high amounts of solutes could rapidly diffuse out from the matrices of egesta (Jumars et al. 1989). Thus the release of DOM by protozoa (10–30% of ingestion, Tables 2 and 3) appears to be fully accounted for by egestion.

The release of DOM by protozoa varies depending on food abundance (Nagata and Kirchman 1991; Ferrier-Pagès et al. 1998). Jumars et al. (1989) proposed a model in which grazers maximize the gain (total amount of assimilated organic matter) by changing τ, the retention time of prey in a digestive system; τ drops when prey abundance is high for processing ingested prey rapidly. This "optimum digestion" model predicts that the release of DOM relative to ingestion increases with the increase of prey abundance. This is consistent with observations by Nagata and Kirchman (1991), who found that flagellates release higher amounts of dissolved free amino acids relative to ammonium when food abundance is higher. However, other studies have found that the retention time of prey in food vacuoles in ciliates varies little depending on food abundance (Sherr et al. 1988; Capriulo and Degnan 1991; Dolan and Simek 1997). These results suggest that digestion variables other than vacuole retention time (e.g., availability of membranes to form food vacuoles; type and amounts of digestive enzymes; chemical condition of vacuoles) are important in determining the ingestion/egestion mass budget of protozoa (Dolan 1997; Strom et al. 1998).

In addition to the release of DOM during egestion, protozoa excrete organic metabolites such as urea and purines (Caron and Goldman 1990). Goldman et al. (1985) estimate that urea contributes less than 15% of total N excretion (ammonium + urea) by flagellates grazing on diatom and bacteria, representing less than 5% of total N ingestion.

Release of DOM by Zooplankton

In coastal areas, the contribution of zooplankton to the DOM flux is potentially large because these grazers consume a large fraction of primary production dominated by large phytoplankton (White and Roman 1992; Dagg 1993). This idea is supported by indirect evidence showing a positive correlation between DOM concentrations (or fluxes) and zooplankton abundance (Eppley et al. 1981; Riemann et al. 1986; R. Williams and Poulet 1986; Fuhrman 1987; Roman et al. 1988; Peduzzi and Herndl 1992) and enhancement of bacterial production in the presence of zooplankton (Eppley et al. 1981; Riemann et al. 1986; Roman et al. 1988; Peduzzi and Herndl 1992). Although quantitative data that relate the release of DOM to ingestion are quite limited (Table 4), the available results are consistent with the foregoing notion; the release of DOC by zooplankton can represent roughly 10–20% of ingestion (Conover 1966; Copping and Lorenzen 1980; Strom et al. 1997). In environments where phytoplankton growth is balanced by zooplankton grazing, the rate of DOC release by zooplankton may account for 10–20% of primary production.

Table 4. Release of DOM by crustacean zooplankton

Zooplankton	Prey	DOM	Quantity and Remarks	Ref[a]
Copepod (*Calanus hyperborus*)	Diatom (*Thalassira fluviatilis*)	DOC	15% of particulate C removed[b]	1
Freshwater cladocera (*Daphnia pulex*)	Six species of freshwater algae	DOC	4–17% of ingested C was released, possibly owing to sloppy feeding	2
Copepod (*Calanus pacificus*)	Diatom (*Thalassiosira fluviatilis*)	DOC	19% of C ingestion	3
Copepod (*Calanus pacificus*)	Diatom (*Coscinodiscus angstii*)	DOC	21% of C ingestion	3
Copepod (*Calanus pacificus*)	Natural phytoplankton	DOC	9% of C ingestion	3
Copepod (*Calanus pacificus*)	Dinoflagellate (*Prorocentrum minimum*)	DOC	16% of C ingestion	4
Copepod (*Calanus pacificus*)	Flagellate (*Oxyrrhis marina*)	DOC	29% of C ingestion	4

[a] 1, Hellebust and Conover [unpublished data cited in Conover (1966)]; 2, Lampert (1978); 3, Copping and Lorenzen (1980); 4, Strom et al. (1997).
[b] This value has been quoted in the literature as an estimate of DOC release attributable to sloppy feeding (e.g., Strom et al. 1997). But Conover (1966) ascribed the observed release of DOC to "leaching from damaged cells and excretion by copepods."

There are at least four modes of DOM release by zooplankton: sloppy feeding, excretion, egestion, and release from fecal pellets. The relative contribution of each mechanism to the bulk DOM release by zooplankton is not clear. Sloppy feeding causes the leakage of DOM from prey cells broken by mouthparts of zooplankton (Conover 1966). Currently, the only study to have determined DOM production by sloppy feeding is that of Lampert (1978). By using ^{14}C-labeled tracer cells, the author found that freshwater cladocera (*Daphnia*) release 4–17% of ingested prey carbon as DOC during initial stage of feeding. This release was ascribed to sloppy feeding. Consistent with this hypothesis, the release from small cells swallowed whole was low, whereas the release from large cells (more susceptible to mechanical damage) was high. The application of this observation to marine zooplankton species remains to be examined. Behavioral and morphological phenotype of zooplankton should be critical in determining the degree of "sloppiness" in feeding.

Another mechanism of DOM production by zooplankton is excretion of organic metabolites such as urea and dissolved free amino acids, which generally represent less than 10–20% of total N excretion (Webb and Johanness 1967; Corner and Davies 1971; Bidigare 1983), or less than 8% of total N ingestion, assuming that the ratio of N excretion relative to N ingestion is 40% in well-fed zooplankton (Gardner and Paffenhöfer 1982). The excretion mechanism of dissolved free amino acids is not known. Gardner and Paffenhöfer (1982) found that the release by copepods of dissolved free amino acids occurs as occasional discrete events, while ammonium excretion is continuous. This observation is consistent with the hypothesis that the release of dissolved free amino acids is related to egestion events rather than excretion, as suggested for protozoa (Nagata and Kirchman 1991; Nagata in press).

Unassimilated remains of food materials are egested as DOM and particles (fecal pellets). Subsequently, DOM is released from pellets by diffusion, mechanical breakage and enzymatic hydrolysis (Jumars et al. 1989; Lampitt et al. 1990; Strom et al. 1997). A practical, rigorous distinction of these two modes of DOM production is difficult, and quantitative data on rates of DOM release by these processes are absent. Jumars et al. (1989) hypothesized that rapid diffusional release of DOM from pellets should be important in DOM production. In contrast, Strom et al. (1997) suggested that mechanical breakage of pellets is critical; the authors observed that pellets enhanced bacterial growth only when pellets were broken by copepods. In any event, unassimilated prey materials are a potentially important source of DOM. Data on assimilation efficiency of zooplankton vary widely with a typical range of 60–80% (Corner and Davies 1971; Gaudy 1974), implying that 20–40% of ingested carbon can be egested. Assuming that half of egesta is DOM (including DOM released from pellets) and half is pellets, the release of DOM by egestion-related mechanisms accounts for 10–20% of total ingestion, in good agreement with data presented in Table 4. We may hypothesize that a dominant control of DOM release by zooplankton is egestion and rapid dissolution of solutes from pellets (Jumars et al. 1989).

RELEASE OF DOM MEDIATED BY VIRAL INFECTION

Several studies have suggested that viruses can be a major cause of mortality for bacteria and phytoplankton (Proctor and Fuhrman 1990; Suttle 1994; Fuhrman and Noble 1995; Chapter 11) and play a role in DOM fluxes within oceanic environments (Bratbak et al. 1990; Gobler et al. 1997). It is likely that the primary step of the release of DOM is the burst event: the release of viral particles from host cells, which results in the release of cellular solutes and fragments of cell structures. This event may be followed by the release of DOM from host cell fragments by diffusion and enzymatic breakdown. In support of this hypothesis, experiments have revealed that viral infection of marine microorganisms significantly enhances the release of DOM (Weinbauer and

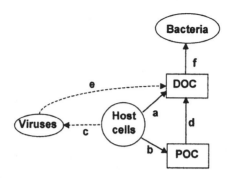

Figure 3. Carbon flows within a food web system consisting of host cells, viruses, and bacteria. Solid and broken lines indicate major and minor flows, respectively. During viral infection, host cell material is transformed to DOC (a), POC (b), and viral biomass (c). DOC is released from POC by dissolution and enzymatic hydrolysis (d). Viral biomass contributes to the formation of DOC pool (e). DOC is consumed by bacteria (f). Note that the viral system provides a highly efficient mechanism of DOC production in seawater because respiration and biomass production are minimal. (From Gobler et al. 1997, with modifications.)

Peduzzi 1995) and detrital colloids (Shibata et al. 1997) from host cells. The mass balance of viral infection was investigated by Gobler et al. (1997), who suggested a remarkable role of viruses in DOM production during the declining phase of phytoplankton blooms. By using a model system consisting of phytoplankton (chrysophyte, *Aureococcus anophageﬀerens*), algal viruses, and bacteria (Figure 3), the authors found that, of the total carbon of algal cells that were lysed by viruses, 15–22% was rapidly transformed to DOC. Also, it was estimated that 22% of the lysed carbon was incorporated into bacterial biomass. Thus, about 40% of the host cell carbon was channeled through the DOC–bacterial pathway as a result of viral infection. Although the authors' mass balance was incomplete (respiration was not measured), the substantial transfer of host cell carbon to DOC and bacteria strongly suggests a major role of viral infection in enhancing DOC flux and microbial loop during the breakdown of algal blooms. In support of this notion, recent evidence suggests that viruses that infect phytoplankton are ubiquitous (Cottrell and Suttle 1991) and can contribute to terminating algal blooms (Nagasaki et al. 1994). The new knowledge just described demands reevaluation of the conventional explanation that the high flux of DOC during declining phase of blooms is due solely to algal extracellular release (Lancelot 1983; Azam et al. 1983; Norrman et al. 1995).

WHICH MECHANISMS ARE IMPORTANT FOR BACTERIA?

The review of potential mechanisms of DOM production strongly suggests that the rates of bulk DOM inputs to seawater are controlled by multiple trophic level processes. This view differs greatly from the original proposition of the microbial loop (Azam et al. 1983), where the major process producing DOC is algal extracellular release. The revised view, however, is consistent with a growing consensus that heterotrophs play an important role in DOM production in marine systems (P. J. le B. Williams 1981; Hagström et al. 1988; Jumars et al. 1989; Fuhrman 1992; Nagata and Kirchman 1992a; Gobler et al. 1997; Strom et al. 1997). Here I try to integrate elementary mechanisms into the framework of DOM–food web dynamics in the oceans.

Oligotrophic oceanic environments are dominated by small phytoplankton which are under the tight grazing control by protozoa. This type of ecosystem characterizes a large part of the world's oceans including equatorial (Landry et al. 1997), subtropical (Liu et al. 1995), and subpolar (Miller et al. 1991) regions and plays a substantial role in global carbon cycles. My approach is to construct an "average" food flow diagram for open oceans with reasonable assumptions that do not seriously conflict with published data. This model is used to examine the extent by which each mechanism can contribute to DOC production. Given that open-ocean ecosystems are highly dynamic and heterogeneous, what we can derive from a set of fixed variables and coefficients is a hypothetical, if not imaginary, model at best. The hypothetical model, however, provides a useful background for identifying important features of the DOM production system in the oceans.

We first set the total primary production as 100 and then evaluate the flow of carbon relative to this value (Figure 4). The primary production is delivered to four compartments (Z, P, V, DOC) based on observational data (Table 5). The outflows from each heterotrophic compartment (Z, P, B, V) are distributed to different destinations (Table 5). The protozoan food chain is considered by dividing protozoa into three omnivorous components (Figure 4b). Similarly, zooplankton are divided into herbivores and carnivores (Figure 4c). For each grazer subsystem, a flow diagram is constructed, and integrated outflows are presented in Figure 4a. Finally, the flows of organic carbon out of the system (export: Legendre and Rassoulzadegan 1996), including food web, sinking POC, and semilabile (and refractory) DOC, are assigned according to empirical data.

The resultant model (Figure 4a) shows that grazers are a dominant source of DOC; 65% of total DOC production is accounted for by the release of DOC by grazers (Table 6). Among grazers, protozoa play a major role in DOC production; more than half DOC production is attributable to protozoa. The validity of this conclusion primarily depends on two assumptions: that protozoa consume a major fraction (80% in the model) of primary production, and that protozoa release a significant portion (25% in the model) of ingested prey as DOC. Both assumptions are supported by data. Field measurements have demonstrated that "microzooplankton," mostly protozoa, can consume more than 80% of daily primary production in many regions of open oceans (Miller et al. 1991; Liu et al. 1995; Landry et al. 1997). As I have already discussed, studies have increasingly revealed that protozoa can release large amounts of

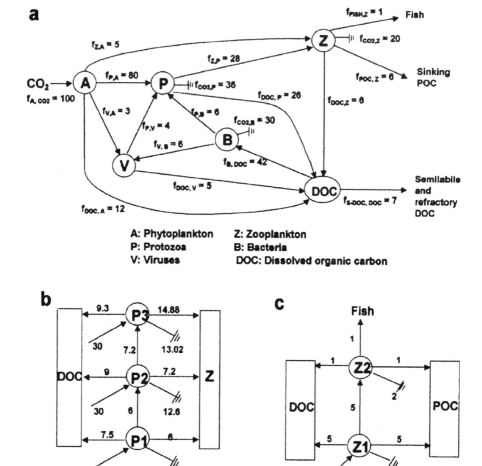

Figure 4. (a) Hypothetical food flow diagram in oligotrophic open oceans. Sources of data for parameterization are summarized in Table 5. Respiration (dissipation of energy) is symbolized by —|||. This model intends to examine production processes of DOC. For clarity, the flow of carbon through detritus (nonliving particulate organic carbon) is not diagrammed. This omission affects only slightly my major arguments concerning DOC. Outflows from protozoan and zooplankton compartments are integrated outflows from subsystem analyses (see below). (b) The protozoan community is divided into three omnivorous subcompartments (P1, P2, P3) connected by a prey–predator chain. Total food intake to protozoa ($f_{P,A} + f_{P,V} + f_{P,B} = 90$) is evenly distributed to three compartments. Higher trophic level compartments gain additional food intake from lower trophic level compartments. For each compartment, fractions of food intake allocated to CO_2, biomass, and DOC are 0.35, 0.40, and 0.25, respectively, which is consistent with published data of protozoan gross growth efficiency (40–50%; Caron and Goldman 1990; cf. Straile 1997) and DOC release (10–30%, Tables 2, 3). For P1 and P2, it is assumed that half the biomass production is consumed by zooplankton and the remaining half consumed by protozoa. The biomass production of P3 is totally consumed by zooplankton. (c) Zooplankton subsystem consists of herbivores (Z1) and carnivores (Z2). The fractions of food intake allocated for CO_2, egesta (DOC:pellets = 1:1) and biomass are 0.55, 0.30, and 0.15 for Z1 and 0.4, 0.4, and 0.2 for Z2. These values are within the ranges of published values of zooplankton gross growth efficiency (10–30%; Corner and Davies 1971) and DOC release (10–20% of ingestion; Table 4).

DOC, which account for 20–30% of ingestion (Tables 2 and 3; Nagata and Kirchman 1992a; Strom et al. 1997; Pelegrí et al. 1998). This conclusion is consistent with the model proposed by Jumars et al. (1989), who suggested that DOM production by grazers is the dominant pathway that fuels the microbial loop in the oceans. Also, on the basis of rate measurements of microbial food web processes, Hagström et al. (1988) suggested that protozoa are the major contributor to DOC production in oligotorophic Mediterranean water. Thus, ingestion, digestion, and egestion of protozoa (Figure 2) appear to be a major regulatory mechanism of DOM dynamics in oceanic waters.

The role of viruses is less clear than other aspects of DOC production. In my model, I assigned 3% of primary production and 50% of bacterial production to viral consumption (Figure 4a). These values are from estimates in coastal environments, and no data are available in oceanic waters (Table 5). It is probably surprising that viruses account for a significant fraction (10%) of total DOC production in my model (Table 6), despite modest assumptions about infection rates. When viruses cause substantial mortality for phytoplankton (Suttle et al. 1990; Suttle 1994), the role of viruses in DOC dynamics could be substantial (Gobler et al. 1997). Much remains to be examined about virus-induced mortality in oceanic environments.

My review emphasizes food web controls rather than algal physiology in determining DOC production in the oceans. In fact, results from the model suggest that algal extracellular release need not be a dominant mechanism of DOC production to explain bacterial carbon demand (Table 6). However, I note that data on direct measurements of algal extracellular release in oligotrophic waters are scarce. The traditional ^{14}C technique is extremely sensitive to experimental artifacts (Sharp 1977; P. J. le B. Williams 1990), particularly when this technique is applied to oligotrophic waters. Also, it is now clear that results from these experiments cannot be taken at face value because of possible contributions of viruses and protozoa to $[^{14}C]DOC$ production in incubation bottles.

In the model, total heterotrophic consumption far exceeds primary production (Table 7). This is counterintuitive but arises because organic carbon is recycled primarily by DOC–microbial food chains (Scavia 1988; Strayer 1988). Note that inflows and outflows of all compartments and the whole system are completely in balance (Figure 4); that is, the second thermodynamic law is not violated. This phenomenon has been theoretically analyzed by Higashi et al. (1993a), who called it "network virtual amplification." It is "virtual" because no new energy is created by recycling; but it should be "real" for oceanic microbial life, which largely depends on recycled energy. Theory predicts that a food web with many recycling flows tends to homogenize the distribution of energy among food web participants; that is, energy is distributed more or less evenly among populations (Higashi et al. 1993a, b). If it applies to microbial food webs in the oceans, intensive recycling by DOC pathways would influence microbial diversity and community structure.

In summary, I conclude that grazers, particularly protozoa, are the dominant source of DOC for bacteria in oligotrophic oceanic waters. This

Table 5. Summary of data sources for parameterization of food flows presented in Figure 3

Process	Flow in the model (% total primary production)	Sources and remarks
Protozoan grazing	$f_{P,A} = 80$	25–>100 (subarctic Pacific, Landry et al. 1993) 90–104 (equatorial Pacific, Landry et al. 1997) 20–>100 (subtropical Pacific, Liu et al. 1995)
Zooplankton grazing	$f_{Z,A} = 5$	5–15 (equatorial Pacific, Roman and Gauzens 1997) 6–15 (subarctic Pacific, Dagg 1993) 9–26 (Sargasso Sea, Roman et al. 1993)
Algal extracellular release of DOC	$f_{DOC,A} = 12$	11 (oligotrophic Mediterranean, Hagström et al. 1988) 13 (average of many aquatic systems, Baines and Pace 1991)
Viral infection	$f_{V,A} = 3$	2–3 (*Synechococcus* community in coastal systems, Suttle 1994)
Bacterial production	$f_{P,B} + f_{V,B} = 12$	10–15 (subarctic Pacific, Kirchman et al. 1993) 10–15 (equatorial Pacific, Ducklow et al. 1995) 10–30 (many marine systems, Ducklow and Carlson 1992)
Protozoan removal and DOC production		See legend of Figure 4b
Zooplankton removal and DOC production		See legend of Figure 4c
Bacterial removal and respiration	$f_{P,B} = 6$ $f_{V,B} = 6$ $f_{CO_2,B} = 30$	Bacterial production is assumed to be balanced by protozoan grazing and viral infection, and these processes are assumed to contribute equally to the total mortality (Fuhrman and Noble 1995). Bacterial gross growth efficiency in the model (0.29) is within the range of current estimates (0.2–0.3, del Giorgio et al. 1997; see also Chapter 10).

Table 5. (*Continued*)

Process	Flow in the model (% total primary production)	Sources and remarks
Virus + host removal and DOC production	$f_{P,V} = 4$ $f_{DOC,V} = 5$	It is assumed that 56% of total host cell carbon is converted to DOC. This value is higher than that reported in a phytoflagellate–virus system (40%: Gobler et al. 1997; see the text). The higher value is used because many colloids (DOC) could be produced during viral infection of bacteria and cyanobacteria. Remains of host cells and some viral particles are assumed to be consumed by protozoa (Gonzales and Suttle 1993).
Food web export	$f_{FISH,Z} = 1$	0.5 (open oceans, Iverson 1990)
POC export (sinking flux)	$f_{POC,Z} = 6$	5 (Sargasso Sea, Carlson et al. 1994) 3 (equatorial Pacific, Murray et al. 1994) 14 (open ocean composite, Martin et al. 1987)
DOC export (advective flux)	$f_{R-DOC,DOC} = 7$	7 (Sargasso Sea, Carlson et al. 1994) 8 (equatorial Pacific, Peltzer and Hayward 1996)

Table 6. Relative contribution of different planktonic organisms to total DOC production in the food web depicted in Figure 4

Organisms	DOC Flux[a] (% of primary production)	Percent of Total DOC Production
Phytoplankton	12	25
Zooplankton	6	12
Protozoa	26	53
Viruses	5	10
Total	49	100

[a] $f_{DOC,X}$.

Table 7. Consumption of organic carbon by heterotrophs in the food web depicted in Figure 4[a]

Heterotrophs	Carbon Consumption (% of primary production)
Bacteria	42
Protozoa	90
Viruses	9
Zooplankton	33
Total	174

[a]Note that summed carbon consumption far exceeds primary production (100%) because of recycling of organic carbon within the food web (see text).

conclusion appears to be general to the extent that protozoa dominate herbivory. In coastal environments, different parameterization is required for modeling the DOC flux, which may result in different conclusions, probably with more emphasis on zooplankton grazers (P. J. le B. Williams 1981). I expect dynamic changes of flow patterns during phytoplankton blooms. Although the traditional paradigm of bloom–DOC dynamics has been strictly framed with the concept of "algal extracellular release," we should reevaluate this model by taking into account the contribution of viruses to DOC production during blooms (Bratbak et al. 1990; Gobler et al. 1997).

NEW PERSPECTIVES ON PRODUCTION PATHWAYS OF REFRACTORY DOM

Although a major part (> 90%) of DOM released to seawater is consumed and respired on a time scale of days or less, a certain component of DOM escapes rapid microbial degradation and accumulates in seawater. The accumulated pool of DOM is available for export by mixing and advection, influencing carbon mass balance of the oceans (Copin-Montegut and Avril 1993; Carlson et al. 1994; Pelzer and Hayward 1996; P. J. le B. Williams 1995). Also, oceanic DOC is one of the largest active reservoirs of organic carbon on earth (Hedges 1992). To better understand global carbon cycles, therefore, increasing attention has been paid to modeling DOM dynamics in the oceans (Siegenthaler and Sarmiento 1993; Yamanaka and Tajika 1997). Investigators are keen to characterize the DOM pool and to understand source and sink mechanisms of DOM. A large part of the DOM pool still remains to be identified, but recent advances in techniques of analyzing DOM have begun to provide remarkable information that suggest possible source mechanisms of the DOM pool in seawater. This section introduces emerging ideas on these topics within the scope of microbial food web mechanisms. Other perspectives of this exciting issue are discussed in Chapter 6 (see also Lee and Wakeham 1992; Benner 1998).

The oceanic DOC pool is dominated by highly refractory organic carbon, with an average age estimated to exceed 1000 years (P. M. Williams and Druffel 1987). Although there is a pool of labile DOC (turnover time < days) that supports bacterial production, it does not represent even 1% of total DOC (Carlson and Ducklow 1995). Traditionally, it has been assumed that refractory DOM consists mainly of complex macromolecules resulting from abiotic condensation (Harvey et al. 1983). This view has been challenged by recent findings that a significant portion of oceanic DOC is composed of unmodified forms of compounds with distinctive biochemical signatures (Aluwihare et al. 1997; McCarthy et al. 1997). These results suggest that selected components of biochemical compounds persist after more labile components have been degraded. In support of this hypothesis, Tanoue and coworkers (Tanoue et al. 1995, 1996; Suzuki et al. 1997) demonstrated that bacterial membrane proteins dominate dissolved protein molecules in a wide variety of oceanic waters, including surface and intermediate layers. Furthermore, McCarthy et al. (1998) have suggested that remnants of bacterial cell wall components, specifically peptidoglycan, constitute a major fraction of high molecular weight dissolved organic nitrogen in both surface and deep oceanic waters. These surprising new findings suggest that bacteria are a major proximate "source" of DOM in the oceans and that bacteria-derived organic matter can be preserved for periods longer than the time scales of the circulation of intermediate and deep water masses (> 100 years). How can we reconcile this view with the observation that bacteria are the "sink" of DOM? What mechanisms are involved in the production and preservation of bacterial organic matter? Although much remains to be done before these important questions can be answered, experimental studies have begun to uncover critical processes governing bacteria–refractory DOM cycles.

Structural components of bacterial cells including membranes and peptidoglycan can be introduced to seawater as DOM during bacterial death due to protozoan grazing and viral infection (Nagata and Kirchman 1999). More precisely, the released materials are most likely part of small particles (colloids), but they are operationally classified as DOM in oceanography (Nagata and Kirchman 1997). Nagata and Kirchman (1992b) suggested that flagellates grazing on bacteria release a part of prey cell components as liposome-like particles (i.e., phospholipid particles with an aqueous center). The authors hypothesized that these particles are an important source of refractory DOM in seawater. In support of this hypothesis, experiments have revealed that organic matter associated with structurally complex, liposome-like particles are protected from rapid enzymatic attack in seawater (Nagata et al. 1998; Borch and Kirchman 1999). Liposome-like particles are likely formed during lysis of bacteria by viruses (Shibata et al. 1997), providing a further source mechanism of refractory DOM derived from bacteria. If the preceding scenario is generally correct, the intensive cyclic flows of organic matter along DOC → B →

$P \rightarrow DOC$ and $DOC \rightarrow B \rightarrow V \rightarrow DOC$ pathways (Figure 4a) could be regarded as a "concentrating" mechanism of refractory DOM in oceanic waters. Furthermore, photosynthetic prokaryotes (*Synecchococcus* and prochlorophytes), which have the same cell structures as heterotrophic bacteria, dominate primary production in some oceanic regions such as subtropical and equatorial waters (Campbell and Vaulot 1993), providing additional routes ($A \rightarrow P \rightarrow DOC$ and $A \rightarrow V \rightarrow DOC$ in Figure 4a) for the production of refractory DOM bearing a bacterial signature.

We still know very little about the production mechanisms of refractory DOM in the oceans, particularly about how low molecular weight refractory DOM is produced and preserved (Amon and Benner 1994). It should be pointed out that besides the source mechanisms already discussed, the formation of refractory DOM can be affected by several biotic and abiotic processes, including the direct release by bacteria and phytoplankton of refractory materials (Brophy and Carlson 1989; Tranvik 1993; Aluwihare et al. 1997; Stoderegger and Herndl 1998), diagenetic reactions of labile precursor molecules (Harvey et al. 1983; Hedges 1988; Keil and Kirchman 1994; Nagata and Kirchman 1996), and photochemical reactions of organic matter (Keil and Kirchman 1994; Benner and Biddanda 1998). However, in any event, the production, preservation and degradation of DOM in the oceans cannot be fully understood apart from the framework of the microbial food web, which is the central regulatory element of oceanic DOM dynamics (Figure 4).

CONCLUSIONS

The discussion in this chapter has been guided by the spirit of Pomeroy (1974), P. J. le B. Williams (1981), and Azam et al. (1983), which suggests a sound combination of reductionism and holism in this field of research. That my conclusions sometimes deviate substantially from earlier studies reflects our evolving understanding of microbial processes in the oceans.

1. In cultures, exponentially growing phytoplankton release on average 5% of total primary production as DOC. This value tends to be higher when algal cells have been exposed to suboptimum conditions. In natural environments, the extracellular release of DOC typically accounts for about 10% of primary production, which is not sufficient to fulfill bacterial carbon demand.

2. Protozoan grazers can release 20–30% of ingested prey organic carbon as DOC. The corresponding values for metazoan zooplankton are 10–20%. Egestion of unassimilated prey material appears to be a major mechanism of DOM release by grazers.

3. Viral infection of host cells (phytoplankton and bacteria) may result in substantial release of DOM. Because viruses can be a significant cause of

mortality for bacteria and phytoplankton, the contribution of viruses to DOM production in seawater is potentially large.

4. A trophic flow model for oligotrophic open oceans suggests that grazers, particularly protozoa, are the major contributor to DOM production. This type of oceanic ecosystem is characterized by intensive recycling of organic carbon through DOM–microbial food chains.

5. Geochemical studies have suggested that a substantial portion of high molecular weight refractory DOM in oceanic water is derived from bacteria. The release into seawater of structural components of bacterial cells during protozoan grazing and viral infection could be a key mechanism that controls refractory DOM cycles in the oceans.

REFERENCES

Al-Hasan, R. H., and Fogg, G. E. (1987) Glycollate concentrations in relation to hydrography in Liverpool Bay. *Mar. Ecol. Prog. Ser.* 37:305–307.

Aluwihare, L. I., Repeta, D. J., and Chen, R. F. (1997) A major biopolymeric component to dissolved organic carbon in surface sea water. *Nature* 387:166–169.

Ammerman, J. W., and Azam, F. (1985) Bacterial 5'-nucleotidase in aquatic ecosystems: A novel mechanism of phosphorus regeneration. *Science* 227:1338–1340.

Amon, R. M. W., and Benner, R. (1994) Rapid cycling of high-molecular-weight dissolved organic matter in the ocean. *Nature* 369:549–552.

Andersen, O. K., Goldman, J. C., Caron, D. A., and Dennett, M. R. (1986) Nutrient cycling in a microflagellate food chain. III. Phosphorus dynamics. *Mar. Ecol. Prog. Ser.* 31:47–55.

Anderson, G. C., and Zeutschel, R. P. (1970) Release of dissolved organic matter by marine phytoplankton in coastal and offshore areas of the northeast Pacfic Ocean. *Limnol. Oceanog.* 15:402–407.

Andersson, A., Lee, C., Azam, F., and Hagström, Å. (1985) Release of amino acids and inorganic nutrients by heterotrophic marine microflagellates. *Mar. Ecol. Prog. Ser.* 23:99–106.

Azam, F., Fenchel, T., Field, J. G., Gray, J. S., Meyer-Reil, L. A., and Thingstad, F. (1983) The ecological role of water-column microbes in the sea. *Mar. Ecol. Prog. Ser.* 10:257–263.

Baines, S. B., and Pace, M. L. (1991) The production of dissolved organic matter by phytoplankton and its importance to bacteria: Patterns across marine and freshwater systems. *Limnol. Oceanogr.* 36:1078–1090.

Beardall, J. (1989) Photosynthesis and photorespiration in marine phytoplankton. *Aquat. Bot.* 34:105–130.

Benner, R. (1998) Cycling of dissolved organic matter in the ocean. In D. Hessen and L. Tranvik, eds., *Aquatic Humic Substances: Ecology and Biogeochemistry.* Springer-Verlag, Berlin, pp. 317–331.

Benner, R., and Biddanda, B. (1998) Photochemical transformations of surface and deep marine dissolved organic matter: Effects on bacterial growth. *Limnol. Oceanogr.* 43:1373–1378.

Berman, T., and Holm-Hansen, O. (1974) Release of photoassimilated carbon as dissolved organic matter by marine phytoplankton. *Mar. Biol.* 28:305–310.

Biddanda, B., and Benner, R. (1997) Carbon, nitrogen, and carbohydrate fluxes during the production of particulate and dissolved organic matter by marine phytoplankton. *Limnol. Oceanogr.* 42:506–518.

Bidigare, R. R. (1983) Nitrogen excretion by marine zooplankton. In E. J. Carpenter and D. G. Capone, eds. *Nitrogen in the Marine Environment.* Academic Press, New York, pp. 385–409.

Bjørnsen, P. K. (1988) Phytoplankton exudation of organic matter: Why do healthy cells do it? *Limnol. Oceanogr.* 33:151–154.

Borch, N. H., and Kirchman, D. L. (1999) Protection of protein from bacterial degradation by submicron particles. *Aquat. Microb. Ecol.* 16:265–272.

Bratbak, G., Heldal, M., Norland, S., and Thingstad, T. F. (1990) Viruses as partners in spring bloom microbial trophodynamics. *Appl. Environ. Microbiol.* 56:1400–1405.

Bronk, D. A., and Glibert, P. M. (1993) Contrasting patterns of dissolved organic nitrogen release by two size fractions of estuarine plankton during a period of rapid NH_4^+ consumption and NO_2^- production. *Mar. Ecol. Prog. Ser.* 96:291–299.

Brophy, J. E., and Carlson, D. J. (1989) Production of biologically refractory dissolved organic carbon by natural seawater microbial populations. *Deep-Sea Res.* 36:497–507.

Campbell, L., and Vaulot, D. (1993) Photosynthetic picoplankton community structure in the subtropical North Pacific Ocean near Hawaii (Station ALOHA). *Deep-Sea Res.* 40:2043–2060.

Capriulo, G. M., and Degnan, C. (1991) Effect of food concentration on digestion and vacuole passage time in the heterotrichous marine ciliate *Fibrea salina. Mar. Biol.* 110:199–202.

Carlson, C. A., and Ducklow, H. W. (1995) Dissolved organic carbon in the upper ocean of the central equatorial Pacific Ocean. *Deep-Sea Res.* 42:639–656.

Carlson, C. A., Ducklow, H. W., and Michaels, A. F. (1994) Annual flux of dissolved organic carbon from the euphotic zone in the northwestern Sargasso Sea. *Nature* 371:405–408.

Caron, D. A., and Goldman, J. C. (1990) Protozoan nutrient regeneration. In G. M. Capriula, ed., *Ecology of Marine Protozoa.* Oxford University Press, New York, pp. 283–306.

Caron, D. A., Goldman, J. C., Andersen, O. K., and Dennett, M. R. (1985) Nutrient cycling in a microflagellate food chain. II. Population dynamics and carbon cycling. *Mar. Ecol. Prog. Ser.* 24:243–254.

Caron, D. A., Porter, K. G., and Sanders, R. W. (1990) Carbon, nitrogen and phosphorus budgets for the mixotrophic phytoflagellate *Poterioochromonas malhamensis* (Chrysophyceae) during bacterial ingestion. *Limnol. Oceanogr.* 35:433–443.

Chase, Z., and Price, N. M. (1997) Metabolic consequences of iron deficiency in heterotrophic marine protozoa. *Limnol. Oceanogr.* 42:1673–1684.

Cole, G. T., and Wynne, M. J. (1974) Endocytosis of *Microcystis aeruginosa* by *Ochromonas danica. J. Phycol.* 10:397–410.

Cole, J. J., Findlay, S., and Pace, M. L. (1988) Bacterial production in fresh and saltwater ecosystems: A cross-system overview. *Mar. Ecol. Prog. Ser.* 43:1–10.

Conover, R. J. (1966) Feeding on large particles by *Calanus hyperboreus* (Kroyer). In H. Barnes, ed., *Some Contemporary Studies in Marine Science*. Allen & Unwin, London, pp. 187–194.

Copin-Montegut, G., and Avril, B (1993) Vertical distribution and temporal variation of dissolved organic carbon in the north-western Mediterranean Sea. *Deep-Sea Res.* 40:1963–1972.

Copping, A. E., and Lorenzen, C. J. (1980) Carbon budget of a marine phytoplankton–herbivore system with carbon-14 as a tracer. *Limnol. Oceanogr.* 25:873–882.

Corner, E. D. S., and Davies, A. G. (1971) Plankton as a factor in the nitrogen and phosphorus cycles in the sea. *Adv. Mar. Biol.* 9:101–204.

Cottrell, M. T., and Suttle, C. A. (1991) Wide-spread occurrence and clonal variation in viruses which cause lysis of a cosmopolitan, eukaryotic marine phytoplankter, *Micromonas pusilla*. *Mar. Ecol. Prog. Ser.* 78:1–9.

Dagg, M. J. (1993) Grazing by the copepod community does not control phytoplankton production in the subarctic Pacific Ocean. *Prog. Oceanogr.* 32:163–183.

Del Giorgio, P. A., Cole, J. J., and Cimbleris, A. (1997) Respiration rates in bacteria exceed phytoplankton production in unproductive aquatic systems. *Nature* 385:148–151.

Dolan, J. R. (1997) Phosphorus and ammonia excretion by planktonic protists. *Mar. Geol.* 139:109–122.

Dolan, J. R., and Simek, K. (1997) Processing of ingested matter in *Strombidium sulcatum*, a marine ciliate (Oligotrichida). *Limnol. Oceanogr.* 42:393–397.

Ducklow, H. W., and Carlson, C. A. (1992) Oceanic bacterial production. *Adv. Microb. Ecol.* 12:113–181.

Ducklow, H. W., Quinby, H. L., and Carlson, C. A. (1995) Bacterioplankton dynamics in the equatorial Pacific during the 1992 El Niño. *Deep-Sea Res.* 42:621–638.

Eppley, R. W., and Sloan, P. R. (1965) Carbon balance experiments with marine phytoplankton. *J. Fish. Res. Board Can.* 22:1083–1097.

Eppley, R. W., Horrigan, S. G., Fuhrman, J. A., Brooks, E. R., Price, C. C., and Sellner, K. (1981) Origins of dissolved organic matter in southern California coastal waters: Experiments on the role of zooplankton. *Mar. Ecol. Prog. Ser.* 6:149–159.

Fenchel, T. (1982) Ecology of heterotrophic microflagellates. II. Bioenergetics and growth. *Mar. Ecol. Prog. Ser.* 8:225–231.

Fenchel, T. (1986) The ecology of heterotrophic microflagellates. *Adv. Microb. Ecol.* 9:57–97.

Fenchel, T. (1987) *Ecology of Protozoa: The Biology of Free-Living Phagotrophic Protists*. Springer-Verlag, Berlin.

Ferrier-Pagès, C., Karner, M., and Rassoulzadegan, F. (1998) Release of dissolved amino acids by flagellates and ciliates grazing on bacteria. *Oceanol. Acta* 21:485–494.

Fogg, G. E. (1983) The ecological significance of extracellular products of phytoplankton photosynthesis. *Bot. Mar.* 26:3–14.

Fuhrman, J. (1987) Close coupling between release and uptake of dissolved free amino acids in seawater studied by an isotope dilution approach. *Mar. Ecol. Prog. Ser.* 37:45–52.

Fuhrman, J. (1992) Bacterioplankton roles in cycling of organic matter: The microbial food web. In P. G. Falkowski and D. Woodhead, eds., *Primary Productivity and Biogeochemical Cycles in the Sea.* Plenum Press, New York, pp. 361–383.

Fuhrman, J., and Noble, R. T. (1995) Viruses and protists cause similar bacterial mortality in, coastal seawater. *Limnol. Oceanogr.* 40:1236–1242.

Gardner, W. S., and Paffenhöfer, G. A. (1982) Nitrogen regeneration by the subtropical marine copepod *Eucalanus pileatus. J. Plankton Res.* 4:725–734.

Gaudy, R. (1974) Feeding four species of pelagic copepods under experimental conditions. *Mar. Biol.* 25:125–141.

Gobler, C. J., Hutchins, D. A., Fisher, N. S., Cosper, E. M., and Sanudo-Wilhelmy, S. A. (1997) Release and bioavailability of C, N, P, Se., and Fe following viral lysis of a marine chrysophyte. *Limnol. Oceanogr.* 42:1492–1504.

Goldman, J. C., Caron, D. A., Andersen, O. K., and Dennett, M. R. (1985) Nutrient cycling in a microflagellate food chain: I. Nitrogen dynamics. *Mar. Ecol. Prog. Ser.* 24:231–242.

Gonzales, J. M., and Suttle, C. A. (1993) Grazing by marine nanoflagellates on viruses and virus-sized particles: Ingestion and digestion. *Mar. Ecol. Prog. Ser.* 94:1–10.

Gosselin, M., Levasseur, M., Wheeler, P. A., Horner, R. A., and Booth, B. C. (1998) New measurements of phytoplankton and ice algal production in the Arctic Ocean. *Deep-Sea Res.* 44:1623–1644.

Gowning, M. M., and Silver, M. W. (1985) Minipellets: A new and abundant size class of marine fecal pellets. *J. Mar. Res.* 43:395–418.

Guillard, R. R. L., and Hellebust, J. A. (1971) Growth and the production of extracellular substances by two strains of *Phaeocystis poucheti. J. Phycol.* 7:330–338.

Guillard, R. R. L., and Wangersky, P. J. (1958) The production of extracellular carbohydrates by some marine flagellates. *Limnol. Oceanogr.* 3:449–454.

Hagström, Å., Azam, F., Andersson, A., Wikner, J., and Rassoulzadegan, F. (1988) Microbial loop in an oligotrophic pelagic marine ecosystem: Possible roles of cyanobacteria and nanoflagellates in the organic fluxes. *Mar. Ecol. Prog. Ser.* 49:171–178.

Harvey, G. R., Boran, D. A., Chesal, L. A., and Tokar, J. M. (1983) The structure of marine fulvic and humic acids. *Mar. Chem.* 12:119–132.

Hedges, J. I. (1988) Polymerization of humic substances in natural environments. In F. H. Frimmel and R. F. Christman, eds., *Humic Substances and Their Role in the Environment,* Wiley, New York, pp. 45–58.

Hedges, J. I. (1992) Global biogeochemical cycles: Progress and problems. *Mar. Chem.* 39:67–93.

Hellebust, J. A. (1965) Excretion of some organic compounds by marine phytoplankton. *Limnol. Oceanogr.* 10:192–206.

Higashi, M., Burns, T. P., and Patten, B. C. (1993a) Network trophic dynamics: The modes of energy utilization in ecosystems. *Ecol. Modelling* 66:1–42.

Higashi, M., Burns, T. P., and Patten, B. C. (1993b) Network trophic dynamics: The tempo of energy movement and availability in ecosystems. *Ecol. Modelling* 66:43–64.

Hutchins, D. A. (1995) Iron and the marine phytoplankton community. *Prog. Phycol. Res.* 11:1–48.

Hutchins, D. A., and Bruland, K. W. (1994) Grazer-mediated regeneration and assimilation of Fe, Zn and Mn from planktonic prey. *Mar. Ecol. Prog. Ser.* 110:259–269.

Ignatiades, L., and Fogg, G. E. (1973) Studies on the factors affecting the release of organic matter by *Skeletonema costatum* (Greville) Cleve in culture. *J. Mar. Biol. Assoc. U.K.* 53:937–956.

Ittekkot, V., Brockmann, U., Michaelis, W., and Degens, E. T. (1981) Dissolved free and combined carbohydrates during a phytoplankton bloom in the northern North Sea. *Mar. Ecol. Prog. Ser.* 4:299–305.

Iverson, R. L. (1990) Control of marine fish production. *Limnol. Oceanogr.* 35:1593–1604.

Jensen, L. M. (1983) Phytoplankton release of extracellular organic carbon, molecular weight composition, and bacterial assimilation. *Mar. Ecol. Prog. Ser.* 11:39–48.

Johanness, R. E. (1965) Influence of marine protozoa on nutrient regeneration. *Limnol. Oceanogr.* 10:434–442.

Jones, D. R., Karl, D. M., and Laws, E. A. (1996) Growth rates and production of heterotrophic bacteria and phytoplankton in the North Pacific subtropical gyre. *Deep-Sea Res.* 43:1567–1580.

Jumars, P. A., Penry, D. L., Baross, J. A., Perry, M. J., and Frost, B. W. (1989) Closing the microbial loop: Dissolved carbon pathway to heterotrophic bacteria from incomplete ingestion, digestion and absorption in animals. *Deep-Sea Res.* 36:483–495.

Keil, R. G., and Kirchman, D. L. (1991) Contribution of dissolved free amino acids and ammonium to the nitrogen requirements of heterotrophic bacterioplankton. *Mar. Ecol. Prog. Ser.* 73:1–10.

Keil, R. G., and Kirchman, D. L. (1994) Abiotic transformation of labile protein to refractory protein in seawater. *Mar. Chem.* 45:187–196.

Kirchman, D. L., Keil, R. G., Simon, M., and Welschmeyer, N. A. (1993) Biomass and production of heterotrophic bacterioplankton in the oceanic subarctic Pacific. *Deep-Sea Res.* 40:967–988.

Lampert, W. (1978) Release of dissolved organic carbon by grazing zooplankton. *Limnol. Oceanogr.* 23:831–834.

Lampitt, R. S., Noji, T., and von Bodungen, B. (1990) What happens to zooplankton faecal pellets? Implication for material flux. *Mar. Biol.* 104:15–23.

Lancelot, C. (1979) Gross excretion rates on natural marine phytoplankton and heterotrophic uptake of excreted products in the southern North Sea, as determined by short-term kinetics. *Mar. Ecol. Prog. Ser.* 1:179–186.

Lancelot, C. (1983) Factors affecting phytoplankton extracellular release in the Southern Bight of the North Sea. *Mar. Ecol. Prog. Ser.* 12:115–121.

Landry, M. R., Monger, B. C., and Selph, K. E. (1993) Time-dependency of microzooplankton grazing and phytoplankton growth in the subarctic Pacific. *Prog. Oceanogr.* 32:205–222.

Landry, M. R., Barber, R. T., Bidigare, R. R., Chai, R., Coale, K. H., Dam, H. G., Lewis, M. R., Lindley, S. T., McCarthy, J. J., Roman, M. R., Stoecker, D. K., Verity, P. G., and White, J. R. (1997) Iron and grazing constraints on primary production in the central equatorial Pacific: An EqPac synthesis. *Limnol. Oceanogr.* 42:405–418.

Larsson, U., and Hagström, Å. (1982) Fractionated phytoplankton primary production, exudate release and bacterial production in a Baltic eutrophication gradient. *Mar. Biol.* 67:57–70.

Laws, E., and Caperon, J. (1976) Carbon and nitrogen metabolism by *Monochrysis lutheri*: Measurement of growth-rate-dependent respiration rates. *Mar. Biol.* 36:85–97.

Leboulanger, C., Oriol, L., Jupin, H., and Descolas-Gros, C. (1997) Diel variability of glycolate in the eastern tropical Atlantic Ocean. *Deep-Sea Res.* 44:2131–2139.

Lee, C., and Wakeham, S. G. (1992) Organic matter in the water column: Future research challenges. *Mar. Chem.* 39:95–118.

Legendre, L., and Rassoulzadegan, F. (1996) Food-web mediated export of biogenic carbon in oceans: Hydrodynamic control. *Mar. Ecol. Prog. Ser.* 145:179–193.

Lignell, R. (1990) Excretion of organic carbon by phytoplankton: Its relation to algal biomass, primary productivity and bacterial secondary productivity in the Baltic Sea. *Mar. Ecol. Prog. Ser.* 68:85–99.

Liu, H., Campbell, L., and Landry, M. R. (1995) Growth and mortality rates of *Prochlorococcus and Synechococcus* measured with a selective inhibitor technique. *Mar. Ecol. Prog. Ser.* 116:277–287.

Mague, T. H., Friberg, E., Hughes, D. J., and Morris, I. (1980) Extracellular release of carbon by marine phytoplankton: A physiological approach. *Limnol. Oceanogr.* 25:262–279.

Martin, J. H., Knauer, G. A., Karl, D. M., and Broenkow, W. W. (1987) VERTEX: carbon cycling in the Northeast Pacific. *Deep-Sea Res.* 34:267–285.

Martin, J. H., Gordon, R. M., and Fitzwater, S. E. (1991) The case for iron. *Limnol. Oceanogr.* 36:1793–1802.

McCarthy, M., Pratum, T., Hedges, J., and Benner, R. (1997) Chemical composition of dissolved organic nitrogen in the ocean. *Nature* 390:150–154.

McCarthy, M. D., Hedges, J. I., and Benner, R. (1998) Major bacterial contribution to marine dissolved organic nitrogen. *Science* 281:231–234.

Miller, C. B., Frost, B. W., Wheeler, P. A., Landry, M. R., Welschmeyer, N., and Powell, T. M. (1991) Ecological dynamics in the subarctic Pacific, a possibly iron-limited ecosystem. *Limnol. Oceanogr.* 36:1600–1615.

Murray, J. W., Barber, R. T., Roman, M. R., Bacon, M. P., and Feely, R. A. (1994) Physical and biological controls on carbon cycling in the equatorial Pacific. *Science* 266:58–65.

Myklestad, S., Holm-Hanse, O., Varum, K. M., and Volcani, B. E. (1989) Rate of release of extracellular amino acids and carbohydrates from the marine diatom *Chaetoceros affinis*. *J. Plankton Res.* 11:763–773.

Nagasaki, K., Ando, M., Itakura, S., Imai, I., and Ishida, Y. (1994) Viral mortality in the final stage of *Heterosigma akashiwo* (Raphidophyceae) red tide. *J Plankton Res.* 16:1595–1599.

Nagata, T. (in press) "Picopellets" produced by phagotrophic nanoflagellates: Role in the material cycling within marine environments. *Dynamics and Characterization of Marine Organic Matter*. In N. Handa, E. Tanoue, and T. Hama, eds., Terra Scientific Publishing, Tokyo.

Nagata, T., and Kirchman, D. L. (1990) Filtration-induced release of dissolved free amino acids: Application to cultures of marine protozoa. *Mar. Ecol. Prog. Ser.* 68:1–5.

Nagata, T., and Kirchman, D. L. (1991) Release of dissolved free and combined amino acids by bacterivorous marine flagellates. *Limnol. Oceanogr.* 36:433–443.

Nagata, T., and Kirchman, D. L. (1992a) Release of dissolved organic matter by heterotrophic protozoa: Implications for microbial food webs. *Arch. Hydrobiol. Beih. Ergebn. Limnol.* 35:99–109.

Nagata, T., and Kirchman, D. L. (1992b) Release of macromolecular organic complexes by heteterotrophic marine flagellates. *Mar. Ecol. Prog. Ser.* 83:233–240.

Nagata, T., and Kirchman, D. L. (1996) Bacterial degradation of protein adsorbed to model submicron particles in seawater. *Mar. Ecol. Prog. Ser.* 132:241–248.

Nagata, T., and Kirchman, D. L. (1997) Roles of submicron particles and colloids in microbial food webs and biogeochemical cycles within marine environments. *Adv. Microb. Ecol.* 15:81–103.

Nagata, T., and Kirchman, D. L. (1999) Bacterial mortality: A pathway for the formation of refractory DOM? *New Frontiers in Microbial Ecology: Proceedings of the Eighth International Symposium on Microbial Ecology.* In M. Brylinsky, C. Bell, and P. Johnson-Green, eds., Atlantic Canada Society for Microbial Ecology, Halifax.

Nagata, T., Fukuda, R., Koike, I., Kogure, K., and Kirchman, D. L. (1998) Degradation by bacteria of membrane and soluble protein in seawater. *Aquat. Microb. Ecol.* 14:29–37.

Norrman, B., Zweifel, U., Li, W., Hopkinson, C. S., and Fry, B. (1995) Production and utilization of dissolved organic carbon during an experimental diatom bloom. *Limnol. Oceanogr.* 40:898–907.

Obernosterer, I., and Herndl, G. J. (1995) Phytoplankton extracellular release and bacterial growth: Dependence on the inorganic N:P ratio. *Mar. Ecol. Prog. Ser.* 116:247–257.

Passow, U., Alldredge, A. L., and Logan, B. E. (1994) The role of particulate carbohydrate exudates in the flocculation of diatom blooms. *Deep-Sea Res.* 41:335–357.

Peduzzi, P., and Herndl, G. J. (1992) Zooplankton activity fueling the microbial loop: Differential growth response of bacteria from oligotrophic and eutrophic waters. *Limnol. Oceanogr.* 37:1087–1092.

Pelegrí, S. P., Christaki, U., Dolan, J., and Rassoulzadegan, F. (1999) Particulate and dissolved organic carbon production by the hetrotrophic nanoflagellate *Pteridomonas danica*, Patterson and Fenchel 1985, fed *Escherichia coli*. *Microb. Ecol.* 37:276–284.

Peltzer, E. T., and Hayward, N. A. (1996) Spatial and temporal variability of total organic carbon along 140°W in the equatorial Pacific Ocean in 1992. *Deep-Sea Res.* 43:1155–1180.

Pomeroy, L. R. (1974) The ocean's food web, a changing paradigm. *BioScience* 36:310–315.

Proctor, L. M., and Fuhrman, J. A. (1990) Viral mortality of marine bacteria and cyanobacteria. *Nature* 343:60–62.

Riemann, B., Jørgensen, N. O. G., Lampert, W., and Fuhrman, J. A. (1986) Zooplankton-induced changes in dissolved free amino acids and in production rates of freshwater bacteria. *Microb. Ecol.* 12:247–258.

Roman, M. R., and Gauzens, A. L. (1997) Copepod grazing in the equatorial Pacific. *Limnol. Oceanogr.* 42:623–634.

Roman, M. R., Ducklow, H. W., Fuhrman, J. A., Garside, C., Glibert, P. M., Malone, T. C., and McManus, G. B. (1988) Production, consumption and nutrient cycling in a laboratory mesocosm. *Mar. Ecol. Prog. Ser.* 42:39–52.

Roman, M. R., Dam, H. G., Gauzens, A. L., and Napp, J. M. (1993) Zooplankton biomass and grazing at the JGOFS Sargasso Sea time series station. *Deep-Sea Res.* 40:883–901.

Scavia, D. (1988) On the role of bacteria in secondary production. *Limnol. Oceanogr.* 33:1220–1224.

Sellner, K. G. (1981) Primary productivity and the flux of dissolved organic matter in several marine environments. *Mar. Biol.* 65:101–112.

Sharp, J. H. (1977) Excretion of organic matter by marine phytoplankton: Do healthy cells do it? *Limnol. Oceanogr.* 22:381–399.

Sherr, E. B., and Sherr, B. F. (1994) Bacterivory and herbivory: Key roles of phagotrophic protists in pelagic food webs. *Microb. Ecol.* 28:223–235.

Sherr, B. F., Sherr, E. B., and Rassoulzadegan, F. (1988) Rates of digestion of bacteria by marine phagotrophic protozoa: Temperature dependence. *Appl. Environ. Microbiol.* 54:1091–1095.

Shibata, A., Kogure, K., Koike, I., and Ohwada, K. (1997) Formation of submicron colloidal particles from marine bacteria by viral infection. *Mar. Ecol. Prog. Ser.* 155:303–307.

Siegenthaler, U., and Sarmiento, J. L. (1993) Atmospheric carbon dioxide and the ocean. *Nature* 365:119–125.

Smith, R. E. H., and Platt, T. (1984) Carbon exchange and ^{14}C tracer methods in a nitrogen-limited diatom, *Thalassiosira pseudonana*. *Mar. Ecol. Prog. Ser.* 16:75–87.

Smith, W. O., Barber, R. T., and Huntsman, S. A. (1977) Primary production off the coast of northwest Africa: Excretion of dissolved organic matter and its heterotrophic uptake. *Deep-Sea Res.* 24:35–47.

Stoderegger, K., and Herndl, G. J. (1998) Production and release of bacterial capsular material and its subsequent utilization by marine bacterioplankton. *Limnol. Oceanogr.* 43:877–884.

Stoecker, D. K. (1984) Particle production by planktonic ciliates. *Limnol. Oceanogr.* 29:930–940.

Straile, D. (1997) Gross growth efficiencies of protozoan and metazoan zooplankton and their dependence on food concentration, predator–prey weight ratio, and taxonomic group. *Limnol. Oceanogr.* 42:1375–1385.

Strayer, D. (1988) On the limits to secondary production. *Limnol. Oceanogr.* 33:1217–1220.

Strom, S. L., Benner, R., Ziegler, S., and Dagg, M. J. (1997) Planktonic grazers are a potentially important source of marine dissolved organic carbon. *Limnol. Oceanogr.* 42:1364–1374.

Strom, S. L., Morello, T. A., and Bright, K. J. (1998) Protozoan size influences algal pigment degradation during grazing. *Mar. Ecol. Prog. Ser.* 164:189–197.

Suttle, C. A. (1994) The significance of viruses to mortality in aquatic microbial communities. *Microb. Ecol.* 28:237–243.

Suttle, C. A., Chan, A. M., and Cottrel, M. T. (1990) Infection of phytoplankton by viruses and reduction of primary productivity. *Nature* 347:467–469.

Suzuki, S., Kogure, K., and Tanoue, E. (1997) Immunochemical detection of dissolved proteins and their source bacteria in marine environments. *Mar. Ecol. Prog. Ser.* 158:1–9.

Tanoue, E., Nishiyama, S., Kamo, M., and Tsugita, A. (1995) Bacterial membranes: Possible source of a major dissolved protein in seawater. *Geochim. Cosmochim. Acta* 59:2643–2648.

Tanoue, E., Ishii, M., and Midorikawa, T. (1996) Discrete dissolved and particulate proteins in oceanic waters. *Limnol. Oceanogr.* 41:1334–1343.

Taylor, G. T., Iturriaga, R., and Sullivan, C. W. (1985) Interactions of bacterivorous grazers and heterotrophic bacteria with dissolved organic matter. *Mar. Ecol. Prog. Ser.* 23:129–141.

Thomas, J. P. (1971) Release of dissolved organic matter from natural populations of marine phytoplankton. *Mar. Biol.* 11:311–323.

Tortell, P. D., Maldonado, M. T., and Price, N. M. (1996) The role of heterotrophic bacteria in iron-limited ocean ecosystems. *Nature* 383:330–332.

Tranvik, L. J. (1993) Microbial transformation of labile dissolved organic matter into humic-like matter in seawater. *FEMS Microbiol. Ecol.* 12:177–183.

Turk, V., Rehnstam, A.-S., Lundberg, E., and Hagström, Å. (1992) Release of bacterial DNA by marine nanoflagellates, an intermediate step in phosphorus regeneration. *Appl. Environ. Microbiol.* 58:3744–3750.

Twiss, M. R., and Campbell, P. G. C. (1995) Regeneration of trace metals from picoplankton by nanoflagellate grazing. *Limnol. Oceanogr.* 40:1418–1429.

Verity, P. G. (1981) Effects of temperature, irradiance, and daylength on the marine diatom *Leptocylindrus danicus* Cleve. II. Excretion. *J. Exp. Mar. Biol. Ecol.* 55:159–169.

Watanabe, Y. (1980) A study of the excretion and extracellular products of natural phytoplankton in Lake Nakanuma, Japan. *Int. Rev. Ges. Hydrobiol.* 65:809–834.

Webb, K. L., and Johannes, R. E. (1967) Studies of the release of dissolved free amino acids by marine zooplankton. *Limnol. Oceanogr.* 12:376–382.

Weinbauer, M. G., and Peduzzi, P. (1995) Effect of virus-rich high molecular weight concentrates of seawater on the dynamics of dissolved amino acids and carbohydrates. *Mar. Ecol. Prog. Ser.* 127:245–253.

White, J. R., and Roman, M. R. (1992) Seasonal study of grazing by metazoan zooplankton in the mesohaline Chesapeake Bay. *Mar. Ecol. Prog. Ser.* 86:251–261.

Williams, P. J. le B. (1981) Incorporation of microheterotrophic processes into the classical paradigm of the planktonic food web. *Kieler Meeresforsch.* 5:1–28.

Williams, P. J. le B. (1990) The importance of losses during microbial growth: Commentary on the physiology, measurement and ecology of the release of dissolved organic material. *Mar. Microb. Food Webs* 4:175–206.

Williams, P. J le B. (1995) Evidence for the seasonal accumulation of carbon-rich dissolved organic material, its scale in comparison with changes in particulate material and consequential effect on net C/N assimilation ratios. *Mar. Chem.* 51:17–29.

Williams, P. J. le B., and Yentsch, C. S. (1976) An examination of photosynthetic production, exrection of photosynthetic products, and heterotrophic utilization of dissolved organic compounds with reference to results from a coastal subtropical sea. *Mar. Biol.* 35:31–40.

Williams, P. M., and Druffel, E. R. M. (1987) Radiocarbon in dissolved organic matter in the central North Pacific Ocean. *Nature* 330:246–248.

Williams, R., and Poulet, S. A. (1986) Relationship between the zooplankton, phytoplankton, particulate matter and dissolved free amino acids in the Celtic Sea. *Mar. Biol.* 90:279–284.

Wolter, K. (1982) Bacterial incorporation of organic substances released by natural phytoplankton populations. *Mar. Ecol. Prog. Ser.* 7:287–295.

Wood, A. M., and Van Valen, L. M. (1990) Paradox lost? On the release of energy-rich compounds by phytoplankton. *Mar. Microb. Food Webs* 4:103–116.

Yamanaka, Y., and Tajika, E. (1997) Role of dissolved organic matter in the marine biogeochemical cycle: Studies using an ocean biogeochemical general circulation model. *Global Biogeochem. Cycles* 11:599–612.

Zlotnik, I., and Dubinsky, Z. (1989) The effect of light and temperature on DOC excretion by phytoplankton. *Limnol. Oceanogr.* 34:831–839.

6

HETEROTROPHIC BACTERIA AND THE DYNAMICS OF DISSOLVED ORGANIC MATERIAL

Peter J. le B. Williams

School of Ocean Sciences,
Marine Sciences Laboratory,
University of Wales, Bangor, United Kingdom

WHAT IS ORGANIC AND WHAT IS DISSOLVED

The term *organic* is not well defined. Historically the term meant "pertaining to plant or animal organisms" (Feiser and Feiser 1956). Lavoisier's pioneering work on combustion in the late 1700s established that carbon is present in all organic compounds. Until this point, a better understanding of microbial processes should have led to the placing of ammonia in the organic category. Wöhler's synthesis of urea in 1828 (incidentally, not the first organic compound to be synthesized: that was oxalic acid) was critical in refuting the concept of "vital force," a property believed to be unique to biological production. With the advent of the synthesis of carbon-containing compounds that were new to nature, the original definition of an organic compound as synthesized by organisms became obsolete; clearly we do not question that DDT is organic. If this earlier definition can no longer stand, then what can? An organic compound must, of course, contain carbon, and the crux of the definition probably lies in the state of this carbon.

All compounds that we do not hesitate to call organic contain a covalent carbon–hydrogen bond, that is, a reduced carbon atom. A definition based on

Microbial Ecology of the Oceans, Edited by David L. Kirchman.
ISBN 0-471-29993-6   Copyright © 2000 by Wiley-Liss, Inc.

the state of the carbon atom would thus include all the conventional biological organic compounds and most of the synthesized ones. The reduced carbon is biochemically important as a proton and electron source, hence as a source of metabolic energy. The abstraction of a proton and electron from water during the first steps in the photosynthetic formation of organic material may be seen as the creation of the "vital force," and so the original concept is retained in a modern context. The definition would exclude urea and carbon tetrachloride and the freon group. Significantly, these compounds share the property of nonflammability; carbon tetrachloride and the freons have been used as fire extinguishing fluids and urea is a fire retardant. Thus, the common concept (e.g., Libes 1992, and others) that urea is organic does not seem to be justified. Urea is better regarded, along with ammonia, as an organic excretion product; its ecology is consistent with this. For example, urea occurs at micromolar concentrations similar to inorganic nitrogen nutrients and very different from individual organic compounds, which are present in nanomolar concentrations.

The distinction between particulate and dissolved material has no fundamental basis and must be seen as operational. Sharp (1973) illustrated this point by showing the continuous nature of the size distribution of organic material in the sea. Nearly all reports effectively define dissolved material as that passing a glass fiber filter (Whatman GF/F filter, with a operational pore size of 0.7 μm). Microorganisms will pass these filters: perhaps 40–90% in the case of marine bacteria (Lee et al. 1995) and probably 100% in the case of marine viruses. Thus, there is a general mismatch between the chemical and microbial definitions of "dissolved." This may render suspect some measurements as some microbial components have been analyzed along with truely dissolved organic material. The use of ultrafilters with submicrometer pore sizes is giving a much more accurate description of the size distribution of organic material.

SOURCES AND CATEGORIES OF COMPOUNDS PRESENT IN THE SEA

The productivities of the oceans and the land may be regarded to be roughly comparable (Field et al. 1998), each equal to about 50 Pg (4 Pmol) of organic material per year.* Both systems are open, and there is transport of organic material, although substantially one way, between the two. The main external organic flux to the marine system is river input. (See Table 1). A figure of 34 Tmol C y^{-1} can be put to this, of which much is estimated to be deposited along the oceanic margins (Hedges and Keil 1995). There are uncertainties over the aeolian input to the open oceans; a figure of 2 T C y^{-1} is probably a maximum (Buat-Ménard et al. 1989).

*The prefixes (T) tera and (P) peta indicate 10^{12} and 10^{15}, respectively.

Table 1. Estimates of gross oceanic organic imports and exports to the oceanic system

Fluxes	Data Quality	Rate (Tmol C y^{-1})	% of Annual Oceanic Production[a]	Ref.
Inputs				
River input of DOC	Quite well constrained	17	0.43	Smith and Hollibaugh (1993)
River input of POC	Quite well constrained	17	0.43	Smith and Hollibaugh (1993)
Aeolian input of POC	Poorly constrained	<8	<0.2	Hedges and Keil (1995)
		2	0.05	Buat-Ménard et al. (1989)
Outputs				
Organic gases	Relatively well constrained	3	0.08	Libes (1992)
Burial				
Coastal	Quite well constrained	9	0.23	Smith and Hollibaugh (1993)
		12	0.3	Hedges and Keil (1995)
Oceanic	Relatively well constrained	2	0.05	Smith and Hollibaugh (1993)
		0.5	0.001	Hedges and Keil (1995)
Total inputs		36– <42	0.85– <1.1	
Total outputs		12.5–17	0.3–0.4	

[a] Assumed oceanic production rate = 4000 Tmol C y^{-1} (5×10^{16} g C y^{-1}).

Primary Biochemicals

Marine algal production is the overriding source of organic material to the oceans. A number of studies have estimated that about 50% of algal production passes into the dissolved organic carbon (DOC) pool. This is discussed later in this chapter and in greater detail in Chapter 5. Thus, the annual production of DOC is in the region of 2–3 Pmol. Assuming the global ocean to contain some 55 Pmol of DOC (see below), this gives a mean turnover time of approximately 25 years. Since much of this turnover will be restricted to the surface ocean, the mean turnover of the DOC pool in these waters will be about a year.

Secondary Chemical Products

The conventional view is that most of the dissolved organic matter (DOM) produced will be used by microorganisms, either in its original form or following enzyme hydrolysis. However, there is a long-standing recognition that purely chemical modification will also occur. This was discussed by Duursma (1965), who considered various chemical oxidations and condensations giving rise to complex, stable products and believed that these eventually lead to the formation of "humic" acids. There is some more or less direct evidence for these purely chemical modifications of biologically produced organics. Keil and Kirchman (1993) demonstrated the abiotic formation of microbiologically resistant glucosylated protein and speculated that the complexation of labile compounds with existing DOC may be a critical first step in the formation of refractory organic material. Interest has reawakened in this area, arising from the growing awareness of the importance of photochemical reactions, which are discussed in detail in Chapter 7. Lara and Thomas (1995) showed progressive changes over time in the physicochemical nature of organic material from a diatom culture. These reactions have important consequences when the mechanisms for decomposition and persistence of organic material are considered.

Concentration and State

Since production in the sea is localized at small point sources, such as phytoplankton cells and fecal pellets, there has been a long debate about the heterogeneous distribution of organic matter on small size scales. It has been speculated that microzones occur as oases of high concentration of organic material separated by a desert of barren dilute water (Krogh 1934; Azam 1998). This has led to the speculation that heterotrophic bacterial activity is restricted to areas of intense activity associated with these microzones, and consequently bulk analyses of seawater may be profoundly misleading.

The counterview is that molecular diffusion is a powerful corrosive force at the low Reynolds numbers characteristic of the microbial environment (see

Purcell 1977). Many of the structures that create and delimit microzones (e.g., mucilages) would not be barriers to molecular diffusion. Mitchel et al. (1985) calculated the likelihood of microzones in the vicinity of phytoplankton cells. They concluded that even with a very weak definition of substrate enrichment (10% above ambient) such microzones could exist only under very special circumstances (e.g., the thermocline). Kaplan and Wofsy (1985) calculated that at typical oxygen concentrations in the sea and characteristic metabolic rates, particles greater than $600-800\,\mu m$ in diameter would be anoxic (i.e., would have passed the limit to active metabolism), thus giving an upper limit to microzone size. Müller-Niklas et al. (1996) counted bacterial numbers in replicate $100\,nL$ volumes (linear dimension ca. $200\,\mu m$) and found that they were normally distributed. These authors concluded that microzones in the waters studied would need to be smaller than $200\,\mu m$. This study defines an upper limit for organic-rich microzones. The lower limit is less well understood, but it will be set by the very high metabolic rates required to compensate for the rapidly increasing impact of molecular diffusion as size is reduced. Thus, there seems to be a narrow window for microzones containing elevated substrate concentrations and activity.

Surface-Absorbed Organics

Certain classes of organic compounds (e.g., lipids, polymers) will have an affinity for surfaces and will exist bound to particles to varying degrees. Certain extraction procedures used for seawater analysis — notably solvent extraction (for lipids) and also ligand complexation procedures (used in early analysis of amino acids) — could well remove the sorbed material. Thus, the free concentration of the molecular species may bear a complex relationship to the determined volumetric concentration.

What is far from clear is the consequence of the binding to surface on the acquisition and decomposition of organic material by microorganisms. Although there is, as a result of sorption, a greater abundance of material at the site (the concept of concentration is not applicable in this circumstance), the activity of the sorbed molecules, in the thermodynamic sense, is reduced and so too is its reactivity. When a molecule is sorbed, its mobility will be reduced, impeding its availability to the active site of an enzyme. This effect would probably manifest itself in an increase in the activation energy of the reaction with a consequent reduction in reaction rate. Indeed, some studies have demonstrated that sorption can inhibit degradation of organic material (e.g., Nagata and Kirchman 1996), and "sorptive preservation" has emerged as one mechanism to explain the presence of apparently labile organic material in sediments (Keil et al. 1994).

Inhibition of hydrolysis by sorption may not be universal, however. Taylor (1995) showed experimentally that at low solute concentrations, bound protein was hydrolyzed 6–10 times faster than that in the dissolved fraction. He suggested that the sorption process denatures proteins and thus removes the steric hindrance, which otherwise slows down proteolytic enzymes.

Complexed Organics

Early work (P. M. Williams 1969; Foster and Morris 1971) showed that the quantity available for analysis of certain metals, notably copper, increases when the organic material present in the sample is destroyed by photooxidation. Useful reviews of the organic complexation of metals in seawater are by Burton and Statham (1990) and Donat and Bruland (1994). In surface oceanic water copper is extensively, if not almost exclusively, bound up in organic complexes (characteristically <99%). Out of the euphotic zone (i.e., at depths >200 m) the extent of complexation decreases to 50% and, as a consequence, the free copper ion concentration increases some three orders of magnitude, from 10^{-13} M to 10^{-10} M. Zinc and iron exhibit much the same pattern. The general increase in the free ion concentration with depth leads to the expectation that microbiological processes are removing the organic chelators and releasing the free ions. The interaction between the decomposition of organic chelates and the consequential toxic effect of the released metals such as copper would be interesting to explore and is returned to in a later section.

HIGHWAY OF DECOMPOSITION

Most microbial mineralization processes (e.g., the production of the inorganic forms of carbon and nitrogen) involve oxidation, the general exception being the release of inorganic phosphate, which is hydrolytic for the most part (but not exclusively: see Clark et al. 1998). Biological oxidation requires a highly structured system, and this level of organization and the necessary physicochemical environment can be realized only within the confines of the cell.

Generally biopolymers must be hydrolyzed down to monomers or oligomers before they can be transported into the cell and acted on by enzymes, effecting mineralization. Amino acids can be deaminated by hydrolysis, and so, in principle, this reaction can occur outside the cell. Pantoja and Lee (1994) have demonstrated cell surface oxidation of amino acids and make the point that this form of oxidation may, on occasion, account for as much as 40% of the total oxidation rate. The conventional view has been that low molecular weight compounds are most reactive and as a group are turning over more rapidly than their high molecular counterparts. Although it is most likely correct for the normal suite of biochemicals, it seems to be not to be entirely so for the DOM present in seawater (Amon and Benner 1994). What this may mean is that a small fraction of the low molecular weight compounds is turning over very rapidly, whereas the bulk is resistant and cycling very slowly.

THE BASIS OF BIOLOGICAL STABILITY AND RESISTANCE

There are paradoxical problems surrounding the removal and persistence of DOM in seawater. In part, the system appears to be very dynamic, with material turning over actively with exceptionally low concentrations maintained. At the same time, ^{14}C dating of DOC from surface and deep water gives ages of 2000 years for the former and 4000–6000 years for the latter (see below: Figure 6, and P. M. Williams and Druffel 1987; Druffel et al. 1992); that is, the mean residence time of the refractory deep water DOC is 3 to 4 oceanic cycling times. It is not an idle question to ask why only 3 to 4 cycling times; if it is indeed refractory, why not much longer? This leads to the question of what comprises resistance to decomposition. We may usefully separate the concepts of stability and resistance; the former may be regarded to be an absolute property and the latter an ephemeral one, determined by prevailing circumstances.

In-Built Stability

The dramatic production of foam at sewage outfalls, following the introduction of synthetic detergents, brought to light the resistance of branched hydrocarbon chains to microbial decomposition. Studies on the resistance of synthetic organic chemicals to biodecomposition have given rise to the concept of xenophores—substitutions into molecules that impart biological stability. This concept, and the evidence surrounding it, is discussed by Alexander (1994). The more common groups are Cl, Br, NO_3, SO_4, and CH_3. Two important features are to be noted at this stage. First, most of the substituted elements are present, if not abundant, in seawater in ionic form. Second, whereas these structures impart biological resistance, they will, without exception, increase the chemical reactivity of the carbon atom to which they are attached. This asymmetry has relevance in constructing a model to account for the cycling of DOC in oceanic waters (see later section).

It is axiomatic that compounds produced enzymatically (i.e., biologically) can be likewise broken down enzymatically. This thinking derives (probably wrongly) from the assumption that enzymatic reactions are reversible. If we wish to retain this axiom, yet at the same time recognize that compounds with xenophoric structures are produced in nature, we are then forced to conclude that these structures must be produced chemically from otherwise labile molecules. It is not difficult to anticipate the reactions that might be involved. These chemical substitutions could occur just as easily within the cell at the time and at the site of production of a biochemical and would be mistaken as a biological product, thus constituting a departure from the axiom above. However, there are signs that the axiom may not be wholly robust—there is, for example, unequivocal evidence for the enzymatic formation of biologically resistant halocarbons by an enzyme-catalyzed halogen transferase reaction (Wousmaa and Hager 1990).

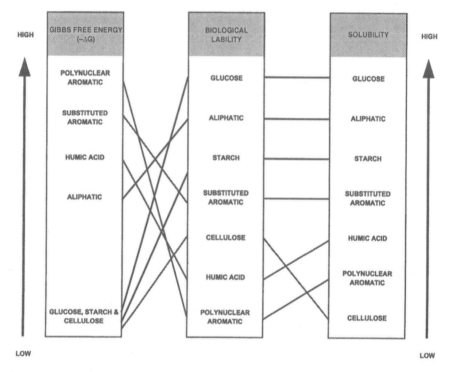

Figure 1. Qualitative ranking of change in Gibb's free energy, stability, and solubility.

Inherent Resistance

Upon ranking microbiological stability of organic compounds in a very simple manner against the potential driving force for oxidation (the change in Gibb's free energy), one sees almost an inverse correlation (left-hand side of Figure 1). The reason in part probably lies in the extent of polarity of the molecule. In essence, the greater the extent of reduction and aromaticity, the greater the change in free energy on oxidation, but the lower the solubility and reactivity. Other molecular features affect the solubility and reactivity of biological products, notably the extent of polymerization and the nature of the polymer. The nicest example comes from glucose and two of its polymers (cellulose and starch). Because biological polymers are formed by condensation, their changes in free energy (on a carbon basis) are very similar, but glucose, the starches, and cellulose are very different in biological reactivity, and in solubility. The latter would appear to be a major determinant of their potential biological reactivity (see right-hand side of Figure 1).

Ephemeral Resistance: Thermodynamic and Kinetic Constraints to Acquisition

As a resource becomes sparse, two factors arise that may set constraints on a material's exploitation: the economics of acquiring it and its frequency of occurrence (i.e., whether it occurs frequently enough to satisfy requirements to keep the system operational). These factors are relevant to the question of whether there is some minimum concentration of an organic substrate below which acquisition by heterotrophic bacteria is thermodynamically and kinetically constrained. The answers are not necessarily intuitive.

The question of economics is akin to asking, Why is there still gold in the Welsh hills? That is, at what point does the cost of acquiring a resource fall below its value? In metabolic terms, the question may be recast as follows: at what point does the energy gained from the oxidation of an organic substrate (i.e., the change in Gibb's free energy $-\Delta G$) equal the thermodynamic energy required to so increase its concentration inside the cell that metabolism of it may proceed at a rate that will enable the organism to survive and compete? We may assume that this latter concentration is in the same region as the Michaelis–Menten constant of enzymes; for this we can use an estimate of 10^{-4} M. It transpires that this value is not critical to the course of the argument. Box 1 gives a calculation for glucose using the thermodynamic equation for the heat of dilution. The calculation shows that the purely thermodynamic constraints to acquisition are of no significance. Indeed, it is difficult to see how this could be accommodated biochemically, for it would require a variable number of ATP molecules to be used for substrate transport into the cell.

Regardless, we can safely assume that the arrival rate of glucose molecules from 10^{476} oceans would not be frequent enough to sustain the observed growth rate of bacteria. This leads to a second calculation (Box 2). There are assumptions embedded in this calculation that make it less secure than the preceding one. They surround simplifying assumptions made for the calculation of the arrival rate of molecules in solution and the fraction of successful collisions. Nonetheless, presuming that they are not too much awry, we end up with a minimum dilution of the order of 10^{-12} M. This is about two to three orders of magnitude below our minimum sensitivity for the measurement of organic nutrient molecules such as sugars and amino acids. Thus, when we currently report undetectable concentrations we cannot eliminate the possibility that these molecules will have been run down to this minimum concentration. If, on the other hand, we are able to provide unequivocal demonstration of their presence in deep water of, say, nanomolar concentrations of individual organic molecules, we may eliminate diffusion as a constraint to decomposition. We would then need to search for other controlling factors — such as the cell maintenance requirements (see Chapter 10). Furthermore, the concentration calculated above is a maximum one, because it assumes that the organic molecule in question is the only one

---

Box 1. Thermodynamic constraints to acquisition.

Question: Is it economical to concentrate dilute material to respire it?

HEAT OF DILUTION CALCULATION

$$-\Delta F = RT \ln(C_1/C_2)$$

When $-\Delta F = -\Delta G$, there is no surplus energy.

The $-\Delta G$ for the oxidation of glucose is 2870 kJ mol^{-1}.

Then:

$$-2870 \times 10^3 = 8.3 \times 290 \times 2.303 \times \log(C_1/C_2) = -5.5 \times 10^3 \times \log(C_1/C_2)$$

$$\therefore \log(C_1/C_2) = \frac{-2870 \times 10^3}{5.5 \times 10^3} = -517$$

where C_1 and C_2 are, respectively, the external and internal concentrations of the substrate.

Thus

$$C_1/C_2 = 10^{-517}$$

Assume $C_2 = 10^{-4}$ M, then $C_2 = 10^{-521}$ M.

Take Avogadro's number as 6×10^{23} and the oceans as 10^{21} liters.

Thus ONE molecule in the whole of the oceans would have a concentration of $\sim 10^{-45}$ M, that is, 10^{476} times greater than the minimum concentration.

Thus, by these calculations it would be economical to acquire ONE molecule of glucose from 10^{476} oceans!

---

being used and that the growth rate is sustained at these depleted concentrations. Neither assumption is likely to be correct.

Ephemeral Resistance: Stoichiometric Control—the C/N Cell Quota Constraint

Because of their high protein and nucleic acid content, bacteria need to sustain a low cell C/N ratio (see also Chapter 9). If a percentage of the carbon is lost

---

Box 2. Kinetic constraints to acquisition.

Question: At low concentrations do molecules arrive at a sufficient rate for use by bacteria?

Consider bacteria at a concentration of $12 \, \text{mg C m}^{-3}$ ($1 \, \mu\text{M C}$) growing at $1 \, \text{div/d}^{-1}$ with a growth efficiency of 25%.

This carbon demand would require $1 \times 4 \times 10^{-6} \times 6 \times 10^{23} \, \text{C atoms d}^{-1}$ (i.e., $2.78 \times 10^{13} \, \text{C atoms s}^{-1}$).

The total bacterial surface area in seawater is approximately $1 \, \text{cm}^2 \, \text{dm}^{-3}$ (Williams, 1981a, 1984).

The minimum arrival must be $3 \times 10^{13} \, \text{C atoms/s}^{-1}$ or 5×10^{12} glucose $\text{s}^{-1} \, \text{cm}^{-1}$. Such a collision rate would be sustained by 10^9 molecules dm^{-3}, or $\sim 10^{-15} \, \text{M}$.

Assume that the successful collision is 1 in 10^3. Then the minimum concentration of the organic substrate would need to be $10^{-12} \, \text{M}$, that is, 10^{510} times greater than the calculated thermodynamic constraint but still less than the observed minimum concentration ($10^{-9} \, \text{M}$).

Thus, the arrival rate of molecules at a bacterium's surface does not seem to set a limiting concentration to acquisition of DOM.

---

to respiration during assimilation, then a minimum C/N ratio may be estimated for the organic substrate to sustain the bacterial C/N quota. The concept defines a threshold for the C/N ratio of substrates for balanced growth below which they will mineralize nitrogen as ammonia and above which they will need to assimilate inorganic nitrogen, characteristically ammonia. Arithmetically the threshold is the C/N quota divided by the fractional growth yield. Taking the bacterial C/N ratio to be 6 (see discussion in Chapter 9) and a bacterial growth yield* of 15%, we obtain a threshold value of 40. This "stoichiometric" model was developed by Goldman et al. (1987) and Anderson (1992). There is plenty of evidence that heterotrophic bacteria in the sea can use ammonia even in the presence of competition with the algae (Kirchman et al. 1994; Wheeler and Kirchman 1986). However they are believed to be poorer competitors for nitrate (although see Chapter 9).

Embedded in the stoichiometric model is the concept that both the bacterial cell quota and the carbon growth yield are constant properties of the organism and the molecule. Whereas the need to metabolize and divide will put certain

---

*Bacterial growth yield and its variability are discussed in Chapter 10. The discussion in that chapter makes it clear that there is no fixed value. For the sake of consistency I use the mean value of 15% for oceanic bacteria throughout this chapter.

minimum constraints on the nitrogen content of the cell and thus to a degree stabilize the bacterial C/N and the C/P quotas, there is no obvious factor stabilizing the carbon growth yield in the same manner. In Chapter 10, del Giorgio and Cole argue that the growth yield is related to growth rate, so this will add a degree of fuzziness to any stoichiometric model.

Experimental studies have revealed that labile organic additions to natural marine communities do not necessarily stimulate growth. Sometimes the stoichiometric model could be used to explain the observations, but at other times (e.g., in the absence of stimulation of bacterial growth by an added nitrogen-containing organic compound such as glycine: Thingstad et al. 1999), it could not.

THE ORGANIC ENVIRONMENT

Away from the immediate influence of rivers, the autotrophic and hetero-trophic microorganisms are major agents shaping the organic composition of the oceans. Their role is not exclusive, the growing awareness of the importance of the role of photochemical reactions (see Chapter 7) now means that we need to think of chemical, photochemical and microbiological processes in parallel or, more likely, in concert.

The total concentration of DOC generally falls in the range 40–80 μM. Deep-water values are typically 40–50 μM (Sharp et al. 1995). Surface waters vary seasonally by undetectable amounts to 30 μM (Carlson et al. 1994; P. J. le B. Williams 1995; Carlson et al. 1998, Peltzer and Hayward 1996), superim-posed on a baseline figure of some 50–60 μM. Regional variations may be seen in DOC (Duursma 1961; Peltzer and Hayward 1996) and dissolved organic nitrogen (DON) (Duursma 1961; Libby and Wheller 1997). We have a great deal less information on the concentration and distribution of individual organic compounds. Concentrations of compounds such as monosaccharides and amino acids fall typically in the range 0.1–50 nM (Fuhrman 1987; Rich et al. 1996; Borch and Kirchman 1997; Skoog and Benner 1997). The extent to which there are seasonal and depth-related patterns in the concentration of individual molecules is not clear, but such information is critical to our understanding of the factors controlling the dynamic of these molecules. Skoog and Benner (1997) found no compelling evidence for systematic depth-related differences in concentration of monosaccharides; in general, the concentrations at 4000 m were comparable to those in the first 200 m. There are very few seasonal studies of individual organic compounds; those that exist (Andrews and Williams 1971; Fuhrman 1987) give no sign of systematic seasonal changes in concentration, such as those characteristic of the inorganic nutrients and total DOC. It was the conclusion of Andrews and P. J. le B. Williams (1971) and Fuhrman (1987) that the system was tightly coupled: that is, whereas there were seasonal changes in the rate of turnover, there was no systematic change in concentration.

Early views were that the deep-water DOM consisted principally of complex heterogeneous macromolecules (see, e.g., Libes 1992, p. 411). This view is undergoing a major revision in two respects. First, the macromolecular structures do not appear to be amorphous but more restricted and conservative in their composition (Aluwihare et al. 1997). Second, and surprisingly, a substantial fraction of the DOC is below 1000 Da (Guo et al. 1994, 1995; Amon and Benner 1996). Aluwihare et al. (1997) conclude that a significant fraction of deep-water DOM is structurally related to biosynthetically derived acyl oligosaccharides that persist after more labile organic matter has been degraded. Certainly, polysaccharides appear to be a more important component of DOM and of products released by algae (Benner et al. 1992; Pakulski and Benner 1994; McCarthy et al. 1996) than hitherto recognized. The major structural polysaccharide in bacteria, peptidoglycan, may also be a large component of the DOM pool (McCarthy et al. 1998).

Rates of DOM Fluxes

Examining DOM fluxes presents many challenges. At the gross level we are studying the turnover of one of the major carbon reservoirs in the biosphere and the largest exchangeable pool of organic carbon on earth. At the detailed level we are trying to determine the fluxes of organic compounds at extreme dilution (i.e. $\sim 10^{-9}$ M or less). We can anticipate the very general boundaries to DOM fluxes: an upper level set by primary plankton production ($5-35$ mmol C m^{-3} d^{-1}) and a lower level set by the cell maintenance of microbes. The aim of the following sections is to assess the scale of DOM utilization by bacterial heterotrophs, ultimately in relation to overall planktonic organic turnover.

Rates Inferred from Respiration   It is convenient to start with chemically determined rates of plankton respiration, since the database is comparatively homogeneous and the respiration data are frequently associated with measurements of plankton photosynthesis, which provide a logical reference parameter.

The full data set of published chemically determined respiration measurements amounts to some $400-600$ observations, minute compared with the number of ^{14}C-determined photosynthetic rate measurements. The available data are summarized in Table 2. Almost without exception, the respiration measurements are restricted to the top 100 m or less of the water column. The mean rates for particular studies in the upper water column generally lie in the range $1-4$ mmol O$_2$ m^{-3} d^{-1} (Table 2). With the striking exception of the data reported by Griffith et al. (1900) and Griffith and Pomeroy (1995), most individual observations are between 0.2 and 10 mmol O$_2$ m^{-3} d^{-1}. The lower limit has little significance other than as an indication of the limit of sensitivity of the rate measurement.

Table 2. Rates for offshore plankton metabolism inferred from changes in oxygen concentration

Location	Approach	Respiration Rate ($mmol\ O_2\ m^{-3}\ d^{-1}$)	Heterotrophic Carbon Utilization ($mmol\ C\ m^{-3}\ d^{-1}$)[a]	Ref.
Gulf of Maine	ΔO_2	4.1 ± 2.6 ($n = 23$)	1.7	Packard and P. J. le B. Williams (1981)
U.S. outer continental shelf				Griffith and Pomeroy (1995)
December–January	ΔO_2	2.4–17	1–6.9	
April–September	ΔO_2	17–60	6.9–24	
>100 km offshore	ΔO_2	88–120	36–49	
Northeastern Atlantic	ΔO_2 and ^{18}O	3.9 ± 2.3 ($n = 30$)	1.6	Griffith et al. (1990)
Northeastern Atlantic	ΔO_2	3.4 ± 2.6 ($n = 130$)	1.4	Kiddon et al. (1995)
Southern Ocean	ΔO_2	1.6 ± 1.3 ($n = 38$)	0.65	P. J. le B. Williams (1998)
Southern Ocean	ΔO_2	1.3 ± 1.1 ($n = 13$)	0.53	Blight (unpublished)
Southern Ocean	ΔO_2	1.9 ± 1.0 ($n = 30$)	0.77	Bouquegneau et al. (1992)
				Robinson and P. J. le B. Williams (1993)
Southern Ocean	ΔO_2	2.8 ± 2.2 ($n = 11$)	1.1	Aristegui et al. (1996)
Southern Ocean	ΔO_2	2.3 ± 2.0 ($n = 28$)	0.93	Boyd et al. (1995)
Mediterranean	ΔO_2	2.3 ± 2.1 ($n = 104$)	0.93	Lefèvre et al. (1998)
Mediterranean	ΔO_2	1.7 ± 1.2 ($n = 9$)	0.69	Aristigui (unpublished)
North Central Pacific Gyre	ΔO_2	0.9 ± 0.5 ($n = 22$)	0.36	P. J. le B. Williams and Purdie (1991)
Sargasso Sea	ΔO_2	1.8 ± 0.3 ($n = 3$)	0.72	P. J. le B. Williams and Jenkinson (1982)

[a] Calculated as: O_2 rate $\times 0.4 \times 1.18 \times 0.88$; where 0.4 is the fraction of community respiration due to bacteria, 0.88 is the respiration coefficient, and 1.18 allows for carbon incorporation (15%) during growth. Thus total carbon consumption equals $R[1/(1 - Y)]$, where R is the respiratory carbon dioxide production and Y is the fractional growth yield.

The question arises of how much of overall respiration can be attributed to bacterial metabolism of DOC. We can ignore any bacterial metabolism associated directly with particulate organic carbon because attached bacterial biomass and growth appear to be negligible compared to free-living levels (e.g., Alldredge et al. 1986).

Low flow, reverse filtration fractionation procedures have been used to separate various size fractions as a means of estimating the microbial component of overall metabolism (P. J. le B. Williams 1981b). This technique was dismissed by Sherr et al. (1988) because of the general problem of the disruption of feedback controls due to separation of predator from prey. However, time courses of oxygen consumption are usually linear in cool, temperate waters (see, e.g., P. J. le B. Williams 1981b; Harrison 1986; Blight et al. 1995), an indication that the approach yields reliable data. Anomalously higher rates in the small size fractions, probably due to bacterial growth, are characteristically found with samples from warm water environments. Hopkinson et al. (1989), who examined respiration in size fractions and analyzed the results in relation to location and temperature, give the clearest picture of events. The anomalies occurred in the inshore stations when temperatures exceeded 20°C, presumably giving rise to high bacterial growth rates. In all other circumstances there was no sign of anomalous behavior. Thus, it would appear that with the possible exception of inshore warm waters, the size fractionation procedure does give sensible results. Figure 2 shows a summary

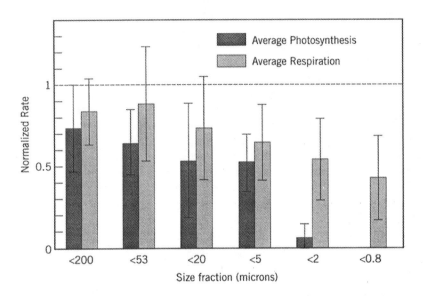

Figure 2. The size distribution of respiration and photosynthesis, normalized to that of the unfiltered sample; bars show standard errors. Data from Harrison (1986), Williams (1981c), Williams and Bentley (unpublished), Blight et al. (unpublished).

taken from some 30 observations in the northeastern Atlantic, the Canadian Arctic, and the Southern Ocean. The mean contribution of bacterial (strictly the <0.8 or 1 μm fraction) to overall metabolism is about 40%. We return to this value in a later section.

The foregoing generalizations allow us some insight into the overall distribution of metabolism among the autotrophs, the microheterotrophs, and the metazoan heterotrophs. Duarte and Cebrián (1996) suggested that algal respiration on average was equal to 35% of primary production. A similar figure was obtained by Robinson et al. (1999). Taking this figure in conjunction with the 40% for microheterotroph respiration estimated above, we are left with 25% or so of production to be consumed by the metazoan heterotrophs, giving a figure for the ratio of microheterotrophy to metazoan heterotrophy above unity. Values above unity for this ratio were also obtained by P. J. le B. Williams (1984) from biomass projections.

Rates Determined from DOC Concentration Change   The first attempt to estimate DOC turnover rate from time course studies was by Duursma (1963). He calculated an annual figure for DOC production from its net seasonal accumulation by making corrections for simultaneous removal, obtaining a figure of $52\,\mathrm{g\,C\,m^{-2}}$ for the annual production of DOC (Table 3). The calculation was repeated by Banoub and P. J. le B. Williams (1973) for DOC at station E_1 in the English Channel. They estimated a similar rate of $34\,\mathrm{g\,C\,m^{-2}}$ (i.e., $\sim 30\%$ of photosynthetic production, equivalent to rates of about $1\,\mathrm{mmol\,C\,m^{-3}\,d^{-1}}$). Carlson et al. (1994) analyzed the time course of the seasonal decrease in the concentration of DOC at the Bermuda Atlantic Time Series Station (BATS). They calculated the loss due to biological utilization in the top 100 m during the period of stratification and also the gain in the depth range of 100–250 m from the water above, once water column stability had broken down. The sum of these two estimates was $1.5\,\mathrm{mol\,C\,m^{-2}\,y^{-1}}$. Assuming an annual production of $8\,\mathrm{mol\,C\,m^{-2}}$ for the area, implies that the observed accumulation of DOC was about 20% of primary production. Hansell and Carlson (1998b) very thoroughly reviewed DOC production in relation to net community production for the Sargasso Sea, the Ross Sea, and the equatorial Pacific. Their generalizations for these areas have been incorporated into Table 3. Additionally they project their findings of a global assessment of DOC production (which they place in the semilabile category). They estimated global new production to be $0.6\,\mathrm{Pmol\,y^{-1}}$ and DOC production to be $0.1\,\mathrm{Pmol\,y^{-1}}$; that is, DOC production is 17% of new production and a mere 2% of gross production.

These estimates, which can be considered to be "net" production of semilabile DOC, are well below those estimated from microbial growth and respiration (this chapter and also Chapter 5); "gross" DOC production would be DOC accumulated in the water column, as estimated by Hansell and Carlson (1998b), plus the produced DOC immediately consumed by microbes. The fraction of material routed through the DOC pool estimated from microbial parameters approaches (or exceeds) 50% of primary production (i.e.,

~2.5 Pmol y^{-1}). Andrews and P. J. le B. Williams (1971) attempted to estimate this fraction by measuring uptake and respiration of glucose and amino acids. Based on these data, they estimated an annual overall turnover of DOC of some 8.3 mol C m^{-2} (40% of primary production), equivalent to a figure of 3.3 mmol C m^{-3} d^{-1}. Assuming a respiratory coefficient of 0.88 (Robinson et al. 1999) this would convert to 3.8 mmol O$_2$ m^{-3} d^{-1} (i.e., within the ranges reported in Table 2).

Estimates derived from seasonal in situ changes in DOC concentration are open to a number of cautions, one being that only net changes can be measured and uptake of DOC is not included. Perhaps more seriously, advection is ignored. An alternative is to derive rates from in vitro incubations. This puts considerable pressures on the precision of the DOC analysis if long incubations are to be avoided; projections from the respiratory measurements suggest a mean turnover rate of DOC in surface waters of about 5% per day. The present generation of DOC analyzers can achieve precisions of better than 1% (see Carlson et al. 1994).

Early studies of the time course of DOC removal were by Ogura (1973, 1975). He used first-order kinetics to analyze the concentration changes and noted two phases in decomposition — an initial rapid phase, with rate constants 0.033 to 0.095 d^{-1}, followed by a slower phase with rate constants about an order of magnitude lower (0.003–0.004 d^{-1}). In inshore waters, the two phases were quite distinct, the initial phase lasting about 5 days. In offshore waters, the separation was less clear but the initial rapid phase lasted much longer — about 50 days. Rates of removal of DOC calculated from Ogura's reported rate constants and the DOC concentrations are given in Table 3. Those for the offshore station are comparable to the estimates derived by Banoub and P. J. le B. Williams (1973) and Andrews and Williams (1971). The small number of short-term measurements of DOC flux is summarized in Table 3. The fluxes show a great deal of scatter, with high values for the Gulf of Mexico and low values for the Ross Sea. The unusually high value for the Gulf of Mexico seems to be associated with very high respiration and bacterial production rates. When the estimates of the annual production of DOC are normalized by the presumed rates of primary production, the DOC fluxes appear to be in the range of 15–40% of the photosynthetic input. The rates will omit rapidly cycling material and would probably fall into the semilabile fraction of Kirchman et al. (1993).

Rates Inferred from Bacterial Growth Rate Measurements   The problem of assessing overall DOM metabolism by microbes may be approached from an alternative direction; it may be derived from [^{3}H]thymidine- and [^{3}H]leucine-determined estimates of bacterial production. There are a number of steps, with associated uncertainties, in the calculations. Both methods require a conversion of the tracer uptake rates to cell carbon production and then a further major correction to account for the material lost to respiration. Aspects of this are discussed by Ducklow and Carlson (1992), by Cole and Pace (1995), and in Chapters 4 and 10.

Table 3. Rates of DOC metabolism inferred from changes in DOC concentration over time[a]

Study Area	Approach	Rate $(\text{mmol C m}^{-3}\,\text{d}^{-1})$	Rate $[\text{mol C m}^{-2}\,\text{y}^{-1}\,(\%\text{PP})]$	Ref.
Coastal waters				
Southern North Sea	Analysis of seasonal DOC curve, with correction for concurrent removal	—	4.3 (17%)	Duursma (1963)
English Channel	As Duursma (1963)	1.1[b]	2.8 (13%)	Banoub and P. J. le B. Williams (1973)
English Channel	As above, plus the flux of amino acids and glucose	3.3[b]	8.3 (40%)	Andrews and P. J. le B. Williams (1971)
Inshore	As Ogura (1973): initial rate constant = 0.095 d^{-1}, subsequent rate constant = 0.004 d^{-1}	11.8	—	Ogura (1975)
Gulf of Mexico	From DOC concentration in in situ incubations HMW: DOC turnover time 6 days LMW: DOC turnover time 26 days	HMW: 35 LMW: nd		Amon and Benner (1994)
Oceanic waters				
Sargasso Sea	Total DOC production	—	1.5 (17%)	Carlson et al. (1994)
Sargasso Sea	Accumulation over spring increase	0.72 ± 0.04	(≈60% NCP)	Carlson et al. (1998); Hansell and Carlson (1998b)

			Rate (mmol C m^{-3} y^{-1})	
Ross Sea	Accumulation over spring increase	0.126	(8–14% NCP)	Carlson et al. (1998); Hansell and Carlson (1998b)
Equatorial Pacific	Various approaches		(10–30% NCP)	Hansell and Carlson (1998b)
Offshore Pacific	Rates of concentration	2.75	—	Ogura (1973)
Deep water				
North Atlantic Ocean	Spatial distribution of deep-water DOC		0.05	Hansell and Carlson (1998a)
Central North Pacific	Calculation from vertical diffusion model from bacterial growth rate measurement		0.05–0.15 0.003–0.05	P. J. le B. Williams and Carlucci (1975) P. J. le B. Williams and Carlucci (1975)

[a] Abbreviations: HMW, LMW, high and low molecular weight; nd, not detectable; PP, primary production; BP, bacterial protection; NCP, net community production.
[b] Calculated assuming that the process continues over 6 months of the year and is restricted to the top 15 m of the water column, constrained by the seasonal thermocline.

171

In their review, Ducklow and Carlson (1992) compiled estimates of bacterial carbon production as a percentage of primary production. The figures they obtained for oceanic environments ranged from 56 to 153%. Del Giorgio and Cole's figure of 15% (Chapter 10) for the growth yield of oceanic bacteria would give bacterial carbon assimilation ranging from 370 to 1000% of primary production — quite unacceptable values. In a subsequent review, Ducklow (1999) obtained values for bacterial production in the region of 20% primary production, giving the bacterial carbon assimilation as roughly 130% of primary carbon production, a high, but more acceptable, figure. In passing it is worth noting that the observations for the Ross Sea stand out as being much lower than the others. The 4% value for the Ross Sea does not seem to be consistent with observations of respiration rates (see Table 2), which are only marginally lower than areas like the North Atlantic. This, however, may be a problem arising because of too few samples.

Rates Estimated from Selected DOM Components The three approaches described above give no insight into the flux of individual organic molecules in the sea — which is the fundamental basis for understanding the dynamics of microbial growth. The difficulty of measuring the concentration of individual organic compounds in oceanic waters and the dynamic nature of the system has meant that no direct measurements of time-dependent changes, analogous to the direct studies of overall DOC fluxes (as discussed above), have been undertaken. Radioisotope procedures provide the required sensitivity. In part, it was these studies that laid the basis for the paradigm shift from the old notions of the bacterial component of the marine food web turning over slowly and contributing little to overall plankton metabolism (<1% was suggested by Zobell 1976) and mainly located in close proximity to the phytoplankton (Krogh 1934), to a highly dynamic dispersed system turning over 20–50% or more of plankton production, that is, the paradigm of Pomeroy (1974). Sieburth (1977) noted the reluctance of many biological oceanographers to accept this paradigm shift, and in that respect the arguments put forward by Pomeroy (1974), P. J. le B. Williams (1981a, 1984), and Azam et al. (1983) were crucial to the acceptance of the overall case.

Parsons and Strickland (1962) introduced the use of radiotracers to determine the flux of organic material in seawater. They used tracer quantities of [^{14}C]-glucose to measure its uptake. The technique was taken up by others. John Hobbie (e.g., Wright and Hobbie, 1965) continued its use to measure the uptake and assimilation of glucose and other substrates. Williams and Askew (1968) used a tracer procedure to determine the respiration of glucose. The dilemma facing these and other workers at that time was that the tracer uptake at best could provide estimates for the turnover time of the substrate and kinetic parameters: that is, the maximum uptake velocity and the sum of the half-saturation constant and the natural substrate concentration (Wright and Hobbie 1965). What this approach could not do by itself was provide the rate of mass utilization; this required a concurrent measurement of the substrate concentration in the sample. Although attempts were made to measure the concentration of individual substrates, this was close to or beyond the capability of the techniques available at that point in time. These early attempts to measure organic metabolism and substrate concentration were reviewed by Williams (1975).

The flux rates of individual molecules range from below detection (i.e., generally $<0.1\,\mu\text{mol}\,\text{C}\,\text{m}^{-3}\,\text{d}^{-1}$) to $100\,\mu\text{mol}\,\text{C}\,\text{m}^{-3}\,\text{d}^{-1}$. These compounds would constitute the labile fraction of Kirchman et al. (1993). Commonly, the highest rates for single molecular species are obtained for glucose. However, since now that there is evidence for substantial photochemical production of organic acids, the early observations of Billen et al. (1980) of very high fluxes of organic acids such as lactate and glycollate strongly merit further examination. These compounds have been largely ignored in recent studies. Attempts have been made to scale these observed rates to bacterial or primary production. P. J. le B. Williams et al. (1976) estimated the amino acid flux to be from 0.5 to 10% of primary production. Rich et al. (1996) concluded that glucose uptake could account for 27–35% of bacterial metabolism. Chapter 9 discusses the quantitative importance of amino acids in overall bacterial N acquisition.

Whereas the early work gave faster turnover times (days to tens of days) in coastal waters than in offshore waters, more recent work does not suggest any systematic difference. In both cases turnover times are characteristically a day or so or less; in some cases fractions of an hour (e.g., Fuhrman 1987; Fuhrman and Ferguson 1986). In surface waters the system seems to be very tightly coupled. The variation in turnover time with depth is not clear. The early work of Banoub and P. J. le B. Williams (1972) on glucose and amino acid turnover in the western Mediterranean and that of Azam and Holm-Hansen (1973) for amino acids in the North Pacific suggested substantial increases in the turnover times with depth. In contrast are the remarkable observations of Craven and Carlucci (1989), who found little change over a 900 m profile with turnover times in the range 0.5–5 days.

Synthesis of Observations

The objective of the preceding four sections has been to establish the magnitude of DOC flux and total bacterial carbon demand; the tacit assumption is that they are one and the same process. There have been earlier attempts to do this. P. J. le B. Williams (1981a) examined the division of flows at various points in plankton metabolism and from this analysis obtained a figure of 56% for flow of DOC to bacteria and the rest of the microbial loop, scaled against primary production. From a completely different set of measurements, Azam et al. (1992), using what they regarded as conservative respiration losses of 50%, estimated bacterial carbon demand from thymidine-based bacterial production to be 30–60% of primary production; a growth yield of 15% would increase these estimates to 100–200%.

The present account has considered four very different sets of measurements, with different assumptions and uncertainties embedded in each one. Before any comparison can be made, the observations need to be converted to a common currency, total bacterial carbon demand is the logical one. To do this, a figure for the growth yield is required. I have used 15% for the growth yield, while recognizing that it is almost certainly a variable property and presently not well constrained (see Chapter 10).

Projections from overall plankton respiratory metabolism are, in principle, the simplest. The calculation relies upon knowing two properties of the microbial system: its contribution to overall metabolism and its growth yield. The greatest uncertainty centers on the fraction of overall respiration that can be attributed to bacteria. Early views (e.g., ZoBell) were that it was very small, less than 1%. Current thinking is that the percentage is substantially higher; the calculation to follow uses a figure in the region of 40%, as suggested by the analysis in the section on respiration. However, it must be stressed that the data set is small and there are potential artifacts. A starting point for the calculation is the assumption that planktonic photosynthesis and respiration are equal and that net production on an annual time scale is very close to zero. From an analysis of field observations of respiration and photosynthesis, P. J. le B. Williams (1998) concluded that the upper water column was close to being in balance. Smith and Hollibaugh (1993) inferred (from mass balance considerations) that the open oceans are marginally (0.4%) heterotrophic. Given the two assumptions above, we may infer that if bacterial respiration is equal to 40% of overall planktonic respiration, it must be equal to the same percentage of planktonic gross photosynthesis. This being the case, given a growth yield of 15%, the total DOM flux and thus total bacterial carbon utilization would amount to $\approx 50\%$ (40% × 1.18) of plankton production.

In an earlier section, I derived a figure of 130% for bacterial carbon demand as a percentage of photosynthetic carbon production from reports of bacterial production. There are more uncertainties in this latter calculation than with the projection from respiration. First, the thymidine method requires an estimate of the carbon content of an individual bacterial cell (see Chapter 4 by Ducklow) to allow bacterial growth to be expressed in mass flux units. Ducklow and Carlson (1992) show the spread of values used to be 10-fold. Second, estimates from bacterial production also requires a further correction for respiratory losses; the correction* in this case is $1/Y$ (where Y is the fractional growth yield), whereas the correction for respiration measurements is $1/(1 - Y)$. If the value of Y is substantially less than 50%, then it becomes a major correction term in the case of estimates from growth rates and very sensitive to the value of G, which in itself is variable and difficult to determine. Low values of Y gives a smaller and correspondingly less uncertain correction in the case of respiration. This in part was the debate between Jahnke and Craven (1995) and Cole and Pace (1995).

Of the two estimates of overall metabolism, that derived from respiration measurements is presently the more robust and does not lead to the perplexing

*Given that $Y = G/(G + R)$, where R = respiration, G = growth (more precisely, biomass production), and Y = growth yield, then $G = R[Y/(1 - Y)]$; thus total carbon demand = G/Y, i.e. $R/(1 - Y)$.

situation in which bacterial production exceeds primary production—which can be achieved only if Y is greater than 50%. In this respect the arguments of Jahnke and Craven (1995) should be given serious consideration.

The rates derived from the other two sets of measurements (i.e., those derived from the measured changes in DOC and the rates of uptake and respiration of radiolabeled compounds) give insights into two different aspects of DOM flux. Kirchman et al. (1993) and Carlson and Ducklow (1995) separated DOM into three metabolic categories (labile, semilabile, and refractory) based on their rates of recycling. Measuring the metabolism of radiolabeled organic additions usually describes the flux of the first of the three categories. As discussed before, measurements of the time-dependent changes in DOC concentration probably fail to include the rapidly cycling (labile) fraction of the DOC and deal with time scales far below that of the refractory DOG; thus they may be considered to be an estimate of the flux of second of the three categories—the semilabile fraction. The rates of measured DOC flux (see Table 3) fall in the range of 13–40% of local plankton production, compared with estimates of the order 50% or so for bacterial DOC utilization derived from respiration measurements. If we make the reasonable assumption that the respiration measurements include the metabolism of both the labile and the semilabile fractions, but not that of the refractory material, then by difference the turnover of the labile material would be 10–37% of photosynthetic carbon fixation and 20–75% of estimated DOC turnover.

The estimate of the turnover of the labile fraction, as a percentage of overall metabolism, is not inconsistent with the estimate of Rich et al. (1996) from measurements with labeled monosaccharides or the correction made by Andrews and P. J. le B. Williams (1971) to calculations based on time-dependent changes in DOC (see Table 3). They are higher than the estimate of P. J. le B. Williams et al. (1976) of 0.5–10% of primary production. It may be notable that the estimates of Rich et al. (1996) were based on observations on sugars, whereas those of Williams et al. were based on amino acids. There is a prevailing view that sugars dominate DOM fluxes (Benner et al. 1992).

Thus, in a very tentative way it has been possible to assemble these various approaches to measuring microbial metabolism into a framework. There is a measure of consistency in the estimates of bacterial carbon demand, with rates equal to 50% or more of phytoplankton carbon production. The rates derived from bacterial production determinations are uncomfortably high in that they exceed phytoplankton carbon production, which can occur only if the bacterial growth yields are well in excess of 50%, whereas the modern evidence (Chapter 10) is that they are low: 15% in the case of oceanic bacteria. Projections from production are very sensitive to the value adopted for growth yield. For example, the figure of 130% bacterial carbon demand versus photosynthetic carbon production derived earlier in this chapter using a bacterial growth yield of 15% would be reduced to 80% if 25% is used for growth yield and 60% if 33% were used. In addition to uncertainties over the details of mass balance, we are certainly not in a position to make firm generalizations about the changes in the scales of these various processes temporally and spatially, in the way that it can be done for planktonic photosynthesis.

In many respects this exercise reveals more about what we do not know than about our knowledge of the details and scales of the metabolism of planktonic heterotrophs. Probably to the disadvantage of understanding, plankton studies over the past four decades have been dominated by a single technique — the ^{14}C technique for measuring photosynthetic production. As a consequence, we have a massive, though generally poorly organized database of some a quarter- to half-million ^{14}C-derived measurements of production, while possessing only about half-a-thousand direct observations of the counter-balancing process — respiration. Since a major proportion of primary production does appear to be passing through the microbial consortium, microbial heterotrophy needs better quantification. There is undoubtedly an urgent need to increase the database of heterotrophic measurements, especially for oceanic environments — the point made by Ducklow and Carlson (1992). Equally, there is a need to redirect it. In the past, heterotrophic studies have been commonly embedded in productivity programs, and the sampling strategy determined more often than not by the need to get a satisfactory profile of autotrophy rather than of heterotrophy. The depths out of the euphotic zone are undersampled in standard productivity programs. The paper of Biddanda and Benner (1997) sounds the warning bell that roughly half of bacterial biomass and metabolism appears to occur out of the euphotic zone.

TROPHIC CONSEQUENCES: THE LINK, SINK, AND FUNCTIONALITY DEBATES

Historically, bacteria and microorganisms in general were perceived as mineralizers — "decomposers" and little else. The change in perspective came from litter ecology, where bacteria were seen to be recyclers of organic material because their capability to assimilate inorganic nitrogen and phosphorus enriches the quality of otherwise low grade material. This paradigm is only slowly being assimilated into marine science. The matter of the dual properties of bacteria as mineralizers and recyclers was the basis of the "link or sink" debate (Ducklow et al. 1986, 1987; Sherr et al. 1987). A number of experiments have been undertaken in mesocosms to determine the yield of organic material at the zooplankton level originating from DOM (Parsons et al. 1981; Gamble and Davies 1982; Ducklow et al. 1986, 1987). They generally demonstrated low yields (2 to <0.5%). These studies led to the view that microorganisms were mainly acting as a sink for DOM, passing only a small fraction of their ultimate food source onto the higher trophic levels. This conclusion was sharply criticized by Sherr et al. (1987) on the basis of the design and representativeness of the experiments.

The "link or sink" debate has a number of facets. First, the yield from DOM, assimilated by the bacteria and passed on to the zooplankton, will depend on the number of intervening trophic steps and the conversion efficiency. If there were a single conversion efficiency, then the yield would be n^e (where n is the number of steps and e is the efficiency). In such a simple case, the question of whether the microorganisms are either a link or a sink may have some significance and usefulness. However, most likely neither the number of trophic steps nor the conversion efficiencies is a fixed property. In this respect, as pointed out by Sherr et al., caution is appropriate.

The perspective of the argument is critical, for although the efficiency of transfer may be low, it still may at times represent a significant supplementation to the metazoan food supply. P. J. le B. Williams (1981a) argued that the microheterotrophic organisms (meaning bacteria) could pass on significant quantities of organic material to the next trophic level—he estimated that bacterial production was perhaps twice herbivorous zooplankton production. This was based on an assumed growth yield of 70% for the bacteria, a figure that would now be regarded as high. If one adopts a more conservative figure in the region of 15–25%, then the figures for production at the level of the microheterotrophs and the herbivorous zooplankton become comparable. In the case of the zooplankton, the timing of any transfer of organic material originating from the DOM pool transferred to the zooplankton via the microorganisms is very likely of considerable importance. At certain times of the year, when plankton photosynthetic production is low, the supplementation of the zooplankton food supply from DOC, via the microorganisms, may be crucial to the survival of the former, despite the absolute yields being low. The dynamics of DOC production and removal was postulated by P. J. le B. Williams (1995) to introduce a critical time delay: the DOM pool being slowly charged up in the spring–midsummer period, when phytoplankton production exceeds the needs of the zooplankton, and discharged during the late summer–autumn period when the reverse situation applies. This is much like a stock farm that produces silage from excess grass production through the spring and early summer, then feeds it to livestock over the winter—subsequent to its enrichment by bacteria. Anderson and P. J. le B. Williams (1998) examined this effect, the transfer of DOM from bacteria to zooplankton, in a model for DOC and DON set in the context of the English Channel. They observed that 40–50% of the dietary nitrogen of the zooplankton (microzooplankton in the study) originated from DON, assimilated initially by heterotrophic bacteria during some seasons (Figure 3). In this context the proposition that the temporary accumulation of DOC is caused by a "malfunctioning" microbial loop (Thingstad et al. 1997) is puzzling. Malfunction must imply a knowledge of design function, which for natural systems calls for divine insight. Rather than a malfunctioning microbial food web, it seems more likely that this is a malfunctioning hypothesis, arising from a view of the role of the marine food web in the microbial community that is acknowledged to be restricted.

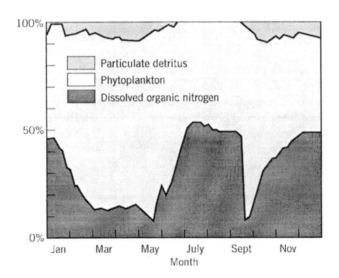

Figure 3. Relative contribution of various sources to the nitrogen diet of zooplankton. From Anderson and P. J. le B. Williams (1998).

The Control of Microbial Growth and the Effect of DOM Additions

In comparison with the soil and many fresh and brackish environments, the sea is organically poor. This was certainly the view of the early workers speculating on the nutrient status of heterotrophic microorganisms in the ocean (e.g., Krogh 1934). Subsequent analytical work confirmed these expectations, revealing that the concentration of individual molecules was characteristically at the nanomolar level. By contrast, much classical microbiology is carried out at the grams per liter (i.e., millimolar) level: that is, some million times greater. It therefore was natural to expect that the metabolism and growth of heterotrophic microorganisms in the oceans would be limited by the availability (i.e., the concentration) of organic substrates and that they would be energy-limited or in "carbon limitation." This view has been challenged by Thingstad and Lingell (1997) who, from theoretical analysis, argue that the whole microbial system (autotrophs plus heterotrophs) is nutrient-limited. They consider phosphate to be the limiting nutrient, but the concept could just as well apply to nitrogen or presumably in the broad sense iron. Upon analyzing the kinetics of the equilibrium between a nutrient, algae, bacteria, and protozoa, they concluded that bacterial abundance and metabolism are limited by the total nutrient present in the system and that in theory, adding more organic material would not result in an increase in bacterial biomass or metabolism.

Interestingly, in theory, it matters not whether the substrate contains the nutrient in organic form. The analysis, thus, appears to exclude the intermediate hypothesis that the severity of the competition for nutrients will be ameliorated when the organic molecule contains the nutrient. The theoretical background is discussed in Chapter 8. Here I shall simply review the field experiments of how heterotrophic bacteria responded to organic and inorganic additions.

Table 4 summarizes most of the field experiments involving dosing with organic and inorganic substrates undertaken in coastal and offshore environments and analyses them in the context of two apparently alternative hypotheses. The first hypothesis (hypothesis of organic limitation) proposes that microbial metabolism is controlled by the concentration of organic material (either containing nitrogen or not). It leads to the prediction that following an organic addition there is an increase in one or more of the following parameters: substrate uptake, respiration, microbial growth, and biomass accumulation. The alternative hypothesis (hypothesis of nutrient limitation) is that the addition of an organic compound alone will not cause an increase in any of the foregoing properties; the increase is seen only if the addition is accompanied with an inorganic nutrient addition, or the nutrient addition alone is made. The studies in Table 4 are organized broadly in relation to the trophic status of the environment. While noting the caution that the data are few and analysis is confounded by the great variety of approaches used, some trends seem to be apparent.

Most striking is the pattern observed for offshore mesotrophic areas, where there is no observation in support of the nutrient limitation hypothesis. Some type of organic addition stimulated microbial addition; but significantly, there was no consistency over the effective molecules. Amino acids often were the most effective (Kirchman 1990), although Cherrier et al. (1996) found that they only had a small effect on growth. The nutrient limitation hypothesis seems to be more frequently sustained in coastal regions. In many respects this is surprising, considering that these regions receive allochthonous inputs of nutrients and also regenerated nutrients from the underlying sediments. The data are too few to draw any reliable conclusion for oligotrophic regions. Elser et al. (1995) made an interesting comparison of the effect of organic and inorganic additions on bacterial growth rates in Wisconsin lakes and the open ocean. The pattern of response was similar in both environments: stimulation both by nutrient (usually nitrate and/or phosphate, but not both) and organic material (glucose) occurred, but surprisingly there was no additive effect when more than one compound was added. The extent of stimulation was manyfold greater in the freshwater environment.

Dose–response studies are in many ways unsatisfactory because they allow time-dependent shifts and adjustments in the populations and, as a result, they are as often as not open to a number of alternative explanations. Since probes now exist to establish nutrient (presently phosphate) control (Scanlan et al. 1997), we may expect an expansion of this type of methodology that will enable

Table 4. Short-term dose–response studies testing hypotheses about controls on bacterial activity[a]

Location and Reference	Nutrient or Organic Status	Additions	Organic Limitation Hypothesis	Nutrient Limitation Hypothesis
Coastal regions				
English Channel, Williams and Gray (1970)	Mid-summer, high NO_3, DOC, and DON	Glucose, amino acids	Supported	Not supported
Norwegian Fjord, Thingstad et al. (1999)	Low nutrients	PO_4, glycine	Not supported	Supported
Bothnian Sea, Zweifel et al. (1995)		PO_4, $PO_4 + NH_3$	Not supported	Supported
Conception Bay, Newfoundland, Pomeroy et al. (1991)	Spring bloom	Glucose + peptone	Supported	Not supported
Norwegian Fjord, Sondegaard et al. (in prep.)	Low nutrients	NO_3, PO_4, glucose	Partial support	Some support
Mesotrophic Offshore Regions				
Equatorial Pacific, Kirchman and Rich (1996)	HNLC area	Glucose, glucose + NH_3 amino acids, NH_3	Supported, with caveat[b]	Not supported

Subarctic Pacific, Kirchman (1990)	HNLC area	Glucose, glucose + NH_3, amino acids	Supported, with caveat[b]	Not supported
Subarctic Pacific, Keil and Kirchman (1991)		Glucose	Supported	Not supported
Northeastern Pacific, Cherrier et al. (1996)	June and October	Algal extract, NH_3, glucose, glucose + NH_3, PO_4, amino acids, urea	Partial support	Not supported
Gerlache Straits, Antarctic, Pakulski et al. (1996)		Fe	Not supported	Supported
Oligotrophic regions				
Sargasso Sea, Carlson and Ducklow (1996)	July; low N and P concentration and DOC accumulation	Glucose, NH_3, amino acids	Supported	Not supported
Gulf Stream and Central Pacific, Elser et al. (1995)		Glucose, NH_3, PO_4	Partial support[c]	Partial support[c]

[a] An expanded version of this table is given at the book's Web site:http://www.wiley.com/products/subject/life/kirchman.
[b] Organic nitrogen seems more important than carbon in limiting growth.
[c] Effects are very small.

more direct, as well as more extensive studies on the nutrient control of microorganisms.

Cross-Shelf Transport of Organic Material and the Balance of Overall Plankton Metabolism in the Open Oceans

The oceans are open systems and a priori we may expect transfers into and out of these systems. Rivers introduce organic material into the coastal region. Part of this material will enter the organic cycle of these areas, and part will simply pass through the system to enter the oceanic organic cycle. Since there is a general decrease in productivity offshore, there is the possibility of transfer of excess production from the shelf system into the open oceans.

Although these transfers are known in principle and organic material of terrestrial origin such as lignin may be detected in open ocean DOC (Opsahl and Benner 1997), we have only the most approximate notion of the quantities transferred. It is thought (see Hedges et al. 1997) that 70% or so of riverine particulate organic material is respired or deposited in estuarine or shelf regions. Opsahl and Benner (1997) concluded that the terrestrial material was rapidly decomposed in the marine system, where its residence time is short (21–132 years) compared with the refractory component of marine DOC (4000–6000 years). From budgets of biological production and utilization, Walsh et al. (1981) inferred the existence of extensive export of coastal production across the shelf system. A subsequent study by Falkowski et al. (1988) led to the conclusion that the shelf system was much more closely balanced, with only a small fraction (10–20%) of production unaccounted for. Curiously, neither of these studies considered water column oxidation of organic material by the microorganisms, and the scale of microbial utilization of plankton production identified in the earlier sections of this chapter would easily close these budgets without the need to attribute the surplus to export. The extent of net production in the coastal oceans and its export to the open oceans must be regarded as unknown at this time.

Working from a geochemical standpoint, Smith and Hollibaugh (1993) compiled a comprehensive mass balance for both the coastal and the open ocean systems. A subsequent study by Hedges and Keil (1995) would suggest minor corrections to the sedimentary loses. Smith and Hollibaugh's flow sheet, with minor revisions, is given in Figure 4; data on inputs and outputs are given in Table 1. The major uncertainty in the budget is the transfer from the coastal to the open-ocean system. This derives from the uncertainty of the net community production in both coastal and open-ocean waters. Smith and Hollibaugh estimate a net production of $7\,\text{Tmol}\,\text{C}\,\text{y}^{-1}$ for the world's shelf–estuarine ecosystem (a factor of 10 less than the lower estimate derived by Falkowski et al. 1988), with an uncertainty in the region of 50%. There appears to be an uncertainty in the region of $\pm 1.5\,\text{Tmol}\,\text{C}\,\text{y}^{-1}$ over coastal sedimentation. Thus the resultant export figure for the coastal regions ($15\,\text{Tmol}\,\text{C}\,\text{y}^{-1}$) has an overall uncertainty of about $\pm 4\,\text{Tmol}\,\text{C}\,\text{y}^{-1}$. The oceanic budget has

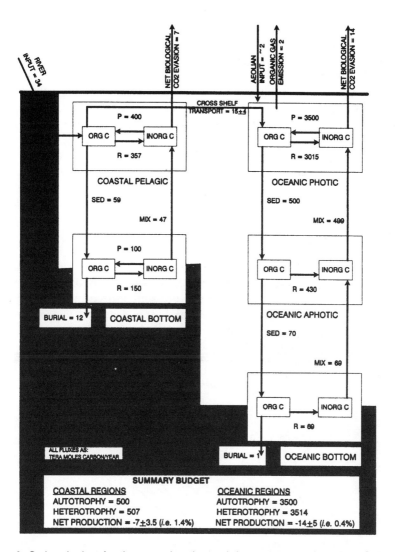

Figure 4. Carbon budget for the coastal region and the open ocean based on Smith and Hollibaugh (1993), with modifications from Hedges and Keil (1995): SED, sedimentation; ORG C, organic carbon; INORG C, inorganic carbon; R, respiration; P, photosynthesis; MIX, mixing rate between the deep and surface ocean.

further uncertainties with regard to the estimates for the annual rates of gas exchange, aeolian input, and net sedimentation. The combined error of these fluxes is probably about $1–2\,\mathrm{Tmol\,C\,y^{-1}}$.

The overall budget leads to the conclusion that the open oceans are net heterotrophic and that net oceanic consumption is $14 \pm 5\,\mathrm{Tmol\,C\,y^{-1}}$. This is also an estimate of the net emission of carbon dioxide, originating from biological processes, from the open oceans. For the marine system as a whole, net production is estimated as $21 \pm 5\,\text{Tmol}\,\text{C}\,\text{y}^{-1}$. (Note that the error remains the same, since the error in the coastal budget is included in that of the oceanic budget.) Although by themselves these figures are substantial, the carbon budget for the oceans is in fact very finely balanced, with only a 0.4% difference between autotrophy and heterotrophy. From an analysis of field observations of respiration and photosynthesis Williams (1998) came to the view that there were no grounds for large mismatches regionally or globally between autotrophy and heterotrophy. There are, however, dissenting views; from similar data sets, del Giorgio et al. (1997) and Duarte and Agusti (1998) found major regional deficits.

THE OCEANIC CYCLE OF DOC

The aim of this section is to explore the nature of the processes that shape the distribution of DOC and its properties on the time scale of oceanic circulation. The essential properties to be explained are summarized in Figures 5 and 6. The work of the late Peter M. Williams and his colleagues has provided much of the basic data for this discussion, and their seminal work on the ^{14}C-determined age of DOC is vital in constraining the system. The mean age of the deep-water DOC is equal to three to four times the cycling times of the

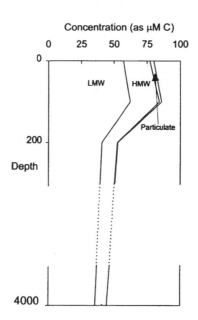

Figure 5. Generalized oceanic profile of DOC molecular fractions: LMW, low molecular weight; HMW, high molecular weight. Redrawn from Benner (1998).

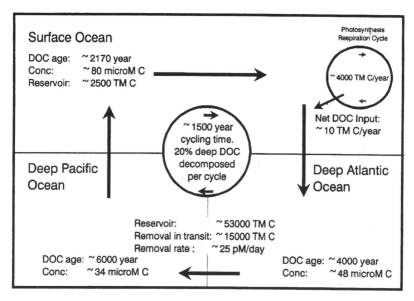

Figure 6. Reservoirs, concentrations, fluxes, and ^{14}C-determined ages for DOC in the oceans. Data principally from Druffel et al. (1992).

ocean. As a consequence, the surface water will contain a mixture of very old upwelled DOC plus young DOC, the latter produced during the time the water had been in the euphotic zone or in contact with it. Unfortunately, the ^{14}C age of the DOC in surface waters cannot be measured directly because these waters contain ^{14}C-labeled carbon originating from nuclear bomb testing. This results in the loss of a valuable constraint. The age of surface water DOC can, however, be deduced with apparent reliability (P. M. Williams and Druffel 1987; Druffel et al. 1992). The two matters that are poorly understood are the time course of the decomposition of the old refractory DOC and the processes giving rise to the apparently selective utilization of molecules of different composition and molecular weight. There are, in principle, three fundamentally different modes of decomposition: during the cycling of water between the surface and deep ocean, decomposition may be continuous, periodic, or intermittent. These three modes are not mutually exclusive.

Continuous Decomposition

The conventional view is that microbial decomposition is continuous and may be treated as a parallel series of first-order reactions, modeled with a spread of first-order rate constants. Equations of this type give a satisfactory account of short-term decomposition of DOC (e.g., Ogura 1973, 1975) and can indeed give satisfactory description of the type of vertical DOC profile (Figure 6). The question is whether the approach is sound for the implied long time scales of decomposition of DOC in deep water. The difficulty is the very low rate constants required to account for DOC with ages in the region of 4000–6000 years. The mean rate constant for deep water decomposition will be approximately the reciprocal of the ^{14}C-age (i.e., 1/6000 to 1/4000 y^{-1}). Whereas there are no particular constraints to purely chemical reactions (e.g., radioactive decay constants for the isotopes of thorium range from 10^{-7} second to 10^{10} years) there are upper and lower limits to microbiological reactions. The upper is set by diffusion and the processing rate of enzyme sites, the lower by the need for a constant input of energy to maintain the system against the forces of entropy. It is difficult to estimate this minimum rate with any exactness, but in an oxygenated aquatic environment it is likely to be not a great deal less than reciprocals of years to tens of years and unlikely to be in reciprocals of millennia or tens of millennia. Thus while we may have no particular problem accepting some form of conceptual model for DOC decomposition based on first-order kinetics during the period of time the DOC is in surface water, it is hard to envisage this simply continuing on for a millennium or so once the DOC has been advected into the deep ocean. A second problem is that this type of model offers no explanation for the predominance of low molecular weight material in deep water.

Periodic Decomposition

An alternative possibility, namely, that decomposition terminates in deep water, avoids the problem of ultraslow decomposition rates. The essence of this model is that decomposition occurs mainly in surface water and is halted in deep water. There has to be a sequel to this; otherwise the oceans would perpetually accumulate organic material. We can envisage a number of mechanisms that switch off decomposition in deep water and surface water. Two possible "on/off" switches related to concentration of individual molecules were discussed and eliminated in an earlier section (see Boxes 1 and 2). Other possible switches are given in Table 5.

Table 5. Potential "off" and "on" Switches of Heterotrophic Metabolism in Deep Water

Potential Stop/Start Mechanism	Deep Water "Off" Switch	Surface Water "On" Switch
1. Fe-limitation	Exhaustion of available Fe	Photochemical production of Fe^{2+}
2. Release of toxic metal	Chelating properties of DOM reduced in deep water	Surface water cycling of DOC resulting in complexation of toxic metals
3. Co-metabolism	Exhaustion of co-metabolites	Formation of co-metabolites
4. Accumulation of stable molecules	Rundown of DOC cycling results in accumulation of resistant DOC	Stable structures slowly photodegraded

The first three switches would not require deep-water DOC to be refractory, inasmuch as the cessation of decomposition is instigated by some other factor. This gives rise to the fatal consequence that on arrival at the surface, decomposition would be universally switched on, so thus the deep-water DOC would enter the planktonic cycle, resulting in extensive or complete rejuvenation of the DOC—which is contrary to observation. The fourth "on/off" switch offers a solution to this problem by incorporating the concept of substitution of xenophoric groups—the "off" switch. Essential to the argument is the noted asymmetry between chemical and biochemical reactivities of the xenophore substitutions. The "on" switch would be the photochemical removal of the xenophore group, freeing up the remainder of the molecule to microbial decomposition. The photochemical production of labile from refractory DOC has been discussed by Mopper et al. (1991); see also Chapter 7. The elegant study by Cherrier et al. (1999) has provided direct demonstration of the uptake and incorporation of old DOC in surface water by marine bacteria. The foregoing type of mechanism requires the photochemical reaction to be slow (i.e., far from complete during the time the water remains in the illuminated part of the oceans)—as would appear to be the case—otherwise, we return to the problem above. That is, the deep-water DOC would be rapidly removed by decomposition along with the recycling of surfaced-produced DOC, and so the DOC would be rejuvenated.

The explanation based on photochemical decomposition of xenophoric structures would account for the observed accumulation of low molecular material in the refractory pool, since high molecular weight conjugated structures would be more effective targets for photons and thus more susceptible to photochemical attack. This type of model for the cycling of oceanic DOC was used by Anderson and P. J. le B. Williams (1999), who successfully reproduced the characteristic profile of DOC (e.g., Figure 5), based on rate constants taken from the literature. By itself the explanation would restrict decomposition to the upper waters of the ocean and thus could not explain the 29% reduction in deep-water DOC concentrations in transit from the North Atlantic to the Pacific Ocean (see Hansell and Carlson 1998a and Figure 6). This gives a minimum estimate of the in situ removal of DOC, and it exceeds the calculated figure of 20% per oceanic cycle derived by Druffel et al. (1992) from steady state considerations. Thus, it would appear that only a small fraction of the decomposition or removal of refractory DOC can occur in surface water by whatever mechanism—co-metabolism or photochemical removal of xenophoric substitutions.

Intermittent Decomposition

There is a continual rain of particulate material through the deep water during its transit in the deep oceans. Some or many of these particles will contain attached bacteria. The transit time of these particles is perhaps less than a year to a few years (i.e., a thousandfold less than the transit time of the deep water

---

Box 3. The flux of particles during the passage of water through the deep ocean.

Question: What is the total flux of particles through a 1 cm^2 section during the time water circulates through the deep ocean?

Assume a mean gross loss by sedimentation into the deep ocean to be between 1 and 10% of surface production, with 2% as a conservative figure.

Given a mean global production rate of $150 \, \text{g C m}^{-2} \, \text{y}^{-1}$, this gives a flux rate of $150 \times 0.02/12 \, \text{mol C m}^{-2} \, \text{y}^{-1}$ or, during the 1500-year transit in the deep oceans, $375 \, \text{mol C m}^{-2}$ or $\sim 400{,}000 \, \mu\text{mol C cm}^{-2}$ over 1500 years.

Assume that a 100 μm diameter particle contains 0.001 μmol C; then over the deep-water transit time $400{,}000/0.001 = 4 \times 10^8$ 100 μm diameter particles would pass through each 1 cm^2 section of deep water.

Similarly, $400{,}000/1 = 4 \times 10^5$ 1 mm particles will pass through each 1 cm^2 section.

---

itself). Thus if decomposition of deep-water DOC took place in short bursts during the passage of a particle rather than as a continuous process, we get away from the problem of long-time constants required in the case of continuous in situ decomposition. In Box 3, the number of particles passing through a 1 cm^2 section of deep-ocean water is estimated. This size scale is chosen because it has a characteristic diffusion time of about 10 minutes, which scales to the possible particle settling speeds of 0.1–1 cm min^{-1}. Given the assumptions in Box 3, over the 1500-year passage through the deep oceans, some 4×10^8 particles of 100 μm diameter or 4×10^5 millimeter-sized particles will pass within the diffusion distance of every deep-water DOC molecule. It would be surprising if during this high number of opportunities, microbial assimilation or chemical absorption of dissolved organics (either or a combination of the two mechanisms would suffice) did not occur. Evidence for adsorption of DOC by deep-water particles is given by Druffel et al. (1996). Absorption may be expected to give rise to preferential removal of high molecular weight material and so satisfy this requirement.

In summary, of the three mechanisms considered, the third involving particle-facilitated removal appears to be the only one to satisfy all the requirements set out in Figures 5 and 6. The second is unable to account for the observed reduction in DOC decomposition in deep water when in transit, but it is able to account for the observations of Cherrier et al. (1999) that deep-water DOC is utilized by bacterial in the surface oceans. Further, with rates abstracted from the literature, it gives a satisfactory vertical profile of DOC. The first explanation, continuous first-order decomposition, is seen to be unsatisfactory on a number of accounts.

CONCLUSIONS

The review has built on the existing concepts of the bacterial flux of DOC in the pelagic system in the oceans and, in addition, new avenues have been explored. I have approached the quantification of the flux of DOC through the microbial system from four different directions. Projections downward from overall measurements of plankton community respiration and upward from bacterial growth rate measurements appear to give answers of the same order, but when looked at hard, not much better. Calculations from respiration and measured DOC flux give figures for total bacterial carbon demand or total DOC flux (they are tacitly taken to be the same) in the range of 15–50% of phytoplankton carbon fixation (this chapter and Chapter 5). The point was made earlier that some of the estimates will be low because they will fail to include the rapidly cycling fraction. The impression one gains is that these figures suggest rates much the same as the earlier, largely intuitive estimates (e.g., P. J. le B. Williams 1981a) and continue to substantiate the claims of this and other authors (Pomeroy 1974; Azam et al. 1983). However, when other calculations are included, the situation is seen to be far from tidy. Hansell and Carlson's careful study implies a global rate of net (semilabile) DOC production of 0.1 Pmol y^{-1} (Hansell and Carlson 1998b), that is, only 2% of global primary production; but this is net DOC production, not gross. Conversely, projections from microbial growth measurements are in the region of 130–1000% (see calculations in this chapter) and give rise to a mass balance problem. Given the bacterial growth yield of 15% for oceanic bacterial estimated by del Giorgio and the need to sustain micro- and mesozooplankton metabolism (a sum of 123% of phytoplankton production is estimated in Chapter 5), it is difficult to explain these high percentages as a consequence of recycling of organic material within the food web.* This may be due to an overestimate of bacterial production from the leucine and thymidine techniques or (and) an overestimate of bacterial carbon demand from production because of a low value used for growth yield. The arithmetic is very sensitive to the figure used for growth yield. The increase in carbon flow due to recycling and the projection of bacterial carbon demand from bacterial production respond oppositely to the figure adopted for growth yield; the problem disappears with growth yields above 25–30%. It would be easy to accept the foregoing implications and discard the low growth yields were it not for the thoroughness of the del Giorgio and Cole review (Chapter 10). Thus, it seems that we must bite the bullet and acknowledge that we face a mass balance problem when we attempt to embed the bacterial production measurements into autochthonous plankton carbon flux. One could solve the problem by attributing the imbalance to input of allochthonous carbon (Duarte and Agustí 1998), but that simply

---

*Maximum yield as a result of recycling is equal to $Y^0 + Y^1 + Y^2 + Y^3 + Y^4 \cdots = 1/(1 - Y)$, where Y is the growth yield (as a fraction). Thus for $Y = 0.15, 0.2, 0.3,$ and 0.5, the maximum yields are respectively 1.18, 1.25, 1.43, and 2.0.

substitutes one mass balance problem for another (P. J. le B. Williams and Bowers 1999).

The two other approaches have been analyzed within the generalization (Kirchman et al. 1993) of labile and semilabile fractions in the DOC assemblage of molecules. Whereas it has been possible from these four approaches to make broad statements of the general flux of DOC and its appointment, we have a woefully incomplete understanding of the temporal and spatial aspects of microbial metabolism. One must reinforce the observation of Ducklow and Carlson (1992) that there is a serious imbalance between our knowledge of heterotrophic and autotrophic processes. If we aspire to understand and quantify processes such as carbon balance in the oceans, it is both bad economics and virtually pointless to perpetuate the present situation of having an enormously detailed record of one side of the balance sheet (autotrophy) and a rudimentary understanding of the other (heterotrophy).

The persistence of dissolved organic compounds in the oceans has been examined from various aspects and a number of causes have been discussed and examined. It has been shown from thermodynamic and kinetic calculations that concentration (more correctly, dilution) would not be a likely cause of persistence at concentrations above 10^{10} molar — presently beyond the limit of detection for compounds such as monosaccharides and amino acids. The creation of resistance in organic compounds by the substitution of various groups to give so-called xenophoric molecules has been discussed in conjunction with the axiom of universal degradability of biochemical products. The concept is used as a means of accounting for the persistence of DOC material in seawater and, in principle, the predominance of low molecular weight DOC material in deep-ocean water. Other ephemeral forms of resistance have been discussed.

SUMMARY

1. We lack a clear understanding why organic material persists in seawater, and until this situation improves, models of global DOC distribution must be seen to be tentative. Obtaining a more detailed picture of the concentrations of individual low molecular weight organic compounds in deep water is an achievable short-term target that would help resolve between alternative hypotheses.

2. Whereas a figure of 50% is a commonly encountered one for the fraction of primary production that passes through DOM prior to decomposition, careful contemporary studies give rise to values as low as 2% and as high as 130%. In part this variability reflects the insufficient database on heterotrophic processes in the global ocean. That the oceans are grossly undersampled for microbial rates cannot be overstressed.

3. A variety of suitable methods to study microbial processes are now available. Whereas most marine microbiologists would be inclined to use either thymidine or leucine incorporation to measure bacterial growth rates, because of uncertainties over conversion factors (see item 4), presently projections from bacterial carbon metabolism are probably most accurately made from respiration measurements.

4. A major factor limiting our understanding of the exact role of bacteria in ocean carbon flux is the uncertainty over the carbon growth yield of oceanic bacteria; its value will determine the potential scale of organic recycling and the balance between planktonic bacteria as mineralizers or recyclers of organic material. The general downward revision of the value for bacterial growth yield in recent years swings the balance toward bacteria as a sink of organic material rather than a link.

5. We are going through a revision of our expectation of the factors that limit the growth rate of heterotrophic bacteria. Simple carbon limitation no longer appears to be the exclusive mechanism.

6. Perhaps the single most striking thing to come of this review (and also Chapter 7 by Moran and Zepp) is the rapidly growing awareness of the importance of chemical, and particularly photochemical, reactions. Two things are very clear. First, the scale of photochemical reactions is significant in the overall turnover of DOC. In one instance (Moran and Zepp, 1997), rates of photochemical production of carbon dioxide from DOC are reported that would be comparable to rates of microbial respiration. Second, all the signs are that chemical (both dark and photochemical) and microbiological transformations of dissolved material cannot be considered as isolated routes of decomposition. There is now compelling evidence to show that they operate in concert. This is the new paradigm emerging for the organic decomposition in marine and fresh waters (where interest in photochemical work is equally if not more active), and we can expect the next 3–5 years to result in a major revision of our perception of the functioning of the organic cycle in aquatic systems. This will raise some interesting questions concerning organic decomposition in particular areas of the oceans such as the Antarctic and the deep ocean, which are subject to prolonged periods of darkness and the consequence of an increasing N-flux to high latitude waters.

ACKNOWLEDGMENTS

I am pleased to acknowledge Drs. Rubina Rodrigues, Carol Robinson, and David Thomas for their thoughtful and constructive comments on the text and David Kirchman for his rigorous editing. I am greatly indebted to Lynda Williams for typing the first draft of the manuscript.

REFERENCES

Alexander, M. (1994) *Biodegradation and Bioremediation.* Academic Press, San Diego, CA.

Alldredge, A. L., Cole, J. J., and Caron, D. A. (1986) Production of heterotrophic bacteria inhabiting macroscopic organic aggregates (marine snow) from surface waters. *Limnol. Oceanogr.* 31:68–78.

Aluwihare, L. I., Repeta, D. J., and Chen, R. F. (1997) A major biopolymeric component to dissolved organic carbon in surface sea water. *Nature* 387:166–169.

Amon, R. M. W., and Benner, R. (1996) Bacterial utilization of different size classes of dissolved organic matter. *Limnol. Oceanogr.* 41:41–51.

Amon, R. M. W., and Benner, R. (1994) Rapid cycling of high-molecular-weight dissolved organic matter in the ocean. *Nature* 369:549–552.

Anderson, T. R. (1992) Modelling the influence of food C:N ratio, and respiration on growth and nitrogen excretion in marine zooplankton and bacteria. *J. Plankton Res.* 14:1645–1671.

Anderson, T. R., and Williams P. J. le B. (1998) Modelling the seasonal cycle of dissolved organic carbon at Station E_1 in the English Channel. *Estuarine. Coastal and Shelf Science* 46:93–109.

Anderson, T. R., and Williams, P. J. le B. (1999) A one-dimensional model of dissolved organic carbon cycling in the water column incorporating combined biological-photochemical decomposition. *Glob. Biogeochem. Cycle* 13:337–349.

Andrews, P., and Williams, P. J. le B. (1971) Heterotrophic utilisation of dissolved compounds in the sea. III. Measurements of the oxidation rates and concentrations of glucose and amino acids in sea water. *J. Mar. Biol. Assoc. UK* 51:111–125.

Aristegui, J., Montero, M. F., Ballesteros, S., Basterretxea, and van Lenning, K. (1996) Planktonic primary production and microbial respiration measured by ^{14}C assimilation and dissolved oxygen changes in coastal waters of the Antarctic Peninsula during the austral summer: Implications for carbon flux studies. *Mar. Ecol. Prog. Ser.* 132:191–201.

Azam, F. (1998) Microbial control of oceanic carbon flux: The plot thickens. *Science* 280:694–696.

Azam, F., and Holm-Hansen, O. (1973) Use of tritiated substrates in the study of heterotrophy in seawater. *Mar. Biol.* 23:191–196.

Azam, F., Fenchel, T., Field, J. G., Gray, J. S., Meyer-Reil, L., and Thingstad, F. (1983) The ecological role of water column microbes in the sea. *Mar. Ecol. Prog. Ser.* 10:257–263.

Azam, F., Smith, D. C., and Carlucci, A. F. (1992) Bacterial transformation and transport of organic matter in the Southern California Bight. *Prog. Oceanog.* 30:151–166.

Banoub, M. W., and Williams, P. J. le B. (1972) Measurement of microbial activity and organic material in the western Mediterranean Sea. *Deep-Sea Res.* 19:433–443.

Banoub, M. W., and Williams, P. J. le B. (1973) Seasonal changes in the organic forms of carbon, nitrogen and phosphorous in the English Channel in 1968. *J. Mar. Biol. Assoc. UK* 53:695–703.

Benner, R. H. (1998) Cycling of dissolved organic matter in the oceans. *Ecol. Stud.* 133:317–331.

Benner, R., Pakulski, J. D., McCarthy, M., Hedges, J. I., and Hatcher, P. G. (1992) Bulk chemical characteristics of dissolved organic matter in the ocean. *Science* 255:1561–1564.

Biddanda, B., and Benner, R. (1997) Major contribution from mesopelagic plankton to heterotrophic metabolism in the upper ocean. *Deep-Sea Res. I* 44:2069–2085.

Billen, G., Joiris, C., Wijnant, I., and Gillain, G. (1980) Concentration and microbiological utilization of small organic molecules in the Scheld Estuary, the Belgian Coastal Zone of the North Sea and the English Channel. *Estuarine Coastal Mar. Sci.* 11:279–294.

Blight, S. P., Bentley, T. L., Lefèvre, D., Robinson, C., Rodrigues, R., Rowlands, J., and Williams, P. J. le B. (1995) The phasing of autotrophic and heterotrophic plankton metabolism in a temperate coastal ecosystem. *Mar. Ecol. Prog. Ser.* 128:61–74.

Borch, N. H., and Kirchman, D. L. (1997) Concentration and composition of dissolved combined neutral sugars (polysaccharides) in seawater determined by HPLC-PAD. *Mar. Chem.* 57:85–95.

Bouquegneau, J. M., Gieskes, W. W. C., Kraay, G. W., and Larsson, A. M. (1992) Influence of physical and biological processes on the concentration of O_2 and CO_2 in the ice-covered Weddell Sea in the spring of 1988. *Polar Bio.* 12:163–170.

Boyd, P., Robinson, C., Savidge, G., and Williams, P. J. le B. (1995) Water column and sea ice primary production during austral spring in the Bellingshausen Sea. *Deep-Sea Res. II* 42:1177–1200.

Buat-Ménard, P., Cachier, H., and Chesselet, R. (1989) Sources of particulate carbon in the marine atmosphere. In J. P. Riley and R. Chester, eds., *Chemical Oceanography*, Vol. 10. Academic Press, London, pp. 251–279.

Burton, J. D., and Statham, P. J. (1990) Trace metals in sea water. In R. Furness and P. S. Rainbow, eds., *Heavy Metals in the Marine Environment*. CRC Press, Boca Raton, FL, pp. 5–25.

Carlson, C. A., and Ducklow, H. W. (1995) Dissolved organic carbon in the upper ocean of the Central Equatorial Pacific Ocean, 1992: Daily and finescale vertical variations. *Deep-Sea Res. II* 42:639–656.

Carlson, C. A., and Ducklow, H. W. (1996) Growth of bacterioplankton and consumption of dissolved organic carbon in the Sargasso Sea. *Aquat. Microb. Ecol.* 10:69–85.

Carlson, C. A., Ducklow, H. W., and Michaels, A. F. (1994) Annual flux of dissolved organic carbon from the euphotic zone in the northwestern Sargasso Sea. *Nature* 371:405–408.

Carlson, C. A., Ducklow, H. W., Hansell, D. A., and Smith, W. O. (1998) Carbon dynamics during spring blooms in the Ross Sea polynya and the Sargasso Sea: Contrasts in dissolved and organic carbon partitioning. *Limnol. Oceanogr.* 43:375–386.

Carlucci, A. F., Craven, D. B., Robertson, K. J., and Henrichs, S. M. (1986) Microheterotrophic utilization of dissolved free amino acids in depth profiles of Southern California Borderland basin waters. *Oceanol. Acta* 9:89–96.

Carlucci, A. F., Wolgast, D. M., and Craven, D. B. (1992) Microbial populations in surface films: Amino acid dynamics in nearshore and offshore waters off Southern California. *J. Geophy. Res. C* 97:5271–5280.

Cherrier, J., Bauer, J., and Druffel, E. R. M. (1996) The utilization and turnover of labile dissolved organic matter by bacterial heterotrophs in eastern North Pacific waters. *Mar. Ecol. Prog. Ser.* 139:267–279.

Cherrier, J., Bauer, J., Druffel, E. R. M., Coffin, R. B., and Chanton, J. P. (1999) Radiocarbon in marine bacteria: Evidence for the ages of assimilated carbon. *Limnol. Oceanogr.* 44:730–736.

Clark, L. L., Ingall, E. D., and Benner, R. (1998) Marine phosphorus is selectively mineralised. *Nature* 393:426.

Cole, J. J., and Pace, M. L. (1995) Why measure bacterial production? *Limnol. Oceanogr.* 40:441–444.

Craven, D. B., and Carlucci, A. F. (1989) Spring and fall microheterotrophic utilization of dissolved free amino acids in a Californian Borderland basin. *Mar. Ecol. Prog. Ser.* 51:229–235.

del Giorgio, P. A., Cole, J. J., and Cimbleris, A. (1997) Respiration rates in bacteria exceed plankton production in unproductive aquatic systems. *Nature* 385:148–151.

Donat, J. R., and Bruland, K. W. (1994) Trace elements in the ocean. In B. Salbu and E. Steinnes, eds., *Trace Elements in Natural Waters.* CRC Press, Boca Raton, FL, pp. 247–281.

Druffel, E. R. M., Williams, P. M., Bauer, J. E., and Ertel, J. R. (1992) Cycling of dissolved and particulate organic matter in the open ocean. *J. Geophys. Res. Oceans* 97:15639–15659.

Druffel, E. R. M., Bauer, J. E., Williams, P. M., Griffin, S., and Wolgast, D. (1996) Seasonal variability of particulate organic radiocarbon in the northeast Pacific Ocean. *J. Geophys. Res.* 101:20543–20552.

Duarte, C. M., and Agustí, S. (1998) The CO_2 balance of unproductive aquatic ecosystems. *Science* 281:234–236.

Duarte, C. M., and Cebrián, J. (1996) The fate of marine autotrophic production. *Limnol. Oceanogr.* 41:1758–1766.

Ducklow, H. W. (1999) The bacterial component of the oceanic euphotic zone. *FEMS Microb. Ecol.* 30:1–10.

Ducklow, H. W., and Carlson, C. G. (1992) Oceanic bacterial production. *Adv. Microb. Ecol.* 12:113–180.

Ducklow, H. W., Purdie, D. A., Williams, P. J. le B., and Davies, J. H. (1986) Bacterioplankton: A sink for carbon in a coastal marine plankton community. *Science* 232:865–867.

Ducklow, H. W., Purdie, D. A., Williams, P. J. le B., and Davies, J. H. (1987) Bacteria. Link or sink? *Science* 235:88–89.

Duursma, E. K. (1961) Dissolved organic carbon, nitrogen and phosphorus in the sea. *Neth. I. Sea Res.* 1:1–147.

Duursma, E. K. (1963) The production of dissolved organic matter in the sea, as related to the primary production of organic matter. *Neth. J. Sea Rers.* 2:85–94.

Duursma, E. K. (1965) In J. P. Riley and G. Skirrow, eds., *Chemical Oceanography,* Vol. 1. Academic Press, London, pp. 433–475.

Elser, J. J., Stabler, B. L., and Hassett, R. P. (1995) Nutrient limitation of bacterial growth and rates of bactivity in lakes and oceans: A comparative study. *Aquat. Microb. Ecol.* 9:105–110.

Falkowski, P. G., Flagg, C. N., Rowe, G. T., Smith, S. L., Whitledge, T. E., and Wirick, C. D. (1988) The fate of a spring phytoplankton bloom: Export or oxidation. *Cont. Shelf Res.* 8:457–484.

Feiser, L., and Feiser, M. (1956) *Organic Chemistry*. Reinhold, New York.

Field, C. B., Behrenfeld, M. J., Randerson, J. T., and Falkowski, P. (1998) Primary production of the biosphere: Integration terrestrial and oceanic components. *Science* 291:237–240.

Foster, P., and Morris, A. W. (1971) The seasonal variation of dissolved ionic and organically associated copper in the Menai Strait. *Deep-Sea Res.* 18:231–236.

Fuhrman, J. A. (1987) Close coupling between release and uptake of dissolved free amino acids in seawater studied by an isotope dilution approach. *Mar. Ecol. Prog. Ser.* 37:45–52.

Fuhrman, J. A., and Ferguson, R. L. (1986) Nanomolar concentrations and rapid turnover of dissolved free amino acids in seawater: Agreement between chemical and microbiological measurements. *Mar. Ecol. Prog. Ser.* 33:237–242.

Gamble, J. C., and Davies, J. M. (1982) Application of enclosures to the study of marine pelagic systems. In G. D. Grice and M. R. Reeve, eds., *Marine Mesocosms*. Springer-Verlag, New York.

Goldman, J. C., Caron, D. A., and Dennett, M. R. (1987) Regulation of gross efficiency and ammonium regeneration in bacteria by substrate C:N ratio. *Limnol. Oceanogr.* 32:1239–1252.

Griffith, P. C., and Pomeroy, L. R. (1995) Seasonal and spatial variations in pelagic community respiration on the southeast U.S. continental shelf. *Cont. Shelf Res.* 15:815–825.

Griffith, P. C., Douglas, D. J., and Wainwright, S. C. (1990) Metabolic activity of size-fractionated microbial plankton in estuarine, near shore, and continental shelf waters of Georgia. *Limnol. Oceanogr.* 59:263–270.

Guo, L., Coleman, C. H., and Santschi, P. H. (1994) The distribution of colloidal and dissolved organic carbon in the Gulf of Mexico. *Mar. Chem.* 45:105–119.

Guo, L., Santschi, P. H., Warnken, K. W. (1995) Dynamics of dissolved organic carbon (DOC) in oceanic environments. *Limnol. Oceanogr.* 40:1392–1403.

Hansell, D. A., and Carlson, C. A. (1998a) Deep-ocean gradients in the concentration of dissolved organic carbon. *Nature* 395:263–266.

Hansell, D. A., and Carlson, C. A. (1998b) Net community production of dissolved organic carbon. *Global Geochem. Cycles* 12:443–453.

Harrison, W. G. (1986) Respiration and its size-dependence in microplankton populations from surface waters of the Canadian Arctic. *Polar Biol.* 6:145–152.

Harvey, H. W. (1950) On the production of living matter in the sea. *J. Mar. Biol. Assoc. UK* 29:97–138.

Hedges, J. I., and Keil, R. G. (1995) Sedimentary organic matter preservation: An assessment and speculative synthesis. *Mar. Chem.* 49:81–115.

Hedges, J. I., Keil, R. G., and Benner, R. (1997) What happens to terrestrial organic matter in the ocean? *Org. Geochem.* 27:195–212.

Hopkinson, C. S., Sherr, B., and Wiebe, W. J. (1989) Size fractionated metabolism of coastal microplankton. *Mar. Ecol. Prog. Ser.* 51:155–166.

Jahnke, R. A., and Craven, D. B. (1995) Quantifying the role of heterotrophic bacteria in the carbon cycle: The need for respiration rate measurements *Limnol. Oceanogr.* 40:436–441.

Kaiser, D. (1996) Bacteria also vote. *Science* 272:1598–1599.

Kaplan, W. A., and Wofsy, S. C. (1985) The biogeochemistry of nitrous oxide: A review. *Adv. Aquat. Ecol.* 3:181–203.

Keil, R. G., and Kirchman, D. L. (1993) Dissolved combined amino acids: Chemical form and utilization by marine bacteria. *Limnol. Oceanogr.* 38:1256–1270.

Keil, R. G., and Kirchman, D. L. (1994) Abiotic transformation of labile protein to refractory protein in sea water. *Mar. Chem.* 45:187–196.

Keil, R. G., and Kirchman, D. L. (1991) Contribution of dissolved free amino acids and amonium to the nitrogen requirements of heterotrophic bacterioplankton. *Mar. Ecol. Prog. Ser.* 73:1–10.

Keil, R. G., Montlucon, D. B., Prahl, F. G., and Hedges, J. I. (1994) Sorptive preservation of labile organic matter in marine sediments. *Nature* 370:549–552.

Kiddon, J., Bender, M. L., and Marra, J. (1995) Production and respiration in the 1989 North Atlantic spring bloom: An analysis of irradiance-dependent changes. *Deep-Sea Res.* 42:553–576.

Kieber, D. J., and Mopper, K. (1987) Photochemical formation of glyoxylic and pyruvic acids in seawater. *Mar. Chem.* 21:135–149.

Kieber, D. J., McDaniel, J., and Mopper, K. (1989) Photochemical source of biological substrates in sea water: Implications for carbon cycling. *Nature* 341:637–639.

Kirchman, D. L. (1990) Limitation of bacterial growth by dissolved organic matter in the subarctic Pacific. *Mar. Ecol. Prog. Ser.* 62:47–54.

Kirchman, D. L., and Rich, J. H. (1997) Regulation of bacterial growth rates by dissolved organic carbon and temperature in the Equatorial Pacific Ocean. *Microb. Ecol.* 33:22–30.

Kirchman, D. L., Keil, R. G., Simon, M., and Welschmeyer, N. A. (1993) Biomass and production of heterotrophic bacterioplankton in the oceanic subarctic Pacific. *Deep-Sea Res. I* 40:967–988.

Kirchman, D. L., Ducklow, H. W., McCarthy, J. J., and Garside, C. (1994) Biomass and nitrogen uptake by heterotrophic bacteria during the spring phytoplankton bloom in the North Atlantic Ocean. *Deep-Sea Res.* 41:879–895.

Krogh, A. (1934) Conditions of life in the ocean. *Ecol. Monogr.* 4:421–429.

Lara, R. J., and Thomas, D. N. (1995) Formation of recalcitrant organic matter: Humification dynamics of algal-derived dissolved organic carbon and its hydrophobic fractions. *Mar. Chem.* 51:193–199.

Lee, S. H., Kang, Y.-C., and Fuhrman, J. A. (1995) Imperfect retention of natural bacterioplankton cells by glass fibre filters. *Mar. Ecol. Prog. Ser.* 119:285–290.

Lefèvre, D., Minas, H. J., Minas, M., Robinson, C., Williams, P. J. le B., and Woodward, E. M. S. (1997) Review of gross community production, primary production, net community production and dark community respiration in the Gulf of Lions. *Deep Sea Res. II* 44:801–832.

Libby, P. S., and Wheeler, P. A. (1997) Particulate and dissolved organic nitrogen in the central and eastern equatorial Pacific. *Deep-Sea Res.* 44:354–361.

Libes, S. (1992) *An Introduction to Marine Biogeochemistry.* Wiley, New York.

McCarthy, M., Hedges, J., and Benner, R. (1996) Major biochemical composition of dissolved high molecular weight organic matter in seawater. *Mar. Chem.* 55:281–297.

McCarthy, M. D., Hedges, J. I., and Benner, R. (1998). Major bacterial contribution to marine dissolved organic nitrogen. *Science* 281:231–234.

Mitchell, J. M., Okubo, A., and Fuhrman, J. (1985) Microzones surrounding phytoplankton form the basis for a stratified marine microbial ecosystem. *Nature* 316:58–59.

Moffett, J. W., and Brand, L. E. (1996) Production of strong extracellular Cu chelators by marine cyanobacteria in response to Cu stress. *Limnol. Oceanogr.* 41:388–395.

Mopper, K., Zhou, X., Kieber, R. J., Kieber, D. J., Sikorski, R. J., and Jones, R. D. (1991) Photochemical degradation of dissolved organic carbon and its impact on the oceanic carbon cycle. *Nature* 353:60–62.

Moran, M. A., and Zepp, R. G. (1997) Role of photoreactions in the formation of biologically labile compounds from dissolved organic matter. *Limnol. Oceanogr.* 42:1307–1316.

Müller-Niklas, G., Agis, M., and Herndl, G. J. (1996) Microscale distribution of bacterioplankton in relation to phytoplankton: Results from 100 nL samples. *Limnol. Oceanogr.* 41:1577–1582.

Nagata, T., and Kirchman, D. L. (1996) Bacterial degradation of protein adsorbed to model submicron particles in sea water. *Mar. Ecol. Prog. Ser.* 132:241–248.

Ogura, N. (1973) Rate and extent of decomposition of dissolved organic matter in surface seawater. *Mar. Biol.* 13:89–93.

Ogura, N. (1975) Further studies on decomposition of dissolved organic matter in coastal seawater. *Mar. Biol.* 13:101–111.

Opsahl, S., and Benner, R. (1997) Distribution and cycling of terrigenous dissolved organic matter in the ocean. *Nature* 386:480–482.

Packard, T. T., and Williams, P. J. le B. (1981) Rates of respiratory oxygen consumption and electron transport in surface seawater from the northwest Atlantic. *Oceanol. Acta* 4:351–358.

Pakulski, J. D., and Benner, R. (1994) Abundance and distribution of carbohydrates in the ocean. *Limnol. Oceanogr.* 39:930–940.

Pakulski, J. D., Coffin, R. B., Kelley, C. A., Holder, S. L., Downer, R., Aas, P., Lyons, M. M., and Jeffrey, W. H. (1996) Iron stimulation of Antarctic bacteria. *Nature* 383:133–134.

Palenik, B., Kieber, D. J., and Morel, F. M. M. (1989) Dissolved organic nitrogen use by phytoplankton: The role of cell-surface enzymes. *Biol. Oceanogr.* 6:347–354.

Pantoja, S., and Lee, C. (1994) Cell-surface oxidation of amino acids in seawater. *Limnol. Oceanogr.* 39:1718–1726.

Parsons, T. R., and Strickland, J. D. H. (1962) On the production of particulate organic carbon by heterotrophic processes in sea water. *Deep-Sea Res.* 8:211–222.

Parsons, T. R., Albright, L. J., Whitney, E., Wong, C. S., and Williams, P. J. le B. (1981) The effect of glucose on the productivity of seawater: An experimental approach using controlled aquatic ecosystems. *Mar. Environ. Res.* 4:229–242.

Peltzer, E. T., and Hayward, N. A. (1996) Spatial and temporal variability of total organic carbon along 140°W in the equatorial Pacific. *Deep-Sea Res.* 43:1155–1180.

Pomeroy, L. R. (1974) The oceans' food web, a changing paradigm. *BioScience* 24:499–504.

Pomeroy, L. R., Wiebe, W. J., Deibel, Thompson, R. J., Rowe, G. T., and Pakulski, J. D. (1991) Bacterial responses to temperature and substrates concentration during the Newfoundland spring bloom. *Mar. Ecol. Prog. Ser.* 75:143–159.

Purcell, E. M. (1977) Life at low Reynolds numbers. *Am. J. Phys.* 45:3–11.

Rich, J. H., Ducklow, H. W., and Kirchman, D. L. (1996) Concentrations and uptake of neutral monosaccharides along 140 degrees W in the Equatorial Pacific: Contribution of glucose to heterotrophic bacterial activity and the DOM flux. *Limnol. Oceanogr.* 41:595–604.

Robinson, C., and Williams, P. J. le B. (1993) Temperature response of Antarctic plankton respiration. *J. Plankton Res.* 15:1035–1051.

Robinson, C., and Williams, P. J. le B. (1999) Plankton net community production and dark respiration in the Arabian Sea during September 1994. *Deep-Sea Res. II* 46:745–765.

Robinson, C., Archer, S. D., and Williams, P. J. le B. (1999) Microbial dynamics in coastal waters of East Antarctica: Plankton production and respiration. *Mar. Ecol. Prog. Ser.* 180:23–36.

Scanlan, D. J., Silman, N. J., Donald, K. M., Wilson, W. H., Carr, N. G., Jiont, I., and Mann, N. H. (1997) An immunological approach to detect phosphate stress in populations and single cells of photosynthetic picoplankton. *Appl. Environ. Microbial.* 63:2411–2420.

Sharp, J. H. (1973) Size classes of organic carbon in sea water. *Limnol. Oceanogr.* 18:441–447.

Sharp, J. H., Benner, R., Bennett, L., Carlson, C. A., Fitzwater, S. E., Peltzer, E. T., and Tupas, L. M. (1995) Analysis of dissolved organic carbon in seawater: The JGOF Eq Pac methods comparison. *Mar. Chem.* 48:91–108.

Sherr, B. F., Sherr, E. B., and Hopkinson, C. S. (1988) Trophic interactions within pelagic microbial communities: Indications of feedback regulation of carbon flow. Special issue: The role of microorganisms in aquatic environments. *Hydrobiologia* 159:19–26.

Sherr, E. B., Sherr, B. F., and Albright, L. J. (1987) Bacteria: Link or sink. *Science* 235:288.

Sieburth, J. McN. (1977) International Helgoland Symposium: Convener's report. *Helgoland Wiss. Meeresunters.* 30:565–574.

Skoog, A., and Benner, R. (1997) Aldolose in various size fractions of marine organic matter. Implications for carbon cycling. *Limnol. Oceanogr.* 42:1803–1813.

Smith, S. V., and Hollibaugh, J. T. (1993) Coastal metabolism and the oceanic organic carbon balance. *Rev. Geophys.* 31:75–89.

Taylor, G. T. (1995) Microbial decomposition of sorbed and dissolved protein in seawater. *Limnol. Oceanogr.* 40:875–885.

Thingstad, T. F., and Lignell, R. (1997) Theoretical models for the control of bacterial growth rate, abundance, diversity and carbon demand. *Aquat. Microb. Ecol.* 13:19–27.

Thingstad, T. F., Hagström, Å., and Rassoulzadegan, F. (1997) Accumulation of degradable DOC in surface waters: Is it caused by a malfunctioning microbial loop? *Limnol. Oceanogr.* 35:424–433.

Thingstad, T. F., Havskum, H., Kaas, H., Lefèvre, D., Nielsen, T. G., Reimann, B., and Williams, P. J. le B. (1999) Bacteria–protist interactions and organic matter degradation under P-limited conditions: Analysis of an enclosure experiment using a simple model. *Limnol. Oceanogr.* 44:62–79.

Walsh, J. J., Rowe, G. L., Iverson, R. L., and McRoy, C. P. (1981) Biological export of shelf carbon is a neglected sink of the global CO_2 cycle. *Nature* 291:196–201.

Wheeler, P. A., and Kirchman, D. L. (1986) Utilization of inorganic and organic nitrogen by bacteria in marine systems. *Limnol. Oceanogr.* 31:998–1009.

Williams, P. J. le B. (1975) Biological and chemical aspects of dissolved organic material in sea water. In J. P. Riley and G. Skirrow, eds., Vol. 2. Academic Press, London, pp. 301–363.

Williams, P. J. le B. (1981a) Incorporation of microheterotrophic processes into the classical paradigm of the food web. *Kieler Meeresforsch. Sonderh.* 5:1–28.

Williams, P. J. le B. (1981b) Microbial contribution to overall marine plankton metabolism: Direct measurements of respiration. *Oceanol. Acta* 4:359–364.

Williams, P. J. le B. (1984) Bacterial production in the marine food chain: The emperor's new suit of clothes. In M. J. R. Fasham, ed., *Flows of Energy and Material in Marine Ecosystems: Theory and Practice*. Plenum Press, New York, pp. 271–299.

Williams, P. J. le B. (1995) Evidence for the seasonal accumulation of carbon-rich dissolved organic material, its scale in comparison with changes in particulate material and the consequential effect on net C/N assimilation ratios. *Mar. Chem.* 51:17–29.

Williams, P. J. le B. (1998) The balance of plankton respiration and photosynthesis in the open oceans. *Nature* 394:55–57.

Williams, P. J. le B., and Askew, C. (1968) A method of measuring the mineralization by micro-organisms of organic compounds in sea water. *Deep-Sea Res.* 15:365–375.

Williams, P. J. le B., and Bowers, D. G. (1999) Major regional carbon imbalances in the oceans. *Science* 284:1735b.

Williams, P. J. le B., and Gray, R. W. (1970) Heterotrophic utilization of dissolved organic compounds in the sea. II. Observations on the responses of heterotrophic marine populations to abrupt increase in amino acid concentration. *J. Mar. Biol. Assoc. UK* 50:871–8881.

Williams, P. J. le B., and Jenkinson, N. W. (1982) A transportable micro-processor controlled Winkler titration suitable for field station and shipboard use. *Limnol. Oceanogr.* 27:576–584.

Williams, P. J. le B., and Purdie, D. A. (1991) *In vitro* and *in situ* derived rates of gross production, net community production and respiration of oxygen in the oligotrophic subtropical gyre of the North Pacific Ocean. *Deep-Sea Res.* 38:891–910.

Williams, P. J. le B., Berman, T., and Holm-Hansen, O. (1976) Amino acid uptake and respiration by marine heterotrophs. *Mar. Biol.* 35:41–47.

Williams, P. M. (1969) The association of copper with dissolved organic matter in seawater. *Limnol. Oceanogr.* 14:156–158.

Williams, P. M, and Carlucci, A. F. (1975) Bacterial utilization of organic matter in the deep sea. *Nature* 262:810–811.

Williams, P. M., and Druffel, E. R. M. (1987) Radiocarbon in dissolved organic matter in the North central Pacific Ocean. *Nature* 330:246–248.

Wright, R. T., and Hobbie, J. E. (1965) The uptake of organic solutes in lakes. *Limnol. Oceanogr.* 10:22–28.

Wuosmaa, A. M., and Hager, A. M. (1990) Methyl chloride transferase: A carboxylation route for the synthesis of halometabolites. *Science* 249:160–162.

ZoBell, C. E. (1976) Discussion. In C. Litchfield, ed., *Marine Microbiology*. Academic Press, London, p. 198.

Zweifel, U. L., Wikner, J., Hagström, Å, Lundberg, E., and Norman, B. (1995) Dynamics of dissolved organic carbon in a coastal ecosystem. *Limnol. Oceanogr.* 40:299–305.

7

UV RADIATION EFFECTS ON MICROBES AND MICROBIAL PROCESSES

Mary Ann Moran

*Department of Marine Sciences,
University of Georgia,
Athens, Georgia*

Richard G. Zepp

*Ecosystems Research Division,
U.S. Environmental Protection Agency,
Athens, Georgia*

INTRODUCTION

The ultraviolet (UV) region of solar radiation is defined as wavelengths in the range of 200–400 nm. In contrast to visible radiation (400–800 nm), which has a well-defined role as the energy source for most of the earth's primary production, the effects of UV radiation on biological processes in the ocean include an assortment of both positive and negative aspects, the balance of which is not readily understood. Recent increases in components of UV radiation reaching the ocean surface as a result of stratospheric ozone depletion makes comprehending, and ultimately predicting, the effects of UV radiation on the ocean ecosystem an especially important task.

Understanding the ecological effects of UV radiation in the ocean requires sufficient knowledge of at least three factors. First, one needs information on

Microbial Ecology of the Oceans, Edited by David L. Kirchman.
ISBN 0-471-29993-6   Copyright © 2000 by Wiley-Liss, Inc.

the number of photons of light reaching the surface of the ocean for each individual wavelength in the UV range. Since photons at 310 nm are likely to inflict microbial damage whereas photons at 380 nm can stimulate repair of such damage, the spectrum of UV radiation reaching the seawater surface is an important determinant of its effect on marine microbes. Second, information is required on the penetration of UV radiation into seawater at each individual wavelength. UV wavelengths are differentially absorbed by seawater constituents and therefore penetrate (and have the potential to affect biological processes) to different depths within the water column. Third, information is required on the effectiveness of each photon surviving to a given depth to bring about photobiological or photochemical changes. Molecules in both living cells and nonliving organic matter are susceptible to modification by UV light, and the outcome is highly dependent on wavelength.

This chapter is organized as a stepwise progression through each of these important factors. The first two, light reaching and penetrating into seawater, are straightforward processes that can be readily described with currently available instruments, although certainly more knowledge in both areas is needed. The third factor, effects on molecules within living cells or as part of the nonliving organic matter pool, is a complex issue, but one that has been the subject of considerable recent research. Ultimately, however, the challenge is to understand how UV-mediated changes at the cellular and molecular levels, occurring at various depths in the water column, are reflected in the microbial processes of the ocean, including such critical functions as the workings of marine microbial food webs and the rates of microbially mediated biogeochemical processes. This is a much more speculative area, where hard data are difficult to acquire, but it is also a critical one. Thus the final topic in this chapter is a discussion of what is known about the interface of UV radiation and the microbial ecology of the ocean.

THE SOLAR SPECTRUM AND LIGHT PENETRATION IN SEAWATER

Sunlight at the Ocean Surface

UV radiation is divided into three wavelength bands that differ with regard to their potential to impact biological processes in the ocean. UV-C radiation, ranging from 200 to 280 nm, is efficiently absorbed by the atmosphere and therefore is selectively removed from solar radiation before it reaches the ocean surface; these wavelengths are of little concern for ecological issues and are not considered further in this chapter. UV-B radiation, ranging from 280 to 315 nm (although sometimes defined as 280–320 nm), is strongly absorbed by stratospheric ozone (O_3) and therefore is partially removed from the solar radiation that reaches the ocean surface. But the UV-B that remains (primarily the 300–315 nm wavelengths) can have pronounced effects on living organisms and biological processes and is a major focus of this chapter. UV-A radiation, ranging from 315 to 400 nm, is not as efficiently absorbed by the atmosphere and reaches the surface of the ocean relatively undiminished. Like UV-B, UV-A radiation also has important effects on many biological processes.

The energy associated with photons of UV radiation (calculated as $\varepsilon = 1988/\lambda \times 10^{19}$ J, where λ is wavelength expressed in nanometers: Kirk 1994b) diminishes with increasing wavelength. Photons of UV-B light at 300 nm, for example, have 5% more energy than photons at 315 nm (the wavelength at the boundary between UV-B and UV-A radiation), and 33% more energy than photons at 400 nm (at the boundary between UV-A and visible light). While the energy per photon decreases as wavelength increases, the number of photons exhibits the opposite pattern.

The integrated effects of wavelength-related patterns in both energy per photon and photon number can be viewed as a plot of wavelength versus energy flux, in units of watts per square meter (W m^{-2}) or joules per second per square meter (J s^{-1} m^{-2}), as in Figure 1. For example, the midday solar irradiance at the ocean surface at the equator in mid-January integrated over UV-B, UV-A, and visible wavelengths (280–700 nm) is approximately 524 W m^{-2}. UV-B radiation accounts for only 0.5% of this energy because the total number of photons in this range is low (although each photon is relatively higher in energy); UV-A radiation accounts for 11.7%, while visible light accounts for the remaining 87.8%.

Such quantitative estimates of solar irradiance, which provide important starting points for estimating the underwater irradiance in the ocean, can be calculated from analytical models that account for solar zenith angle (the displacement of the sun relative to the direct vertical position), ozone thickness, aerosol thickness, and surface albedo (the ratio of reflected energy to incident energy) (Green et al. 1980; Frederick et al. 1989; Madronich 1993; Herman et al. 1998). These models account for variance in irradiance with latitude, season, and time of day, all of which affect intensity of sunlight. As a pronounced example of the combined effect of diurnal and latitudinal gradients, integrated solar irradiance (280–700 nm) at 60°N latitude in mid-January at 4:00 P.M. is only 1.5% of the irradiance at midday, mid-July at latitude 30°N. Latitudinal gradients (i.e., decreasing irradiance with increasing latitude) are most pronounced during the spring, fall, and especially winter months, being much less noticeable during the summer (Figure 1A).

Ozone has a very characteristic effect on the spectrum of sunlight reaching the earth's surface, absorbing sunlight primarily in the UV-B region with relatively little effect on longer wavelength radiation (Madronich et al. 1998; Herman et al. 1998). Thus the recently documented decreases in ozone concentration on a global scale are acting to increase the flux of UV-B radiation relative to other components of sunlight (Smith et al. 1992; Madronich et al. 1998; Herman et al. 1998), (Figure 1B). Clouds and aerosols also have important effects on the transmission of UV radiation (Green and Schippnick 1982; Madronich et al. 1998; Herman et al. 1998), but these appear to be less selective; that is, they also affect UV-A and visible radiation.

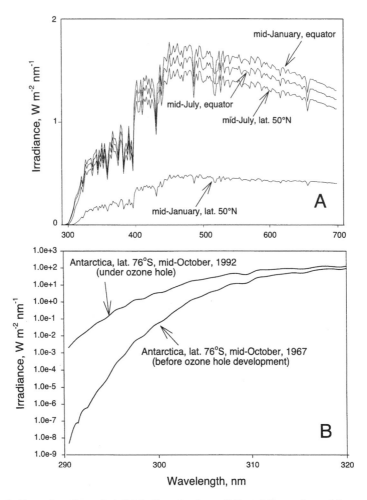

Figure 1. Examples of spectral distribution of solar radiation at the surface of the ocean. (A) Midday irradiance during mid-January and mid-July at the equator and at latitude 50°N. (B) UV-B irradiance during mid-October at latitude 76°S (Antarctic coast) before (1967) and after (1992) development of the "ozone hole"; ozone data were obtained from Harris et al. (1995). Irradiance was computed using the TUV computer program developed by Madronich (1993).

Nonetheless, because clouds and aerosols are likely to be strongly affected by changes in climate, future research will likely focus more closely on their effects on solar UV radiation reaching the sea surface.

Sunlight Penetration in Seawater

Once an adequate model of the spectral distribution of sunlight at the ocean surface has been established, we must try to understand how sunlight is attenuated as it travels downward through seawater. Recent research has resulted in the development of a variety of numerical models that are being applied in optical oceanography; these have been reviewed in some detail by several authors (Mobley et al. 1993; Mobley 1994; Kirk 1994b). Our purpose here is to provide an introduction to some of the basic concepts and terms that are commonly used in estimating radiative transfer in the sea.

The transmittance of light through seawater at a particular wavelength λ can be described in terms of a diffuse attenuation coefficient, $K_d(\lambda)$. The coefficient $K_d(\lambda)$ is referred to as an "apparent" optical water property (e.g., Jerlov 1976; Preisendorfer 1976; Smith and Tyler 1976; Kirk 1994a) and is derived from underwater light measurements. For an environmental situation in which light is present as a mixture of diffuse light and direct light entering the water at a variety of angles, the irradiance at depth z can be related to the downwelling irradiance immediately below the sea surface, $E_d(0, \lambda)$ by:

$$E_d(z, \lambda) = E_d(0, \lambda)\exp[-K_d(\lambda)z] \tag{1}$$

Because the irradiance immediately beneath the surface can be related to that reaching the surface (e.g., using equations that describe reflective loss and refraction of light at the air−sea interface), equation (1) is of great importance in quantitatively relating the irradiance reaching the sea surface to underwater solar spectral irradiance at any given depth.

Calculations of K_d are made using measurements of downwelling irradiance, $E_d(z, \lambda)$, at various depths z:

$$K_d(\lambda) = \frac{1}{E_d(z, \lambda)} \frac{dE_d(z, \lambda)}{dz} \tag{2}$$

Values of K_d indicate that shorter wavelength UV light is attenuated most rapidly and that the rate of attenuation decreases throughout the UV region and into the visible (Figure 2a). Not only does K_d vary with wavelength, it also exhibits wide variation (at any given wavelength) across marine environments (Figure 2b). Coastal marine systems typically have high diffuse attenuation coefficients (i.e., shallow light penetration), while open-ocean waters have low diffuse attenuation coefficients. Depths at which irradiance in the mid-UV (using 320 nm as an index) drops to 1% of surface values vary from 6.8 m for a typical coastal system (Figure 2b) to 49 m in the clearest seawater; longer wavelengths in the UV region show similar patterns across systems, although the depth to 1% of surface levels is shifted downward in the water column.

The constituents in seawater that affect absorption and scattering of UV light ultimately determine the value of $K_d(\lambda)$. The "colored" component of dissolved organic matter (DOM), referred to as "CDOM," makes the greatest contribution to the absorption of UV radiation in ocean water, especially in regions close to the coast (Kirk 1994b; DeGrandpre et al. 1996) (Figure 3).

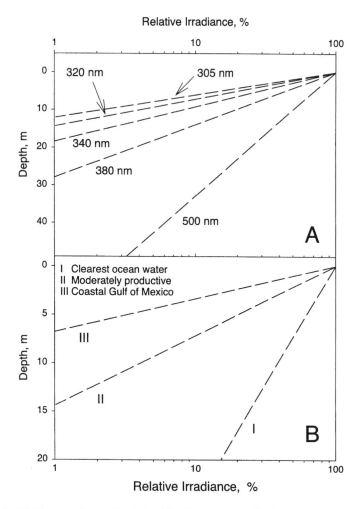

Figure 2. (A) Wavelength-specific plots of irradiance versus depth for moderately productive ocean water (0.5 mg chlorophyll a m^{-3}; Smith and Baker 1979), expressed as a percentage of surface irradiance. The y-axis intercept is the depth at which irradiance is 1% of surface. (B) Plots of irradiance versus depth for three marine environments, illustrating variations in K_d (slope) and depth to 1% of surface irradiation for UV (320 nm) light: I, clearest ocean water (Smith and Baker, 1981); II, moderately productive ocean water; III, Gulf of Mexico coastal waters (Smith and Baker 1979).

Even in highly oligotrophic ocean water, absorption by CDOM appears to be mainly responsible for UV and short-wavelength visible light attenuation (Siegel and Michaels 1996), based on comparisons of the absorptivities of pure water (Quickenden and Irvin 1980; Kirk 1994b) to those of the clearest ocean water (Smith and Baker 1981). Absorption by CDOM typically decreases in an approximately exponential fashion with increasing wavelength (Jerlov 1976;

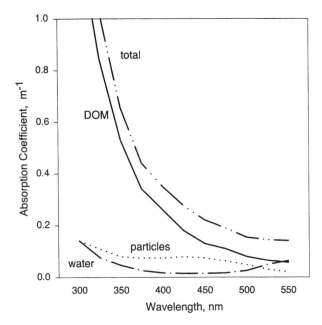

Figure 3. Absorption spectra for DOM, particulate matter, water, and total for coastal surface water (Mid-Atlantic Bight near the United States). (From DeGrandpre et al. 1996, copyright American Geophysical Union.)

Bricaud et al. 1981; Zepp and Schlotzhauer 1981; Baker and Smith 1982; Carder et al. 1989; Blough et al. 1993).

Particulate detritus has a similar absorption spectrum to CDOM and also contributes to UV attenuation (Siegel and Michaels 1996). Likewise, chlorophyll contributes to absorption of UV light, and thus the biomass of phytoplankton affects attenuation (although the strongest light absorption by chlorophyll is in the visible region). Particulate material, both living and nonliving, also contributes to attenuation by scattering UV light; the importance of scattering is not well characterized for marine environments, but it may be particularly important in coastal ecosystems (Nelson and Guarda 1995; DeGrandpre et al. 1996).

Vertical mixing also has important effects on photobiological and photochemical processes in natural waters and may ultimately determine the exposure of microorganisms to UV radiation (Morowitz 1950; Miller and Zepp 1978; Plane et al. 1987; Najjar et al. 1995). Under the simplifying assumption that the water column is mixed more rapidly than the photobiological effect occurs, the depth-averaged dose rate $v(z, \lambda)$ at depth z can be expressed approximately by:

$$v(z, \lambda) = \frac{v(0, \lambda)\{1 - \exp[-K_d(\lambda)z]\}}{zK_d(\lambda)} \tag{3}$$

where $v(0, \lambda)$ is the near-surface rate (Morowitz 1950; Zepp and Cline 1977). At depths sufficient to ensure the absorption of all the active radiation, $v(z, \lambda)$ approximately equals $v(0, \lambda)/[zK_d(\lambda)]$. In the clearest ocean waters, for example, the average rate of a photoreaction involving UV-B radiation in the top hundred meters is calculated to be about 10% of the near-surface rate. However, this simplified analysis is complicated by the failure of wavelengths that induce damage to living cells to penetrate as efficiently as wavelengths that repair biological damage (see below).

EFFECTS OF UV RADIATION ON LIVING CELLS

Molecular-Level Effects

Photons of light that penetrate into seawater and impinge on living cells have the potential to interact with chemical bonds and excite electrons to higher energy states. Visible light interacts with pigment molecules in this manner, thus forming the basis for light-driven photosynthesis. UV radiation, however, is generally not involved in energy-generating mechanisms within the cell, and as these higher energy photons are dissipated, they can modify the chemical structure of macromolecules. Thus for many organisms, exposure to UV light is largely an unavoidable consequence of having access to visible (photosynthetically active) radiation, or of being in an environment where other organisms are actively photosynthesizing.

There has been considerable research aimed at identifying the molecules within living cells that are most susceptible to UV damage, and to understanding consequences to the cell of accumulating damaged molecules. UV damage to living cells falls for the most part into two categories: damage to DNA, which is considered to be the primary lethal effect of UV radiation, and damage to other molecules, including proteins, RNA, and membrane-associated molecules.

DNA Damage from UV Radiation

Although DNA absorbs maximally at 260 nm (a wavelength in the UV-C region of the solar spectrum that does not survive passage through the atmosphere), significant absorption continues through the UV-B region to 315 nm. Photons absorbed directly by DNA can modify covalent bonds within the macromolecule, forming two characteristic types of cross-links between adjacent pyrimidine (thymine or cytosine) bases (Karentz et al. 1991; Jeffrey et al. 1996b). The biological consequences of the formation of these cross-links between adjacent bases is interference with DNA replication and transcription, since DNA polymerase and reverse transcriptase enzymes are stalled or physically blocked by the structural changes in the DNA molecule.

Along with direct absorption of UV-B radiation by DNA, indirect mechanisms that contribute to DNA damage in living cells are at work, as well. Other chromophoric molecules within the cell can absorb UV radiation, particularly in the UV-A region where DNA does not absorb directly. These chromophores include NADH, NADPH, flavins, proteins, unsaturated lipids, and other molecules with π-electron systems (Cockell 1998). Once the chromophores or "photosensitizer" molecules have absorbed energy, they can either pass the energy directly to DNA or react with an intermediary molecule to generate reactive intermediates. Photosensitized reactions are thought to result in single- and double-strand DNA breaks, DNA–protein cross-links, and photohydrates (the addition of a water molecule to a cytosine or thymine molecule) (Mitchell 1995; Peak and Peak 1995). The combined result of accumulated DNA damage in living cells from UV radiation, whether by direct or indirect (photosensitized) mechanisms, is that DNA replication or transcription is disrupted, cell division may cease, and mutations in essential genes may cause cell death (Mitchell 1995).

Other Damaging Effects of UV Radiation

A host of other effects of UV radiation on living cells have also been identified. These effects are sometimes referred to as "sublethal" in recognition of the extensive evidence that DNA damage is the primary mechanism for UV-induced cell death (Jagger 1981). Although these non-DNA injuries may not lead directly to cell death, they can significantly affect the growth and reproduction of cells, and they can act synergistically with the effects of DNA damage to increase lethality. In any case, they cost the organism energy to repair or overcome.

Cell membranes are one of the important secondary targets of UV radiation effects. Exposure to UV radiation affects permeability of membranes and transport of molecules (including amino acids and sugars) into the cell (Kubitschek and Doyle 1981; Moss and Smith 1981). In addition, many of the electron transport chain components (located within the cell membrane) including menaquinone, riboflavin, and porphyrins, can absorb UV radiation (Eisenstark 1987). Damage to these molecules interrupts the electron transport chain and disrupts proton gradients across the cell membrane (Jagger 1981). UV radiation also affects cell membranes by damaging permease enzymes located within the membranes, with subsequent inactivation of membrane transport functions (Jagger 1981). RNA can also be photodamaged by UV radiation, through the formation of cross-links between adjacent uracil molecules (Cockell 1998) and the formation of cross-links with tRNA molecules (Jagger 1981). In photosynthetic organisms, UV radiation damages the photosynthetic apparatus by affecting photosystem II energy transfer and by bleaching photosynthetic pigments (Neale 1987; Sinha et al. 1995).

Photoprotection and Repair

Although susceptible to damage by UV radiation, microbes are not defenseless. As a first line of defense, cells can limit the exposure of DNA and other chromophoric molecules to damaging radiation using photoprotection mechanisms. Compounds such as scytoenmin (Garcia-Pichel 1994) and mycosporine-like amino acids (Karentz et al. 1991; Mitchell and Karentz 1993), which are synthesized by marine phytoplankton (but not heterotrophic bacteria), are examples of UV screening compounds. These compounds have absorption peaks in the UV (λ_{max} values ranging from 310 to 360 nm; Mitchell and Karentz 1993) and therefore intercept damaging photons before they reach DNA and other cellular components. Screening can also occur by placement of DNA where it is least susceptible to UV radiation (e.g., shielded by organelles) or where UV light must first travel a long distance through the cytoplasm (e.g., in the center of a large cell) (Mitchell and Karentz 1993). Bacteria have been found to be relatively more susceptible to UV damage than eukaryotic microorganisms (Jeffrey et al. 1996b), presumably because of their small size and ineffective cellular shading of DNA (Garcia-Pichel 1994).

An additional defense against exposure to UV radiation includes the production of enzymes (e.g., superoxide dismutases, catalases, and peroxidases) and scavenger molecules (e.g., vitamins C, B, and E; cysteine, glutathione), which can quench or scavenge UV-produced excited states and reactive oxygen species within the cell (Karentz et al. 1994; Peak and Peak 1995). Behavioral defenses have also been described in which diurnal rhythms can restrict UV radiation exposure in eukaryotic microorganisms during times of high UV flux or at susceptible stages in cell development (Karentz et al. 1994; Häder and Worrest 1991).

If such protective mechanisms are insufficient to prevent damage to macromolecules, repair mechanisms are also available to the cell for reversing UV radiation damage. One of the two dominant repair mechanisms for DNA damage is a light-induced repair process, also referred to as photoreactivation. Repair involves the binding of a repair enzyme (photolyase) to a damaged strand of DNA, absorption of a photon of UV-A or blue light (380–450 nm) by the bound enzyme, and subsequent utilization of the absorbed energy to repair the DNA damage (Jeffrey et al. 2000). The second repair mechanism is a dark (light-independent) process, also referred to as nucleotide excision repair. In this case, repair relies on the redundant information in the non-damaged complementary DNA strand to repair damage. Dark repair involves the recognition of structural distortions of DNA, excision of the damaged region, and resynthesis of the excised area using the undamaged strand as the template (Mitchell 1995). Repair of UV-induced damage to other (i.e., non-DNA) cell components has not been well studied, but living cells can repair and/or replace damaged proteins and membrane components as well.

The balance between injury and repair ultimately determines how much UV-induced damage accumulates in living microorganisms, although understanding the integrated effects of injury and repair over time and space in oceanic surface waters is a difficult undertaking. Jeffrey et al. (1996a) and Herndl (1997) point out that light responsible for DNA photorepair processes (UV-A and visible wavelengths) penetrates deeper into the water column than does damaging wavelengths (primarily UV-B) (Figure 2). Thus depending on the mixing process operating in the water column at any particular time and place in the surface ocean, microorganisms can be carried into regions of the water column where they realize net repair of DNA damage (Kaiser and Herndl 1997). The variation in the concentration and spectral character of CDOM also has important effects on the balance between UV-A and UV-B radiation in the sea; analysis of CDOM spectral properties suggest that open-ocean CDOM removes a greater proportion of damaging wavelengths relative to repair wavelengths, and thus that net DNA damage may be less severe in the open ocean than in coastal waters, given the same dose of visible light. Other types of cell damage that are UV-A mediated, however, will be occurring simultaneously with DNA repair.

Organism-Level Effects

Cellular-level effects of UV radiation on nucleic acids, proteins, and other macromolecules are ultimately manifested at the organismal level. A number of excellent reviews summarize UV radiation effects on the physiology of marine microorganisms (e.g., Holm-Hansen et al. 1993; Karentz et al. 1994; Häder et al. 1995), effects that include reduction in survival rates (Karentz et al. 1991), changes in rates of primary and secondary production (Worrest 1983; Döhler 1984; Sieracki and Sieburth 1986; Smith 1989; Hebling et al. 1992; Aas et al. 1996; Herndl 1997; Kaiser and Herndl 1997; Sommaruga et al. 1997; Pakulski et al. 1998), modifications in rates of nitrogen fixation and assimilation (Kumar et al. 1996; Döhler and Hagmeier 1997), changes in rates of microbial metabolism (Pakulski et al. 1998), effects on orientation and motility (Häder and Worrest 1991), changes in microbial community composition (Herndl et al. 1997; Pakulski et al. 1998), effects on extracellular enzyme activity (Herndl et al. 1993; Müller-Niklas et al. 1995), and effects on the infectivity of bacterial and algal viruses (Wommack et al. 1996; Wilhelm et al. 1998; Jeffery et al. 2000). These studies cover a wide array of organisms, light exposure regimes, seasons, depths, latitudes, and mixing conditions that reflect the vast matrix of possible situations under which UV radiation will influence living cells in the ocean.

   In some cases, however, effects of UV radiation on marine microorganisms have been described through the development of "action spectra," quantitative descriptions of the relationship between the dose of UV radiation at a specific wavelength and subsequent measurable effects on organisms. These studies, which are particularly valuable from the perspective of modeling or predicting effects of UV radiation on the biology of marine microorganisms, are the focus of the next section.

Action Spectra for Biological Damage

The effect of radiation on a biological target (e.g., macromolecules, single cells, or whole organisms) can be measured as a function of wavelength, using either a series of individual single-wavelength exposures that span the UV region of the solar spectrum, or using multiple-wavelength (broad band) exposures that incrementally add back portions of UV-A and UV-B to a constant background of visible light (Cullen and Neale 1994). The latter case produces a coarser but perhaps more ecologically relevant action spectrum (or "biological weighting function"), since complex interactions between regions of the solar spectrum (e.g., reversal of DNA damage by wavelengths active in light repair) are integrated into a single function (Cullen and Neale 1994).

Much of the recent research on action spectra has focused on UV inhibition of photosynthesis in marine phytoplankton (Figure 4), using data based on both detailed laboratory studies of cultured marine organisms (Cullen et al. 1992) and on broadband field studies of natural phytoplankton communities (Boucher and Prézelin 1996; Neale et al. 1994, 1998). Considerable variability is evident among the photosynthesis inhibition action spectra, particularly with regard to the effect of UV-A radiation, and they differ considerably from the action spectrum of DNA damage (Figure 4).

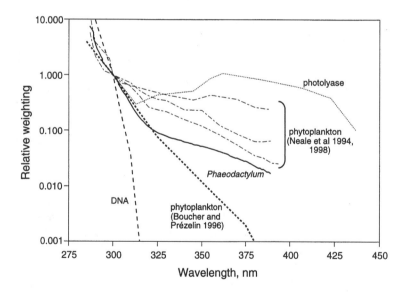

Figure 4. Photon-based action spectra for biological damage expressed in units normalized to 1.0 at 300 nm. Measured responses include inhibition of photosynthesis in a diatom culture (*Phaeodactylum* sp.; Cullen et al. 1992) and in natural phytoplankton communities (Neale et al. 1994, 1998; Boucher and Prézelin 1996), and DNA damage (Setlow 1974). An action spectrum for the DNA photorepair enzyme (photolyase) in *E. coli* is shown for comparison (Sancar 1994).

While action spectra are acknowledged to be critical for predicting the effect of UV radiation on biological processes (Cullen and Neale 1994; Coohill 1997), there are a number of concerns related to constructing and interpreting these plots. First, measurements are usually made under constant exposure to a given dose of UV radiation, a situation much different from what would be experienced in a well-mixed water column. Action spectra are therefore likely to misrepresent UV effects (probably by overestimating) unless mixing effects are considered (Karentz et al. 1994; Jeffrey et al. 2000). Second, underlying assumptions about "reciprocity" of response (i.e., whether a short exposure to highly damaging radiation is equivalent to a longer exposure to less damaging radiation, provided the cumulative dose is the same) and other aspects of the kinetics of damage and recovery of biological systems have yet to be resolved (Cullen and Neale 1997); the action spectra collected in Figure 4 reflect different underlying assumptions about the kinetics of UV inhibition. Third, action spectra have been found to vary considerably with taxonomy and physiology of organisms and with environmental conditions, suggesting there may be a need to determine multiple ecosystem-specific action spectra that match the time and space scales of microbial community dynamics (Jagger 1981; Boucher and Prézelin 1996; Neale et al. 1998; Jeffrey et al. 2000); this would obviously be an enormous undertaking. Fourth, measurements used to construct action spectra are often made over relatively short periods of time (e.g., hours), a practice that eliminates consideration of adaptation to UV radiation exposure by microorganisms, either through physiological mechanisms or through shifts in community composition (Karentz et al. 1994; Herndl 1997; Pakulski et al. 1998). The variability in action spectra evident from the literature may therefore simply reflect, at least to some extent, differences in the degree to which microbial communities have adapted to UV radiation effects (Smith and Cullen 1995; Neale et al. 1998).

Despite the differences among action spectra for inhibition of marine microorganisms (Figure 4) and the caveats detailed above, the development of these relationships have been critical for some of the first quantitative estimates of the effects of UV radiation on microbial processes in the ocean, as well as for predicting the effects of changing UV levels. For marine phytoplankton, for example, a number of consensus points have emerged from comparisons of the action spectra quantifying inhibitory effects. First, UV radiation is responsible for the bulk of photoinhibition of phytoplankton activity, with little or no involvement of visible wavelengths (although this does not seem to be the case for heterotrophic bacteria: Aas et al. 1996). Second, significant inhibition of phytoplankton occurs even under ambient levels of UV radiation. Third, while UV-B radiation is more damaging on a per-photon basis, UV-A is more abundant in the solar spectrum and can therefore also play a significant role in photoinhibition (Cullen and Neale 1994; Karentz et al. 1994) although not all action spectra–based studies agree on this point (Boucher and Prézelin 1996).

EFFECTS OF UV RADIATION ON NONLIVING ORGANIC MATTER

Despite intense interest in the effects of UV radiation on marine organisms, most of the UV radiation reaching the surface of the ocean interacts not with living cells but with nonliving organic matter. Absorption by CDOM is the primary fate of UV radiation in the ocean, with secondary importance attaching to absorption and scattering by particulate material [i.e., both nonliving particulate organic material (POM) and living cells]. As the dominant player in UV light absorption, nonliving organic matter plays a critical role as a natural UV screen in the ocean. But equally important, nonliving organic matter that intercepts UV radiation is itself susceptible to photochemical alterations. Changes that occur in DOM following exposure to sunlight include the degradation of larger molecules into smaller ones (i.e., reduction in average molecular weight of DOM) and specific destruction of the light-absorbing components (i.e., loss of absorptivity and fluorescence efficiency). Such alterations can ultimately have far-reaching effects on the standing stocks and turnover of major organic matter pools in the ocean. This section discusses the changes that occur in nonliving organic matter during exposure to UV light and explores what is known about the modified compounds (photoproducts) that result.

Photomineralization

The energy in a photon of UV radiation has the potential to convert DOM into inorganic photoproducts, including dissolved inorganic carbon (DIC: CO_2, HCO_3^-, and CO_3^{2-}), carbon monoxide, and carbonyl sulfide. The inorganic product formed most efficiently from DOM by solar radiation is DIC; current data suggest that at least 15% of the carbon in marine DOM may be susceptible to conversion to DIC by exposure to natural sunlight (Miller and Zepp 1995; Amon and Benner 1996). This conversion represents a direct, nonbiological mineralization that removes organic carbon from the ocean ecosystem.

The second most abundant photoproduct resulting from the interaction of UV radiation with DOM is CO, which is produced at rates about 15- to 20-fold lower than DIC (Jones 1991; Mopper et al. 1991; Schmidt and Conrad 1993; Miller and Zepp 1995). Like DIC, CO photoproduction represents a direct loss of organic carbon from oceanic surface waters; but unlike DIC, it also represents a potential substrate for marine bacteria. CO can be utilized for growth by chemoautotrophic bacteria termed "carboxidotrophs" (Mörsdorf et al. 1992), while other bacteria (ammonia oxidizers and methylotrophs) oxidize CO as a result of broad substrate specificity of their enzymes but obtain no energy from the process. It is questionable whether CO can be utilized for growth at the concentrations measured in oceanic surface waters, however (Schmidt and Conrad 1993), and oceanic bacteria utilizing CO must compete with volatilization to the atmosphere (Zuo and Jones 1995).

Formation of Biologically Labile Substrates

Along with the conversion of DOM to inorganic forms, absorption of photons of UV radiation also results in the cleavage of DOM into smaller organic compounds, many of which are more biologically labile than the parent DOM. Identification of these lower molecular weight organic photoproducts indicates a predominance of carbonyl compounds, primarily fatty acids and keto acids (Moran and Zepp 1997). Evidence also exists that DOM not converted to compounds small enough and simple enough to be readily identified by standard chemical methods may nonetheless exhibit enhanced biological lability (Miller and Moran 1997; Mopper and Kieber 2000). Organic photoproducts have been shown to significantly stimulate activity of marine bacteria, and increases in bacterial cell numbers or secondary production following exposure of DOM to UV radiation have been documented for both coastal and oceanic systems (Kieber et al. 1989; Miller and Moran 1997; Benner and Biddanda 1998); deep ocean DOC may be particularly susceptible to photochemical modification (Benner and Biddanda 1998).

The nitrogenous components of DOM are also photochemically modified by exposure to UV radiation. Both ammonium (Bushaw et al. 1996) and amino acids (Amador et al. 1989; Bushaw-Newton and Moran 1999) are released from DOM by photochemical processes, resulting in the conversion of organic nitrogen to a form that is more easily assimilated by marine microorganisms.

Formation of Recalcitrant Compounds

In keeping with the complex nature of the effects of UV radiation in the ocean, DOM can also be modified into more biologically refractory forms by UV light, essentially the opposite effect of that described in the preceding section. Individual compounds (e.g., proteins; Keil and Kirchman 1994) or complex mixtures of algal-derived material (Tranvik and Kokalj 1998; Pausz and Herndl in press) can be converted to more biologically refractory forms by sunlight, presumably with significant UV involvement; the higher molecular weight fraction of DOM is potentially important in catalyzing this transformation (Tranvik and Kokalj 1998). The UV-mediated formation of recalcitrant organic matter may be an important mechanism for the production of the refractory DOM that persists in the deep ocean for decades and longer (Benner and Biddanda 1998), and it certainly points out the complexity of interactions between UV radiation and the biogeochemistry of marine DOM.

Formation of Reactive Intermediates

A final category of photoproducts produced by UV radiation from DOM consists of reactive oxygen intermediates. As has been demonstrated for organic matter inside living cells, nonliving organic matter can also act as photosensitizers, absorbing UV radiation and subsequently reacting with molecular oxygen to form reactive oxygen intermediates (Blough and Zepp 1995; Mopper and Kieber 2000). These reactive oxygen species can react with constituents of living cells, interacting with cell-surface-associated proteins, lipids, or polysaccharides to bring about cell damage (Mill et al. 1990; Palenik et al. 1991). Alternatively, reactive intermediates may interact with nonliving organic matter, serving as an important indirect route for the formation of carbon gases or biologically labile photoproducts from DOM (Mopper and Kieber 2000).

Action Spectra for DOM Photoproduct Formation

As was the case for UV effects on living cells, effects on nonliving organic matter have been quantified through the construction of action spectra. A number of studies have measured the formation of DOM photoproducts, focusing on CO (Valentine and Zepp 1993), biologically available carbonyl compounds (Kieber et al. 1989), and reactive oxygen species (OH and H_2O_2: Weiss et al. 1995; Zafiriou personal communication in press; Yocis et al. in press) (Figure 5). Like the action spectra for living organisms (Figure 4), incident photons in the lower wavelength (more energetic) regions of the solar spectrum are most effective in bringing about photochemical transformations.

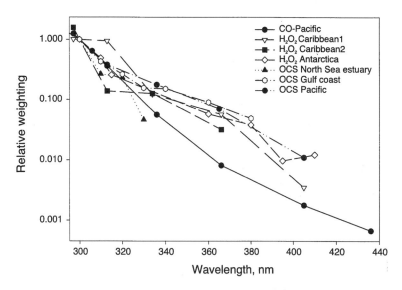

Figure 5. Photon-based action spectra for photochemical production of DOM photoproducts, normalized to 1.0 at 300 nm. Photoproducts include carbon monoxide (CO) in the Pacific Ocean (Zafiriou, personal communication, 1997), hydrogen peroxide (H_2O_2) in the Caribbean (Moore et al. 1993; Blough et al. 1993) and Antarctic (Yocis et al. in press), and carbonyl sulfide (OCS) in the North Sea estuary, coastal Gulf of Mexico (Zepp and Andreae 1994), and Pacific Ocean (Weiss et al. 1995).

Interpretation of action spectra for nonliving organic matter is somewhat less problematic than for living organisms. Effects are typically limited to simple photoreactions, rather than the higher order, cumulative effects that occur in living cells. Because of this, mixing conditions in the upper ocean are a less serious issue for determining the rate of formation of photoproducts, although mixing may influence the depth at which photoproducts are formed. Variability among action spectra is considerably lower for nonliving organic matter than for living cells (Figure 5), suggesting that one or a few spectra might suffice for most DOM photoproducts (Moran and Zepp 1997). However, assumptions about the kinetics underlying photochemical reactions are still problematic. Because most empirical studies of photoproduct formation are conducted over short time periods (hours to days), rates must be extrapolated to the longer time intervals that are relevant to ecological processes. Underlying assumptions about kinetics of photoproduct formation, and specifically whether rates are linear with cumulative exposure of DOM to UV radiation, will have important effects on predicted results.

SCALING UP: ECOLOGICAL IMPLICATIONS OF UV RADIATION

The potential roles of UV radiation in regulating marine biogeochemical processes and the implications of increasing UV-B levels related to stratospheric ozone depletion are important issues that motivated many of the studies discussed above. However, extrapolating results of studies conducted on single populations under highly defined conditions to questions about complex natural systems is a difficult undertaking. This section focuses on scaling up current understanding of the role of UV radiation in the ocean by applying information gathered at the molecular and organismal levels to identify points of interaction at the ecosystem and global scales.

Autotrophic Processes

As predicted from the well-documented effects on phytoplankton at the organismal level, UV radiation adversely affects rates of carbon fixation in the ocean. In studies of the Southern Ocean (which have dominated research on this topic in this area), ambient UV levels (i.e., those typical of a nondiminished ozone layer) decrease rates of primary production. The amount of inhibition reported varies among studies, which is to be expected, but all values are relatively high ($\leqslant 50\%$) (Cullen et al. 1992; Arrigo 1994; Boucher and Prézelin 1996). When UV effects are integrated throughout the water column, inhibitory effects of UV radiation decrease in quantitative importance, bringing the decreases into the range of 25% or less relative to values recorded when only visible light is present (Boucher and Prézelin 1996).

Recent changes in the level of UV radiation reaching the ocean surface due to stratospheric ozone depletion raise questions about the effects of increasing

UV, and in particular increasing UV-B, on carbon fixation in marine eco-systems. A number of studies have addressed this issue for the Southern Ocean, where ozone depletion is most severe. Despite differences in action spectra among studies, ozone-related increases in UV radiation are consistently pre-dicted to decrease rates of primary production in the Southern Ocean by a maximum of 15% in near-surface waters (Figure 6) (Cullen et al. 1992; Smith et al. 1992; Arrigo 1994; Boucher and Prézelin 1996). When effects are integrated throughout the water column and over seasons (the ozone hole is present for only about three months of the year, during the austral spring), the annual decrease in carbon fixation in the Southern Ocean is predicted to amount to only 1–4% of historical values (Smith et al. 1992; Arrigo 1994; Helbling et al. 1994).

From a global perspective, even small decreases in rates of primary production might affect atmospheric CO_2 concentrations and the sequestering of fixed carbon in the deep ocean. Decreases in primary production for the Southern Ocean, however, are predicted to be approximately 7×10^{12} g of carbon per year, a value about three orders of magnitude lower than estimates of global primary production by marine phytoplankton (Smith and Cullen 1995). Likewise, models predict that if UV effects halted phytoplankton production completely in the Southern Ocean, the resulting effect on atmo-spheric CO_2 concentrations would be relatively small (Peng 1992). Nonethe-less, these scenarios are based on the assumption that ozone depletion is occurring only in the Southern Ocean; current satellite data indicate ozone-

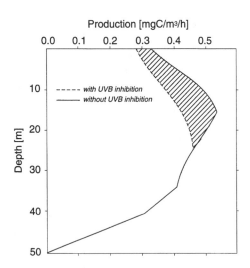

Figure 6. Average values for in situ phytoplankton productivity versus depth within the marginal ice zone of the Bellingshausen Sea, Antarctica, comparing rates inside and outside the ozone hole. Higher UV-B levels inside the ozone hole are consistently associated with reduced levels of primary production. (From Smith and Cullen 1995; copyright American Geophysical Union.)

related increases in UV-B at midlatitudes in both hemispheres (Herman et al. 1998), and ozone depletion from pre 1980s levels is expected to continue to the mid-2000s (Madronich et al. 1998).

The concentration and spectral properties of CDOM effect the extent of UV inhibition of carbon fixation in the ocean. First, CDOM absorbs damaging UV wavelengths, acting as a protective screen for marine phytoplankton. Second, CDOM absorbs photosynthetically active radiation, essentially competing with phytoplankton for visible wavelengths and reducing rates of primary productivity. Arrigo and Brown (1996) developed a model that integrates these two opposing effects and predicts that carbon fixation in surface waters is generally enhanced by the presence of CDOM. However, depth-integrated gross primary production under most scenarios (varying concentrations of CDOM, chlorophyll a, and ozone; varying solar zenith angle; varying sensitivity of phytoplankton to UV radiation) is reduced by the presence of CDOM, essentially because the relative balance of protection versus competition shifts with depth.

Heterotrophic Processes and the Microbial Loop

Like effects on autotrophic processes, UV effects on heterotrophic activity are complex and possibly conflicting. While relatively few estimates of system-level effects of UV radiation on heterotrophic microorganisms currently exist, it is clear that UV simultaneously inhibits activity (by damaging DNA and other cell components, and possibly by forming recalcitrant photoproducts) and stimulates activity (by providing labile photoproducts).

Inhibition by UV radiation of bacterial secondary production, measured as uptake of [^{3}H]thymidine and [^{3}H]leucine, has been demonstrated in surface waters of a number of coastal marine environments, with inhibition of 40–70% being typical (Herndl et al. 1993; Aas et al. 1996; Sommauga et al. 1997). These estimates, although limited in number, suggest that short-term effects of UV radiation on heterotrophic production may be greater than effects on primary production. Measures of DNA damage accumulating in the heterotrophic bacteria size fraction ($<0.8\,\mu$m) relative to the phytoplankton size fraction ($>0.8\,\mu$m) also provide evidence that heterotrophic microorganisms are more susceptible to UV damage than phytoplankton (Jeffrey et al. 1996a). On the other hand, heterotrophic activities occur in the absence of sunlight, and the effect of UV radiation may be primarily that of shifting bacterial activity in time and space (i.e., to lower in the water column or to the nighttime) but not ultimately reducing it. If a significant fraction of the bacterial community is not damaged by UV (Herndl et al. 1997) or if bacteria that are damaged recover rapidly (Kaiser and Herndl 1997), heterotrophic activities are ultimately less likely than autotrophic processes to be directly inhibited by UV radiation when integrated over time and space.

At the same time that UV radiation reduces heterotrophic activity through damage to microorganisms, it also stimulates activity through the formation of labile DOM photoproducts. The ultimate effects of UV radiation on heterotrophic processes in the ocean will therefore depend on the balance between cell damage and photoproduct formation (Figure 7), with conversion of labile DOM into more refractory forms by UV radiation further complicating the equation (Pausz and Herndl in press).

If it can be assumed that damage to heterotrophic microorganisms shifts activity in time and space but does not ultimately reduce it, then the major impact of UV radiation on heterotrophic processes may be the conversion of oceanic DOM into organic or inorganic forms with modified susceptibility to heterotrophic transformation. Two global-scale estimates have been made of the amount of marine DOM converted annually by UV radiation to either inorganic compounds (DIC and CO) or microbial substrates. Moran and Zepp (1997), who used a solar irradiance model (Valentine and Zepp 1993) and recent total ozone data (WMO 1994), applied the CO action spectrum to all classes of DOM photoproducts to estimate an annual photochemical transformation of 14×10^{15} g of DOC into DIC $(12 \times 10^{15} \, \text{g C y}^{-1})$, CO $(0.8 \times 10^{15} \, \text{g C y}^{-1})$, and biologically labile photoproducts $(0.9 \times 10^{15} \, \text{g C y}^{-1})$. Mopper and Kieber (2000) used the lower end of the range of global CO photoproduction rates to calculate an annual transformation of 3×10^{15} g of DOC into DIC $(2.7 \times 10^{15} \, \text{g C y}^{-1})$ and CO $(0.1 \times 10^{15} \, \text{g C y}^{-1})$. These estimates differ by almost five-fold; but both

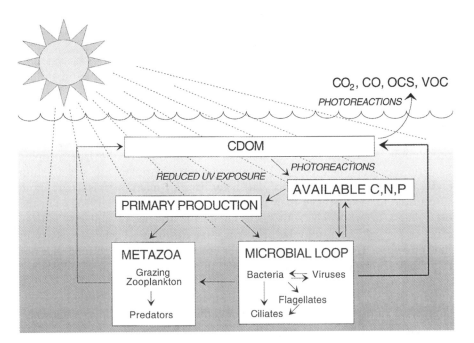

Figure 7. Interactions of UV radiation, CDOM, and microorganisms in the surface ocean: OCS, carbonyl sulfide; VOC, volatile organic compounds. (After Pomeroy and Wiebe 1988.)

suggest an important role for photochemical modification of DOM, with the primary effect coming from the UV region of the solar spectrum. For example, these estimated annual conversions of marine DOM account for up to 3% of the oceanic DOC pool, and are 100-fold greater than the estimated annual burial of carbon in marine sediments (Hedges 1992; Mopper and Kieber 2000). Global-scale transformations of marine DOM via UV-mediated photodegradation can therefore be viewed both as potential substrates for the microbial loop and as a mechanism for the turnover of oceanic DOM, some components of which might otherwise have extremely long biological turnover times.

SUMMARY

1. CDOM is responsible for most of the absorption of UV and short-wavelength visible light in seawater.

2. UV radiation effects on marine microorganisms at the molecular level are also manifested at the organismal level, evident as reductions in survival rates, changes in primary and secondary productivity, and alterations in other aspects of the physiology of marine microorganisms.

3. UV radiation effects on nonliving organic matter include photochemical mineralization (to DIC and CO), modification of biological availability, and alterations in optical properties; these changes can affect turnover rates of nonliving organic matter in seawater and alter the penetration of UV radiation into the upper ocean.

4. Action spectra provide a quantitative description of the relationship between the dose of UV radiation at a specific wavelength and subsequent measurable effects on living organisms and nonliving organic matter.

5. Effects of UV radiation on the microbial ecology of the ocean include reductions in carbon fixation rates in the surface ocean and complex influences on the turnover of oceanic DOM pools. The ultimate effects on heterotrophic processes in the ocean depend on the balance between cell damage and photoproduct formation.

REFERENCES

Aas, P., Lyons, M. M., Pledger, R., Mitchell, D. L., and Jeffrey W. H. (1996) Inhibition of bacterial activities by solar radiation in nearshore waters and the Gulf of Mexico. *Aquat. Microb. Ecol.* 11:229–238.

Amador, J. A., Alexander, M., and Zika, R. G. (1989) Sequential photochemical and microbial degradation of organic molecules bound to humic acid. *Appl. Environ. Microbiol.* 55:2843–2849.

Amon, R. M. W., and Benner, R. (1996) Photochemical and microbial consumption of dissolved organic carbon and dissolved oxygen in the Amazon River system. *Geochim. Cosmochim. Acta* 60:1783–1792.

Arrigo, K. R. (1994) The impact of ozone depletion on phytoplankton growth in the Southern Ocean: Large-scale spatial and temporal variability. *Mar. Ecol. Prog. Ser.* 114:1–12.

Arrigo, K. R., and Brown, C. W. (1996) Impact of chromophoric dissolved organic matter on UV inhibition of primary productivity in the sea. *Mar. Ecol. Prog. Ser.* 140:207–216.

Baker, K. S., and Smith, R. C. (1982) Bio-optical classification and model in natural waters. II. *Limnol. Oceanog.* 27:500–509.

Benner, R., and Biddanda, B. (1998) Photochemical transformations of surface and deep marine dissolved organic matter: Effects on bacterial growth. *Limnol. Oceanogr.* 43:1373–1378.

Blough, N. V., and Zepp, R. G. (1995) Reactive oxygen species in natural waters. In C. S. Foote et al., eds., *Active Oxygen: Reactive Oxygen Species in Chemistry*. Chapman & Hall, London, pp. 280–333.

Blough, N. V., Zafiriou, O. C., and Bonilla, J. (1993) Optical absorption spectra of waters from the Orinoco River outflow: Terrestrial input of colored organic matter to the Caribbean. *J. Geophys. Res.* 98:2271–2278.

Boucher, N. P., and Prézelin, B. B. (1996) Spectral modeling of UV inhibition of in situ Antarctic primary production using a field-derived biological weighting function. *Photochem. Photobiol.* 64:407–418.

Bricaud, A., Morel, A., and Prieur, L. (1981) Absorption by dissolved organic matter of the sea (yellow substance) in the UV and visible domains. *Limnol. Oceanogr.* 26:43–53.

Bushaw, K. L., Zepp, R. G., Tarr, M. A., Schulz-Jander, D., Bourbonniere, R. A., Hodson, R. E., Miller, W. L., Bronk, D. A., and Moran, M. A. (1996) Photochemical release of biologically available nitrogen from dissolved organic matter. *Nature* 381:404–407.

Bushaw-Newton, K. L., and Moran, M. A. (1999) Photochemical formation of biologically available nitrogen from dissolved humic substances in coastal marine environments. *Aquat. Microb. Ecol.* 18:285–292.

Carder, K. L., Steward, R. G., Harvey, G. R., and Ortner, P. B. (1989) Marine humic and fulvic acids: Their effects on remote sensing of ocean chlorophyll. *Limnol. Oceanogr.* 34:68–81.

Cockell, C. S. (1998) Ultraviolet radiation, evolution and the π-electron system. *Biol. J. Linn. Soc.* 63:449–457.

Coohill, T. P. (1997) UV action spectra for marine phytoplankon. *Photochem. Photobiol.* 65:259–260.

Cullen, J. J., and Neale, P. J. (1994) Ultraviolet radiation, ozone depletion, and marine photosynthesis. *Photosynth. Res.* 39:303–320.

Cullen, J. J., and Neale, P. J. (1997) Effect of UV on short-term photosynthesis of natural phytoplankton. *Photochem. Photobiol.* 65:264–266.

Cullen, J. J., Neale, P. J., and Lesser, M. P. (1992) Biological weighting function for the inhibition of phytoplankton photosynthesis by ultraviolet radiation. *Science* 258:646–650.

DeGrandpre, M. D., Vodacek, A., Nelson, R. K., Bruce, E. J., and Blough, N. V. (1996) Seasonal seawater optical properties of the U.S. Middle Atlantic Bight. *J. Geophys. Res.* 101:722–736.

Döhler, G. (1984) Effect of UV-B radiation on the marine diatoms *Lauderia annulata and Thalassiosira rotula* grown in different salinities. *Mar. Biol.* 83:247–253.

Döhler, G., and Hagmeier, E. (1997) UV-effects on pigments and assimilation of ^{15}N-ammonium and ^{15}N-nitrate by natural marine phytoplankton of the North Sea. *Bot. Acta* 11:481–488.

Eisenstark, A. (1987) Mutagenic and lethal effects of near-ultraviolet radiation (290–400 nm) on bacteria and phage. *Environ. Mol. Mutagen.* 10:317–337.

Frederick, J. E., Snell, H. E., and Haywood, E. K. (1989) Solar ultraviolet radiation at the Earth's surface. *Photochem. Photobiol.* 50:443–450.

Garcia-Pichel, F. (1994) A model for internal self-shading in planktonic organisms and its implications for the usefulness of ultraviolet sunscreens. *Limnol. Oceanogr.* 39:1704–1717.

Green, A. E. S., and Schippnick, P. F. (1982) UV-B Reaching the surface. In: J. Calkins, ed., *The Role of Ultraviolet Radiation in Marine Ecosystems*. NATO Conference Series, IV *Marine Sciences*, Vol. 7. Plenum Press, New York, pp. 5–27.

Green, A. E. S., Cross, K. R., and Smith, L. A. (1980) Improved analytic characterization of ultraviolet light. *Photochem. Photobiol.* 31:59–65.

Häder, D.-P., and Worrest, R. C. (1991) Effects of enhanced solar ultraviolet radiation on aquatic ecosystems. *Photochem. Photobiol.* 53:717–725.

Häder, D.-P., Worrest, R. C., Kumar, H. D., and Smith, R. C. (1995) Effects of increased solar ultraviolet radiation on aquatic ecosystems. *Ambio* 24:174–180.

Harris, N. R. P. (1995) Oxone measurements. In D. L. Albritton and R. T. Watson, eds., *Scientific Assessment of Ozone Depletion:1994*. Global Ozone Research and Monitoring Project, Report No. 37. World Meteorological Organization, Geneva, pp. 1.1–1.48.

Hedges, J. I. (1992) Global biogeochemical cycles: Progress and problems. *Mar. Chem.* 39:67–93.

Helbling, E. W., Villafañe, V., Ferrario, M., and Holm-Hansen, O. (1992) Impact of natural ultraviolet radiation on rates of photosynthesis and on specific marine phytoplankton species. *Mar. Ecol. Prog. Ser.* 80:89–100.

Helbling, E. W., Villafañe, V., and Holm-Hansen, O. (1994) Effects of ultraviolet radiation on Antarctic marine phytoplankton photosynthesis with particular attention to the influence of mixing. In C. S. Weiler and P. A. Penhale, eds., *Ultraviolet Radiation in Antarctica: Measurements and Biological Effects*. Antarctic Research Series, Vol. 62, pp. 207–227.

Herman, J. R., McKenzie, R. L., Diaz, S., Kerr, J., Madronich, S., and Seckmeyer, G. (1998) Ultraviolet radiation at the earth's surface. In D. Albritton, P. Aucamp, G. Megie, and R. Watson, eds., *Scientific Assessment of Ozone Depletion: 1998*. World Meteorological Organization, Geneva, pp. 9.1–9.46.

Herndl, G. J. (1997) Role of ultraviolet radiation on bacterioplankton activity. In D.-P. Häder, ed., *The Effects of Ozone Depletion on Aquatic Ecosystems*. R. G. Landes, San Diego. p. 9.1–9.46.

Herndl, G., Müller-Niklas, G., and Frick, J. (1993). Major role of ultraviolet-B in controlling bacterioplankton growth in the surface layer of the ocean. *Nature* 361:717–719.

Herndl, G. J., Brugger, A., Hager, S., Kaiser, E., Obernosterer, I., Reitner, B., and Slezak, D. (1997) Role of ultraviolet-B radiation on bacterioplankton and the availability of dissolved organic matter. *Plant Ecol.* 128:42–51.

Holm-Hansen, O., Lubin, D., and Helbling, E. W. (1993) Ultraviolet radiation and its effects on organisms in aquatic environments. In A. R. Young, L. Björn, J. Moan, and W. Nultsch, eds., *Environmental UV Photobiology*. Plenum Press, New York, pp. 379–425.

Jagger, J. (1981) Near-UV radiation effects on microorganisms. *Photochem. Photobiol.* 34:761–768.

Jeffrey, W. H., Aas, P., Lyons, M. M., Coffin, R. B., Pledger, R. J., and Mitchell, D. L. (1996a) Ambient solar radiation-induced photodamage in marine bacterioplankton. *Photochem. Photobiol.* 64:419–427.

Jeffrey, W. H., Pledger, R. J., Aas, P., Hager, S., Coffin, R. B., Von Haven, R., and Mitchell, D. L. (1996b) Diel and depth profiles of DNA photodamage in bacterioplankton exposed to ambient solar ultraviolet radiation. *Mar. Ecol. Prog. Ser.* 137:283–291.

Jeffrey, W. H., Kase, J. P., and Wilhelm, S. W. (2000) Ultraviolet radiation effects on bacterioplankton and viruses in marine ecosystems. In S. J. deMora, S. Demers, and M. Vernet, eds., *The Effects of UV Radiation in the Marine Environment*. Cambridge University Press, Cambridge.

Jerlov, N. G. (1976) *Marine Optics*. Elsevier, New York.

Jones, R. D. (1991) Carbon monoxide and methane distribution and consumption in the photic zone of the Sargasso Sea. *Deep-Sea Res.* 38:625–635.

Kaiser, E., and Herndl, G. J. (1997) Rapid recovery of marine bacterioplankton activity after inhibition by UV radiation in coastal waters. *Appl. Environ. Microbiol.* 63:4026–4031.

Karentz, D., Cleaver, J. E., and Mitchell, D. L. (1991) Cell survival characteristics and molecular responses of Antarctic phytoplankton to ultraviolet-B radiation. *J. Phycol.* 27:326–341.

Karentz, D., Bothwell, M. L., Coffin, R. B., Hanson, A., Herndl, G. J., Kilham, S. S., Lesser, M. P., Lindell, M., Moeller, R. E., Morris, D. P., Neale, P. J., Sanders, R. W., Weiler, C. S., and Wetzel, R. G. (1994) Impact of UV-B radiation on pelagic freshwater ecosystems: Report of working group on bacteria and phytoplankton. *Ergebn. Limnol.* 43:31-69.

Keil, R. G., and Kirchman, D. L. (1994) Abiotic transformation of labile protein to refractory protein in seawater. *Mar. Chem.* 45:187–196.

Kieber, D. J., McDaniel, J., and Mopper, K. (1989) Photochemical source of biological substrates in sea water: Implications for carbon cycling. *Nature* 341:637–639.

Kirk, J. T. O. (1994a) Optics of UV-B radiation in natural waters. *Ergebn. Limnol.* 43:1–16.

Kirk, J. T. O. (1994b) *Light and Photosynthesis in Aquatic Ecosystems*, 2nd ed. Cambridge University Press, Cambridge.

Kubitschek, H. E., and Doyle, R. J. (1981) Growth delay induced in *Escherichia coli* by near-ultraviolet radiation: Relationship to membrane transport functions. *Photochem. Photobiol.* 33:695–702.

Kumar, A., Sinha, R. P., and Häder, D.-P. (1996) Effect of UV-B on enzymes of nitrogen metabolism in the cyanobacterium *Nostoc calcicola. J. Plant Physiol.* 148:86–91.

Madronich, S. (1993) The atmosphere and UV-B radiation at ground level. In A. R. Young, L. O. Björn, J. Moan, and W. Nultsch, eds., *Environmental UV Photobiology.* Plenum Press, New York, pp. 1–39.

Madronich, S., McKenzie, R. L., Björn, L. O., and Caldwell, M. M. (1998) Changes in biologically active ultraviolet radiation reaching the Earth's surface. In *Environmental Effects of Ozone Depletion: 1998 Assessment.* United Nations Environmental Programme, Nairobi, pp. 1–27.

Mill, T., Haag, W., and Karentz, D. (1990) Estimated effects of indirect photolysis on marine organisms. In N. B. Blough and R. G. Zepp, eds., *Effects of Solar Ultraviolet Radiation on Biogeochemical Dynamics in Aquatic Environments.* Woods Hole Oceanographic Institution Technical Report WHOI-90-09, pp. 89–93.

Miller, G. C., and Zepp, R. G. (1978) Effects of suspended sediments on photolysis rates. *Water Res.* 13:453–459.

Miller, W. L., and Moran, M. A. (1997) Interaction of photochemical and microbial processes in the degradation of refractory dissolved organic matter from a coastal marine environment. *Limnol. Oceanogr.* 42:1317–1324.

Miller, W. L., and Zepp, R. G. (1995) Photochemical production of dissolved inorganic carbon from terrestrial organic matter: Significance to the oceanic organic carbon cycle. *Geophys. Res. Lett.* 22: 417–420.

Mitchell, D. L. (1995) DNA damage and repair. In W. M. Horspool and P. Song, eds., *CRC Handbook of Organic Photochemistry and Photobiology.* CRC Press. Boca Raton, FL, pp. 1326–1331.

Mitchell, D. L., and Karentz, D. (1993) The induction and repair of DNA photodamage in the environment. In A. R. Young, L. Björn, J. Moan, and W. Nultsch, eds., *Environmental UV Photobiology.* Plenum Press, New York, pp. 345–377.

Mobley, C. D. (1994) *Light and Water: Radiative Transfer in Natural Waters.* Academic Press, San Diego, CA.

Mobley, C. D., Gentili, B., Gordon, H. R., Jin, Z., Kattawar, G. W., Morel, A., Reinersman, P. S., Stamnes, K., and Stavn, R. H. (1993) Comparison of numerical models for computing underwater light fields. *Appl. Opt.* 32:7484–7504.

Moore, C. A., Farmer, C. T., and Zika, R. G. (1993) Influence of the Orinoco River on hydrogen-peroxide distribution and production in the eastern Caribbean. *J. Geophys. Res.* 98:2289–2298.

Mopper, K., and Kieber, D. J. (2000) Marine photochemistry and its impact on carbon cycling. In S. J. deMora, S. Demers, and M. Vernet, eds., *The Effects of UV Radiation in the Marine Environment.* Cambridge University Press, Cambridge.

Mopper, K., Zhou, X., Kieber, R. J., Kieber, D. J., Sikorski, R. J., and Jones, R. D. (1991) Photochemical degradation of dissolved organic carbon and its impact on the oceanic carbon cycle. *Nature* 353:60–62.

Moran, M. A., and Zepp, R. G. (1997) Role of photoreactions in the formation of biologically labile compounds from dissolved organic matter. *Limnol. Oceanogr.* 42:1307–1316.

Morowitz, H. J. (1950) Absorption effects in volume irradiation of microorganisms. *Science* 111:229–231.

Morris, D. P., Zagarese, H., Williamson, C. E., Balseiro, E. G., Hargreaves, B. R., Modenutti, B., Moeller, R., and Queimalinos, C. (1995) The attenuation of solar UV radiation in lakes and the role of dissolved organic carbon. *Limnol. Oceanogr.* 40:1381–1391.

Mörsdorf, G., Frunzke, K., Gadkari, D., and Meyer, O. (1992) Microbial growth on carbon monoxide. *Biodegradation* 3:61–82.

Moss, S. H., and Smith, K. D. (1981) Membrane damage can be a significant factor in the inactivation of *Escherichia coli* by near-ultraviolet radiation. *Photochem. Photobiol.* 33:203–210.

Müller-Niklas, G., Heissenberger, A., Puskaric, S., and Herndl, G. J. (1995) Ultraviolet-B radiation and bacterial metabolism in coastal waters. *Aquat. Microb. Ecol.* 9:111–116.

Najjar, R. G., Erickson, D. J., and Madronich, S. (1995) Modeling the air–sea fluxes of gases formed from the decomposition of dissolved organic matter: Carbonyl sulfide and carbon monoxide. In R. G. Zepp and C. Sonntag, eds., *The Role of Nonliving Organic Matter in the Earth's Carbon Cycle.* Wiley, New York, pp. 107–132.

Neale, P. J. (1987) Algal photoinhibition and photosynthesis in the aquatic environment. In D. J. Kyle, C. B. Osmond, and C. J. Arntzen, eds., *Photoinhibition.* Elsevier Science Publishers, Amsterdam, pp. 39–65.

Neale, P. J., Cullen, J. J., and Davis, R. F. (1998) Inhibition of marine photosynthesis by ultraviolet radiation: Variable sensitivity of phytoplankton in the Weddell–Scotia Confluence during the austral spring. *Limnol. Oceanogr.* 43:433–448.

Neale, P. J., Lesser, M. P., and Cullen, J. J. (1994) Effects of ultraviolet radiation on the photosynthesis of phytoplankton in the vicinity of McMurdo Station, Antarctica. *Antarctic Res. Ser.* 62:125–142.

Nelson, J. R., and Guarda, S. (1995) Particulate and dissolved spectral absorption on the continental shelf of the southeastern United States. *J. Geophys. Res.* 100:8715–8732.

Pakulski, J. D., Aas, P., Jeffrey, W., Lyons, M., Von Waasenbergen, L., Mitchell, D., and Coffin, R. (1998) Influence of light on bacterioplankton production and respiration in a subtropical coral reef. *Aquat. Microb. Ecol.* 14:137–148.

Palenik, B., Price, N. M., and Morel, F. M. M. (1991). Potential effects of UV-B on the chemical environment of marine organisms: A review. *Environ. Pollut.* 70:117–130.

Pausz, C., and Herndl, G. J. (2000) Role of ultraviolet radiation on phytoplankton extracellular release and its subsequent utilization by marine bacterioplankton. *Aquat. Microb. Ecol.* (in press).

Peak, M. J., and Peak, J. G. (1995) Photosensitized reaction of DNA. In W. M. Horspool and P. Song, eds., *CRC Handbook of Organic Photochemistry and Photobiology.* CRC Press, Boca Raton, FL, pp. 1318–1325.

Peng, T.-H. (1992) Possible effects of ozone depletion on the global carbon cycle. *Radiocarbon* 34:772–779.

Plane, J. M. C., Zika, R. G., Zepp, R. G., and Burns, L. A. (1987) Photochemical modeling applied to natural waters In R. G. Zika and W. J. Cooper, eds., *Photochemistry of Environmental Aquatic Systems*, ACS Symposium Series 327. American Chemical Society, Washington, DC, pp. 250–267.

Pomeroy, L. R., and Wiebe, W. J. (1988) Energetics of microbial food webs. *Hydrobiologia* 159:7–18.

Preisendorfer, R. W. (1976) *Hydrologic Optics*. U.S. Department of Commerce, Washington, DC.

Quickenden, T. I., and Irvin, J. A. (1980) The ultraviolet absorption spectrum of liquid water. *J. Chem. Phys.* 72:4416–4428.

Sancar, A. (1994) Structure and function of DNA photolyase. *Biochemistry* 33:2–9.

Schmidt, U., and Conrad, R. (1993) Hydrogen, carbon monoxide, and methane dynamics in Lake Constance. *Limnol. Oceanogr.* 38:1214–1226.

Setlow, R. B. (1974) The wavelengths in sunlight effective in producing skin cancer: A theoretical analysis. *Proc. Natl. Acad. Sci. USA* 71:3363–3366.

Sharma, R. C., and Jagger, J. (1979) Ultraviolet (254–405 nm) action spectrum and kinetic studies of analine uptake in *Escherichia coli* B/R. *Photochem. Photobiol.* 30:661–666.

Siegel, D. A., and Michaels, A. F. (1996) Quantification of non-algal light attenuation in the Sargasso Sea: Implications for biogeochemistry and remote sensing. *Deep-Sea Res.* 43:321–346.

Sieracki, M. E., and Sieburth, J. McN (1986) Sunlight-induced growth delay of planktonic marine bacteria in filtered seawater. *Mar. Ecol. Prog. Ser.* 33:19–27.

Sinha, R. P., Kumar, H. D., Kumar, A., and Häder, D.-P. (1995) Effects of UV-B irradiation on growth, survival, pigmentation and nitrogen metabolism enzymes in cyanobacteria. *Acta Protozool.* 34:187–192.

Smith, R. C. (1989) Ozone, middle ultraviolet radiation and the aquatic environment. *Photochem. Photobiol.* 50:459–468.

Smith, R. C., and Baker, K. S. (1979) Penetration of UV-B and biologically effective dose-rates in natural waters. *Photochem. Photobiol.* 29:311–323.

Smith, R. C., and Baker, K. S. (1981) Optical properties of the clearest natural waters (200–800 nm). *Appl. Opt.* 20:177–184.

Smith, R. C., and Cullen, J. J. (1995) Effects of UV radiation on phytoplankton. *Rev. Geophys.* 33(suppl.):1211–1223.

Smith, R. C., and Tyler, J. E. (1976) Transmission of solar radiation into natural waters. *Photochem. Photobiol. Rev.* 1:117–155.

Smith, R. C., Prézelin, B. B., Baker, K. S., Bidigare, R. R., Boucher, N. P., Coley, T., Karentz, D., MacIntyre, S., Matlick, H. A., Menzies, D., Ondrusek, M., Wan, Z., and Waters, K. J. (1992) Ozone depletion: Ultraviolet radiation and phytoplankton biology in Antarctic waters. *Science* 255:952–959.

Sommaruga, R., Obernosterer, I., Herndl, G. J., and Psenner, R. (1997) Inhibitory effect of solar radiation on thymidine and leucine incorporation by freshwater and marine bacterioplankton. *Appl. Environ. Microbiol.* 63:4178–4184.

Tranvik, L., and Kokalj, S. (1998) Decreased biodegradability of algal DOC due to interactive effects of UV radiation and humic matter. *Aquat. Microb. Ecol.* 14:301–307.

Valentine, R. L., and Zepp, R. G. (1993) Formation of carbon monoxide from the photodegradation of terrestrial dissolved organic carbon in natural waters. *Environ. Sci. Technol.* 27:409–412.

WMO (World Meteorological Organization). (1994) Scientific assessment of ozone deletion: 1994. In D. L. Albritton, R. T. Watson, and P. J. Aucamp, eds., Report No. 37, World Meteorological Organization, Geneva.

Weiss, P. S., Andrews, S. S., Johnson, J. E., and Zafiriou, O. C. (1995) Photoproduction of carbonyl sulfide in South Pacific ocean waters as a function of irradiation wavelength. *Geophys. Res. Lett.* 22:215–218.

Wilhelm, S. W., Weinbauer, M. G., Suttle, C. A., Pledger, R. J., and Mitchell, D. L. (1998) Measurements of DNA damage and photoreactivation imply that most viruses in marine surface waters are infective. *Aquat. Microb. Ecol.* 14:215–222.

Wommack, K. E., Hill, R. T., Muller, T. A., and Colwell, R. R. (1996) Effects of sunlight on bacteriophage viability and structure. *Appl. Environ. Microbiol.* 62:1336–1341.

Worrest, R. C. (1983) Impact of solar ultraviolet-B radiation (290–320 nm) upon marine microalgae. *Physiol. Plant* 58:428–434.

Yocis, B. H., Kieber, D. J., and Mopper, K. Photochemical production of hydrogen peroxide in Antarctic waters. *Deep-Sea Res.* In press.

Zepp, R. G., and Andreae, M. O. (1994) Factors affecting the photochemical production of carbonyl sulfide in sea water. *Geophys. Res. Lett.* 21:2810–2816.

Zepp, R. G., and Cline, D. M. (1977) Rates of direct photolysis in the aquatic environment. *Environ. Sci. Technol.* 11:359–366.

Zepp, R. G., and Schlotzhauer, P. F. (1981) Comparison of photochemical behavior of various humic substances in water. III. Spectroscopic properties of humic substances. *Chemosphere* 10:479–486.

Zuo, Y., and Jones, R. D. (1995) Formation of carbon monoxide by photolysis of dissolved marine organic material and its significance in the carbon cycling of the oceans. *Naturwissenschaften* 82:474–474.

8

CONTROL OF BACTERIAL GROWTH IN IDEALIZED FOOD WEBS

T. Frede Thingstad

Department of Microbiology, University of Bergen,
Bergen, Norway

INTRODUCTION

Studies on the ecology of heterotrophic bacteria can focus on different levels of organization. One can focus on the ecophysiology of community members or on the internal structure of the bacterial community by looking at properties such as diversity; or one can take a more "black box" approach, trying to unravel the role of heterotrophic bacteria by regarding them as one functional unit with a role to play in food webs and in the rest of the ecosystem. Here we concentrate on the second perspective. To restrict the discussion, we will assume an idealized world in which there are only five possible interactions between heterotrophic bacteria and their trophic neighbors in the food web (Figure 1). This includes two types of loss processes (bacterivory and viral lysis) and three types of substrate: (labile dissolved organic carbon, L-DOC; dissolved inorganic nutrients, e.g., PO_4^{3-} and NH_4^+); and organic forms of mineral nutrients, e.g., amino acid bound nitrogen) (Thingstad and Lignell 1997).

The basic question we ask here is, How will this idealized bacterial community influence, and be influenced by, the rest of the photic zone pelagic food web? To approach an answer, we need also to idealize the complex set of trophic interactions and trophic strategies forming this web. Most such idealizations are based on the idea that food webs in the pelagic environment are highly size-structured (Sheldon et al. 1972; Azam et al. 1983; Fenchel 1987;

Microbial Ecology of the Oceans, Edited by David L. Kirchman.
ISBN 0-471-29993-6   Copyright © 2000 by Wiley-Liss, Inc.

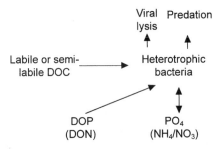

Figure 1. Idealized view of the interactions between heterotrophic bacteria and their immediate neighbors in the trophic web.

Moloney and Field 1991; Thingstad 1998). One way of representing such an idealization is shown in Figure 2. The flow of carbon through the food web presented here proceeds from phytoplankton of different size classes toward predators of increasing size. There is also a series of mechanisms by which organic matter is lost from this "upward" flux toward larger particles, to the pool of dissolved organic material (DOM). Parts of this DOM may be reincorporated into particles via uptake into heterotrophic bacteria, closing what is sometimes termed the "microbial loop" (Azam et al. 1983; Azam et al. 1994). Apart from respiratory losses in the photic zone, organic carbon may also be lost from this system via export to the aphotic zone. This may be caused by sinking of particles, a process mainly linked to large particles in the "classical" food chain at the right side in the food web of Figure 2. In addition, organic material may be exported by transport of DOC by downwelling and diffusion (Legendre and Gosselin 1989; Copin-Montegut and Avril 1993; Carlson et al. 1994). Carbon is also lost presumably from pools of organic matter with low bacterial degradability (e.g., Brophy and Carlson 1989). In the representation in Figure 2, DOC consumption and export processes are linked to the left "microbial" side of the food web.

We know today that there are aspects complicating this idealized view. Examples of complications include mixotrophic protists (Riemann et al. 1995) that combine the roles of osmo- and phagotrophs, parasites (Pan et al. 1997; Rice et al. 1998), pallium (Buskey 1997), peduncle feeders (Calado et al. 1998) or small organisms that feed on larger organisms, and "baleen whales" of the microbial world such as appendicularians (King 1982), which feed on particles several orders of magnitude smaller than themselves. Still it is probably fair to claim that most of the research effort put into the microbial part of the marine pelagic food web over the last 25–30 years has confirmed or added new aspects to the picture in Figure 2 rather than seriously challenging its general validity.

The advancement over the last 25 years in our description of *how* the system in Figure 2 is structured and *how much* material it processes may perhaps seem impressive and exciting. One could, however, claim that the advancement in our understanding of *why* the system works as observed has not advanced with

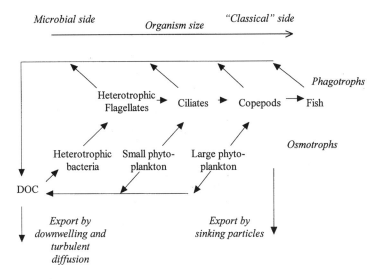

Microbial side   *Organism size*   *"Classical" side*

Figure 2. Idealized scheme of carbon flow through the pelagic food web. Lower line-unicellular microorganisms taking up dissolved organic and inorganic substances (osmotrophs) from the water; upper line: organisms feeding on other organisms by ingesting particles (phagotrophs). There is a general increase in size from left to right and an "upward" flux of carbon toward larger organisms, but also a "microbial loop" in which organic material is lost from this upward flux to a pool of dissolved organic carbon (DOC) from which (parts of) it may be reincorporated into the particulate food chain by heterotrophic bacteria.

the same speed. As a consequence, there is no well-established consensus regarding how heterotrophic bacteria in the lower left corner of Figure 2 influence, and how they are influenced by, the rest of the food web structure. Here we explore the question of whether existing information can be assembled into one framework consistent both in its internal logic and with observational and experimental data. Although several different forms of control on bacterial growth are discussed, phosphorus is given some prominence because it appears to be an important factor limiting bacterial growth even in marine systems (but see Chapter 6).

TOP-DOWN OR BOTTOM-UP CONTROL?

Correlating Bacterial Biomass to Bacterial Production

The discussion of control in food webs is often centered on the concepts of "bottom-up" versus "top-down" control. "Bottom-up" refers to the causal chain in which resource limitation influences the consumer and the consumer's predators, and so on, up the food chain. "Top-down" refers to the cascading effects of predators controlling their prey, which again may control their prey and so on, down the food chain (e.g., McQueen et al. 1989). These concepts are, however, not quite as straightforward as they may seem, since causal relationships in the pelagic food web are not linear, but circular (nutrients → phytoplankton → zooplankton → remineralized nutrients) and multibranched (nutrients → size classes of phytoplankton → size classes of zooplankton → remineralized nutrients). Also, the criterion that growth must balance loss for any component in steady state means that bottom-up and top-down forces in this sense must be of exact equal importance, at least when considered over sufficiently long time scales.

Inspired by Steele's (Steele 1974) question of how the food chain produces enough food for fish, much of the discussion concerned with the food web role of heterotrophic bacteria has had a bottom-up perspective. A central question asked has been whether carbon and energy reassimilated by bacteria from dissolved material into the particulate food web can be transferred upward in the food web, potentially reaching trophic levels of commercial interest such as fish (Azam et al. 1983; Ducklow et al. 1986). The reciprocal top-down perspective of the extent to which fish predation has a cascading effect downward on bacterial biomass and production has received some attention in limnology (Pace and Cole 1996; Jeppesen et al. 1998), but less in the marine field.

As already discussed in Chapter 4, bacterial carbon demand (BCD) is bacterial production ($\mu_B B$) divided by the carbon yield (Y_{BC}):

$$BCD = \frac{\mu_B B}{Y_{BC}} \qquad (1)$$

where μ_B is specific growth rate and B is bacterial biomass. The problem of what controls BCD can thus be split into a discussion of control of each of the three factors μ_B, B, and Y_{BC} (Table 1). Growth rate μ_B and yield Y_{BC} are affected by food availability, as opposed to control of B, which is believed to be influenced by predators. Control of μ_B and control of Y_{BC}, versus control of B, thus correspond to bottom-up and top-down control, respectively. As we shall discuss more closely, there is no reason to assume that the control is by one of the three factors alone; all may be influenced by mechanisms in the rest of the food web.

A simple and elegant approach to the question of what varies most, B or μ_B, was introduced by Billen et al. (1990). If measurements of bacterial biomass (B) are plotted versus bacterial production ($\mu_B B$) in a log–log plot, the variation in biomass can be compared to that in production. A small variation in biomass compared to that in production indicates a large variation in bacterial growth rate. In environments ranging from the Southern Ocean to highly eutrophic freshwater environments, Billen et al. (1990) found production to vary over approximately five orders of magnitude. Total variation in biomass covered a somewhat smaller range, leaving a variation in growth rate

Table 1. Symbols used in this chapter

Symbol	Description	Numerical Value Used	Units
State variables			
P	Free mineral nutrient		nmol P L^{-1}
B	Biomass of heterotrophic bacteria		nmol P L^{-1}
A	Biomass of autotrophic flagellates		nmol P L^{-1}
D	Biomass of diatoms		nmol P L^{-1}
H	Biomass of heterotrophic flagellates		nmol P L^{-1}
C	Biomass of ciliates		nmol P L^{-1}
M	Biomass mesozooplankton		nmol P L^{-1}
Other variable			
BCD	Bacterial carbon demand	Computed	nmol C h^{-1}
Parameters			
Affinities and clearance rates:			
α_B	Bacterial affinity for free mineral nutrients	0.1	L nmol P h^{-1}
α_A	Autotrophic flagellate affinity for free mineral nutrients	0.05	L nmol P h^{-1}
α_D	Diatom affinity for free mineral nutrients		L nmol P h^{-1}
α_H	Heterotrophic flagellate clearance rate for bacteria	0.008	L nmol P h^{-1}
α_C	Ciliate clearance rate for flagellates	0.001	L nmol P h^{-1}
α_M	Mesozooplankton clearance rate for ciliates and diatoms	0.00025	L nmol P h^{-1}
Yields			
Y_H	Heterotrophic flagellate yield on mineral nutrient	0.20	
Y_C	Ciliate yield on mineral nutrient	0.15	
Y_M	Mesozooplankton yield on mineral nutrient	0.10	
Y_{BC}	Bacterial yield on organic carbon	0.008	nmol P (nmol C)$^{-1}$
Loss rates			
δ_B	Bacterial loss rate (apart from predation)		h^{-1}
δ_H	Heterotrophic flagellate loss rate		h^{-1}
δ_M	Mesozooplankton loss rate	Computed	h^{-1}
δ_A	Autotrophic flagellates loss rate		h^{-1}
Others			
P_T	Total P in planktonic food web	20–5000	nmol P L^{-1}
k	Specific excretion rate of L-DOC from phytoplankton biomass	0.02	h^{-1}
r	Fraction of L-DOC respired by bacteria	0.6	
K_1	Fixed value of δ_M for $P_T < K_2$	0.00025	h^{-1}
K_2	Value of P_T at which δ_M becomes a function of P_T	75	nmol P L^{-1}
K_3	Exponent giving the dependence of δ_M on P_T	0.85	
Conversion factors			
C:P (molar) in protist biomass		106	
C:P (molar) in bacterial biomass		50	

of two orders of magnitude. These investigators concluded that in this set of data, there was a tighter control (smaller variation) of growth rate than of biomass. The slope of the regression line of log biomass (on the y axis) versus log production (on the x-axis) in these data (Billen et al. 1990) was 0.7. Using data from Chesapeake Bay and open-ocean locations, Ducklow (1992) found a slope of 0.46, while when open-ocean data only were used, the slope decreased to 0.28, suggesting less bottom-up and more top-down control in oceanic environments. These data, comprising together about 2800 individual measurements of bacterial biomass and production, thus suggest a range in slope of 0.5 ± 0.2.

In this simple manner, a lot of information can be derived from otherwise isolated measurements of bacterial biomass and production, but such an analysis does not really answer our *why* question. What mechanisms determine the slope in such plots?

To approach the question of control, we need briefly to consider the question of steady states in food webs. Control of a process in a dynamic system is obviously different when the system is in steady state, such as might be expected in a climax community, and in a transient phase (e.g., in a spring bloom). The question is, however, slightly more complex, since organisms in the pelagic food web have widely different potential generation times: from the order of hours for heterotrophic bacteria to the order of weeks for copepods to years for higher organisms. Therefore food webs can be considered to be a system of nested processes with different response times. The part containing microorganisms may reach internal equilibrium faster than the part including copepods; the part containing microorganisms and copepods may approach steady state faster than the system containing fish, and so on. Presumably there is some level at which the steady state approximation becomes invalid because both frequency and amplitude of disturbances are faster and stronger than the relaxation time at that level in the system. In such a picture, steady state in the microbial part will be a function of (among other factors) copepod abundance, and thus change with changing copepod population. For the steady state assumption to be a valid approximation, the changes in copepod population would need to be slow compared to the response times of the microbial part of the system. In the sections that follow, we explore the controls in the microbial food web under the assumption that the examined part is in approximate steady state.

Control of Bacterial Biomass

With potential generation times of around 5 hours (Fenchel 1982), heterotrophic flagellates have a growth potential allowing a very rapid response in flagellate abundance when the abundance of their bacterial prey increases. Predatory control has thus to a large extent taken over the role as the standard explanation for control of bacterial biomass, where a more traditional argument was that resource limitation had to be the controlling mechanism. The argument was that bacteria, with their short generation time and exponential growth, have the potential to build up a biomass larger than the mass of the earth in the time span of days. This does not happen because, it was then suggested, bacterial growth rapidly exhausts the resources available for growth or produces an excess of toxic substances (see, e.g., Stanier et al. 1963, p. 329).

An expression for the expected level of bacterial abundance can be obtained by using the steady state requirement for a nonselective heterotrophic flagellate at abundance H, eating bacterial prey at abundance B with a specific clearance rate of α_H, a yield of Y_H, and subject to a specific loss rate δ_H. The clearance rate is the volume swept for prey per unit predator per unit time, and the yield is units of predator formed per units of prey caught. The δ_H parameter contains at this stage any kind of loss of flagellates (cell death, viral lysis, predation from organisms higher up in the food chain, etc.). The number of bacterial prey caught per unit time per unit predator would be given by $\alpha_H B$, and the total loss due to predation would be $\alpha_H BH$. The number of predators formed would thus be $Y_H \alpha_H BH$. For the predator population to remain in steady state, formation of predators must balance their total loss $\delta_H H$, that is:

$$Y_H \alpha_H B^* H^* = \delta_H H^* \qquad (2)$$

where the asterisk denote steady state concentrations of prey and predator. Solving for B^* gives (provided $H^* \neq 0$):

$$B^* = \frac{\delta_H}{Y_H \alpha_H} \qquad (3)$$

The reader will recognize this as the steady state solution to the simplest form of the Lotka–Volterra equations for predators and their prey (see, e.g., Lynch and Hobbie 1988). When resource availability for the bacteria is sufficient to reach the level given by equation (3), more resources will lead to higher B^* only if more resources also leads to an increase in δ_H. This would be the case if the added resources led to more biomass of organisms preying on H (or if, for some less obvious reason, Y_H or α_H were to decrease).

If we assume a turnover of the flagellate population once per day, $\delta_H = 1.0 \, d^{-1}$ ($0.04 \, h^{-1}$), a "typical" flagellate that needs in the order of 200 bacteria to divide ($Y_H = 0.005$), and has a maximum clearance rate of 10^{-5} mL h^{-1} (Fenchel 1982), equation (3) gives an estimated steady state population of bacteria of $\approx 8 \times 10^5$ mL^{-1}, a value very close to levels typically observed in marine systems (see Chapter 4). On such a general level, this theory thus seems to be consistent with observations.

Simon et al. (1992) found that freshwater pelagic systems, as a general trend, have higher bacterial abundance than marine systems. In lakes cladocerans are potentially dominant mesozooplankton predators (Pace and Cole 1996). The simple linear food chain given in Figure 3 may then not be a good representa-

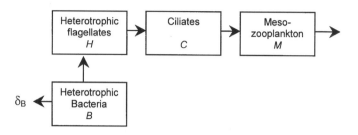

Figure 3. Linear food chain used to discuss the mechanisms controlling bacterial biomass.

tion, and equation (3) may need to be modifed to reflect the observation that mesozooplankton can prey on virtually the entire size range of microbes, from ciliates to bacteria.

Bacterial density in sediment pore water is about 10^9 mL^{-1} (Schmidt et al. 1998). Hence the value of the expression in equation (3) must be at least a factor of 10^3 larger in pore water than in the pelagic zone. The most reasonable difference leading to a sufficient increase in the expression in equation (3) would seem to be a drastically reduced flagellate clearance rate for bacteria in the spatially more complex world of sediment pore waters. In principle, the expression in equation (3) may of course be even greater than 10^9 mL^{-1}, in which case we would have the interesting situation that other balancing mechanisms must have come into play. Respiration balancing a slow supply rate of organic C, stopping increase in bacterial abundance beyond the observed 10^9 mL^{-1}, could be one speculated possibility.

Equation (3) is also relevant to the debate of whether counts of bacteria include dead cells or "ghosts" (Zweifel and Hagström 1995). Since the argument leading to the expression for B^* in equation (3) is based on the food concentration needed by the heterotrophic flagellates to balance their loss, our B^* corresponds to the total count if flagellates eat all bacteria and "ghosts" indiscriminately. As the opposite extreme, it should represent only the live fraction if there is a 100% predator rejection of dead prey cells.

If we introduce an idealized "ciliate" population as the predator on heterotrophic flagellates (Figure 3) and assume predation to be the dominant loss factor, we get:

$$\delta_H = \alpha_C C \qquad (4)$$

where α_C and C are the clearance rate and the biomass, respectively, of ciliates. By insertion into equation (3), bacterial biomass in steady state with a given ciliate biomass becomes:

$$B^* = \frac{\alpha_C}{Y_H \alpha_H} C \qquad (5)$$

For a model with fixed parameters (α and Y values), steady state bacterial biomass will thus be proportional to ciliate biomass.

By adding a ciliate step to our food bacteria–flagellate food chain, the question of bacterial biomass control at steady state has thus been translated into a question of how ciliate biomass is controlled. If we extend the simple linear food chain further by assuming a mesozooplankton population preying on ciliates (Figure 3), and assume the whole food chain to be in steady state, the same argument discussed above will give C proportional to the specific loss rate δ_M of mesozooplankton. Then B^* will be proportional to δ_M. Since δ_M again would be a function of the biomass of the next level of predators preying on mesozooplankton, we see how the variation in B^* becomes a function of how we close the upper end of our food chain model.

Interesting also are some of the processes *not* influencing B^* according to this simple theory. If we introduce an extra specific death rate δ_B to the bacteria (Figure 3), this will not alter our conclusion. If we let δ_B represent the loss of bacteria due to viral lysis, the expression for bacterial biomass in equation (3) will still be valid, and steady state bacterial biomass will be independent of the rate of viral lysis of the bacteria.

Current evidence suggests that the action of bacterial viruses occurs via lytic, rather than lysogenic cycles (e.g. Wilcox and Fuhrman 1994). With an argument similar to that advanced earlier, one can compute the host abundance required to allow existence of a virus. Combining the two, we get a theory according to which predation by an unselective flagellate determines the size of the entire bacterial community, while host-specific viruses will determine the size of each bacterial host subpopulation. The ratio between these types of predation will give an estimate of the number of simultaneously dominant bacterial "species." This theoretical argument leads to the prediction that there are about 100 simultaneously dominant "species" in natural waters (Thingstad et al. 1997). Recent data based on 5S rRNA sequencing (Höfle and Brettar 1995) or denaturing gel gradient electrophoretic analysis of 16S rRNA (Øvreås et al. 1997) suggest that real aquatic systems may contain even fewer (i.e., more on the order of 10) simultaneously dominant "species."

Experimental observations indicate that our assumption of an unselective predation by heterotrophic flagellates is only a rough approximation to reality (Simek and Chrzanowski 1992; del Giorgio et al. 1996; Pernthaler et al. 1996). The consequence would seem to be the possibility for coexisting populations of bacterial prey and thus an increase both in B^* and in bacterial diversity.

Although simple, this theory has quite profound consequences for our understanding of the central mechanisms controlling pelagic food webs. This can be illustrated by asking whether this theory is in accordance with Liebig's law of the minimum (Liebig 1840). Liebig's law predicts an increase in *biomass* if the most limiting element is added. An increase in B^* in the food web of Figure 3 requires that an addition of limiting nutrient leads to increased in mesozooplankton loss rate δ_M. This is at least not the usual way we perceive Liebig's law to function. Also, the virus argument above demonstrates that within this theory, Hutchinson's paradox (Hutchinson 1961) may not be so paradoxical after all. With predators of different selectivity, different prey populations are allowed to coexist, even if they compete for the same resource. No assumptions of heterogeneity in space or time are needed, although such mechanisms of course may add further to the diversity.

Substrate Control of Bacterial Growth Rate

Heterotrophic bacteria obtain energy from the degradation of organic matter. For the formation of new biomass, they also need elements such as C, N, P, and Fe, all of which thus may potentially be in limited supply to the bacteria. In the aphotic zone of the ocean with its measurable concentrations of free phosphate and nitrate, at least N and P are in surplus of what can be incorporated in microbial biomass. Bacterial growth rate in the aphotic zone is thus presumably controlled by the availability of degradable sources of organic carbon. In the surface layer where the heterotrophic bacteria have to compete with phytoplankton for mineral forms of N, P, and Fe, the story is more complicated.

To determine which factor limits bacterial growth rate, various substrates, alone or in different combinations, are added to subsamples and the growth response of the bacteria is measured over time. This general approach gives different answers in different situations. There is an increasing body of evidence that not only in freshwater (Morris and Lewis 1992) but also in estuarine (Thingstad et al. 1993; Zweifel et al. 1993) and marine environments (Pomeroy et al. 1995; Cotner et al. 1997; Thingstad et al. 1998; Zohary and Robarts 1998), where P is deficient relative to N, not only is phytoplankton P-limited but also the bacterial growth rate. We have, however, also found situations in presumably P-deficient estuarine systems where bacterial growth rate could be stimulated by glucose alone (Thingstad, unpublished).

C-limitation has been described in the equatorial Pacific (Kirchman and Rich 1997). In the subarctic Pacific Kirchman (1990) also found cases of C-limitation, but the best response was found when amino acids were added. (see also Kirchman et al. 1990). Lignell et al. (1992) examined a coastal Baltic system in the period after the vernal bloom and found significant stimulation of bacterial activity only when organic C source was added together with the mineral nutrient found to limit phytoplankton. Rivkin and Anderson (1997) found that additions of organic carbon had the largest positive effect on bacterial production in the Caribbean, while in the Gulf Stream and in the Sargasso Sea, phosphate had the largest effect. In Lake Constance, Schweitzer and Simon (1995) observed C-limitation before the spring phytoplankton bloom, then a complex pattern with either P-limitation alone, or C-limitation co-occurring with P-limitation, N-limitation, or both. Reports of N-limited bacterial growth rate seem to be fewer, but Elser et al. (1995) observed that N additions stimulated bacteria in two out of six marine sites and two out of five freshwater sites. While it can be argued theoretically that bacteria should be C-limited in extremely oligotropic situations (Thingstad et al. 1997), it is not intuitively obvious whether high eutrophication levels should favor mineral nutrient or C-limitation of bacteria. The finding of a C-limited bacterial growth rate in the highly eutrophied Danish lake Frederiksborg Slotssø (Kristiansen et al. 1992) at least demonstrates the possibility of C-limitation at high degrees of eutrophication.

If bacterial growth rates can be P- or N-limited, there are no obvious reasons why they should not also be Fe-limited in Fe-deficient areas, in particular since high bacterial Fe content has been shown (Heldal et al. 1996; Tortell et al. 1996). In fact, Pakulski et al. (1996) observed that addition of iron stimulated bacterial growth in Antarctic waters.

Apparently, the system may attain different states in terms of nutrient limitation of heterotrophic bacteria. We obviously need a framework within which we can analyze the mechanisms leading to shifts between C-limited bacterial growth rates and rates limited by mineral nutrients.

The Case of P-Limited Bacterial Growth Rate

Let us first look at the simplest case in which the organic material produced by the system is assumed to be dominated by P-free compounds like carbohydrates and protein.

This situation can be explored by again using the simple Lotka–Volterra formulations (Thingstad et al. 1997). In the simple food web of Figure 4, bacterial biomass B^* at steady state will be given by equation (3). The steady

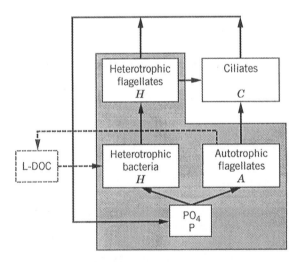

Figure 4. Idealized food web of phosphorus (solid arrows) comprising autotrophic flagellates as bacterial competitors for free phosphate and heterotrophic flagellates as bacterial predators. Ciliates prey on both the competitor and predator. The part of the system inside the shaded area is assumed to be in internal steady state. For illustration of the case with C-limited bacterial growth rate, labile dissolved organic carbon (L-DOC) production is assumed to be proportional to phytoplankton biomass (dotted arrows).

state concentration P^* of phosphate can be obtained from the argument that growth must equal loss for the phytoplankton compartment:

$$\alpha_A P^* A^* = \delta_A A^*, \quad \text{or} \quad P^* = \frac{\delta_A}{\alpha_A}, \tag{6}$$

where α_A is phytoplankton affinity for free phosphate.

With a P-limited bacterial growth rate, bacterial growth rate $\mu_B = \alpha_B P^*$. Insertion of equation (6) then gives:

$$\mu_B = \frac{\alpha_B}{\alpha_A} \delta_A \tag{7}$$

Inserting the expressions for biomass and growth rate [equations (3) and (7), respectively] into the expression for bacterial carbon demand BCD, equation (1), we get an expression for steady state bacterial carbon demand when bacterial growth rate is P-limited:

$$BCD_P^* = (Y_{BC} Y_H)^{-1} \frac{\alpha_B}{\alpha_A \alpha_H} \delta_A \delta_H \tag{8}$$

where subscript P is used to indicate that bacterial growth rate is P-limited. Note how BCD_P^*, through the Y and α-parameters, is a function of the physiology of all the interacting groups of microorganisms, as well as of the loss rates of the competitor (δ_A) and predator (δ_H) populations. If we assume the loss of both autotrophic and heterotrophic flagellates to be dominated by predation from a ciliate population grazing unselectively on the two flagellate populations, we get:

$$\delta_A = \delta_H = \alpha_C C$$

which by insertion into equation (8) gives:

$$BCD_P^* = (Y_{BC} Y_H)^{-1} \frac{\alpha_B}{\alpha_A \alpha_H} \alpha_C^2 C^2 \tag{9}$$

If the system, by autochthonous production alone or in combination with allochthonous import, is supplied with labile DOC at a rate ψ faster than BCD_P^* ($\psi > BCD_P^*$), bacteria will remain P-limited, and DOC will accumulate in the system at the rate $\psi - BCD_P^*$. If, however, $\psi < BCD_P^*$, the pool of degradable organic material will sooner or later be depleted, and bacterial growth rate becomes C-limited; BCD will then be reduced. In a system with C- and P-limitation as the two alternatives, equation (9) gives the maximum BCD attainable for a given value of C.

Note how, in such a model, the ciliates exert *both* top-down and bottom-up control on the heterotrophic bacteria. There is an indirect positive effect of ciliates, both via their predation on bacterial predators (top down), and via their predation on the bacterial competitors for mineral nutrients (the phytoplankton), leading to a bottom-up effect on bacterial growth rate.

By substituting for B^* from equation (3) into equation (9), we can derive the following expression for the relationship between bacterial biomass and production, valid when there is an approximate steady state in a system of bacteria and auto- and heterotrophic flagellates, and the bacterial growth rate is P-limited:

$$\log B^* = \log \left[Y_{BC} Y_H^{-1} \frac{\alpha_A}{\alpha_B \alpha_H} \right]^{1/2} + 0.5 \times \log BCD_P^* \qquad (10)$$

In the simple food web of Figure 4, where there is a common factor (the ciliates) controlling (indirectly) both growth rate and biomass of heterotrophic bacteria, bacterial production spans twice as many orders of magnitude as bacterial biomass, corresponding to the slope of 0.5 in equation (10). This is comfortably in the middle of the previously discussed range of observed slopes of 0.5 ± 0.2, suggesting that some kind of common control of growth rate and biomass may indeed be a feature of the natural system. Not unexpectedly, however, the large range in the observed slopes suggests that additional mechanisms add to the variability in the system.

C-Limited Bacterial Growth Rate

C-limitation of bacterial growth rate is one of the mechanisms that will affect the plots of biomass versus production. When mineral-nutrient-limited bacteria have depleted the pool of labile organic substrates, their growth rate will be reduced until bacterial carbon demand matches the rate at which such material is produced in the system. For a given biomass, points in the plots of biomass versus production will thus be moved toward lower production. This mechanism thus has the potential not only to increase the scatter in such plots, but also to change the slope of the regression line.

There is a multitude of candidates proposed for the processes producing L-DOC in the natural environment (see Chapter 5). These include phytoplankton excretion (Obernosterer and Herndl 1995), which may be passive (Bjørnsen 1988) or active (Myklestad et al. 1989); egestion, excretion, and sloppy feeding from zooplankton (Jumars et al. 1989; Nagata and Kirchman 1992); release by viral lysis (Bratbak et al. 1992; Middelboe et al. 1996; Gobler et al. 1997); and photolysis of humic substances (Lindell et al. 1995; Jorgensen et al. 1998). Probably the production of L-DOC also depends on species composition in the different functional groups of protists, where systems dominated by the carbohydrate-producing alga *Phaeocystis* (Lancelot et al.

1987) may be one example. The one thing these mechanisms have in common is the difficulty in quantifying the process in natural environments. Thus it is currently difficult to make a convincing case for one "correct" model for the production of labile organic substrates for bacterial growth.

For the purpose of illustrating the underlying principles, however, we will analyze the simple case in which L-DOC production in the food web of Figure 4 is proportional to phytoplankton biomass A^*. The mathematical solutions are summarized in Frame 1. C-limitation of bacterial growth rate will, in addition to reducing the bacterial growth rate relative to the corresponding P-limited situation, shift the biomass distribution toward less heterotrophic and more autotrophic flagellates (Frame 1). While bacterial biomass is proportional to C [equation (12)], bacterial production now becomes a function of both the total P-content (P_T) of the system and of the amount of P_T tied up in ciliate biomass (C) [equation (19)]. A one-to-one relationship between bacterial biomass and production can thus be obtained only if additional assumptions are made linking C to P_T. If one considers ciliates to be grazed on by mesozooplankton (as in Figure 5) with a fixed (independent of P_T) loss rate δ_M, the steady state condition for mesozooplankton gives ciliate biomass $C^* = \delta_M / Y_M \alpha_M$; there will be no variation in B^*. Slope in the plots of biomass versus production will thus be 0. Again, this demonstrates the importance of how we choose to close the upper end of our food web model.

One might suspect that addition to the model of a phytoplankton group inedible by ciliates might change the bacterial production versus biomass plots. This is, however, not necessarily the case. The situation can be illustrated by the food web with diatoms (Figure 6) as inedible by ciliates, but serving as food for mesozooplankton. If everything except mesozooplankton is assumed to be in internal steady state, and the growth rates of all osmotrophs are assumed to be P-limited (both L-DOC and silicate available in excess of demand), one can use the same arguments as before to show that steady state bacterial biomass is proportional to mesozooplankton biomass M, while bacterial production is proportional to M^2. Plotting log biomass versus log production will thus again give a slope of 0.5.

Effects of Eutrophication

The main factors identified in the previous analysis affecting the slope in the biomass versus production plots were (1) whether bacteria are limited with respect to mineral nutrient or to carbon; and (2) the manner in which the food web is closed at the upper end. Even with fairly simple models, these two factors can give quite complex relationships between bacterial production and biomass. To illustrate this, the food web of Figure 5 is assumed to be in internal steady state with respect to total phosphorus content P_T and a mesozooplankton loss rate δ_M, which we now assume is an increasing function of P_T. Behind the last assumption lies the idea that there is a relationship between the phosphorus content in this lower part of the food chain and the amount of

Box 1. Steady state solutions for the food web in Figure 4 with P- and C-limited bacterial growth rate.

	Steady State Equations	
Condition	P-Limited Bacteria	C-Limited Bacteria
$dH/dt = 0$	$Y_H \alpha_H B^* = \alpha_C C$	Unaltered
$dB/dt = 0$	$\alpha_B P^* = \alpha_H H^*$	$Y_{BC} k A^* = \alpha_H B^* H^*$
$dA/dt = 0$	$\alpha_A P^* = \alpha_C C$	Unaltered
Mass balance	$P_T = P^* + B^* + A^* + H^* + C$	Unaltered

Solutions for P-Limited Bacterial Growth Rate

$$P^* = \frac{\alpha_C}{\alpha_A} eC \tag{11}$$

$$B^* = \frac{\alpha_C}{Y_H \alpha_H} C \tag{12}$$

$$H^* = \frac{\alpha_B \alpha_C}{\alpha_H \alpha_A} C \tag{13}$$

$$A^* = P_T - \left[1 + \left(\frac{\alpha_C}{\alpha_A} + \frac{\alpha_C}{Y_H \alpha_H} + \frac{\alpha_B \alpha_C}{\alpha_H \alpha_A} \right) \right] C \tag{14}$$

Bacterial production will be

$$BCD_P^* = (Y_{BC} Y_H)^{-1} \frac{\alpha_B}{\alpha_H \alpha_A} \alpha_C^2 C^2 \tag{15}$$

or in terms of B^*:

$$BCD_P^* = Y_{BC}^{-1} Y_H \frac{\alpha_B \alpha_H}{\alpha_A} B^{*2} \tag{16}$$

Solutions for C-Limited Bacterial Rates

Inspection of the steady conditions reveals that, for a given C, neither P^* nor B^* will change relative to the situation with P-limited bacterial growth rate. For the same P_T, the mass balance equation then also implies that $A^* + H^*$ will be the same, and the only change in biomass distributions will be in the balance between A^* and H^*.

Box 1. (*Continued*)

The explicit solutions for A^* and H^* are:

$$A^* = \frac{P_{\mathrm{T}} - \left(1 + \dfrac{\alpha_{\mathrm{C}}}{\alpha_{\mathrm{A}}} + \dfrac{\alpha_{\mathrm{C}}}{Y_{\mathrm{H}}\alpha_{\mathrm{H}}}\right)C}{1 + \dfrac{Y_{\mathrm{BC}}Y_{\mathrm{H}}k}{\alpha_{\mathrm{C}}C}} \qquad (17)$$

and

$$H^* = \frac{Y_{\mathrm{BC}}Y_{\mathrm{H}}k}{\alpha_{\mathrm{C}}C} \frac{P_{\mathrm{T}} - \left(1 + \dfrac{\alpha_{\mathrm{C}}}{\alpha_{\mathrm{A}}} + \dfrac{\alpha_{\mathrm{C}}}{Y_{\mathrm{H}}\alpha_{\mathrm{H}}}\right)C}{1 + \dfrac{Y_{\mathrm{BC}}Y_{\mathrm{H}}k}{\alpha_{\mathrm{C}}C}} \qquad (18)$$

So that C-limited bacterial production will be:

$$BCD_{\mathrm{C}}^* = k \frac{P_{\mathrm{T}} - \left(1 + \dfrac{\alpha_{\mathrm{C}}}{\alpha_{\mathrm{A}}} + \dfrac{\alpha_{\mathrm{C}}}{Y_{\mathrm{H}}\alpha_{\mathrm{H}}}\right)C}{1 + \dfrac{Y_{\mathrm{BC}}Y_{\mathrm{H}}k}{\alpha_{\mathrm{C}}C}} \qquad (19)$$

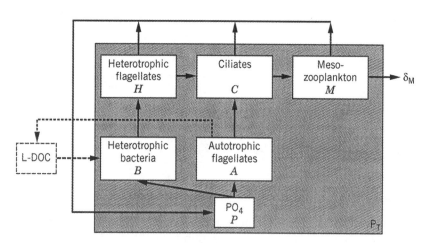

Figure 5. Expansion of the simple food web in Figure 4 to include diatoms and mesozooplankton. The system inside the side shaded area is assumed to be in internal steady state with respect to total phosphorus P_{T} and mesozooplankton biomass M. In this scheme silicate (broken arrow) is required for diatom growth.

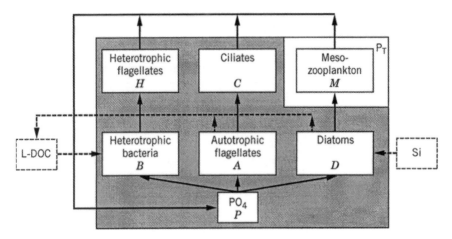

Figure 6. Food web used to explore the effect of total phosphorus P_T and mesozooplankton loss rate δ_M on bacterial growth rate and biomass. System inside shaded area assumed to be in steady state.

planktivorous fish feeding on mesozooplankton. The steady state solutions for the food web in Figure 5, are given in Frame 2.

For illustrative purposes, the function linking δ_M to the phosphorus content P_T of the system (mesozooplankton and lower levels) has somewhat arbitrarily been chosen as:

$$\delta_M = \begin{cases} K_1; & \text{if } P_T < K_2 \\ K_1 \left(\dfrac{P_T}{K_2}\right)^{K_3}; & \text{if } P_T \geqslant K_2 \end{cases} \tag{20}$$

reflecting a mechanism where, beyond a certain P content (K_2) in the system, further enrichment will partially be transferred into biomass of planktivorous fish. Thus δ_M is assumed constant in oligotrophic systems ($P_T < K$) and then increasing with P_T as the power K_3. The corresponding log–log plot of bacterial production versus bacterial biomass is shown in Figure 7 for a range in P_T from 20 to 5000 nmol P L^{-1}. The curve has a flat (slope 0) part in the oligotrophic end corresponding to the range for which δ_M is assumed to be constant, then a part with slope 0.5 corresponding to the range (up to $P_T \approx 1240$ nmol P L^{-1} where bacterial growth rate is P limited), and then finally a part with slope exceeding 0.5 where bacterial growth rate is C-limited. There is also a part at the extreme oligotrophic end (for $P_T < 35$ nmol P L^{-1} with the parameters chosen) where bacterial growth rate is C-limited. For this model, C-limitation of bacterial growth rate thus occurs at both ends of the eutrophication scale.

Box 2. Steady state solutions for the food web in Figure 5 with P- and C-Limited bacterial growth rate

	Steady State Equations	
Condition	P-Limited Bacteria	C-Limited Bacteria
$dH/dt = 0$	$Y_H \alpha_H B^* = \alpha_C C^*$	Unaltered
$dB/dt = 0$	$\alpha_B P^* = \alpha_H H^*$	$Y_{BC} k A^* = \alpha_H B^* H^*$
$dA/dt = 0$	$\alpha_A P^* = \alpha_C C^*$	Unaltered
$dM/dt = 0$	$Y_M \alpha_M C^* = \delta_M$	Unaltered
Mass balance	$P_T = P^* + B^* + A^* + H^* + C^* + M$	Unaltered

Solutions for P-Limited Bacterial Growth Rate

$$P^* = \frac{\alpha_C}{\alpha_A} \frac{\delta_M}{Y_M \alpha_M} \tag{21}$$

$$B^* = \frac{\alpha_C}{Y_H \alpha_H} \frac{\delta_M}{Y_M \alpha_M} \tag{22}$$

$$H^* = \frac{\alpha_B \alpha_C}{\alpha_H \alpha_A} \frac{\delta_M}{Y_M \alpha_M} \tag{23}$$

$$C^* = \frac{\delta_M}{Y_M \alpha_M} \tag{24}$$

$$A^* = R_3 R_1 \left[P_T - \left(\frac{R_4}{R_3 R_1} + R_2 \right) \delta_M \right] \tag{25}$$

$$M^* = R_1 [P_T - R_2 \delta_M] \tag{26}$$

where

$$R_1 = \left(1 + \frac{\alpha_M}{Y_C \alpha_C} \right)^{-1}, \quad R_2 = \frac{1}{Y_M \alpha_M} \left(1 + \frac{\alpha_C}{\alpha_A} + \frac{\alpha_C}{Y_H \alpha_H} \right), \quad R_3 = \frac{\alpha_M}{Y_C \alpha_C},$$

$$R_4 = \frac{\alpha_B \alpha_C}{Y_M \alpha_M \alpha_A \alpha_H}$$

Bacterial production will be:

$$BCD_P^* = (Y_{BC} Y_H)^{-1} \frac{\alpha_B}{\alpha_H \alpha_A} \left(\frac{\alpha_C}{Y_M \alpha_M} \right)^2 \delta_M^2 \tag{27}$$

Box 2. (*Continued*)

or in terms of B^*:

$$BCD_P^* = Y_{BC}^{-1} Y_H \frac{\alpha_B \alpha_H}{\alpha_A} B^{*2} \tag{28}$$

[i.e., identical to equation (15)].

Solutions for C-Limited Bacterial Growth Rate

As above, except equations (23), (25), and (27) are replaced by:

$$A^* = R_1[P_T - R_2\delta_M] \frac{(Y_M Y_H k Y_{BC})^{-1} \delta_M}{Y_C \left[1 + \dfrac{\alpha_C}{\alpha_M} (Y_M Y_H k Y_{BC})^{-1} \delta_M \right]} \tag{29}$$

$$H^* = R_1[P_T - R_2\delta_M] \frac{\alpha_M}{Y_C \alpha_C \left[1 + \dfrac{\alpha_C}{\alpha_M} (Y_M Y_H k Y_{BC})^{-1} \delta_M \right]} \tag{30}$$

$$BCD_C^* = R_1[P_T - R_2\delta_M] \frac{(Y_M Y_H Y_{BC})^{-1} \delta_M}{Y_C \left[1 + \dfrac{\alpha_C}{\alpha_M} (Y_M Y_H k Y_{BC})^{-1} \delta_M \right]} \tag{31}$$

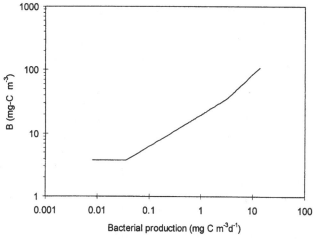

Figure 7. Log–log plot of bacterial biomass versus bacterial production for the model food web in Figure 5 as total P in the planktonic part of the food web (P_T) varies from 20 to 5000 nmol P L^{-1}. Parameter values as in Table 1 and loss rate δM of mesozooplankton assumed to vary with P_T according to equation 20.

The exponent K_3 is a key parameter in determining the range over which bacterial biomass and production varies for a given variation in P_T. The value of 0.85 was chosen as one giving a reasonable range in the variation of bacterial biomass and production. It compares favorably, however, with the value of 0.71 in the empirical relationship $\log Y = 0.71 \log X + 0.774$ between fish crop (Y in kilograms per hectare) and total P (X in micrograms per liter) reported for lakes (Hanson and Leggett 1982).

The Quantitative Importance of Bacteria in Food Webs

Much of the argument around the quantitative importance of bacteria in microbial food webs has been based on comparison of bacterial biomass and production to that of phytoplankton. Combining data from marine and limnetic environments, Simon et al. (1992) found, as a general trend, a positive correlation between bacterial biomass and phytoplankton biomass, but with large scatter in the data. Bacterial biomass covered a range of two orders of magnitude in their data, as compared with more than three orders of magnitude for phytoplankton. The corresponding slope in a log–log plot of bacterial versus phytoplankton biomass was 0.33 ± 0.03 when all data were considered together and 0.23 ± 0.04 when marine data were considered alone. With slopes less than 1, this means that bacterial biomass becomes more dominant relative to phytoplankton biomass as one goes toward oligotrophic environments. If we go back to our model food webs, we see that a constant ciliate biomass (which again was linked to a constant mesozooplankton loss rate δ_M) will make bacterial biomass B^* constant (independent of P_T). Algal biomass A^*, on the other hand, will decrease with decreasing P_T, making bacteria relatively more dominating in oligotrophic environments. The result of plotting B^* versus A^* for the model in Figure 5 is shown in Figure 8A. The line has two parts, one with constant B^* (slope = 0) for the region where mesozooplankton loss rate δ_M is assumed constant, and one where both B^* and A^* as δ_M increase with increasing P_T.

For bacterial production, Cole et al. (1988) found a positive correlation with primary production. The regression line in a log–log plot of bacterial production versus net primary production had a slope of 0.8, corresponding to a smaller variation in bacterial production than in primary production. The model used here gives an average slope around 0.75 (Figure 8B) but consists of parts with different slopes as one goes through four consecutive phases indicated in the figure and characterized by:

1. Constant δ_M, C-limited bacterial growth rate

2. Constant δ_M, P-limited bacterial growth rate

3. δ_M increasing with increasing P_T, P-limited bacterial growth rate

4. δ_M increasing with increasing P_T, C-limited bacterial growth rate

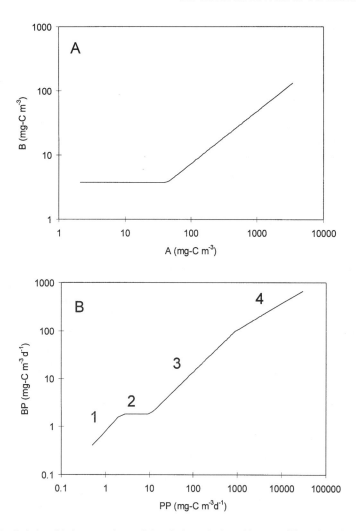

Figure 8. Relationship between bacterial and phytoplankton biomass (A) and production (B) in the food web model of Figure 5 as total P in the planktonic part of the food web (P_T) varies from 20 to 5000 nmol P L^{-1}. Parameter values as in Table 1 and loss rate δM of mesozooplankton assumed to vary with P_T according to equation (20). Numbers 1–4 in (B) indicate different situations discussed in the text.

Note that the positive correlation observed between phytoplankton and bacteria may easily be misinterpreted as demonstrating a bottom-up control of bacteria where more phytoplankton produce more substrates for bacterial growth. As is demonstrated here, such covariation may also happen when bacterial growth rate is P-limited. These plots are thus better understood as demonstrating a common dependence of the independent variable total P.

Patterns of Succession

In one commonly observed pattern of succession, there is a delay between a peak in phytoplankton biomass and the subsequent response in bacterial biomass and production. This can be explained by mechanisms assuming C-limited bacteria, combined with a delay before the photosynthesized organic material becomes available to bacteria. Such a delay can be caused by the need for bacteria to hydrolyze polymers to monomers suitable for bacteria to transport over the cell membrane. Using two classes of polymers of different degradability, Billen (1991) demonstrated the ability of such a model to reproduce observed successions. An alternative explanation is that the production of organic substrates for bacterial growth is mainly a result of activities in the predatory food chain (Jumars et al. 1989), Labile material will then be produced only in the succession phase where grazers consume the phytoplankton.

The model suggested here offers, however, alternative explanations based on a bacterial growth rate limited by the supply of mineral nutrients. If, in Figure 4, the system inside the shaded area is assumed to be in an internal steady state with high P_T and low C, there will be a high phytoplankton biomass A^* [equation (14)], and thus a surplus of food for the ciliates. With a low C, bacterial biomass B^* will be low [equation (12)]. With a surplus of food, the ciliates will grow, and increasing C will lead to a smaller A^* [equation (14)], but to larger B^* [proportional to C equation (12)] and in particular to a larger BCD_P^* [proportional to C^2 in equation (15)]. Thingstad et al. (1999) proposed such a model to explain the pattern of succession observed in a mesocosm experiment performed in the P-deficient brackish layer of a Norwegian fjord. Both the combination of a lack of response to addition of a carbon source (glycine) and a delay in the response of bacteria to addition of phosphate could be explained within the framework of such a model.

Note that when the ciliate biomass increases, the response in bacterial carbon demand will be strong owing to the second power dependence in equation (15). P-limitation will be sustained only if the system's production of L-DOC increases correspondingly. In the absence of such an increase, bacterial consumption would deplete the reservoir of L-DOC, and the system would sooner or later shift to a state with C-limited bacterial growth rate.

The same type of model has also been applied to a microcosm experiment. Here it was suggested to explain a succession from autotrophic flagellates to ciliates as a response to phosphate addition, an increase in bacterial production when phosphate was added without glucose, and a lack of any response to glucose when added without phosphate. A culmination in the bacterial response to phosphate was suggested to be caused by bacteria depleting the pool of degradable of organic C. If this is correct, it means that the system's response in production of L-DOC did not increase as fast as the response in bacterial consumption and P addition thus shifted the system to C-limited bacterial growth rate (Thingstad et al. 1999).

Note that including diatoms as in Figure 6 may strongly influence the effect of phytoplankton–zooplankton succession patterns on bacterial carbon demand. The much longer generation time of mesozooplankton than of ciliates implies that the characteristic time scale expected for a succession from diatoms to mesozooplankton would be much longer than for a succession from autotrophic flagellates to ciliates. With bacterial carbon demand proportional to M^2, the dynamics of a phytoplankton–bacteria succession might therefore be speculated to be slower when available silicate allows a diatom bloom than in the case of a flagellate bloom. This argument would obviously be modified when heterotrophic dinoflagellates with short generation times are the important grazers on diatoms and ciliates.

So far we have not discussed the potential predator–prey oscillations inherent in the type of equations we have used here. From equations (12) and (13), we see that both H^* and B^* are proportional to C. Fluctuations in C so slow that the microbial part of the food web remains in equilibrium with the shifting C should thus move points along a straight line in a plot of H^* versus B^*. In addition, one may have predator–prey oscillations at a smaller time scale between H and B. Rather than a straight line, Tanaka et al. (1997) reported an elliptic seasonal relationship, overlaid by a higher frequency oscillation of smaller amplitude.

Growth on N-Containing Organic Matter

Up to this point, we have considered only the simplified case of phosphorus being the single limiting mineral nutrient potentially limiting bacterial growth rate; we have assumed that labile organic C is released as P-free compounds only and that phosphorus will be released as inorganic orthophosphate. These assumptions may not be too restrictive in a P-deficient environment, where both phytoplankton and heterotrophic bacteria produce enzymes that allow them to split off orthophosphate from dissolved organic phosphates (Jansson et al. 1988). In the case of organic N, amino acid uptake is usually assumed to be dominated by heterotrophic bacteria, although phytoplankton also possesses enzymes for the utilization amino acid N (Pantoja and Lee 1994). In an idealized world where amino acids can be utilized only by bacteria, this changes our simple food web structure, since there will be a pool of dissolved N for which bacteria do not need to compete with the phytoplankton (Figure 1). One effect of this is that, theoretically, the probability for C-limitation of bacteria could be larger in N-deficient regions than in P-deficient regions.

In the case of C-limited bacteria assimilating N-containing organic material, this material may have a surplus of N, and this surplus may be released. Bacteria will then act as remineralizers rather than as competitors, furnishing phytoplankton with the excess N, in the simplest case as NH_4^+. This and the following topics are also discussed in Chapter 9. The stoichiometry of this process can be derived from the requirement that released nitrogen be the difference between nitrogen consumed and nitrogen built into biomass (Parnas 1975):

$$\Delta N = (N{:}C)_S \Delta C - (N{:}C)_B (1 - r) \Delta C$$

or

$$\frac{\Delta N}{\Delta C} = (N{:}C)_S - (N{:}C)_B (1 - r) \tag{32}$$

where ΔN, and ΔC are the amount of nitrogen excreted and the amount of carbon consumed, $(N{:}C)_S$ and $(N{:}C)_B$ are the N:C ratios in the organic substrate and in bacterial biomass, respectively, and r is the fraction respired of the organic carbon. Thus $1 - r$ is the fraction incorporated into biomass. A positive ΔN means that bacteria have a surplus of N, their growth rate will not be N-limited, and they will function as remineralizers supplying inorganic N to phytoplankton. A negative ΔN [e.g., as in the case of growth on carbohydrates: $(N{:}C)_S = 0$] means that bacteria need to take up additional inorganic N from the environment, their growth rate may become N-limited, and they will have to compete with the phytoplankton. From experiments with microbial communities supplied with organic matter of different $(N{:}C)_S$, Billen (1984) reported a very good fit of equation (32) for $(N{:}C)_B (1 - r) = 0.085$ mol $N(\text{mol C})^{-1}$. For a respiration coefficient $r = 0.6$, this corresponds to $(N{:}C)_B = 0.21$, that is, slightly above the Redfield ratio of 0.15, but in good agreement with numbers found for natural communities using X-ray microanalysis (Fagerbakke et al. 1996). On average, bacteria in natural environments seem to have a $(P{:}C)_B$ substantially above Redfield (Fagerbakke et al. 1996; see also Chapter 9). It is possible to interpret this in two quite different ways: either bacteria seldom experience P-limitation and therefore store a surplus of P, or they have a high requirement of P relative to C. The latter interpretation, which is discussed in Chapter 9, would add to the previous arguments for P-limitation of bacteria in P-deficient areas being more probable than N-limited bacteria in N-deficient areas.

The Effects of Variable Yield and Biomass Composition

So far we have discussed factors affecting μ_B and B, assuming that the third factor determining BCD, the bacterial yield on carbon, Y_{BC}, is a constant. If biomass B is in phosphate units, Y_{BC} has units of mol P biomass formed per mol C of substrate consumed. Assuming no excretory losses $Y_{BC} = (1 - r)(P{:}C)_B$. Clearly both a change in respiration coefficient r and a change in biomass composition $(P{:}C)_B$ will affect BCD, and together they will affect the transition between C- and P-limitation of bacteria in analogy with equation (32). Because an increasing fraction of the carbon source is consumed for maintenance purposes at low growth rates (increasing r), chemostat cultures with C-limited bacteria show decreasing biomass yield with decreasing growth rate. Also bacteria in P-limited chemostats have been demonstrated to have a cell quota $(P:C)_B$ that varies with growth rate as described by a Droop model (Vadstein and Olsen 1989). For a fixed growth rate, it has also been shown that there is a range in medium $(P:C)_S$ ratios for which bacteria will consume both the phosphate added as the single source of P, and the glucose added as the single source of organic C (Pengerud et al. 1987), indicating that r and/or $(P:C)_B$ changes with $(P:C)_S$ in the transition zone between C- and P-limited growth rate. This type of simultaneous consumption of two limiting substrates for a range of medium compositions, which has also been reported for nitrogen–carbon limitations and for microorganisms of a number of different types, thus seems to be a general phenomenon (Egli 1995). Looking at the combined possibility of limitation with respect to C, N, and P, a region can also be found in which all three substrates are simultaneously depleted (Martinussen and Thingstad 1987; Thingstad 1987).

The food web also contains other mechanisms that may remove or transform a substrate available in surplus. One example is the conversion by some phytoplankton species of excess mineral nitrogen into dissolved organic N (Collos et al. 1992). Since, in addition, the general N:P ratio of the system over time may be adjusted by nitrogen fixation and denitrification (see Chapter 15), the observation that surface layers depleted of mineral nutrients often seem to be in a state in which both inorganic N and inorganic P are depleted, is perhaps not unexpected. Such mechanisms would mean that within certain limits, excess L-DOC is either respired or stored in biomass of high C:N and/or C:P content. This would reduce the probability of a bacterial growth rate limited by the supply of mineral nutrients leading to accumulation of labile DOC as suggested by Thingstad et al. (1997). Whether labile organic C as a rule is depleted simultaneously with inorganic N and/or P in natural systems remains to be seen. Søndergaard and Middelboe (1995) reported a mean of 19% of the total DOC as being degradable by bacteria.

In addition to these biogeochemically important consequences, a flexible stoichiometry will affect a model's ability to reproduce short time variations where, for example, phytoplankton may continue to grow on internal stores subsequent to the depletion of free mineral nutrients. Adding Droop descriptions to a model simulating nutrient addition experiments in mesocosms improves the fit of the model to nutrient data over the time scale of days (Baretta-Bekker et al. 1998).

There are two traditional ways of introducing flexible stoichiometry. Usually in bacteriology, the Monod formulation is used for growth and a variable yield (Pirt 1982). On the other hand, it is traditional in phytoplankton ecology to use a Droop formulation linking growth with cell quota (Droop 1974). The Droop formulation can be extended to include the case of organic substrates being respired (Thingstad 1987). All these are, however, phenomenological descriptions not really linking the variation at the organism level to an understanding at the lower levels of biochemical processes, electron transfer, and thermodynamics. A model of this type, combined with the assumption that a bacterium optimizes its growth rate, has been suggested by Vallino et al. (1996).

CONCLUSIONS

While studies on the ecological role of pelagic heterotrophic bacteria have been much concerned with bottom-up effects such as supply of organic substrates, phytoplankton–bacteria competition for mineral nutrients, and transfer of organic C upward in the food chain, the models used here highlight the additional top-down effects from a predator food chain, with groups such as ciliates, mesozooplankton, and planktivorous fish playing key roles for bacterial biomass and production. The idealized models discussed here allow a mathematical formulation in which such top-down and bottom-up effects can be analyzed within one unified and consistent description. These models also demonstrate the need to understand how different parts of the food web have different relaxation times (time needed to approach steady state). With a proper understanding of such factors, models of this kind should allow us to analyze food web relationships at different scales, both in large intersystem comparisons and in attempts to understand patterns of succession observed in experimental systems (e.g., mesocosms). Although originally intended as mental guides on the road to understanding principles of control in recycling food webs, these models seem to bear promises as minimum models for actual description and prediction of observed data.

SUMMARY

1. Conditions controlling bacterial growth in the microbial food web are difficult to analyze without some kind of idealized mathematical description of the system.

2. Assuming the Lotka–Volterra type of interactions, a food web structure, and approximate steady state in the microbial part of the food web, one can derive analytical mathematical expressions linking bacterial production and carbon demand to properties of other organisms in the food web and to conditions such as total availability of the limiting element.

3. Organisms in the predator food chain influence bacterial biomass, both via predation on competitors for mineral nutrients. They also influence the growth rate of bacteria when bacteria are limited by mineral nutrients. Since production is the product of growth rate and biomass, predator biomass would be expected to have a squared effect on bacterial production.

4. The concepts of top-down and bottom-up control cannot be readily separated in steady state models of recycling systems. The models used

here do, however, emphasize the role of top predators in controlling the capacity of bacteria to consume organic carbon.

5. The conceptual view of food web control derived in this manner is very different from the one obtained from analysis of C flow alone, which often emphasizes the bottom-up perspective, namely, that the amount of carbon entering lower levels determines the biomass at higher levels.

6. If carbon consumption by bacteria whose growth rate is limited by a mineral nutrient exceeds the supply of organic substrates for growth, the pool of degradable organic carbon will be depleted and bacterial growth will become C-limited. In this case the mechanisms of control shift, and bacterial production will be controlled by the supply rate of organic substrates.

7. Since diatoms are competitors with heterotrophic bacteria for mineral nutrients but are not grazed on by the same predators grazing on bacteria and other picoplankton, the silicate supply would be expected to have a large influence on the magnitude and dynamics of bacterial production.

ACKNOWLEDGMENT

This work was financed by the EU Mast3 program, contract MAS3-CT95-0016 "Medea."

REFERENCES

Azam, F., Fenchel, T., Field, J. G., Gray, J. S., Meyer-Reil, L. A., and Thingstad, T. F. (1983) The ecological role of water-column microbes in the sea. *Mar. Ecol. Prog. Ser.* 10:257–263.

Azam, F., Smith, D., Steward, G., and Hagström, Å. (1994) Bacteria–organic matter coupling and its significance for oceanic carbon cycling. *Microb. Ecol.* 28:167–169.

Baretta-Bekker, H., Baretta, J., Hansen, A., and Riemann, B. (1998) An improved model of carbon and nutrient dynamics in the microbial food web in marine enclosures. *Aquat. Microb. Ecol.* 14:91–108.

Billen, G. (1984) Heterotrophic utilization and regeneration of nitrogen. In J. E. Hobbie and P. J. leB Williams, eds., *Heterotrophic Activity in the Sea.* Plenum Press, New York, pp. 313–356.

Billen, G. (1991) Protein degradation in aquatic environments. In R. Chróst, ed., *Microbial Enzymes in Aquatic Environments.* Springer-Verlag, New York, pp. 123–143.

Billen, G., Servais, P., and Becquevort, S. (1990) Dynamics of bacterioplankton in oligotrophic and eutrophic aquatic environments: Bottom-up or top-down control? *Hydrobiologia* 207:37–42.

Bjørnsen, P. K. (1988) Phytoplankton exudation of organic matter: Why do healthy cells do it? *Limnol. Oceanogr.* 33:151–154.

Bratbak, G., Heldal, M., Thingstad, T. F., Riemann, B., and Haslund, O. H. (1992) Incorporation of viruses into the budget of microbial C-transfer. A first approach. *Mar. Ecol. Prog. Ser.* 83: 273–280.

Brophy, J. E., and Carlson, D. J. (1989) Production of biologically refractory dissolved organic carbon by natural seawater microbial populations. *Deep-Sea Res.* 36:497–507.

Buskey, E. J. (1997) Behavioral components of feeding selectivity of the heterotrophic dinoflagellate *Protoperidinium pellucidum. Mar. Ecol. Prog. Ser.* 153:77–89.

Calado, A., Craveiro, S., and Moestrup, O. (1998) Taxonomy and ultrastructure of a freshwater, heterotrophic *Amphidinium* (Dinophyceae) that feeds on unicellular protists. *J. Phycol.* 34:536–554.

Carlson, C., Ducklow, H. W, and Michaels, A. (1994) Annual flux of dissolved organic carbon from the euphotic zone in the northwestern Sargasso Sea. *Nature* 371:405–408.

Cole, J., Findlay, S., and Pace, M. (1988) Bacterial production in fresh and saltwater ecosystems: A cross-system review. *Mar. Ecol. Prog. Ser.* 43:1–10.

Collos, Y., Dohler, G., and Biermann, I. (1992) Production of dissolved organic nitrogen during uptake of nitrate by *Synedra planctonica:* Implications for estimates of new production in the oceans. *J. Plankton Res.* 14:1025–1029.

Copin-Montegut, G., and Avril, B. (1993) Vertical distribution and temporal variation of dissolved organic carbon in the North-Western Mediterranean Sea. *Deep-Sea Res.* 40:1963–1972.

Cotner, J. B., Ammerman, J. W., Peele, E. R., and Bentzen, E. (1997) Phosphorus-limited bacterioplankton growth in the Sargasso Sea. *Aquat. Microb. Ecol.* 13:141–149.

Del Giorgio, P., Gasol, J., Vaqué, D., Mura, P., Agustí, S., and Duarte, C. (1996) Bacterioplankton community structure: Protists control net production and the proportion of active bacteria in a coastal marine community. *Limnol. Oceanogr.* 41:1169–1179.

Droop, M. R. (1974) The nutrient status of algal cells in continuous culture. *J. Mar. Biol. Assoc. UK* 54:825–855.

Ducklow, H. (1992) Factors regulating bottom-up control of bacterial biomass in open ocean plankton communities. *Arch. Hydrobiol. Beih.* 37:207–217.

Ducklow, H. W., Purdie, D. A., and Williams, P. J. le B. (1986) Bacterioplankton: A sink for carbon in a coastal marine plankton community. *Science* 232:865–867.

Egli, T. (1995) The ecological and physiological significance of the growth of heterotrophic microorganisms with mixtures of substrates. In G. Jones, ed., *Advances in Microbial Ecology.* Vol. 14. Plenum Press, New York, pp. 305–386.

Elser, J. J., Stabler, L. B., and Hasset, R. P. (1995) Nutrient limitation of bacterial growth and rates of bacterivory in lakes and oceans: A comparative study. *Aquat. Microb. Ecol.* 9:105–110.

Fagerbakke, K., Heldal, M., and Norland, S. (1996) Content of carbon, nitrogen, oxygen, sulfur and phosphorus in native aquatic and cultured bacteria. *Aquat. Microb. Ecol.* 10:15–27.

Fenchel, T. (1982) Ecology of heterotrophic microflagellates. II. Bioenergetics and growth. *Mar. Ecol. Prog. Ser.* 8:225–223.

Fenchel, T. (1987) *Ecology–Potentials and Limitations.* Ecology Institute, Oldendorf/ Luhe, Germany.

Gobler, C. J., Hutchins, D. A., Fisher, N. S., Cosper, E. M., and SanudoWilhelmy, S. A. (1997) Release and bioavailability of C, N, P, Se, and Fe following viral lysis of a marine chrysophyte. *Limnol. Oceanogr.* 42:1492–1504.

Hanson, J. M., and Leggett, W. C. (1982) Empirical prediction of fish biomass and weight. *Can. J. Fish. Aquat. Sci.* 39:257–263.

Heldal, M., Fagerbakke, K., Tuomi, P., and Bratbak, G. (1996) Abundant populations of iron and manganese sequestering bacteria in coastal water. *Aquat. Microb. Ecol.* 11:127–133.

Hutchinson, G. E. (1961) The paradox of the plankton. *Am. Nat.* 95:137–145.

Höfle, M., and Brettar, I. (1995) Taxonomic diversity and metabolic activity of microbial communities in the water column of the central Baltic Sea. *Limnol. Oceanogr.* 40:868–874.

Jansson, M., Olsson, H., and Petterson, K. (1988) Phosphatases: Origin, characteristics and function in lakes. *Hydrobiologia* 170:157–175.

Jeppesen, E., Søndergaard, M., Jensen, J. P., Mortensen, E., Hansen, A. M., and Jorgensen, T. (1998) Cascading trophic interactions from fish to bacteria and nutrients after reduced sewage loading: An 18-year study of a shallow hypertrophic lake. Ecosystems 1: 250–267.

Jumars, P. A., Penry, D. L., Baross, J. A., Perry, M. J., and Frost, B. W. (1989) Closing the microbial loop: Dissolved carbon pathway from to heterotrophic bacteria from incomplete ingestion, digestion and absorption in animals. *Deep-Sea Res.* 36:483– 495.

Jørgensen, N. O. G., Tranvik, L., Edling, H., Graneli, W., and Lindell, M. (1998) Effects of sunlight on occurrence and bacterial turnover of specific carbon and nitrogen compounds in lake water. *FEMS Microbiol. Ecol* 25:217–227.

King, K. R. (1982) The population biology of the larvacean *Oikopleura dioica* in enclosed water columns. In: G. D. Grice and M. R. Reeve, eds., *Marine Mesocosms.* Springer-Verlag, Berlin.

Kirchman, D. L. (1990) Limitation of bacterial growth by dissolved organic matter in the subarctic Pacific. *Mar. Ecol. Progr. Ser.* 62:47–54.

Kirchman, D. L., and Rich, J. (1997) Regulation of bacterial growth rates by dissolved organic carbon and temperature in the equatorial Pacific Ocean. *Microb. Ecol.* 33:11–20.

Kirchman, D. L., Keil, R. G., and Wheeler, P. A. (1990) Carbon limitation of ammonium uptake by heterotrophic bacteria in the subarctic Pacific. *Limnol. Oceanogr.* 35:1258–1266.

Kristiansen, K., Nielsen, H., Riemann, B., and Fuhrman, J. (1992) Growth efficiencies of fresh-water bacterioplankton. *Microb. Ecol.* 24:145–160.

Lancelot, C., Billen, G., Sournia, A., Weisse, T., Coljin, F., Veldhuis, M., Davies, A., and Wassman, P. (1987) *Phaeocystis* blooms and nutrient enrichment in the continental coastal zones of the North Sea. *Ambio* 16:38–46.

Legendre, L., and Gosselin, M. (1989) New production and export of organic matter to the deep ocean: Consequences of some recent discoveries. *Limnol. Oceanogr.* 34:1374–1380.

Liebig, J. (1840) *Chemistry in Its Application to Agriculture and Physiology.* Taylor and Walton, London.

Lignell, R., Kaitala, S., and Kuosa, H. (1992) Factors controlling phyto- and bacterioplankton in late spring on a salinity gradient in the northern Baltic. *Mar. Ecol. Prog. Ser.* 84:121–131.

Lindell, M. J., Granli, W., and Tranvik, L. J. (1995) Enhanced bacterial growth in response to photochemical transformation of dissolved organic matter. *Limnol. Oceanogr.* 40:195–199.

Lynch, J. M., and Hobbie, J. E. (1988) Microbial population and community dynamics. In J. M. Lynch and J. E. Hobbie, eds., *Micro-organisms in action: Concepts and Applications in Microbial Ecology.* Blackwell Scientific Publishers, Oxford, pp. 70–73.

Martinussen, I., and Thingstad, T. F. (1987) Utilization of N, P, and organic C by heterotrophic bacteria. II. Comparison of experiments and a mathematical model. *Mar. Ecol. Prog. Ser.* 37:285–293.

McQueen, D. J., Johannes, M. R. S., Post, J. R., Steward, T. J., and Lean, D. R. S. (1989) Bottom-up and top-down impacts on freshwater pelagic community structure. *Ecol. Monogr.* 59:289–310.

Middelboe, M., Jørgensen, N. O. G., and Kroer, N. (1996) Effects of viruses on nutrient turnover and growth efficiency of noninfected marine bacterioplankton. *Appl. Environ. Microbiol.* 62:1991–1997.

Moloney, C. L., and Field, J. G. (1991) The size-based dynamics of plankton food webs. 1. A simulation model of carbon and nitrogen flows. *J. Plankton Res.* 13:1003–1038.

Morris, D., and Lewis, W. (1992). Nutrient limitation of bacterioplankton growth in Lake Dillon, Colorado. *Limnol. Oceanogr.* 37:1179–1192.

Myklestad, S., Holm-Hansen, O., Vørum, K. M., and Volcani, B. E. (1989) Rate of release of extracellular amino acids and carbohydrates from the marine diatom *Chaetocheros affinis. J. Plankton Res.* 11:763–773.

Nagata, T., and Kirchman, D. (1992) Release of macromolecular organic-complexes by heterotrophic marine flagellates. *Mar. Ecol. Prog. Ser.* 83:233–240.

Obernosterer, I., and Herndl, G. (1995) Phytoplankton extracellular release and bacterial-growth-dependence on the inorganic N–P ratio. *Mar. Ecol. Prog. Ser.* 116:247–257.

Pace, M. L., and Cole, J. J. (1996) Regulation of bacteria by resources and predation tested in whole-lake experiments. *Limnol. Oceanogr.* 41:1448–1460.

Pakulski, J. D., et al. (1996). Iron stimulation of Antarctic bacteria. *Nature* 383(6596):133–134.

Pan, C. L., Hsu, Y. L., Tsai, G. J., Kuo, H. J., Chang, C. M., Wang, F. J., and Wu, C. S. (1997) Isolation and identification of *Bdellovibrio* from coastal areas of Taiwan. *Fish. Sci.* 63:52–59.

Pantoja, S., and Lee, C. (1994) Cell-surface oxidation of amino acids in seawater. *Limnol. Oceanogr.* 39:1718–1725.

Parnas, H. (1975) Model for decomposition of organic material by microorganisms. *Soil. Biol. Biochem.* 7:161–169.

Pengerud, B., Skjoldal, E. F., and Thingstad, T. F. (1987) The reciprocal interaction between degradation of glucose and ecosystem structure. Studies in mixed chemostat cultures of marine bacteria, algae, and bacterivorous nanoflagellates. *Mar. Ecol. Prog. Ser.* 35:111–117.

Pernthaler, J., Sattler, B., Simek, K., Schwarzenbacher, A., and Psenner, R. (1996) Top-down effects on the size–biomass distribution of a freshwater bacterioplankton community. *Aquat. Microb. Ecol.* 10:255-263.

Pirt, S. J. (1982) Maintenance energy: A general model for energy-limited and energy-sufficient growth. *Arch. Microbiol.* 133:300–302.

Pomeroy, L. R., Sheldon, J. E., Sheldon, W. M. J., and Peters, F. (1995) Limits to growth and respiration of bacterioplankton in the Gulf of Mexico. *Mar. Ecol. Prog. Ser.* 117:259-268.

Rice, T. D., Williams, H. N., and Turng, B. F. (1998) Susceptibility of bacteria in estuarine environments to autochthonous bdellovibrios. *Microb. Ecol.* 35:256-264.

Riemann, B., Havskum, H., Thingstad, T. F., and Bernard, C. (1995) The role of mixotrophy in pelagic environments. In I. Joint, ed., *Molecular Ecology of Aquatic Microbes*, G.38. Springer Verlag, Berlin, pp. 87–114.

Rivkin, R., and Anderson, M. (1997) Inorganic nutrient limitation of oceanic bacterioplankton. *Limnol. Oceanogr.* 42:730–740.

Schmidt, J. L., Deming, J. W., Jumars, P. A., and Keil, R. G. (1998) Constancy of bacterial abundance in surficial marine sediments. *Limnol. Oceanogr.* 43:976–982.

Schweitzer, B., and Simon, M. (1995) Growth limitation of planktonic bacteria in a large mesotrophic lake. *Microb. Ecol.* 30:89–104.

Sheldon, R. W., Prakash, A., and Sutcliffe, W. H. (1972) The size distribution of particles in the ocean. *Limnol. Oceanogr.* 17:327–340.

Simek, K., and Chrzanowski, T. (1992) Direct and indirect evidence of size-selective grazing on pelagic bacteria by fresh-water nanoflagellates. *Appl. Environ. Microbiol.* 58:3715-3720.

Simon, M., Cho, B., and Azam, F. (1992) Significance of bacterial biomass in lakes and the ocean—Comparison to phytoplankton and biogeochemical implications. *Mar. Ecol. Prog. Ser.* 86:103–110.

Slater, J. H. (1988) Microbial population and community dynamics. In J. M. Lynch, and J. E. Hobbie, (eds.) *Micro-organisms in Action. Concepts and Applications in Microbial Ecology.* Oxford, Blackwell Scientific Publishers, Oxford. pp. 51–74.

Stanier, R., Doudoroff, M., and Adelberg, E. (1963) *The Microbial World*, 2nd ed. Prentice-Hall, Englewood Cliffs, NJ.

Steele, J. H. (1974) *The Structure of Marine Ecosystems.* Harvard University Press, Cambridge, MA.

Søndergaard, M., and Middelboe, M. (1995) A cross-system analysis of labile dissolved organic carbon. *Mar. Ecol. Prog. Ser.* 118:283–294.

Tanaka, T., Fujita, N., and Taniguchi, A. (1997) Predator–prey eddy in heterotrophic nanoflagellate–bacteria relationships in a coastal marine environment: A new scheme for predator–prey associations. *Aquat. Microb. Ecol.* 13:249–256.

Thingstad, T. F. (1987) Utilization of N,P, and organic C by heterotrophic bacteria. I. Outline of a chemostat theory with a consistent concept of maintenance metabolism. *Mar. Ecol. Prog. Ser.* 35:99–109.

Thingstad, T. F. (1998) A theoretical approach to structuring mechanisms in the pelagic food chain. *Arch. Hydrobiol.* 363:59–72.

Thingstad, T. F., and Lignell, R. (1997) A theoretical approach to the question of how trophic interactions control carbon demand, growth rate, abundance, and diversity. *Aquat. Microb. Ecol.* 113:19–27.

Thingstad, T. F., Hagström, Å., and Rassoulzadegan, F. (1997) Export of degradable DOC from oligotrophic surface waters: Caused by a malfunctioning microbial loop? *Limnol. Oceanogr.* 42:398–404.

Thingstad, T. F., Skjoldal, E. F., and Bohne, R. A. (1993) Phosphorus cycling and algal–bacterial competition in Sandsfjord, western Norway. *Mar. Ecol. Prog. Ser.* 99:239–259.

Thingstad, T. F., Zweifel, U. L., and Rassoulzadegan, F. (1998) P-limitation of both phytoplankton and heterotrophic bacteria in NW Mediterranean summer surface waters. *Limnol. Oceanogr.* 43:88–94.

Thingstad, T. F., Havskum, H., Kaas, H., Lefevre, D., Nielsen, T. G., Riemann, B., and Williams, P. J. le B. (1999) Bacteria–protist interactions and organic matter degradation under P-limited conditions. Comparison between an enclosure experiment and a simple model. *Limnol. Oceanogr.* 44:236–253.

Thingstad, T. F., Pérez, M., Pelegri, S., Dolan, J., and Rassoulzadegan, F. (1999) Trophic control of bacterial growth in microcosms containing a natural community from northwest Mediterranean surface waters. *Aquat. Microb. Ecol.* 18:145–156.

Tortell, P. D., Maldonado, M. T., and Price, N. M. (1996) The role of heterotrophic bacteria in iron-limited ocean ecosystems. *Nature* 383:330–332.

Vadstein, O., and Olsen, Y. (1989) Chemical composition and phosphate uptake kinetics of limnetic bacterial communities cultured in chemostats under phosphorus limitation. *Limnol. Oceanogr.* 34:939–946.

Vallino, J. J., Hopkinson, C. S., and Hobbie, J. E. (1996) Modeling bacterial utilization of dissolved organic matter: Optimization replaces Monod growth kinetics. *Limnol. Oceanogr.* 41:1591–1609.

Wilcox, R. M., and Fuhrman, J. A. (1994) Bacterial viruses in coastal seawater:Lytic rather than lysogenic production. *Mar. Ecol. Prog. Ser.* 114:35–45.

Zohary, T., and Robarts, R. D. (1998) Experimental study of microbial P limitation in the eastern Mediterranean. *Limnol. Oceanogr.* 43:387–395.

Zweifel, U. L., and Hagström, Å. (1995) Total counts of marine bacteria include a large fraction of non-nucleoid-containing "ghosts." *Appl. Environ. Microbiol.* 61:2180–2185.

Zweifel, U. L., Norrman, B., and Hagström, Å. (1993) Consumption of dissolved organic carbon by marine bacteria and demand for inorganic nutrients. *Mar. Ecol. Prog. Ser.* 101:23–32.

Øvreås, L., Forney, L., Daae, F., and Torsvik, V. (1997) Distribution of bacterioplankton in meromictic Lake Saelenvannet, as determined by denaturing gradient gel electrophoresis of PCR-amplified gene fragments coding for 16S rRNA. *Appl. Environ. Microbiol.* 63:3367–3373.

9

UPTAKE AND REGENERATION OF INORGANIC NUTRIENTS BY MARINE HETEROTROPHIC BACTERIA

David L. Kirchman

Graduate College of Marine Studies,
University of Delaware, Lewes, Delaware

Many different microbes are capable of either taking up or releasing ("regenerating") inorganic nutrients in aquatic ecosystems, but heterotrophic bacteria are uniquely involved in both processes. For the most part phytoplankton are sinks for nutrients, whereas heterotrophic protists grazing on bacteria and small phytoplankton (see Chapter 12) excrete material and are not known to use dissolved compounds at natural concentrations, with the possible exception of high molecular weight biopolymers. Heterotrophic bacteria need the capacity both to take up and release inorganic nutrients because the elemental ratio of organic material used to support bacterial growth often differs from that in bacterial biomass, thus necessitating assimilation or regeneration of elements to maintain a steady state elemental composition. Phytoplankton, on the other hand, never need to excrete inorganic nutrients because they can always adjust CO_2 fixation to the nutrient supply (assuming that they are not in the dark or otherwise stressed) and thus are able to maintain their elemental composition at steady state. Heterotrophic protists may graze on prey deficient (relative to the protist) in a particular element, but apparently they have not devised any strategy to supplement particulate prey by utilizing dissolved nutrients.

Microbial Ecology of the Oceans, Edited by David L. Kirchman.
ISBN 0-471-29993-6   Copyright © 2000 by Wiley-Liss, Inc.

Because of low surface area to volume ratios, heterotrophic protists are at a disadvantage in competing with smaller microbes such as heterotrophic bacteria and many phytoplankton species for dissolved compounds. Only heterotrophic bacteria are potentially involved in both uptake and regeneration of key elements such as N and P, and only they seem to have dual roles in nutrient cycling.

Of course the physiological potential for both taking up and regenerating inorganic nutrients does not mean that heterotrophic bacteria are doing both all the time, in all environments. When and where heterotrophic bacteria take up or release nutrients are the major ecological questions.

The general goal of this chapter is to provide the necessary framework for understanding the processes of nutrient uptake and regeneration by heterotrophic bacteria. Here we focus on the two major nutrients, N and P. I use "bacteria" in this chapter, but in fact most of what is discussed here applies to *Archaea* as well as those prokaryotes in the *Bacteria* domain (Chapter 3).

COMPOSITION OF A BACTERIAL CELL

Examining the elemental composition of microbial cells is an important first step in understanding what compounds are used and excreted by bacteria in the oceans. Figure 1 illustrates how the elements making up cells and the elements available in the earth's crust, the ultimate source of primordial building blocks for life, differ in their relative abundance. The earth's crust has high amounts of Si, but this element is not used by bacteria; diatoms are the most ecologically important user of Si in the oceans. Although Na, Mg, and Ca are abundant in the earth's crust and as major cations in seawater, they (and Cl^-) are not found in any biochemical structure in bacteria. Marine bacteria do require these ions for growth, to maintain osmotic balance; but uptake is insignificant compared to the large concentrations in seawater. Calcium is used only by selected phytoplankton (coccolithophorids), again not by bacteria, except as cationic bridges among polymers.

Several elements occur only in trace amounts in cells and in the earth's crust. Metals like Zn and Co are important cofactors in selected enzymes (e.g., Zn in urease, Co in enzymes requiring vitamin B_{12}), but cells require only small amounts for growth. The most important micronutrient is Fe.

Iron is abundant in the earth's crust and is present in all cells. Although cells need relatively low amounts (C:Fe ratios on the order of 10,000), use by both phytoplankton and heterotrophic bacterial assemblages reduces Fe concentrations to very low (picomolar) levels in the surface layer of many oceans; heterotrophic bacteria appear to account for roughly 40% of total iron use (Tortell et al. 1996), although there is only one published study to date on this important topic. The insolubility of Fe oxides in oxic seawater at near-neutral pH also contributes to low Fe concentrations. In some oceans, most notably in the low-chlorophyll, high-nutrient (HNLC) provinces (e.g., the equatorial

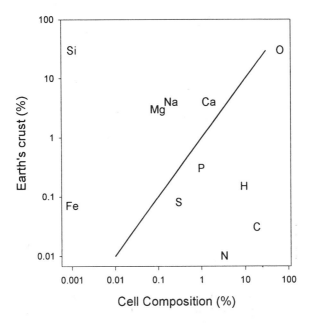

Figure 1. Comparison of the elemental composition of a living cell with some elements in the earth's crust; line indicates equal amounts in both cells and the crust. Note that some elements are highly enriched in cells compared to the inert world, whereas others are present only in low amounts. Silicon is used in cell walls of some algae (diatoms) but not by bacteria. Iron in cells is much lower than indicated in this graph (see text). (Inspired by Brock and Madigan 1991.)

Pacific: Coale et al. 1996) and some upwelling regions (Hutchins and Bruland 1998), iron limits primary production and thus, at the very least, indirectly limits bacterial production. Whether it limits bacterial production directly remains to be seen; there is some evidence that Fe may limit heterotrophic bacterial production in the Southern Ocean (Pakulski et al. 1996), but more work is needed on this topic.

Of the six most abundant elements in bacteria, two (O and H) are readily obtained from water and a third from a major cation in seawater (SO_4^-). Aerobic heterotrophic bacteria easily obtain sufficient S from assimilatory sulfate reduction; the reduced sulfur is used mostly in the synthesis of two sulfur amino acids, methionine and cysteine. The remaining three elements, C, N, and P, are those thought to limit bacterial growth most frequently in marine systems (see Chapter 8).

C:N AND C:P RATIOS FOR HETEROTROPHIC BACTERIA

Heterotrophic bacteria are generally thought to be rich in nitrogen compared with phytoplankton. The C:N of bacteria is often assumed to be below 5, lower (more N) than that of phytoplankton; the phytoplankton or Redfield C:N ratio is 6.7 on a molar basis. Some laboratory studies have measured low C:N ratios [e.g., Goldman and Dennett (1991) reported an average of 4.5], but other laboratory studies have reported higher ratios [e.g., Goldman et al. (1987) found 5.6]. Recently Fukuda et al. (1998) found C:N ratios of 6.8 ± 1.2 and 5.9 ± 1.1 in coastal and oceanic bacterial assemblages, respectively, which do not differ significantly from the Redfield ratio for phytoplankton. In short, bacteria and phytoplankton do not appear to differ substantially in nitrogen content normalized to carbon.

Bacteria do appear to be rich in phosphorus compared with phytoplankton. For example, the C:P ratio of phytoplankton biomass is generally about 106, much higher (much less P) than that for heterotrophic bacteria (ca. 50), as discussed below in more detail. Few investigations have examined the P content of both bacteria and phytoplankton simultaneously in natural communities, but freshwater studies have confirmed the higher P content in natural bacterial assemblages (e.g., Vadstein 1998).

What is not often appreciated, however, is the variation in the C:P ratio. Depending on growth conditions, the bacterial C:P ratio can vary as much as 50-fold in laboratory-grown pure cultures (e.g. Tezuka 1990), much greater than variation in C:N ratios (maximum of three-fold: Tezuka 1990). When all published studies are considered, the C:P ratio varies nearly 60-fold, from 8 (Bratbak 1985) to 464 (Tezuka 1990) for laboratory-grown bacteria, whereas the C:N ratio varies only about four-fold (from 3.8 to 15); the C:N ratio of natural bacteria in various marine environments varies even less, from 3.8 to 9.9 (summarized by Fukuda et al. 1998).

The factors causing such high variation in the C:P ratio are not completely understood, but one factor is apparently the composition of the dissolved compounds in the bacterial growth medium; bacterial C:P ratios increase with the C:P of the compounds used by bacteria for growth (Figure 2). The effect of dissolved N composition on the C:N of bacterial biomass is less clear. Tezuka (1990) found that bacterial C:N increased with media C:N whereas Goldman et al. (1987) found no effect.

Two observations need explanation: Why do bacteria apparently have more P than phytoplankton? And why does C:P vary much more than C:N in bacteria? The size and biochemical composition of bacteria offer clues to answering these questions.

The major constituents of a bacterial cell rapidly growing in pure culture are well known and easily measured, but it is much more difficult to estimate the same parameters for heterotrophic bacteria growing slowly in the oceans. The biggest problem is the presence of other organisms and detritus that cannot be separated easily (if at all) from heterotrophic bacteria. Another problem is that concentrations of various cellular biochemicals are much lower in seawater than in pure cultures. Bacterial abundance in nearly all aquatic environments is roughly 10^6 cells per milliliter, two orders of magnitude lower than that achievable in the lab. Furthermore, slowly growing cells in the oceans

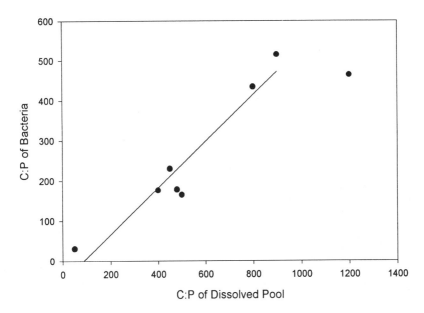

Figure 2. Variation in cellular C:P in pure bacterial cultures as a function of media C:P. (Modified from Tezuka 1990.)

and most natural environments are much smaller than lab-grown cells. Cell sizes can vary greatly, but as a rough guide, natural bacteria are about 0.5 μm in diameter versus cultured bacteria that are 1.5 μm or larger, as illustrated in Figure 3. The threefold difference in diameter gives a nearly 30-fold difference in volume. The small volume but large ratios of surface area to volume of natural heterotrophic bacteria have profound consequences for understanding cell composition and many aspects of microbial life. Some studies have suggested that natural bacteria contain less water per volume than pure culture bacteria (Table 1), which would greatly affect concentrations of various enzymes and intracellular solutes and in turn has profound consequences for all aspects of bacterial metabolism.

The relative amount of two macromolecules in slow-growing natural bacteria appear to differ greatly from rapidly growing lab bacteria. First, slow-growing bacteria have relatively more DNA than rapidly growing bacteria: 10% versus 3% of dry weight for natural and cultured bacteria, respectively (Table 1). The absolute amount of DNA per cell or the genome size of natural bacteria is not any bigger (and perhaps it is a bit smaller), but these cells must put all the genes necessary for independent existence into a smaller cell, typical of slowly growing bacteria (Fuhrman and Azam 1982). So, the ratio of DNA to the rest of cellular biomass has to be higher for slow-growing bacteria.

The second major difference between laboratory and natural bacteria is the RNA content. Bacteria growing at high rates in the lab typically have about

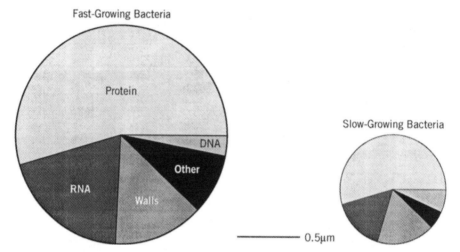

Figure 3. Schematic diagram of bacteria grown in the laboratory on rich media versus bacteria occurring naturally in aquatic ecosystems. See text and Table 1 for more details.

20% of dry weight as RNA, compared to the estimate of 14% for natural bacteria growing slowly. Unfortunately, RNA content of natural bacteria must be estimated by difference (see Table 1); but the decrease if not the absolute value for cellular RNA is consistent with several studies indicating that RNA amount decreases with growth rate (e.g., Kemp et al. 1993). Fast growth requires high rates of protein synthesis, which in turn necessitate large numbers of ribosomes and ribosomal RNA; rRNA is 80% of total RNA.

Variability in the major cellular P-containing components (nucleic acids and phospholipids) explains why C:P ratios can vary greatly in heterotrophic bacteria. Phospholipid amounts per cell also change because of changes in cell size. As cells decrease in size with decreasing growth rate, the ratio of surface area to volume increases, resulting in more lipids per total cellular carbon. Polyphosphate is another P-containing biochemical that may contribute to variability in C:P ratios, although the small cell size of natural bacteria and normally low dissolved P concentrations in seawater suggest that amounts of P stored in polyphosphate bodies will be small (if present at all) relative to nucleic acids and phospholipids. Changes in RNA content may be offset partially by changes in relative DNA and lipid content (Table 1), but variability in all P-containing biochemical compounds especially RNA, leads to variability in C:P ratios.

In contrast, C:N ratios do not vary much because protein as a fraction of cell mass changes little with environmental conditions. Protein is the largest reservoir of N in cells, accounting for about 80% of cellular N. Although nucleic acids are richer in N than protein (2.6 vs 3.8: P. J. le B. Williams, personal communication), changes in nucleic acid content are likely to be small relative to protein and thus should not affect cellular C:N ratios substantially.

Table 1. Biochemical composition of a bacterium growing fast (generation times of ≤ 1 h) and slowly (day time scale)[a]

| Parameter | Growth Rate | | Units |
	Fast[b]	Slow	
Size			
Volume	1.8	0.07	μm^3
Diameter	1.5	0.5	μm
Dry weight per cell	284	24	fg cell^{-1}
Dry weight per volume	161	367	fg μm^{-3}

| | % of Dry Weight | | Comments on Slow-Growth Cells |
	Fast[b]	Slow	
Protein	55	55.0	Simon and Azam (1989)
RNA	20	13.7	Calculated by difference[c]
Lipids	9	12.0	Same lipid/SA ratio[d]
Lipopolysaccharides (LPS)	3.4	3.3	Watson et al. (1977)
Cell wall (peptidoglycan: PG)	2.5	4.1	Same PG/LPS ratio
C storage (glycogen)	2.5	0.0	Assume C-limitation
DNA	3	10.0	Fuhrman and Azam (1982)
Monomers (e.g., metabolic intermediates, inorganic ions)	4	2.1	Set arbitrarily

[a]The dry weight of a slow-growing cell was estimated from the carbon content (12 fg cell^{-1}) measured by Fukuda et al. (1998) and assuming that carbon is 50% of dry weight.
[b]Cell composition for a bacterium growing rapidly in the lab (Ingraham et al. 1983).
[c]Amount of RNA was calculated by subtracting the sum of all other amounts from the total estimated dry weight for a slow-growing natural bacterium.
[d]I assumed that the ratio of lipid or peptidoglycan (PG) per surface area (SA) measured for fast-growing bacteria in the lab can be applied to slow-growing natural bacteria. We do not know lipid nor peptidoglycan amounts of natural bacteria.

The difference in cell size seems to offer the best explanation for why bacteria have relatively more P than phytoplankton. Since heterotrophic bacteria are smaller than phytoplankton, even smaller than cyanobacteria (cyanobacteria are usually 0.8 μm in diameter or bigger), the ratio of surface area to volume is much larger, and thus the amount of membrane per cytoplasmic material is much higher for heterotrophic bacteria than for phytoplankton. Membranes are P-rich due to phospholipids, whereas the cytoplasm is dominated by proteins with little P; the only P-containing proteins are the few enzymes that have been phosphorylated to regulate activity. The net result is a higher P:C ratio for heterotrophic bacteria than for phytoplankton.

STOICHIOMETRY OF NITROGEN UPTAKE AND EXCRETION

The ecological roles of heterotrophic bacteria in the oceans potentially include the uptake of inorganic nitrogen and the release of ammonium during the utilization of dissolved organic nitrogen (DON). Before reviewing what we know about these processes, it is useful to examine a stoichiometric model that aims to explain when and where heterotrophic bacteria take up or excrete inorganic nutrients. Although this model is a good starting point, ultimately it is of limited use for understanding in situ oceanographic processes because we do not know enough about the composition of the DOM used by heterotrophic bacteria for growth. I focus here on N because the high variability of C:P ratios in heterotrophic bacteria greatly complicates using the following model for examining P fluxes.

The model is a fairly simple mass balance for C and N and is a formal way to look at how C:N ratios affect N uptake and excretion. I use the term "C:N mass balance" because the model basically says that the mass of C and N entering the cell must be equal to the C and N exiting the cell. Goldman et al. (1987) explored this model perhaps most thoroughly and also discussed earlier work on it (See also Fenchel et al. 1998). The model relies on two assumptions. First, it assumes that the organic C in DOM either is oxidized to CO_2 or is used for biomass synthesis and that the relationship between the two fates for C (respiration or biomass synthesis) is set by a constant carbon growth efficiency or yield (Y). (See Chapter 4 or 10 for an explicit definition of Y). Likewise, assimilated nitrogen either is used for biomass synthesis or is excreted as ammonium; although the N form is not important to the model, heterotrophic bacteria excrete only ammonium (they do not oxidize it to nitrate) but can take up both ammonium and nitrate (see below). The second assumption is that the N:C ratios of bacterial cells ($N:C_b$) and of the DOM used by bacteria ($N:C_s$) are invariant.

The total amount of N required for biomass synthesis (N demand) is $U_c Y(N:C_b)$ where U_c is DOM uptake in C units. This N demand must be balanced by N assimilated from DOM, that is, $U_c(N:C_s)$, with any imbalances corrected for by excretion or uptake of DIN (F_N); positive F_N indicates net uptake, whereas negative F_N indicates net ammonium excretion. At steady state (constant Y and N:C ratios),

$$U_c Y(N:C_b) = U_c(N:C_s) + F_N \qquad (1)$$

or

$$F_N = U_c[Y(N:C_b) - N:C_s] \qquad (2)$$

This approach for examining ammonium fluxes mediated by heterotrophic bacteria has been validated by several studies mainly using laboratory cultures (e.g., Goldman et al. 1987).

To illustrate the model and to begin to explore ammonium fluxes in the oceans, let us consider the conditions under which ammonium is neither taken up nor excreted (i.e., F_N is zero, the "breakeven point."). These are the curves given in Figure 4. Below these curves, bacteria excrete ammonium, whereas above them, they take up ammonium. We can put further bounds on these conditions by making some assumptions about the parameters in equation (2). At least the ranges in $C:N_b$ and Y are relatively well known, if not the exact values. We can assume that $C:N_b$ will vary between just 4.5 and 6.7 (the Redfield ratio). The two extremes for Y are 0.15 and 0.37, grand averages for the open ocean and estuaries, respectively (Chapter 10). Picking values for $C:N_s$ is the hardest problem because we do not know exactly what DOM is used by bacteria, but we can examine the breakeven points given the various combinations of Y and $C:N_b$ as illustrated in Figure 4; the intersection of the horizontal lines with the $C:N_b$ curves determines the $C:N_s$ for the breakeven point. At one extreme, the highest Y and the lowest $C:N_b$ give a $C:N_s = 12$; at the other extreme of lowest Y and highest $C:N_b$, $C:N_s = 44$. The oceanic values are $Y = 0.15$ (Chapter 10), $C:N_b = 5.9$ (Fukuda et al. 1998), and thus $C:N_s = 29$.

Based on these oceanic values, bacteria using DOM with C:N lower than 29 should excrete ammonium. If bacteria use DOM with the Redfield ratio

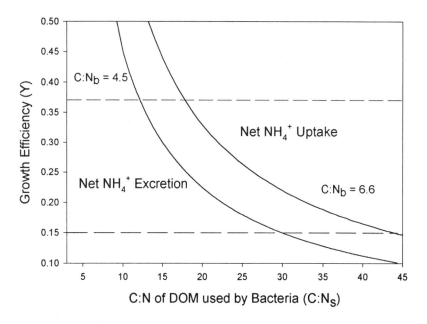

Figure 4. Flux of NH_4^+ (excretion or uptake) for different growth efficiencies and C:N of DOM used by bacteria ($C:N_s$) for two values of C:N of bacterial biomass ($C:N_b$). Curves indicate zero flux for the given $C:N_b$. Any values of Y and $C:N_s$ above the curves indicate net NH_4 uptake, and below the curve, excretion. The top dashed line is set at $Y = 0.37$, the highest average Y for a marine system (estuaries) and the lower dashed line is set at $Y = 0.15$, the lowest average marine Y, which is from open-ocean studies (see Chapter 10).

(6.7), then they should also excrete ammonium. Likewise, they should also excrete ammonium if the $C:N_s$ is equal to the ratio of total DOC to total DON (DOC:DON). Taking DOC and DON concentrations to be 80 μM C and 5μM N, respectively (Hansell and Waterhouse 1997) gives a DOC:DON = $C:N_s$ = 16, which is lower than all the breakeven points except for the case of the highest Y and lowest $C:N_b$. Ignoring that exception, assigning $C:N_s$ values equal to the Redfield ratio or DOC:DON ratio leads to the prediction that heterotrophic bacteria should excrete, ammonium, not take it up.

In fact, we know that heterotrophic bacteria often take up much ammonium, as discussed below in more detail, suggesting that the model is not appropriate for natural bacterial assemblages. It could be argued that we really do not know $C:N_s$ and that the values assumed above are not close to being correct. Since we know that only a small fraction of the total DOM pool is utilized by bacteria on short time scales, we cannot assume $C:N_s$ to be equal to DOC:DON. Unfortunately, we do not know if the Redfield ratio is any closer to being the correct $C:N_s$. In short, the lack of information about $C:N_s$ limits use of the model.

One would think that it would be easier to apply the C:N mass balance model to a case where it is clear what DOM is being used by bacteria. Such was the situation discussed by Tupas and Koike (1990), who examined bacterial assemblages growing on high concentrations of DON.

Tupas and Koike (1990) measured bacterial growth on DON consisting of mostly dissolved combined amino acids (DCAA) released by a mussel. They observed that DON use was very high (70–260% of bacterial N demand), indicating bacteria should excrete ammonium in this experiment, according to the C:N mass balance model. However, Tupas and Koike (1990) observed ammonium assimilation (50–88% of bacterial N demand) simultaneously with DON use. More problematically, even as ammonium was being assimilated, ^{15}N isotope dilution experiments showed that 80% of the assimilated DON was converted to ammonium and excreted. Tupas et al. (1994) went on to demonstrate simultaneous uptake of ammonium and regeneration of $[^{15}N]H_4^+$ from $[^{15}N]$amino acids. In short, these bacterial assemblages seem to be doing everything at the same time: assimilating and releasing ammonium while also using DON.

The C:N mass balance model may not explain the data of Tupas et al. (1994) because of changes in various model parameters not allowed by the model. We would not expect $C:N_b$ to vary much, as argued above, but Y may vary as bacteria adjust their metabolism to changing growth conditions (Chapter 10). Because of changing DOM composition, $C:N_s$ may also vary substantially. Variation in any of these three parameters would greatly complicate predicting whether bacteria make a net contribution to either ammonium mineralization or uptake.

It is also useful to remember that the mass balance model does not consider the internal dynamics of N fluxes, only the end result. For example, there could be much ammonium/ammonia exchange across membranes even if net uptake is zero. Hoch et al. (1992) pointed out that ammonia (NH_3) from intracellular pool could diffuse across cell membranes because intracellular concentrations of this uncharged, small molecule would be much higher than extracellular concentrations. To recapture this lost nitrogen, bacteria need transport mechanisms to bring back ammonium (NH_4^+ being the dominant form at seawater pH) into the cell. Standard ^{15}N techniques would reveal ammonium uptake even if the net effect is excretion of ammonium.

Finally, perhaps the most important consideration is that different groups within the bacterial assemblages may be utilizing different components of the DOM and DIN pools. We know that the community structure of marine bacterial assemblages is quite complex (see Chapter 3). Perhaps C:N mass balance would be more obvious for individual bacterial groups and microniches than for the entire assemblage. Conceivably, some bacteria may be using sugars and ammonium, for example, and other bacterial groups rely more on amino acid pools and end up regenerating ammonium. Still, if each component is in mass balance, the sum should also be in mass balance. Regardless, complexity in the bacterial community and in its environment certainly complicates any simple account of C and N fluxes.

Vallino et al. (1996) present a more complicated budget for bacterial growth that attempts to account not only for C and N stoichiometry but also for electron flow during utilization of DOM and DIN. The basic premise of the model is that bacteria will adjust their metabolism and allocate resources to maximize growth rates, with constraints on energetics, electron balances, and substrate uptake kinetics, in addition to C:N ratios of labile DOM and of bacterial biomass. Unlike the simple C:N stoichiometric model, C:N ratios and Y can vary. The model is too complicated to reproduce here, but the underlying premise of the model is simple: bacteria optimize resource use in order to maximize growth. Of the several simulations presented by Vallino et al. (1996), one is instructive to discuss here.

Figure 5 illustrates the case of bacteria growing with initially high concentrations of a labile DOC component (e.g., glucose) and low concentrations of DON such as amino acids. Under these conditions, amino acids and the glucoselike pool are taken up with no change in ammonium (and nitrate) concentrations, until the amino acids are depleted. Then ammonium is taken up until it is depleted, and only then is nitrate used. Vallino et al. (1996) point out that in this simulation ammonium uptake was zero initially even for $C:N_s = 29$, that is, a $C:N_s$ sufficiently high to lead to predictions of ammonium uptake based on a simple C and N mass balance model. Likewise, in another simulation presented by Vallino et al. (1996) ammonium concentrations remain constant initially, even though $C:N_s = 4.8$, a case in which ammonium excretion would be predicted by the C- and N-only mass balance model. These two examples illustrate how the bioenergetic model of Vallino et al. (1996) predicts a more complicated relationship between DOM and DIN fluxes than apparent from only C and N mass balance.

The problem with applying the bioenergetic model of Vallino et al. (1996) to natural environments is that it requires even more information about the DOM pool than the C:N mass balance model. In addition to C:N ratios of

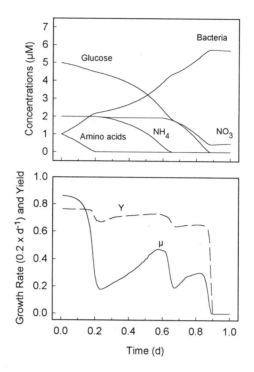

Figure 5. Model results of bacteria growing on high labile DOC (glucose) and low labile DON (amino acids). In the bottom panel, Y is the growth yield and μ the specific growth rate. (Modified from Vallino et al. 1996.)

bacteria and DOM, we need to know the "degree of reduction" (ψ) or the relative oxidation state of DOM components used by heterotrophic bacteria. If we represent the DOM as $C_\alpha H_\beta O_\gamma N_\delta$ with a charge of ξ, then the degree of reduction is:

$$\psi = 4\alpha + \beta - 2\gamma - 3\delta - \xi \tag{3}$$

There are empirical approximations for ψ, but the measurements on naturally occurring DOM are not trivial. Regardless, the model once again points to the importance of information about DOM composition beyond just C:N.

WHAT PROPORTION OF INORGANIC NITROGEN AND PHOSPHORUS UPTAKE IS BY HETEROTROPHIC BACTERIA?

Orthophosphate Uptake

Heterotrophic bacteria appear to account for a large fraction of orthophosphate (P_i) uptake both in the oceans and in freshwater (Kirchman 1994). The median percentage of phosphate uptake attributable to bacteria is 60% when both freshwater and marine studies are considered, but the percentage varies greatly among the many diverse ecosystems examined so far. These percentages are usually estimated from the amount of radiolabeled P_i taken up the small size fraction (< 1.0 or 0.8 μm), that is, the "bacterial" size fraction. Some of the uptake may be due to phytoplankton such as cyanobacteria in the bacterial size fraction. The few studies that corrected for phytoplankton uptake found a lower fraction (24–46%) of P_i uptake attributable to heterotrophic bacteria.

One approach for putting these percentages into perspective is to compare relative P_i uptake with the ratio of bacterial production to primary production (BP:PP). This comparison converts rates of biomass production from the usually reported C units to P units. The maximum expected relative uptake is:

$$\text{Maximum \% uptake} = \text{BP:PP} \times \frac{\text{phytoplankton C:P}}{\text{bacterial C:P}} \times 100 \qquad (4)$$

This estimate is a maximum because any uptake of dissolved organic P (DOP) would lead to lower P_i uptake relative to biomass production. As discussed in Chapter 4 in more detail, biomass production and BP:PP vary greatly with time and space, making it difficult to pick a single "average" value, but here it is instructive to assume BP:PP $= 0.2$. Bacterial C:P ratios also appear to vary greatly, but an average of 53 is not too misleading; the phytoplankton C:P ratio is the Redfield ratio of 106. Given these values, the maximum expected uptake of P_i would be about 40% of total uptake.

The calculated maximum uptake of P_i is about equal to or less than the measured relative uptake of P_i by heterotrophic bacteria, implying that these bacteria obtain much of their P from P_i, not from organic P. Surprisingly, few studies have compared P_i uptake and biomass production directly (e.g., see Fuhrman and Azam 1982).

Uptake of Ammonium, Nitrate, and Urea by Heterotrophic Bacteria

Unlike inorganic P, which occurs exclusively in one oxidation state, there are three oxidation states for dissolved inorganic nitrogen (DIN): ammonium (-3), nitrite ($+3$) and nitrate ($+5$), of which two (ammonium and nitrate) are important N sources for phytoplankton growth. Urea, which is best treated as if it were an inorganic N compound, only occasionally is an important N source for phytoplankton and usually is not thought to be used by bacteria (e.g., Tanmminen and Irmisch 1996), although there have been reports of heterotrophic urea uptake (Kirchman et al. 1991; Jørgensen et al. 1999). Ammonium uptake has been examined in more detail than both nitrate and urea uptake by heterotrophic bacteria, a difference in emphasis that needs to be remembered in looking at previous studies.

Like P_i uptake, ammonium uptake by heterotrophic bacteria varies greatly, from a low of 5% of total uptake in Long Island Sound (Fuhrman et al. 1988) to 78% in Georgia coastal waters (Table 2; see also Kirchman 1994). The overall median is about 40% which is two-fold higher than the maximum expected percentage, assuming a BP:PP of 0.2 and similar C:N ratios for bacteria and phytoplankton (see above). Like P_i uptake, estimates of DIN uptake may be compromised by phytoplankton uptake in the bacterial size fraction, although the proper controls can correct for this uptake. DIN uptake is even more difficult to estimate than P_i uptake because the stable isotope ^{15}N must be used. The overall conclusion that bacteria can be responsible for at least some ammonium uptake has been confirmed by Lipschultz (1995), who used flow cytometry, not size fractionation, to examine the role of various organisms in DIN uptake.

Natural assemblages of heterotrophic bacteria have not been thought to take up much nitrate because the energetic cost of reducing nitrate via assimilatory nitrate and nitrite reductases to ammonium would seem to preclude nitrate uptake by bacteria in natural waters with low DOM concentrations. Use of nitrate requires five NADHs compared to one for ammonium (Vallino et al. 1996), and in laboratory pure cultures, the presence of ammonium often inhibits nitrate reduction and represses the synthesis of assimilatory nitrate reductase. Consequently, one would expect the continuous input of ammonium from grazers would inhibit nitrate uptake by bacteria in the oceans. It was not surprising when initial reports indicated at best low nitrate uptake by heterotrophic bacteria. Eppley et al. (1977) first observed that plots of ammonium uptake versus primary production had a positive vertical intercept, implying uptake independent of phytoplankton, most likely uptake by heterotrophic bacteria. A similar plot for nitrate uptake went through zero, implying negligible uptake by bacteria. Some subsequent studies that directly

Table 2. Summary of studies measuring both ammonium and nitrate uptake by heterotrophic bacteria in marine waters

Location	Total Uptake by Bacteria (%)		Comments	Ref.
	NH_4^+	NO_3^-		
North Atlantic	22–39%	4–14	Spring bloom	Kirchman et al. (1994)
Georgia coastal waters	78	0	Used inhibitors	Wheeler and Kirchman (1986)
Subarctic Pacific	31	32	Station P; 4-month average	Kirchman and Wheeler (1998)
Georges Bank (North Atlantic)	38	27	Coast to open ocean transect	Harrison and Wood (1988)
Boothbay Harbor, ME	34	10	Flow cytometry	Lipschultz (1995)
Average	42	16		

measured uptake also found very low nitrate uptake by heterotrophic bacteria (e.g., Wheeler and Kirchman 1986; Kirchman et al. 1994).

However, there have been a few reports of nitrate uptake by bacteria in the oceans (Table 2). Harrison and Wood (1988) found often substantial nitrate uptake by the < 1.0 μm size fraction along a transect in Georges Bank, but there also was much chlorophyll and CO_2 fixation (i.e., phytoplankton activity) in this size fraction. It is still interesting to note that Harrison and Wood (1988) found more — but not much more — ammonium uptake than nitrate uptake in the small size fraction. Overall, the small size fraction accounted for 27 and 38% of nitrate and ammonium uptake, respectively, (Harrison and Wood 1988). In contrast, Kirchman and Wheeler (1998) found the opposite in the subarctic Pacific: relatively more nitrate than ammonium uptake by the bacterial size fraction (36% vs 18%) when the two processes were measured together. Over the entire data set, however, the fraction taken up by bacteria was about the same (31%) for both nitrate and ammonium.

Assuming that the few studies completed to date are representative of all oceans (a somewhat dubious assumption), it seems that bacteria use more ammonium than nitrate; overall, bacteria account for 42 and 16% of ammonium and nitrate uptake, respectively (Table 2). These unweighted averages, however, cannot be taken too seriously, given that they are based on very different studies of highly diverse marine systems. In fact, the two studies with the largest number of samples found relatively little difference in the amount of ammonium and nitrate taken up by the bacterial size fraction (Harrison and Wood 1988; Kirchman and Wheeler 1998). More work is needed on this topic.

What Nitrogen Sources Support Bacterial Growth in the Oceans?

The comparison of average ammonium uptake with bacterial production indicates that a large fraction of bacterial growth can be supported by ammonium. Direct comparisons of N uptake and bacterial growth reveal a bit more complicated picture. Three conclusions can be drawn from the data given in Table 3.

First, most microbial ecologists would probably guess that dissolved free amino acids (DFAA) supply much N and C for natural bacterial assemblages, and indeed that is often the case. Dissolved protein, measured as dissolved combined amino acids (DCAA), directly supports usually less than 30% of bacterial growth (Table 3), although it may be more important in the undersampled open oceans (e.g., Keil and Kirchman 1999; M. Simon, personal communication). The contribution of DCAA may be underestimated because DFAA release during DCAA hydrolysis is usually not included in estimates of DCAA assimilation. The interactions between DFAA and DCAA (protein) utilization are illustrated in Figure 6. Studies in freshwater lakes have also found that amino acids support a high fraction of bacterial production (e.g., Jørgensen 1987; Rosenstock and Simon 1993). The substantial role of amino

Table 3. Support of heterotrophic bacterial growth by various nitrogen sources in marine systems[a]

Location	Bacterial Growth (%) Supported by[b]				Comments	Ref.
	DFAA	NH_4^+	NO_3^-	DCAA		
Subarctic Pacific	41	105	114	NA	Field	Kirchman and Wheeler (1998)
Subarctic Pacific	24	58	NA	NA	Batch	Keil and Kirchman (1991)
North Atlantic	21	22	5	NA	Field	Kirchman et al. (1994)
North Atlantic	NA	248	355	NA	Batch	Kirchman et al. (1991)
Delaware estuary	5–1000	5	NA	10	Field; seasonal variation	Hoch and Kirchman (1995) Keil and Kirchman (1993)
Delaware estuary	88	73	NA	NA	Field	Keil and Kirchman (1991)
Delaware estuary	44	13	NA	56	Batch 0–31 h	Middleboe et al. (1995)
	7	(−3)[c]		(−40)[c]	(31–55 h)	
Delaware coast	5–100	20–90	NA	0	Field; seasonal variation	Hoch and Kirchman (1995) Keil and Kirchman (1993)
Santa Rosa Sound	82	0	0	28[d]	Batch	Jorgensen et al. (1993)
	24	46	(−30)[c]	124		Kroer et al. (1994)
Flax Pond	120	121	75	10[d]	Batch	Jorgensen et al. (1993)
Sargasso Sea	4	NA	NA	4	Field	Keil and Kirchman (1999)
Gulf of Mexico	14	13	46	50	Batch	Kroer et al. (1994)

[a] Bacterial growth was estimated by radiotracer methods (TdR or Leu incorporation) for "field studies" and for Jorgensen et al. (1993) and by changes in cell abundance for the other "batch" experiments. Field studies examined N uptake and bacterial growth in water samples not manipulated, except for addition of tracers. Batch experiments are incubations with the bacterial size fraction only.

[b] DFAA, dissolved free amino acids; DCAA, dissolved combined amino acids; NA, not analyzed.

[c] Compounds were excreted.

[d] Uptake of DNA in addition to dissolved combined amino acids.

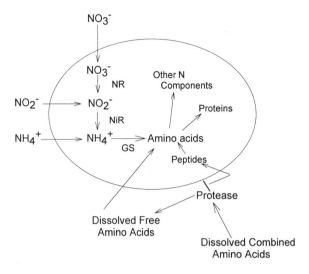

Figure 6. Summary of the physiological mechanisms governing DIN and DON uptake by heterotrophic bacteria: NR, nitrate reductase; NiR, nitrite reductase; GS, glutamine synthetase. Glutamine is the only amino acid synthesized directly by GS; the other amino acids are synthesized from carbon skeletons using glutamine as the nitrogen source. "Protease" is cell-associated and is necessary for hydrolyzing dissolved combined amino acids (DCAA) to amino acids and oligopeptides (< 500 Da) that can be transported into the cell. Some of the amino acids (and oligopeptides) may be released into the free pool.

acids (free and combined forms) is expected, given that the dominant form of nitrogen in cells is protein amino acids.

The second conclusion from Table 3 is that ammonium and nitrate are often as important as amino acids in supporting bacterial growth. The importance of ammonium is not surprising, given that protists and metazoan zooplankton usually excrete ammonium and thus the flux of ammonium is high even when ammonium concentrations are low. Bacteria may use ammonium readily, if appropriate C sources are available, since the difference in energetic costs of using ammonium (and a carbon source like glucose) versus amino acids may be small (Russell and Cook 1995). In contrast, the use of nitrate is energetically expensive (Vallino et al. 1996), and its uptake by bacteria is a bit mysterious. Regardless, bacteria seem to be responsible for similar fractions of total ammonium and nitrate uptake (Table 2) and ammonium and nitrate support roughly the same amount of bacterial growth (Table 3).

The third conclusion to be drawn from Table 3 is that with a few exceptions, bacterial growth can be accounted for by the uptake of DIN and amino acids (either DFAA or DCAA). There is little need to find additional N sources to explain bacterial growth. Organic nitrogen other than amino acids (oDON) comprises a small fraction of cellular N ($\approx 20\%$) and thus the flux of oDON would be low and would not be expected to support much bacterial growth.

The few studies that have directly examined oDON components support this hypothesis. Jørgensen et al. (1993) and Kroer et al. (1994) found that nucleic acids (DNA) supplied little C and N for bacterial growth, although they may contribute much P (Jørgensen and Jacobsen 1996), and Höfle (1984) found that aliphatic amines added to bacterial assemblages were apparently degraded without affecting bacterial growth. The oDON in excess of the little required by cells would need to be catabolized to DOC and ammonium before use in energy production and biomass synthesis.

It is worthwhile emphasizing that oDON is certainly mineralized and utilized by heterotrophic bacteria and that bacterial processing of oDON undoubtedly is important for understanding the flux of the entire DON pool. However, comparisons of N uptake and bacterial growth (e.g., Table 3) indicate that the flux of oDON cannot be large relative to amino acids, unless other mineralization processes (e.g., photochemical degradation) are substantial. Another possibility, that oDON concentrations are increasing in the oceans because oDON production exceeds degradation, seems highly unlikely.

Finding additional N sources to support bacterial growth is not the problem; rather, heterotrophic bacteria appear to be assimilating too much nitrogen, and uptake of sometimes even one nitrogen source frequently exceeds bacterial biomass production. An increase in the N:C ratio of bacterial cells and methodological errors may help explain the imbalance, but N:C ratios do not seem to change enough to account for the imbalance, and there is no obvious experimental error that could explain all experiments. Lysis of cells by viruses may lead to underestimating bacterial production and thus would explain the imbalance in some experiments, but even in situ estimates of bacterial production, which are not affected by viruses, are not in balance with N uptake. Excretion of oDON would help; there is some evidence that heterotrophic bacteria excrete methylamines (Jørgensen et al. 1993) and urea (Berman et al. 1999; Jørgensen et al. 1999), two compounds not measured by other studies of N use by bacteria. Bronk and Gilbert (1993) reported evidence of DON excretion by the < 1.2 μm size fraction, which often is dominated by heterotrophic bacteria.

RELATIONSHIP BETWEEN AMINO ACID AND AMMONIUM UPTAKE

It is worthwhile focusing on the relationships between amino acid and ammonium uptake because these compounds are two of the three most important N sources supporting bacterial growth in the oceans. The third, nitrate, is related to ammonium use via nitrate and nitrite reduction (Figure 6). In addition to ecological connections, amino acid and ammonium uptake are physiologically linked and are coregulated. Ammonium is directly used in the synthesis of two amino acids, glutamine and glutamate, when ammonium concentrations are low and high, respectively. These two amino acids then act as N donors in transamination reactions to synthesize other amino acids and other cellular organic nitrogen. Glutamine and glutamate can inhibit ammonium transport and repress the synthesis of the enzymes involved in ammonium assimilation by pure bacterial cultures in the lab.

Bacteria in pure cultures at least appear to prefer amino acids because, for example, they grow faster in media supplemented with amino acids than in media consisting only of glucose plus ammonium (Russell and Cook 1995). The physiological explanation for this amino acid enhancement effect is not entirely clear as some calculations show that the cost of synthesizing amino acids from glucose and ammonium is only slightly more than the costs of transporting the various amino acids into the cell (e.g. Russell and Cook 1995). Vallino et al. (1996), however, argued that the energetic cost of ammonium use is in fact greater than that for amino acid utilization and that unless ammonium concentrations are exceptionally high, amino acids would be preferred over ammonium. Russell and Cook (1995) hypothesized that amino acids are preferred because anabolic and catabolic rates are more closely coupled and less energy is lost as heat when cells are growing on amino acids.

Some evidence from experiments with mixed bacterial cultures and natural bacterial assemblages can be used to support the hypothesis that amino acids are preferred over ammonium by marine bacterial communities. One approach has been to measure amino acid and ammonium depletion by the bacterial size fraction incubated over a few days. These experiments, which were summarized in Table 3, often show that free amino acids are depleted first, followed by ammonium use (e.g., Kirchman et al. 1989). However, the same type of experiment has also shown simultaneous use of amino acids and ammonium.

Amino acid supply seems to explain some of the ammonium use by natural assemblages of heterotrophic bacteria in estuaries and coastal waters. Hoch and Kirchman (1995) observed a gradient in amino acid concentrations and ammonium uptake in a transect from the Delaware Bay to coastal waters. Free and combined amino acid concentrations and uptake are high in the Delaware Bay, but ammonium use is low. Glibert (1982) also observed low ammonium use by heterotrophic bacteria in the Chesapeake Bay. At the other end of the transect, amino acid concentrations are low in coastal waters and ammonium use is high.

The picture is incomplete, however, because these earlier studies did not examine uptake of all possible DOM components. Goldman and Dennett (1991) pointed out that the supply and concentration of labile organic carbon such as glucose could be what controls ammonium uptake; uptake of glucose and similar compounds has not been measured in experiments examining N uptake (e.g., Kirchman et al. 1989). In fact, ammonium uptake in the Delaware Bay may be relatively low, not because of high amino acid concentrations, but rather because concentrations of glucose and other free neutral sugars are low; these sugars cannot be measured in the Delaware Estuary (< 5 nM; Borch et al., unpublished data). Glucose is an important example and potentially a large part of the DOM pool with high C:N ratios. Glucose is often the only free aldose (or neutral sugar) measurable in oceanic waters, and its uptake can support a substantial fraction (order of 30%) of bacterial production (Rich et al. 1996, 1998).

Another field study failed to find evidence for the hypothesized preference of amino acids over ammonium. That hypothesis and physiological studies lead to the prediction that there should be a negative correlation between amino acid and ammonium uptake. In fact, Kirchman and Wheeler (1998) observed a positive correlation in the subarctic Pacific. They hypothesized that amino acid supply correlated with the DOC supply that in turn drove ammonium uptake.

In short, we have not determined whether the physiological mechanisms observed by pure culture studies can be applied to natural bacterial assemblages. It is conceivable that only bacteria growing in rich media have devised physiological mechanisms to turn off ammonium uptake when presented with selected amino acids. Even in pure cultures, the energetic costs of growing on ammonium versus amino acids are not entirely clear. Few studies have examined bacteria growing on very low concentrations of a complex DOM pool (i.e., closer to real oceanic DOM). Another possibility is that the physiological mechanisms help explain only short-term and small-scale interactions between bacteria and the dissolved pools. Larger scale pattern (e.g., changes in ammonium and amino acid use over seasons and among oceanic regimes), may be best examined by relatively simple stoichiometric models.

EXCRETION OF AMMONIUM BY HETEROTROPHIC BACTERIA AND OTHER ORGANISMS

Ammonium is a crucial nitrogen compound in the oceans because it is the preferred N source of phytoplankton, and a large fraction of primary production usually is supported by ammonium uptake even when nitrate concentrations are much higher than ammonium (e.g., in high-nutrient, low-chlorophyll environments such as the subarctic Pacific: Wheeler and Kokkinakis 1990). For these reasons, we should pay attention to any ammonium excreted by heterotrophic bacteria.

As already observed for ammonium assimilation, there is much variation in the contribution of heterotrophic bacteria to total ammonium mineralization; the bacterial size fraction has accounted for under 10% to 95% of total mineralization (Table 4). As mentioned for ammonium assimilation, some of this variation is probably due to methodological problems associated with measuring regeneration by isotope dilution of added $[^{15}N]H_4^+$ in various size fractions. The most serious problem, perhaps, is that ammonium regeneration by bacteria versus larger organisms must be measured on fractions separated before the incubation ("preincubation"); in contrast, size fractionation can be conducted after the incubation in estimating ammonium uptake ("postincubation"). The problem is that fragile cells can be broken and organic nitrogen released during preparation of the various size fractions (Fuhrman and Bell

Table 4. Summary of studies examining mineralization and uptake of ammonium by bacteria

| | Contribution by Bacteria (%) to NH_4^+ Flux[a] | | |
Location	Mineralization	Uptake[b]	Ref.
Chesapeake	10	10	Glibert (1982)
English Channel	16	10	Corre et al. (1996)
Antarctica	42	16	Tupas et al. (1994)
Lake Biwa	29–51	ND	Haga et al. (1995)
Mississippi River plume	7–50	ND	Cotner and Gardner (1993)
Southern California	39	12	Harrison (1978)
Mid-Atlantic Bight	60–95	53–97	Harrison et al. (1983)
Georgia coast	<10	78	Wheeler and Kirchman (1986)
Median	40	40	

[a]Percentage of total NH_4^+ mineralization or total uptake attributable to bacteria.
[b]See Kirchman (1994) for more complete listing of ammonium uptake by heterotrophic bacteria. The median % NH_4^+ uptake is taken from that study.

1985). The newly released DON is likely to have C:N ratios equal to or less than total cellular C:N, that is, Redfield. This N-rich DON then will stimulate ammonium regeneration and growth by bacteria; the frequently large increase in bacterial abundance following 0.8 or 1.0 μm size fractionation occurs because bacterivorous grazers have been eliminated, but the released DOM also contributes to the rapid growth.

With these caveats in mind, it is still interesting to note that the bacterial size fraction accounts for very roughly 40% of ammonium regeneration (i.e., an amount not greatly different from the bacterial contribution to ammonium uptake). It is worthwhile emphasizing once again the great variability among different studies and locations and in the reported fractions. The temporal variability in ammonium uptake and regeneration is not known at all, again greatly complicating the use of any average to judge these processes. Still, Table 4 is useful in pointing to the potential importance of bacteria in ammonium regeneration.

Phosphate regeneration by bacteria has not been examined at all. As a first approximation, phosphate is probably regenerated mainly by the same organisms (i.e., protists and sometimes zooplankton) that regenerate ammonium. There is some evidence, however, that ammonium and phosphate regeneration is not correlated (e.g., Hassett et al. 1997), but the relationship between the two processes has not been examined extensively.

There has been a tendency to ignore ammonium and phosphate regeneration by heterotrophic bacteria, perhaps in part because uptake is easier to measure and has been studied more frequently. Also, we may tend to minimize the bacterial contribution to nutrient regeneration because that was the old picture of bacteria in aquatic habitats (see Chapters 1 and 2). In fact, there have been very few modern studies of bacterial mineralization. We need more work on these important questions.

CONSEQUENCES OF DIN FLUXES MEDIATED BY HETEROTROPHIC BACTERIA

There are several reasons to learn more about the fluxes of N and P as mediated by heterotrophic bacteria. A simple reason is that if we are to understand what controls bacterial growth and biomass levels in the ocean, we need to consider how fluxes of the major elements, such as N and P, affect bacterial metabolism. A slightly more compelling reason is that we should be interested in any information about factors affecting phytoplankton uptake of at least DIN, the supply of which often limits primary production in many oceans. Bacteria probably never fully "compete" with phytoplankton, in the usual sense of the word, because all heterotrophic microbes ultimately depend on phytoplankton for reduced carbon. Still, heterotrophic bacteria have impacts on DIN uptake by phytoplankton.

We turn now to two more reasons to learn more about DIN fluxes mediated by heterotrophic bacteria The first has to do with how these fluxes affect new production. I focus on DIN and not P fluxes to keep this discussion simple and because N, not P, more often limits biomass production in the oceans.

The uptake of ammonium and nitrate by heterotrophic bacteria has been thought to affect calculations of regenerated and new production, which are important parameters for describing the potential of carbon export out of the surface layer to the deep ocean (Eppley and Peterson 1979). Regenerated production is supported by nitrogen regenerated within the mixed layer and usually is taken as equal to ammonium uptake, the dominant form of regenerated nitrogen. New production, on the other hand, was originally defined as uptake of nitrate (Dugdale and Goering 1967); nitrate is the dominant nitrogenous compound that is "new" or introduced to the surface layer via upwelling and diffusion. Nitrate is assumed to be formed only below the surface layer by nitrification. If nitrogen fixation (Chapter 13) and nitrification within the euphotic zone (Chapter 14) are negligible, as is usually assumed, then nitrate uptake must equal the downward flux of nitrogen ("export production") at steady state. The export of carbon out of the surface layer can then be estimated by converting the nitrogen flux to carbon units using a C:N ratio of the microbes utilizing the nitrate. Suffice it to say that carbon export is of great importance for examining material and energy budgets of oceanic surface layers (Chapter 15).

In the open ocean where outside (allochthonous) inputs of carbon are trivial, the uptake of nitrate by bacteria should not have a large impact on new production calculations. A more complete discussion of this argument can be found in Kirchman et al. (1992). The downward flux of nitrogen would have to equal nitrate uptake at steady state regardless of which group of microbes (phytoplankton or bacteria) is using the nitrate. Substantial uptake of nitrate by bacteria may not complicate converting N fluxes to C fluxes because bacterial C:N ratios may differ little (order of 25%) from the phytoplankton, if at all.

The possibility of ammonium regeneration by bacteria, however, does complicate our interpretation of regenerated production as measured by standard ^{15}N techniques. Kirchman et al. (1992) make the case that only net ammonium uptake by bacteria and phytoplankton should be considered as "regenerated production" because the ammonium excreted by bacteria during degradation of DON initially synthesized by phytoplankton could be used again by bacteria or phytoplankton. This ammonium recycled only within the microbial loop (phytoplankton → DON → bacteria) would be counted twice: once when it is first taken up by phytoplankton and used to synthesize DON, and then again when it is excreted and taken up a second time.

While the calculation of new and regenerated production may not be affected by heterotrophic uptake of DIN, the relationship between new production and export seems more complex at the very least when bacteria take up substantial amounts of nitrate. Uptake of nitrate by small microbes like bacteria would add to a time lag between new and export production. At steady state these two rates must be equal; but often it seems that the oceans are far from steady state, and new and export rates may differ substantially. Nitrate uptake by small microbes would uncouple new and export production because individual bacteria and eukaryotic microbes smaller than 10 μm do not sink and thus need to be "packaged" via grazing (one of several possible processes) into particles large enough to sink. Since the organisms grazing on small microbes are also small and thus unlikely to produce substantial amounts of sinking particles, additional trophic transfers are necessary to package small microbial material into sinking particles. In contrast, uptake of nitrate by large phytoplankton, the prime example being diatoms, is potentially tightly coupled to the sinking flux and export production. Large phytoplankton can sink under some conditions, and their grazers are likely to directly produce fecal pellets and other sinking particles.

Another consequence of DIN uptake by heterotrophic bacteria is the impact of nutrient competition on the community structure of phytoplankton. Small cells like heterotrophic bacteria are presumed to be able to out-compete large cells, including most phytoplankton, for dissolved compounds. The usual argument for the competitive edge of small cells is their high ratio of surface area to volume, although the experimental evidence that the uptake affinity of small cells in natural marine microbial communities is greater than that of large cells is rather slim (Suttle et al. 1990). It is usually argued that large cells are better adapted to utilize pulses of high concentrations. Regardless of whether these generalizations are always applicable, certainly uptake by heterotrophic bacteria will exert some selection pressure on the size if not community structure of phytoplankton assemblages. It is also worthwhile pointing out that the very low concentrations of even nonlimiting inorganic nutrients have forced microbes to devise high affinity mechanisms to transport and assimilate these compounds. Even if the nutrient does not limit production (e.g., phosphate), uptake by heterotrophic bacteria is certainly a process phytoplankton must contend with.

A final, and perhaps most important reason to learn more about N fluxes mediated by heterotrophic bacteria is that the turnover of DOM, the largest exchangeable pool of organic C on the planet, is undoubtedly connected somehow to DIN fluxes. The simplest example is assimilation by bacteria of DIN during consumption of DOM with high C:N ratios. Williams (1995) evoked DIN limitation of heterotrophic bacteria to explain the buildup of DOC during the summer in the North Atlantic and probably elsewhere as well. However, there is no evidence of N limitation of heterotrophic bacteria (see Chapter 8), unlike the evidence indicating N limitation of phytoplankton growth. In contrast, the model of Anderson and Williams (1998) successfully described seasonal changes in DOC and DON when bacteria were net mineralizers; that is, in their model bacteria produce NH_4^+, not consume it, and bacteria do not use NO_3^-. Although the current evidence indicates that bacteria are in fact more likely to assimilate DIN than to excrete NH_4^+, the data are not compelling enough to totally rule out some connection between DIN mineralization and DOC buildup as hypothesized by Anderson and Williams (1998).

SUMMARY

1. The elemental and biochemical composition of microbial cells explains much about the uptake and regeneration of various elements and dissolved compounds by heterotrophic bacteria.

2. Bacteria account for roughly 40% of total uptake of inorganic P, N, and Fe, equal to that expected from the ratio of bacterial production to primary production, corrected for differences in elemental ratios.

3. Generally heterotrophic bacteria seem to take up more ammonium than nitrate, which is expected given the energetic costs associated with using these two DIN sources. However, in the few cases where rates can be compared directly, bacteria accounted for about the same fraction of ammonium and nitrate uptake, indicating at the least that nitrate uptake by heterotrophic bacteria cannot be ignored.

4. Amino acids, ammonium, and nitrate supply all the nitrogen needed for bacterial growth. These data also indicate that amino acids (both free and combined) dominate DON fluxes.

5. Contrary to current models, bacteria appear to be taking up DIN and DON while also simultaneously contributing to ammonium regeneration. One explanation is that different phylogenetic groups within the bacterial assemblage are responsible for these different processes.

6. Excretion of ammonium by heterotrophic bacteria needs to be examined in more detail, because this aspect of N metabolism potentially affects new production the most.

ACKNOWLEDGMENTS

I thank Peter J. le B. Williams, Niels Jørgensen, N. D. Sherry, and Mike Lomas for their comments on this chapter. My work reported on here was supported by the National Science Foundation.

REFERENCES

Anderson, T. R., and Williams, P. J. le B. (1998) Modeling the seasonal cycle of dissolved organic carbon at station E-1 in the English Channel. *Estuarine Coastal Shelf Sci.* 46(1):93–109.

Berman, T., Bechemin, C., and Maestrini, S. Y. (1999) Release of ammonium and urea from dissolved organic nitrogen in aquatic ecosystems. *Aquat. Microb. Ecol.* 16:295–302.

Bratbak, G. (1985) Bacterial biovolume and biomass estimations. *Appl. Environ. Microbiol.* 49(6):1488–1493.

Bronk, D. A. and Glibert, P. M. (1993) Contrasting patterns of dissolved organic nitrogen release by two size fractions of estuarine plankton during a period of rapid NH_4^+ consumption and NO_2^- production. *Mar. Ecol. Progr. Ser.* 96:291–299.

Brock, T. D., and Madigan, M. T. (1991) *Biology of Microorganisms*, 6th ed. Prentice-Hall, Englewood Cliffs, NJ.

Coale, K. H., Johnson, K. S., Fitzwater, S. E., Gordon, R. M., Tanner, S., Chavez, F. P., Ferioli, L., Sakamoto, C., Rogers, P., Millero, F., Steinberg, P., Nightingale, P., Cooper, D., Cochlan, W. P., Landry, M. R., Constantinou, J., Rollwagen, G., Trasvina, A., and Kudela, R. (1996) A massive phytoplankton bloom induced by an ecosystem-scale iron fertilization experiment in the equatorial Pacific Ocean. *Nature* 383:495–501.

Corre, P. L., Wafar, M. Helguen, S. L., and Maguer, J. F. (1996) Ammonium assimilation and regeneration by size-fractionated plankton in permanently well-mixed temperate waters. *J. Plankton Res.* 18:355–370.

Cotner, J. B., Jr., and Gardner, W. S. (1993) Heterotrophic bacterial mediation of ammonium and dissolved free amino acid fluxes in the Mississippi River plume. *Mar. Ecol. Prog. Ser.* 93:75–87.

Dugdale, R. C., and Goering, J. J. (1967) Uptake of new and regenerated forms of nitrogen in primary productivity. *Limnol. Oceanogr.* 12:196–206.

Eppley, R. W., and Peterson, B. J. (1979) Particulate organic matter flux and planktonic new production in the deep ocean. *Nature* 282:677–680.

Eppley, R. W., Sharp, J. H. Renger, E. H., Perry, M. J., and Harrison, W. G.. (1977) Nitrogen assimilation by phytoplankton and other microorganisms in the surface waters of the central North Pacific Ocean. *Mar. Biol.* 39:111–120.

Fenchel, T., King, G. M., and Blackburn, T. H. (1998) *Bacterial Biogeochemistry: The Ecophysiology of Mineral Cycling*, 2nd ed. Academic Press, San Diego, CA.

Fuhrman, J. A., and Azam, F. (1982) Thymidine incorporation as a measure of heterotrophic bacterioplankton production in marine surface waters: Evaluation and field results. *Mar. Biol.* 66:109–120.

Fuhram, J. A., and Bell, T. M. (1985) Biological considerations in the measurement of dissolved free amino acids in seawater and implications for chemical and microbiological studies. *Mar. Ecol. Prog. Ser.* 25:13–21.

Fuhrman, J. A., Horrigan, S. G., and Capone, D. G. (1988) Use of ^{13}N as tracer for bacterial and algal uptake of ammonium from seawater. *Mar. Ecol. Prog. Ser.* 45:271–278.

Fukuda, R., Ogawa, H., Nagata, T., and Koike, I. (1998) Direct determination of carbon and nitrogen contents of natural bacterial assemblages in marine environments. *Appl. Environ. Microbiol.* 64 (9):3352–3358.

Glibert, P. M. (1982) Regional studies of daily, seasonal and size fraction variability in ammonium remineralization. *Mar. Biol.* 70:209–222.

Goldman, J. C., and Dennett, M. R. (1991) Ammonium regeneration and carbon utilization by marine bacteria grown on mixed substrates. *Mar. Biol.* 109:369–378.

Goldman, J. C., Caron, D. A., and Dennett, M. R. (1987) Regulation of gross growth efficiency and ammonium regeneration in bacteria by substrate C:N ratio. *Limnol. Oceanogr.* 32:1239–1252.

Haga, H., Nagata, T., and Sakamoto, M. (1995) Size-fractionated NH_4^+ regeneration in the pelagic nvironments of two mesotrophic lakes. *Limnol. Oceanogr.* 40:1091–1099.

Hansell, D. A., and Waterhouse, T. Y. (1997) Controls on the distributions of organic carbon and nitrogen in the eastern Pacific Ocean. *Deep-Sea Res. I* 44:843–857.

Harrison, W. G. (1978) Experimental measurements of nitrogen remineralization in coastal waters. *Limnol. Oceanogr.* 23(4):684–694.

Harrison, W. G., and Wood, L. J. E. (1988) Inorganic nitrogen uptake by marine picoplankton: evidence for size partitioning. *Limnol. Oceanogr.* 33:468–475.

Harrison, W. G., Douglas, D., Falkowski, P., Rowe, G., and Vidal, J. (1983) Summer nutrient dynamics of the Middle Atlantic Bight: Nitrogen uptake and regeneration. *J. Plankton Res.* 5(4):539–556.

Hassett, R. P., Cardinale, B., Stabler, B., and Elser, J. J. (1997) Ecological stoichiometry of N and P in pelagic ecosystems: Comparison of lakes and oceans with emphasis on the zooplankton–phytoplankton interaction. *Limnol. Oceanogr.* 42:648–662.

Hoch, M. P., and Kirchman, D. L. (1995) Ammonium uptake by heterotrophic bacteria in the Delaware Estuary and adjacent coastal waters. *Limnol. Oceanogr.* 40:886–897.

Hoch, M. P., Fogel, M. L., and Kirchman, D. L. (1992) Isotope fractionation associated with ammonium uptake by a marine bacterium. *Limnol. Oceanogr.* 37:1447–1459.

Höfle, M. G. (1984) Degradation of putrescine and cadaverine in seawater cultures by marine bacteria. *Appl. Environ. Microbiol.* 47:843–849.

Hutchins, D. A., and Bruland, K. W. (1998) Iron-limited diatom growth and S : N uptake ratios in a coastal upwelling regime. *Nature* 393(6685):561–564.

Ingraham, J. L., Maaloe, O. and Neidhardt, F. C. (1983) *Growth of the Bacterial Cell*. Sinauer, Sunderland, MA.

Jørgensen, N. O. G. (1987) Free amino acids in lakes: Concentrations and assimilation rates in relation to phytoplankton and bacterial production. *Limnol. Oceanogr.* 32:97–111.

Jørgensen, N. O. G., and Jacobsen, C. S. (1996) Bacterial uptake and utilization of dissolved DNA. *Aquat. Microb. Ecol.* 11:263–270.

Jørgensen, N. O. G., Kroer, N., Coffin, R. B., Yang, X.-H., and Lee, C. (1993) Dissolved free amino acids, combined amino acids, and DNA as sources of carbon and nitrogen to marine bacteria. *Mar. Ecol. Prog. Ser.* 98:135–148.

Jørgensen, N. O. G., Kroer, N., Coffin, R. B., and Hoch, M. P. (1999) Relationships between bacterial nitrogen metabolism and growth efficiency in an estuarine and an open-water ecosystem. *Aquat. Microb. Ecol.* 18: 247–261.

Keil, R. G., and Kirchman, D. L. (1991) Contribution of dissolved free amino acids and ammonium to the nitrogen requirements of heterotrophic bacterioplankton. *Mar. Ecol. Prog. Ser.* 73:1–10.

Keil, R. G., and Kirchman, D. L. (1993) Dissolved combined amino acids: chemical form and utilization by marine bacteria. *Limnol. Oceanogr.* 38:1256–1270.

Keil, R. G., and Kirchman, D. L. (1999) Utilization of dissolved protein and amino acids in the Northern Sargasso Sea. *Aquat. Microb. Ecol.* 18:293–300.

Kemp, P. F., Lee, S., and LaRoche, J. (1993) Estimating the growth rate of slowly growing marine bacteria from RNA content. *Appl. Environ. Microbiol.* 59:2594–2601.

Kirchman, D. L. (1994) The uptake of inorganic nutrients by heterotrophic bacteria. *Microb. Ecol.* 28:255–271.

Kirchman, D. L., and Wheeler, P. A. (1998) Uptake of ammonium and nitrate by heterotrophic bacteria and phytoplankton in the sub-Arctic Pacific. *Deep-Sea Res. I* 45:347–365.

Kirchman, D. L., Keil, R. G., and Wheeler, P. A. (1989) The effect of amino acids on ammonium utilization and regeneration by heterotrophic bacteria in the subarctic Pacific. *Deep-Sea Res. I* 36:1763–1776.

Kirchman, D. L., Suzuki, Y., Garside, C., and Ducklow, H. W. (1991) High turnover rates of dissolved organic carbon during a spring phytoplankton bloom. *Nature* 352:612–614.

Kirchman, D. L., Moss, J., and Keil, R. G. (1992) Nitrate uptake by heterotrophic bacteria: Does it change the f-ratio? *Arch. Hydrobiol.* 37:129–138.

Kirchman, D. L., Ducklow, H. W., McCarthy, J. J., and Garside, C. (1994) Biomass and nitrogen uptake by heterotrophic bacteria during the spring phytoplankton bloom in the North Atlantic Ocean. *Deep-Sea Res. I* 41:879–895.

Kroer, N., Jørgensen, N. O. G., and Coffin, R. B. (1994) Utilization of dissolved nitrogen by heterotrophic bacterioplankton: A comparison of three ecosystems. *Appl. Environ. Microbiol.* 60:4116–4123.

Lipschultz, F. (1995) Nitrogen-specific uptake rates of marine phytoplankton isolated from natural populations of particles by flow cytometry. *Mar. Ecol. Prog. Ser.* 123:245–258.

Middelboe, M., Borch, N. H., and Kirchman, D. L. (1995) Bacterial utilization of dissolved free amino acids, dissolved combined amino acids and ammonium in the Delaware Bay estuary: Effects of carbon and nitrogen limitation. *Mar. Ecol. Prog. Ser.* 128:109–120.

Pakulski, J. D., Coffin, R. B., Kelley, C. A., Holder, S. L., Downer, R., Aas, P., Lyons, M. M., and Jeffrey, W. H. (1996) Iron stimulation of Antarctic bacteria. *Nature* 383 (6596):133–134.

Rich, J. H., Ducklow, H. W., and Kirchman, D. L. (1996) Concentrations and uptake of neutral monosaccharides along 140°W in the equatorial Pacific: Contribution of glucose to heterotrophic bacterial activity and the DOM flux. *Limnol. Oceanogr.* 41:595–604.

Rich, J., Gosselin, M., Sherr, E., Sherr, B., and Kirchman, D. L. (1998) High bacterial production, uptake and concentrations of dissolved organic matter in the Central Arctic Ocean. *Deep-Sea Res. II* 44:1645–1663.

Rosenstock, B., and Simon, M. (1993) Use of dissolved combined and free amino acids by planktonic bacteria in Lake Constance. *Limnol. Oceanogr.* 38:1521–1531.

Russell, J. B., and Cook, G. M. (1995) Energetics of bacterial growth: Balance of anabolic and catabolic reactions. *Microbiol. Rev.* 59(1):48–62.

Simon, M., and Azam, F. (1989) Protein content and protein synthesis rates of planktonic marine bacteria. *Mar. Ecol. Prog. Ser.* 51:201–213.

Suttle, C. A., Fuhrman, J. A., and Capone, D. G. (1990) Rapid ammonium cycling and concentration-dependent partitioning of ammonium and phosphate: Implications for carbon transfer in planktonic communities. *Limnol. Oceanogr.* 35(2):424–433.

Tamminen, T., and Irmisch, A. (1996) Urea uptake kinetics of a midsummer planktonic community on the SW coast of Finland. *Mar. Ecol. Prog. Ser.* 130:201–211.

Tezuka, Y. (1990) Bacterial regeneration of ammonium and phosphate as affected by the carbon:nitrogen:phosphorus ratio of organic substrates. *Microb. Ecol.* 19:227–238.

Tortell, P. D., Maldonado, M. T., and Price, N. M. (1996) The role of heterotrophic bacteria in iron-limited ocean ecosystems. *Nature* 383(6598):330–332.

Tupas, L., and Koike, I. (1990) Amino acid and ammonium utilization by heterotrophic marine bacteria grown in enriched seawater. *Limnol. Oceanogr.* 35(5):1145–1155.

Tupas, L. M., Koike, I., Karl, D. M., and Holm-Hansen, O. (1994) Nitrogen metabolism by heterotrophic bacterial assemblages in Antarctic coastal waters. *Polar Biol.* 14:195–204.

Vadstein, O. (1998) Evaluation of competitive ability of two heterotrophic planktonic bacteria under phosphorus limitation. *Aquat. Microb. Ecol.* 14:119–127.

Vallino, J. J., Hopkinson, C. S., and Hobbie, J. E. (1996) Modeling bacterial utilization of dissolved organic matter: Optimization replaces Monod growth kinetics. *Limnol. Oceanogr.* 41(8):1591–1609.

Watson, S. W., Novitsky, T. J., Quinby, H. L., and Valois, F. W. (1977) Determination of bacterial number and biomass in the marine environment. *Appl. Environ. Microbiol.* 33(4):940–946.

Wheeler, P. A., and Kirchman, D. L. (1986) Utilization of inorganic and organic nitrogen by bacteria in marine systems. *Limnol. Oceanogr.* 31:998–1009.

Wheeler, P. A., and Kokkinakis, S. A. (1990) Ammonium recycling limits nitrate use in the oceanic subarctic Pacific. *Limnol. Oceanogr.* 35(6):1267–1278.

Williams, P. J. le B. (1995) Evidence for the seasonal accumulation of carbon-rich dissolved organic material, its scale in comparison with changes in particulate material and the consequential effect on net C/N assimilation ratios. *Mar. Chem.* 51:17–29.

10

BACTERIAL ENERGETICS AND GROWTH EFFICIENCY

Paul A. del Giorgio

Horn Point Laboratory,
University of Maryland Center for Environmental Science,
Cambridge, Maryland

Jonathan J. Cole

Institute of Ecosystem Studies,
Millbrook, New York

Marine waters, particularly the open oceans, are extremely dilute environments and represent a considerable challenge for aquatic microorganisms. To survive and grow under natural oceanic conditions, bacteria require the capacity of greatly adjusting cellular structure and metabolic function (Dawes 1985; Poindexter 1987). Experimental work suggests that microbial cells have an enormous catabolic versatility and flexibility, which is probably the key to survival and growth in extremely oligotrophic but variable environments such as the world's oceans (Kovárová-Kovar and Egli 1998). It is clear that free-living bacteria have a remarkable capacity to withstand extreme substrate and nutrient limitation, to scavenge for vanishingly low traces of nutrients, and to modulate their physiology to utilize a wide variety of different substrates. The physiological mechanisms that allow bacteria to perform in this manner are not well understood, but central to the adaptation of bacteria to dilute and variable conditions is an enormous flexibility in the way bacteria partition the

Microbial Ecology of the Oceans, Edited by David L. Kirchman.
ISBN 0-471-29993-6   Copyright © 2000 by Wiley-Liss, Inc.

available energy, so that at any given set of environmental conditions, survival and growth can be maximized (Dawes 1985; Morita 1997).

Aquatic bacteria also exhibit a remarkable diversity in terms of the energy and carbon sources that they can utilize, ranging from phototrophs that utilize light and CO_2, chemolithotrophs that use inorganic compounds for energy and CO_2, and heterotrophs that utilize organic compounds as both the C source and the source of energy. From the point of view of the carbon cycle in marine ecosystems, the organotrophic bacteria are the most important physiological group. These heterotrophic bacteria comprise the bulk of microbial populations inhabiting the water column of the oceans and are responsible for much of the biological transformation of organic matter and the production of CO_2 in the oceans (Sherr and Sherr 1996). The magnitude, regulation, and ecological significance of heterotrophic bacterial biomass production and growth have been extensively studied and are reviewed in Chapters 4 and 6. In this chapter we review the metabolic processes that underlie bacterial growth and production, and we focus on microbial energetics and how bacteria partition the organic matter they take up into anabolic (synthesis) and catabolic (breakdown) pathways.

The distribution of carbon into anabolic and catabolic processes results in bacterial growth efficiency (BGE), defined as the ratio of biomass produced to substrate assimilated (A), where A is the sum of bacterial production (BP) and bacterial respiration (BR), so that BGE $= BP/(BP + BR) = BP/A$. By definition, growth efficiency (or yield — here used interchangeably) is the quantity of biomass synthesized per unit of substrate assimilated (see also Chapter 4). BGE is a fundamental attribute of microbial metabolism, which largely determines the ecological and biogeochemical roles of bacteria in microbial food webs and in aquatic systems (Sherr and Sherr 1996; del Giorgio and Cole 1998). In most studies of carbon flow in aquatic ecosystems, the respiration term is derived from measurements of BP and assumed values of bacterial growth efficiency. For many years, assumed values of BGE had been often based on early measurements using simple radiolabeled organic compounds (Crawford et al. 1974), but these values are now widely regarded as overestimates of the real growth efficiency of natural bacterioplankton utilizing complex natural substrates (Jahnke and Craven 1995; del Giorgio and Cole 1998). Because the magnitude and variation of bacterial growth efficiency are not well understood (del Giorgio et al. 1997), the real magnitude of organic carbon flow through bacterioplankton in the world's oceans remains largely unknown. In addition, the relative paucity of direct measurements of in situ BGE has allowed little progress in the debate of whether bacteria are a significant link from the detrital pool of organic matter to consumers in the pelagic food web (Pomeroy and Wiebe 1993; Jahnke and Craven 1995).

Microbial energetics deals with how bacterial cells partition the energy and carbon they take up from the environment into catabolic and anabolic pathways, as well as with the allocation of resources and metabolizable energy into the components of the cell's energy budget, such as maintenance, transport, growth and reproduction, and the factors that regulate this partition (Wieser 1994; Payne and Wiebe 1978). Growth efficiency (or yield) is the parameter that best integrates these complementary metabolic processes (Dawes 1985) and is the focus of the chapter. This chapter attempts to synthesize the results of research on bacterioplankton growth efficiency done in the last 30 years, focusing on data from natural ecosystems. The reader will note that there are few studies of bacterioplankton growth efficiency relative to the large number of studies of bacterioplankton growth, production, nutrient uptake, and other microbial processes. This is partly because of the technical difficulties associated with the measurements of respiration rates of in situ planktonic assemblages. As a result, energetics of natural bacterial assemblages are still poorly understood, and much remains to be done. Thus we begin by reviewing some basic aspects of microbial energetics, mostly derived from work with cultured organisms. We then attempt to link these concepts to the observed metabolism of natural assemblages.

MICROBIAL ENERGETICS

In the process of growth, various compounds, elements and minerals are converted into cell material at the expense of the energy source (Figure 1). An organic substrate taken up by a bacterial cell will be used partly in catabolic reactions to generate ATP and partly in anabolic reactions for biomass synthesis (Figure 1). The purpose of this scheme is to emphasize that multiple processes determine growth efficiency, and these processes constitute what we call microbial energetics. The overwhelming majority of studies dealing with bacterial energetics have been carried out with single species growing under defined conditions, in batch or continuous cultures. Below we briefly review some of the resulting general principles and experimental results. We cannot summarize all the information on microbial bioenergetics in pure cultures, so we refer the reader to several excellent reviews (Wieser 1994; Battley 1987; Russell and Cook 1995).

There is a long history of studies on microbial energetics in bacterial cultures, with researchers searching for regularities in the relationship between the amount and nature of the organic substrate and how bacteria process this carbon. Among the first to quantify this type of regularity was Monod (1942), who recognized early on that the yield of cells in carbon-limited cultures was constant for any given substrate and growth conditions. But as we will explore, later work has revealed patterns in microbial energetics that are much more complex and much less predictable.

The organic carbon assimilated A can be directed into the production of energy from the oxidation of the substrate R, and into biosynthesis, or production P, so that $A = R + P$. Metabolic energy generated is distributed between demands of two kinds: the demands of biosynthetic processes that produce a net increase in biomass and the demands of processes that do not,

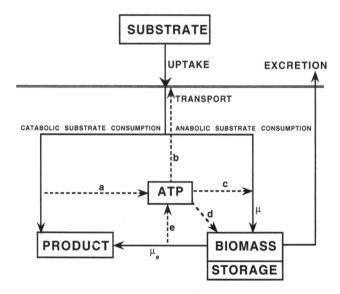

Figure 1. Simplified depiction of catabolic and anabolic pathways that influence growth efficiency in aquatic bacteria. The oxidation of organic compounds contributes to the energy pool as ATP at a rate a. In turn, active transport of substrates into the cell requires energy from this ATP pool at a rate b; anabolic reactions utilize ATP at a rate c, and result in a growth rate, μ. The anabolic pathways result not only in increases in biomass, but also in storage products and organic compounds that may be excreted back to the medium. Maintenance expenditures consume ATP at a rate d. In the absence of exogenous substrates, minimum maintenance energy requirements must be supported by degradation of biomass through endogenous metabolism (μ_e), which supplies ATP at a rate e. Endogenous metabolism is defined here as the state in which no growth is possible, and by definition BGE is 0 under these conditions. (Adapted from del Giorgio and Cole 1998.)

(e.g., regulation of internal pH and osmotic pressure, macromolecular turnover, membrane energization, and motility) (Pirt 1982). The latter are generally referred to as maintenance costs. Therefore a fraction of R is required to meet the expenses of production. In the simplest scenario, these energetic requirements are modeled as a constant proportion of the amount produced (R_p). So in addition to R_p, there are energetic expenditures for maintenance functions (R_m), so that total energy produced can be partitioned as follows:

$$R_t = R_m + R_p$$

The partitioning of metabolizable substrate into R and P (yield or efficiency), as well as into R_m and R_p has been the focus of study for many decades and it is still not well understood. Pirt (1965) conceptualized bacterial growth efficiency or yield Y as a function of the maximum potential yield Y_{max}, the specific uptake rate of energy-yielding substrate used for maintenance K_m, and specific growth rate μ:

$$\frac{1}{Y} = \frac{1}{Y_{\text{max}}} + \frac{K_{\text{m}}}{\mu}$$

It was originally assumed that Y_{max} and K_{m} were invariant with μ, so that a plot of $1/Y$ against $1/\mu$ would give a straight line with slope K_{m} and intercept of $1/Y_{\text{max}}$. This assumption was not supported by experimental data, as we will see.

The Energetic Cost of Growth

The term cost of growth (R_{p}) implies that the metabolic expenditures connected with growth can be separated from the metabolic expenditures of the same organism when it is not growing but still active; in practice, however, this calculation is difficult. The rate of maintenance metabolism may be deduced, for example, by plotting total metabolic rate (R_{t}) of cells growing at different rates against growth rate. If the resulting regression is linear, its intercept characterizes metabolic expenses independent of growth, and the slope of the line would be the net cost of growth.

The ATP requirements for the formation of microbial cells from preformed monomers, with glucose as the sole energy source, was calculated by Stouthamer (1973). The theoretical maximal yield, $Y_{\text{ATP max}}$, amounts to approximately 32 mg of dry biomass per millimple of ATP, which corresponds to roughly a BGE of 0.88, assuming that 1 mol ATP = 1 g C dissimilated. Of the total ATP requirement, about 78% is due to protein synthesis and RNA turnover, and 22% to the transport of ions and monomers. A general pattern found in all microbiological studies is that the realized growth efficiency is (1) extremely variable and (2) always substantially lower than this theoretical value, even in cultures growing presumably under optimum conditions (Russell and Cook 1995). The efficiency of utilization of organic compounds varies greatly, but because the energy content of these compounds also varies, there were early attempts to describe efficiency in terms of mass of cells produced per unit energy (ATP) that could be extracted from the substrate. It was thought originally that whereas the C-based yield should be highly variable because of differences in the energy density, the ATP-based yield should be constant. Later studies, however, have not supported this hypothesis (Wieser 1994). A source of variation in BGE would be changes in cellular composition with growth rate; but these changes in the relative proportion of protein, RNA, and carbohydrate explain only a trivial fraction of the variance in Y_{ATP}, for example (Weiser 1994; Russell and Cook 1995).

Maintenance Requirements

A major source of variability in the apparent cost of growth and in the resulting growth efficiency (or yield) is the large variability in maintenance costs. For example, Tempest and Neijssel (1978) demonstrated that mainten-ance metabolism was dependent not only on experimental conditions but also on the rate of growth itself. It would appear that processes such as the energization of membranes, typically a maintenance function, consume varying amounts of energy depending on the growth stage (Wieser 1994), and that there is a dependence of growth on these maintenance functions, so that stimulation of the former requires the synchronized stimulation of the latter. Attempts to distinguish between these two major components of metabolism have been based on mathematical arguments rather than on biological evi-dence. Pirt (1982) later modified his original model (Pirt 1965) to incorporate these findings in a model that separates total maintenance expenditures (R_m) into a constant component (R_{m_1}) and a component that decreases propor-tionally with growth rate (R_{m_2}). Whether the variable component (R_{m_2}) is considered to be a part of maintenance or of growth is a matter of definition.

Uncoupling Between Catabolism and Anabolism

Results from experimental studies suggest that when growth is unconstrained, as in batch cultures, there is a often a high degree of coupling between catabolism and anabolism. When growth is constrained by the supply of organic substrate or inorganic nutrients, as it is in most chemostat studies and certainly in most natural situations, different degrees of uncoupling are invariably observed. This uncoupling is evidenced in various ways: high rates of oxygen and organic substrate consumption, metabolite overproduction and excretion, excess heat production, and energy-spilling pathways (Battley 1987). All these processes result in reduced growth efficiency. Anomalies in BGE are often found at low growth rates when growth is limited by some substrate other than the energy source. In general, catabolism appears to proceed at the maximum rate at which the organisms are capable in the circumstance, irrespective of whether the energy so produced can be used for biosynthesis. Under conditions of severe constraints to growth, such as under acute lack of mineral nutrients, it has been suggested that maintaining the highest possible flow of energy would be advantageous (Russell and Cook 1995). One of the potential advantages of a high energy flux in the cell may be to maintain the energization of cell membranes and the function of active transport systems, both of which are essential conditions to resume growth whenever environ-mental conditions change (Dawes 1985; Morita 1997). The conclusion that it is advantageous and even necessary for bacteria to maintain a high flow of energy is supported by thermodynamic analysis of microbial energetics, which suggest that microbial growth efficiency is usually low, but optimal for maximal growth rate (Westerhoff et al. 1983).

It is clear that bacteria can alter the coupling between catabolism and anabolism according to the circumstance to maximize survival and growth (Teixeira de Mattos and Neijssel 1997). Although difficult to extrapolate to natural environments, these considerations are relevant in our interpretation of bacterioplankton energetics because planktonic bacteria occupy niches, survive, and grow in an extremely dilute environment where carbon, energy, and other nutrients are often limiting, and growth is usually much slower than in culture. Thus, maintenance energy requirements are expected to play a significant role in determining BGE, and bacterioplankton should generally be in a region of low BGE. In addition, planktonic bacteria are expected to exhibit a relatively large degree of uncoupling between catabolism and anabolism compared to their cultured counterparts. As discussed later, the data from natural aquatic systems generally support these expectations.

Linking Bacterial Energetics to Single-Cell Activity and Growth

Figure 2 attempts to conceptualize the relationship between growth rate and organic carbon consumption, following Harder (1997). In this scheme, we consider the two maintenance components just discussed: one is independent of growth rate (m_e) and the other tends to decline proportionately to growth rate (m_μ). Carbon consumption increases with specific growth rate μ, but at $\mu = 0$ there is still a finite rate of carbon consumption, to meet minimum maintenance energy requirements. A consequence of the maintenance energy concept

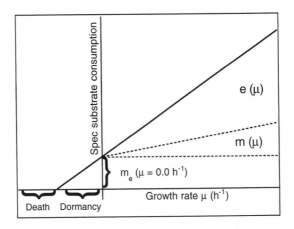

Figure 2. Conceptual relationship between specific substrate consumption, as moles of substrate per cell per hour (*Y* axis) and specific growth rate, following Harder (1997). In this scheme, there are two maintenance components; one is independent of growth rate (m_e) and the other tends to decline proportionately to growth rate (m_μ). The energetic costs directly associated to growth are e_μ. Carbon consumption increases with specific growth rate (μ), but at $\mu = 0$ there is still a finite rate of carbon consumption, to meet minimum maintenance energy requirements. Below a certain minimum level of exogenous substrate consumption, bacteria revert to endogenous metabolism, which results in apparent negative growth, finally leading to a region in which exogenous as well as endogenous substrate metabolism is minimum. These different regions defined by growth rate and substrate consumption in fact define the physiological state of individual cells: moderate to high metabolic activity and growth rates, low growth leading to dormancy (reversible inactivity), and finally death.

is that there can be finite flux of an energy or carbon substrate at zero growth rate (Kovárová-Kovar and Egli 1998). It is clear that although bacteria need to consume carbon to grow, bacteria can consume carbon with little or no resulting growth (Dawes 1985). Below a certain minimum level of exogenous substrate consumption, bacteria revert to endogenous metabolism (i.e., they degrade cellular constituents to obtain energy), finally leading to a region in which both exogenous and endogenous substrate metabolism is minimum. These different regions defined by growth rate and substrate consumption in fact define the physiological state of individual cells: high metabolic activity and growth rates, and low growth leading to dormancy (reversible inactivity) and death (irreversible inactivity). This scheme suggests that (1) there is a positive relationship between growth and growth efficiency, and (2) growth efficiency is intimately linked to the physiological condition of cells, in terms of the level of metabolic activity (Harder 1997). This latter link between single-cell activity or physiological condition and energetics, which is important in understanding the regulation of growth efficiency at the community level, is explored in subsequent sections.

BACTERIAL GROWTH EFFICIENCY IN NATURAL AQUATIC SYSTEMS

Some History

The ecological importance of bacterial growth efficiency was recognized early on by microbial ecologists, although the terminology has differed greatly over the years (Payne and Wiebe 1978). Physiologists and industrial microbiologists, for different reasons, were interested in the yield of microbial biomass per unit of organic substrate utilized largely under controlled conditions. The modern age of aquatic microbiology was ushered in during the mid-1960s, when researchers began adding small amounts of radioactive organic compounds to water samples and examining the uptake (Wright and Hobbie 1965) and respiration as production of labeled CO_2 (Hobbie and Crawford 1969) from labeled organic compounds under ambient conditions by natural assemblages of bacteria. The original goal of this research was largely to see how active bacteria were in various environments (Crawford et al. 1974) and how fast specific substrates were being turned over. However, the ability to directly measure two fates of a given substrate, uptake into biomass and respiration to $[^{14}C]O_2$, offered the possibility of estimating BGE, if only from the point of view of a single, added substrate. The combination of the short incubations used and the simple substrates chosen (acetate, amino acids, glucose) produced a data set in which BGE was quite high, ranging from 0.3 to 0.8. Methodological improvements in measuring bacterial production (by thymidine or leucine incorporation: see Chapter 4), coupled with more sensitive techniques for measuring either oxygen consumption or dissolved organic carbon utilization, allowed BGE to be estimated without depending on the utilization of single substrates. The more modern measurements of ambient BGE are lower than the historical ones based on single substrates, as we will see.

Measuring Bacterioplankton Growth Efficiency

Studies of growth efficiency in laboratory cultures involve detailed energy and carbon budgets during batch or continuous cultures, based on measurements of substrate depletion, cell mass, respiration, and heat production (Battley 1987). The measurement of bacterioplankton growth efficiency, however, continues to challenge microbial ecologists because direct measurements of substrate consumption can seldom be made at realistic time scales, and metabolic rates are often extremely low (Søndergaard and Theil-Nielsen 1997). Early studies followed the uptake, incorporation, and respiration of simple radiolabeled compounds (Crawford et al. 1974). The advantage of this approach is its high sensitivity, which allows us to measure rates of uptake and respiration in short incubations even in the most unproductive aquatic systems. The main disadvantage is that during these short incubation times the intracellular carbon pools often do not attain equilibrium, with the result that respiration is greatly underestimated and BGE grossly overestimated (King and Berman 1985; Bjørnsen 1986). In addition, the single-model compounds may not be representative of the range of substrates utilized by bacteria in nature.

Current Approaches. The use of single radiolabeled compounds has largely been replaced by techniques that attempt to measure growth efficiency of bacteria utilizing the *in situ* pool of organic matter. Two main approaches are used for this purpose:

1. Simultaneous measurements of bacterial respiration and bacterial net production in relatively short (usually < 36 h) incubations, usually using the *in situ* microbial assemblage (Laanbroek and Verplanke 1986; Griffith et al. 1990; Chin-Leo and Benner 1992; Coffin et al. 1993; Biddanda et al. 1994; Daneri et al. 1994; Pomeroy et al. 1995). There are several difficulties here. First, although bacterial production can be measured in an incubation of less than an hour (see Chapter 4), obtaining a measurable change in O_2 or CO_2 can take 24 hours or more, depending on the system. Further, bacteria must be physically separated from other planktonic components. This is usually attempted by filtration in the 0.6–2 µm range. Since complete separation is seldom achieved, we must conclude that a variable fraction of the measured respiration is due to organisms other than bacteria. Bacteria that are attached to particles, which are often the most active of the assemblage, are also eliminated by means of these techniques. In addition, filtration also disrupts the structure of the bacterial assemblage and the supply of substrates. Organic C consumption is approximated as the sum of bacteria production and bacterial respiration. Bacterial respiration is generally measured as O_2 con-

sumption (Griffith et al. 1990; Chin-Leo and Benner 1992; Biddanda et al. 1994) or more rarely, as CO_2 production (Hansell et al. 1995). Bacterial production is generally measured from the rate of protein or DNA synthesis, using radiolabeled leucine or thymidine, although some studies also follow changes in bacterial abundance and size. Chin-Leo and Benner (1992) found that BGE estimates based on leucine incorporation were generally higher than those based on thymidine incorporation, although the differences were minor.

2. Dilution cultures, where filter-sterilized water is reinoculated with a small amount of the native bacterial assemblage, and the changes in dissolved and particulate organic matter (DOC and POC), bacterial biomass, CO_2 and O_2 are followed, generally for days or weeks (Bjørnsen 1986; Tranvik 1988; Kroer 1993; Zweifel et al. 1993; Carlson et al. 1996, in press). Bacterial growth efficiency is then calculated from the rate of decline in substrate (DOC) and the rate of increase in bacterial biomass (POC), for example. The obvious difficulties include the exceedingly long incubation and the depletion of nutrients and substrates; of special concern is the growth of populations that are not representative of the in situ dominant taxonomic groups.

Methodological Problems. There have been no explicit comparisons of bacterial growth efficiency estimated from short- and long-term experiments; both approaches have problems. Whichever approach is taken, bacteria are isolated from their natural sources of organic matter, and separation of bacteria from microbial grazers also uncouples pathways of nutrient regeneration which may be important in maintaining higher growth efficiencies in natural systems. In long-term experiments there may be increasing use of refractory dissolved organic matter and depletion of nutrients, and therefore the resulting estimate of growth efficiency should be generally lower than in short-term incubations, where presumably only the most labile fraction of dissolved organic matter is utilized. Since growth of heterotrophic nanoflagellates often occurs in long-term incubations, the resulting grazing may heavily affect the accumulation of bacterial biomass and the apparent growth efficiency (Johnson and Ward 1997; Linley and Newell 1984). Also in long term experiments the accumulation of toxic metabolic by-products may result in lower growth efficiencies (Landwell and Holme 1979). The actual consumption of organic matter can seldom be directly measured in short-term experiments, and the assumption that the sum of bacterial respiration and production approximates C consumption does not always hold (Cherrier et al. 1996). For example, Pakulski et al. (1995) have shown that nitrification accounted for 20 to > 50% of the oxygen consumption at intermediate salinities in the Mississippi River plume, with the consequent underestimation of the carbon respiration. There can be considerable variation in respiration and production rates even within short incubations (Pomeroy et al. 1991), so that the length of the incubation and the integration method for these rates become critical for the calculation of growth efficiency (Søndergaard and Theil-Nielsen 1997). In general, it is thought that reducing the incubation times to hours results in data that are ecologically more relevant, but in many natural samples this is not possible with current methods.

All the methods used in determining BGE involve assumptions and the application of conversion factors, which surely contribute to the large variability observed in BGE. Some critical assumptions deal with the conversion of bacterial abundance to carbon, and authors use a wide range of factors. Whereas some authors measured bacterial cell to estimate volume (Schwaerter et al. 1988), others assumed a fixed cell size or carbon content per cell (Benner et al. 1988, Sand-Jensen et al. 1990; Kirchman et al. 1991). Likewise, there is variance in the conversion factors used for the calculation of bacterial production (Biddanda et al. 1994; Chin-Leo and Benner 1992). Respiratory quotients (RQ) assumed by authors also vary (Griffith et al. 1990, Middelboe and Søndergaard 1995), although most authors assume a $RQ = 1$ and it is likely that RQ is a minor source of error compared to the problems discussed above.

BGE in Marine Ecosystems

We have compiled from 25 papers a total of 239 direct measurements of BGE from a variety of natural marine ecosystems (Table 1), including 62 points corresponding to open ocean areas, 123 to coastal areas and 54 to estuaries. The range in measured BGE values is large (< 1 to $> 60\%$), but most values are clustered in the 0.05–0.3 range. The median value of BGE for all marine data, excluding estuaries, is 0.20 (mean of 0.23 ± 0.11). The marine data can be further grouped into open-ocean and coastal sites (Figure 3A), and it is clear that there are systematic differences among these areas. The median BGE for open ocean areas is 0.09 (mean 0.15 ± 0.12 SD), lower than the median for coastal areas of 0.25 (mean 0.27 ± 0.18) and for estuaries of 0.34 (mean 0.37 ± 0.15).

These differences in BGE among distinct marine areas are mostly driven by changes in bacterial production (Figure 3B). Average bacterial production in the open-ocean data (mean $= 0.37 \pm 0.054$ (SD) μg C $L^{-1}d^{-1}$) is an order of magnitude lower than bacterial production in coastal areas or estuaries (mean 2.41 ± 0.33 and 5.20 ± 0.97 μg C L^{-1} d^{-1}, respectively), whereas bacterial respiration is less variable among these areas, ranging from 3.13 ± 0.55 μg C L^{-1} d^{-1} in open oceans to 10.24 ± 1.10 and 8.72 ± 1.51 μg C L^{-1} d^{-1} in coastal areas and estuaries, respectively.

Figure 4 shows the relationship between individual measurements of bacterial production and respiration. It is clear that bacterial production has a much larger dynamic range (five orders of magnitude) across marine ecosystems than bacterial respiration (2.5 orders of magnitude). The slope of this log–log relationship is thus extremely low (0.36), whereas the intercept is highly significantly different from 0 ($p < 0.01$). From Figure 4 it is also clear that there is an enormous amount of variance around the production-versus-respiration relationship, and any given value of bacterial production can have an associated respiration ranging around two orders of magnitude. This unexplained

Table 1. Published sources of direct measurements of *in situ* bacterial growth efficiencies (BGE) that appear in Figures 3 and 4, including the range of BGE

System	BGE	Ref.
Marine		
Sargasso Sea	0.04–0.09	Hansell et al. (1995)
Santa Rosa Sound	0.08–0.69	Coffin et al. (1993)
Gulf of Mexico	0.02–0.23	Pomeroy et al. (1995)
North Pacific	0.01–0.33	Cherrier et al. (1996)
Sargasso Sea	0.04–0.30	Carlson and Ducklow (1996)
Baltic Sea	0.31–0.64	Jørgensen et al. (1993)
Weddel Sea and Scotia shelf	0.38–0.40	Bjørnsen and Kuparinen (1991)
North Atlantic	0.04–0.06	Kirchman et al. (1991)
Coastal waters and enclosures	0.07–0.46	Daneri et al. (1994)
Gulf of Mexico	0.26–0.61	Kroer (1993)
Mississippi River plume	0.10–0.32	Chin-Leo and Benner (1992)
Peruvian upwelling	0.30–0.34	Sorokin and Mameva (1980)
Louisiana shelf	0.18–0.55	Biddanda et al. (1994)
Georgia (USA) coastal and open water	0.01–0.25	Griffith et al. (1990)
Belgian coast	0.1–0.3	Billen and Fontigny (1987)
Baltic and Mediterranean Seas	0.21–0.29	Zweifel et al. (1993)
Ross Sea	0.09–0.38	Carlson et al. (1999)
Baltic Sea	0.25	Platpira and Filmanovicha (1993)
Southern Sea	0.26–0.30	Kähler et al. (1997)
Danish fjord	0.38–0.57	Sand-Jensen et al. (1990)
Estuaries		
Florida estuaries	0.11–0.61	Coffin et al. (1993)
Danish fjord	0.22–0.36	Middelboe et al. (1992)
Hudson River	0.18–0.61	Findlay et al. (1992)
Danish fjord	0.19–0.23	Bjørnsen (1986)
Santa Rosa Sound	0.60–0.61	Kroer (1993)
Brackish estuary	0.40	Laanbroek and Verplanke (1986)

variance is linked in part to methodological problems, some outlined above, which contribute to the scatter in the production to respiration relationship. Many different approaches were used, including short- and long-term incubations. But the scatter in the relationship of bacterial production to respiration may also reflect the physiological flexibility of bacteria, and the extent to which catabolism and anabolism can be uncoupled in natural aquatic bacteria to maximize survival and growth.

General Patterns. Collectively, these direct measurements of BGE, although weakened by methodological uncertainties, nevertheless suggest two broad

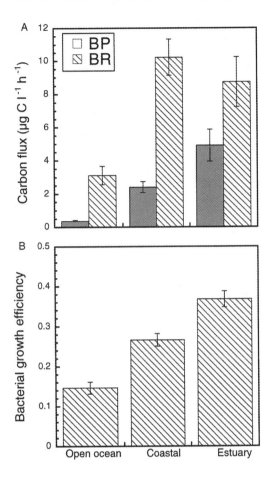

Figure 3. Summary of literature data on direct measurements of bacterial metabolism and growth efficiency in natural aquatic systems, from Table 1. (A) Bacterioplankton production (BP) and respiration (BR) averaged by system (open-ocean, coastal, and estuarine systems). (B) The resulting average bacterial growth efficiency (BGE = $BP/(BR + BP)$] for each system. Bars represent 1 standard error.

patterns: (1) BGE values in natural aquatic systems are generally below 0.4, most often between 0.05 and 0.3, and (2) BGE tends to increase with increasing bacterial production. This latter pattern in fact corresponds to a trend of increasing BGE with overall system productivity, but since there are few actual simultaneous measurements of bacterial growth efficiency and primary production in the sea, we have used the large data set of simultaneous measurements of bacterial production and primary production assembled by Ducklow and Carlson (1992) to explore this relationship. We derived bacterial growth efficiency for each measurement of bacterial production using the relationship

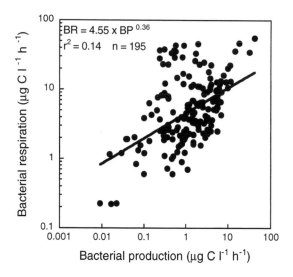

Figure 4. Bacterial respiration as a function of bacterial production in aquatic ecosystems. The data are paired observations of bacterial respiration (*BR*) and production (*BP*); the sources of these data appear in Table 1. The line is the least-squares fit to the log-transformed data.

in Figure 5 and plotted the resulting estimates of BGE as a function of primary production. There is a sharp declining trend of BGE toward the least productive areas (Figure 5), so that the bulk of the world's oceans, which are highly oligotrophic, would appear to be characterized by BGE < 0.15. Next we explore some of the factors that may lead to these patterns and some of their ecological consequences.

Relationship Between BGE and Bacterial Growth

The trends in BGE shown in the preceding section suggest a broad positive relationship between bacterial growth efficiency and growth rate because the latter also tends to increase on average along gradients of primary production (Cole et al. 1988; White et al. 1991). But over smaller scales the relationship between BGE and growth rate does not always hold. For example, Schweitzer and Simon (1995) found no relationship between growth rates and growth efficiencies in natural assemblages of bacterioplankton; Kristiansen et al. (1992) and Middelboe et al. (1992) found a positive relationship between growth rate and BGE in continuos cultures of freshwater bacterioplankton; and Bjørnsen (1986) found a negative relationship between BGE and growth rate. Søndergaard and Theil-Nielsen (1997) found no consistent relation between BGE and growth rates among samples. These data suggest that in any given

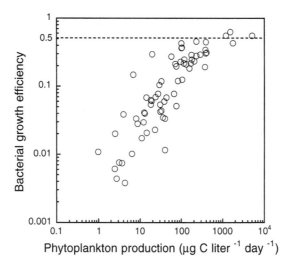

Figure 5. Bacteria growth efficiency as a function of net primary production in marine systems. The large data set on simultaneous measurements of bacterial production and net primary production in Ducklow and Carlson (1992) was used; BGE was calculated for each *BP* point using the model in Figure 4. Data are log-transformed. Dashed line represents asymptotic value.

combination of environmental conditions, BGE may covary with growth rate, but this relationship may be specific for each set of growth conditions; there is no general relationship between bacterial growth rate and efficiency in natural bacterial assemblages.

A factor that uncouples growth rate from growth efficiency is the tendency of bacteria, to maximize growth at the expense of efficiency (Vallino et al. 1996). This maximization is achieved with different energetic costs and varying degrees of uncoupling between catabolism and anabolism. There are clear examples of this type of uncoupling for bacterioplankton. Addition of nutrients sometimes increases substrate consumption with no effect on net growth (Benner et al. 1988). Middelboe et al. (1996) found that viruses decreased BGE in bacterial cultures and at the same increased the growth rate of noninfected bacteria. Zweifel et al. (1993) observed a 70% increase in cell yield (number of cells) and a 20% decrease in BGE after phosphorus was added to the culture media. Robinson et al. (1982) showed that addition of N did not increase BGE but sharply increased the rates of decomposition of detritus, as well as the final yield of bacteria. These examples are important because they suggest that total carbon consumption and BGE may be regulated by factors that are different from those that regulate growth or net bacterial production (Chapters 4, 6, and 8).

REGULATION OF BGE IN NATURAL BACTERIAL ASSEMBLAGES

General problems

Much has been learned from the laboratory studies of microbial energetics, and the mechanisms and general principles that have been identified under laboratory conditions probably also apply to cells inhabiting natural aquatic systems. But there are several fundamental differences between the energetics of single cultures and natural bacterial assemblages. Almost all laboratory studies of bacterial energetics do not consider the following facts: (1) in nature microorganisms grow mostly on mixtures of substrates; (2) growth may not be controlled by a single nutrient alone but by two or more nutrients simultaneously; and (3) adaptation may lead to changes in kinetic and physiologic properties of cells (Kovárová-Kovar and Egli 1998). In addition, one of the difficulties in applying principles derived from single cultures is that natural bacterioplankton assemblages are composed of many species, each with its particular physiological response to environmental conditions (reviewed in Chapter 3). Since the bulk metabolism that we measure from natural assemblages is the summation of the properties of all these coexisting bacterial populations, the study of bacterioplankton energetics elicits questions that laboratory microbiologists have seldom encountered. For example, an increase or a decline of bacterial growth efficiency in an assemblage of planktonic bacteria could result from the depletion of a major nutrient, or a shift to a different organic substrate, and these factors are analogous to conditions encountered in cultures. But it is equally possible that changes in bulk BGE could be the result of a change in the taxonomic composition of the assemblage, or a shift in the relative proportions of cells in different physiologic states (induced, e.g., by viral infection or predation). These last scenarios play a key role in the study of natural assemblages of bacteria but are never significant factors in laboratory cultures.

In the present chapter we have divided the regulation of bacterioplankton energetics, and more precisely of BGE, into two distinct components. The first component groups all the factors that affect the energetics of single cells, such as quality and quantity of organic substrates, availability of inorganic nutrients, and maintenance costs. The second component is related to the factors that affect energetics at the community level, such as the taxonomic composition of the assemblage (reviewed in Chapter 3), as well as the distribution of cells in different physiologic conditions. This second component is an explicit acknowledgment that under a given set of environmental conditions, cells of different species, and cells of the same species but with different physiologic states, will exhibit different growth efficiencies. The bulk measurements that we routinely make in complex assemblages represent the relative contribution of these various taxonomic and physiologic subpopulations. Although these two components of the regulation of bacterioplankton energetics (cellular and community levels) have commonalties, it is conceptually necessary to address them separately.

Regulation of BGE at the Cellular Level

Two main factors can influence BGE at the level of individual cells: the nature and relative availability of organic and inorganic substrates, and the energetic costs associated with survival and growth in the different environments. Traditional kinetics studies are based on the assumption that a single element (e.g., C, P, or N) is limiting; but in contrast to the laboratory, growth in ecosystems proceeds in more complex conditions, where bacteria are faced with mixtures of compounds that can fulfill a particular nutritional function. The growth of bacteria under natural situations can be simultaneously controlled by homologous nutrients (nutrients that fulfill the same function, i.e., various forms of organic carbon), and there is increasing evidence that microbial growth can be stoichiometrically limited at the same time by two or more heterologous nutrients (e.g., C and N) (Kovárová-Kovar and Egli 1998). These factors must be taken into account when we attempt to understand the regulation of BGE in nature.

Regulation of BGE by Nutrient Availability. Many of the studies in natural ecosystems have focused on the dependency of BGE on the relative availability of mineral nutrients and organic carbon. The relationship between BGE, the stoichiometry of N uptake, N excretion, and the C:N ratios of substrate and bacterial biomass is explored in detail by Kirchman in Chapter 9. The basic idea is that bacteria regulate the catabolism of organic substrates to attain the correct intracellular stoichiometry with respect to N (and other nutrients). When the available substrates contain more carbon than nitrogen relative to the needs of the bacterial cells, this excess carbon will be respired or excreted until intracellular stoichiometry with respect to N (or any other nutrient) is attained. Since the elemental composition of bacteria with respect to N is relatively constant (Goldman et al. 1987), BGE should be negatively related to the C:N of the substrate at least in the range of C:N where N, and not C, is limiting. Nitrogen excretion occurs when the carbon in nitrogenous organic compounds is respired and when the resulting N exceeds the stoichiometric cellular quotas (Anderson 1992, Jørgensen et al. 1999).

Regulation of growth efficiency by the availability of mineral nutrients implies that increases in the supply of nutrients should result in increased growth efficiency. Billen (1984) and Goldman et al. (1987) have unequivocally shown that growth efficiency of natural assemblages of marine bacteria grown on a range of substrates is inversely related to C:N ratio of the substrate. But the relationship between BGE and available C:N was considerably weakened when bacteria were exposed to multiple nitrogen and carbon sources (Goldman and Dennett 1991), and under these circumstances the source of the nitrogen (i.e., NH_4^+ or amino acids) becomes important. Bacterial assemblages growing under ambient conditions are exposed to multiple sources of nutrients and organic matter. For example, bacteria take up ammonium along with organic N (see Chapter 9) even when organic N alone can meet the demand (Jørgensen et al. 1999), so it is expected that the relationship between the C:N and BGE may not always hold in natural conditions. Some experimental data support the role of N in regulating BGE (Kroer 1993; Benner et al. 1988; Jørgensen et al. 1999), but most experimental manipulations have showed little or no effect of inorganic nutrient additions in lakes and rivers (Tranvik 1988; Benner et al. 1995), in coastal areas (Robinson et al. 1982; Jørgensen et al. 1993; Zweifel et al. 1993; Daneri et al. 1994), and in open oceans (Carlson and Ducklow 1996; Cherrier et al. 1996; Kirchman and Rich 1997; Carlson et al. in press).

There is thus conflicting evidence regarding the role of N in regulating BGE in marine systems, and no clear pattern emerges along trophic gradients. One reason might be that the bulk C:N is not representative of the substrates actually available and taken up by bacteria (Robinson et al. 1982). It is very likely, however, that nutrients and organic carbon colimit BGE (Egli et al. 1993). Furthermore, there is increasing evidence that P may control BGE not only in freshwater systems (Hessen 1992; Benner et al. 1995) but in marine systems as well (Zweifel et al. 1993; Pomeroy et al. 1995) and that iron deficiency may lower BGE in large areas of the oceans (Tortell et al. 1996).

Regulation of BGE by Energy and Organic Carbon Availability. It has been repeatedly suggested that bacteria in oligotrophic systems are limited primarily by the supply of carbon and energy (Cole et al. 1988; Kirchman 1990; Ducklow and Carlson 1992; Cherrier et al. 1996; Kirchman and Rich 1997; also reviewed in Chapters 4 and 6). The term "limitation" has been used in microbial ecology to describe two completely different phenomena. First, "limitation" is used in a stoichiometric sense to indicate that a certain amount of biomass can be produced from a particular amount of nutrient and that the availability of this nutrient determines the cell density and biomass that can be achieved (following Liebig's law). Second, the term is also used in a kinetic sense to indicate that the microbial growth rate μ is dictated by the concentration of a particular substrate [S], as described, for example, by Monod's equation (Kovárová-Kovar and Egli 1998). These two different types of limitation may have different effects on BGE. In addition, the distinction between energy and carbon limitation is not always fully realized in microbial studies. Growth on relatively oxidized substrates, such as acetate and glycolate, may be energy-limited, and these compounds are typically incorporated into biomass with low efficiency, even if other inorganic nutrients are in excess. This is because the ATP generated during the biological oxidation of this type of compound may be insufficient to reduce all the available carbon in the molecule to the level of bacterial cell carbon (Linton and Stephenson 1978). Thus it is the ratio between the biologically available energy and the carbon content of the organic molecules that determines the maximum growth efficiency (Linton and Stephenson 1978; Connolly et al. 1992).

The Supply of Energy. It is thus important to distinguish between control of BGE arising from the rate of supply of organic matter (and energy) and control from the nature of the available organic matter. Although both may result in low BGE, they are ecologically distinct. If the supply of organic matter is low, regardless of the quality of the substrate, a large fraction of this substrate will be catabolized and used primarily for maintenance energy requirements rather than for growth, with a resulting low growth efficiency (Russell and Cook 1995; Harder 1997). Conversely, there might be an abundant supply of organic substrates that because of their relative energy and carbon content are incorporated with low efficiency even under conditions of excess mineral nutrients. Distinguishing between these two types of limitation in natural situations is difficult, especially because the scenario is further confounded by possible nutrient colimitation. When the actual concentrations of individual dissolved substrates are vanishingly low, bacteria may have to invest additional energy to take up and transport these substrates with an additional lowering of BGE. Williams shows in Chapter 8 that the energy required for work against the thermodynamic gradient might be inconsequential. Williams estimates that heterotrophic bacterial should be able to utilize individual substrates in the oceans at concentrations as low as 10^{-12} M without any significant increase in the cost of acquisition (Chapter 8).

The Supply of Carbon. There is empirical evidence from laboratory cultures that extreme carbon starvation and the resulting slow growth induces the expression of many systems for carbon uptake, transport, and catabolic enzymes, even though the appropriate carbon sources are absent (Sepers 1984). This results in cells that are able to immediately utilize these carbon sources if they become available in the environment (Dawes 1985; Morita 1997). Many studies have further shown that under conditions of severe carbon limitation, heterotrophic bacteria do not restrict themselves to the utilization of a single carbon source but are simultaneously assimilating many of the organic compounds available in their environment, even mixtures of compounds that normally provide diauxic growth at high concentrations (Teixeira de Mattos and Neijssel 1997). For example, *Pseudomonas aeruginosa* was reported to grow with a mixture of 45 carbon compounds, each added to tap water at a concentration of 1 μg C L^{-1}, whereas none of these compounds supported growth on its own at this concentration (van der Kooij et al. 1982).

Regulation of growth efficiency by the supply of organic C implies that increases in the rate of supply of carbon should result in increases in growth efficiency. Empirical and experimental results show that this is not always the case. For example, in the oligotrophic subarctic Pacific, Kirchman (1990) found that growth of planktonic bacteria was not stimulated by the addition of glucose, and others have found similar patterns in other areas (Pomeroy et al. 1995; Carlson et al. in press). Carlson and Ducklow (1996) found that addition of glucose and amino acids resulted in higher BGE; they noted, however, that with glucose addition, cells produced storage carbon and increased in mass rather than in abundance. A similar conclusion was reached by Cherrier et al. (1996). Barillier and Garnier (1993) found a positive relationship between BGE and the concentration of DOC. Our own measurements in the Delaware Bay and surrounding salt marshes in the Atlantic Coast of the United States also suggest a positive relationship between BGE and the concentration of DOC (del Giorgio and Newell, unpublished data), although the best predictor of BGE in these systems is the ratio of inorganic P to DOC rather than DOC alone.

Perhaps the only result common to most addition experiments is that amino acids tend to enhance both BGE and growth of bacteria (Kirchman 1990; Jørgensen et al. 1993; Daneri et al. 1994; Carlson and Ducklow 1996; Cherrier et al. 1996). It has been suggested that it is energetically advantageous to use preformed compounds (Kirchman 1990), although the energetic cost of transporting amino acids across the membranes may offset the advantage of utilizing preformed compounds (Russell and Cook 1995). It is more likely that because amino acids provide both energy and carbon, and are also a source N, they release bacteria from multiple limitation by these factors (i.e., carbon, energy, and N). This experimental evidence suggests that the quality of the organic C, rather than rate of supply of organic matter, may play a large role in regulating BGE in most natural aquatic systems, as has been suggested by Vallino et al. (1996).

The Nature of the Organic Substrate. Qualitative aspects of natural DOC that are relevant to bacterial energetics are difficult to define (Connolly et al. 1992; Vallino et al. 1996). Experiments using different molecular weight fractions of DOC show conflicting results in terms of the resulting BGE (Meyer et al. 1987; Tranvik 1990; Tulonen et al. 1992; Amon and Benner 1996; Middelboe and Søndergaard 1995). Some of these differences can be explained by the C:N of the weight fractions, rather than any qualitative characteristic of the organic carbon itself. Others have measured growth efficiency of bacteria utilizing organic matter derived from specific sources, including phytoplankton (Newell et al. 1981; Bauerfeind 1985; Biddanda 1988) vascular vegetation (Haines and Hanson 1979; Linley and Newell 1984; Moran and Hodson 1989), macroalgae (Lucas et al. 1981; Robinson et al. 1982; Stuart et al. 1982), and feces (Pomeroy et al. 1984; Tupas and Koike 1990). In a recent compilation of these published measurements, del Giorgio and Cole (1998) showed that the efficiency of conversion of detrital organic matter is generally low (< 0.3) regardless of the source of organic matter. Many of these experiments have used organic matter concentrations that are well above normal ambient concentrations, so it is clear that increasing the supply of substrates does not necessarily lead to higher BGE in aquatic systems.

Vallino et al. (1996) have modeled the influence of the bulk properties of DOC on BGE and have concluded that the bulk heat of combustion might provide a good index of the energetic quality of the substrate, but there are few actual measurements that can be used to test this hypothesis. This raises the fundamental question of whether traditional energetic concepts based on cultures with a single species growing on a single substrate can be applied to the environmental conditions where a mixed bacterial assemblage utilizes a variety of C, N, and P compounds simultaneously. It is likely that bacteria utilize some compounds exclusively for energy and others as skeletons and C sources for biosynthesis. For example, nonnitrogenous substrates often contain more energy per unit carbon than nitrogenous ones and therefore on the basis of conserving carbon one might expect the former to be preferentially respired (Anderson 1992). It is at present difficult to differentiate energy from carbon limitation, and it is unclear whether growth efficiency of natural bacterioplankton assemblages is limited by the supply of organic matter, the chemical nature of the organic substrates that are present, or both.

Effect of Temperature, Light, Salinity, and Depth

Laboratory studies show that growth rate declines as temperatures move away from the optima for each type of bacteria, and field studies often show a positive relationship between growth rates and temperature in natural bacterioplankton assemblages (White et al. 1991; Pomeroy et al. 1995). If low temperatures result in lower growth rates, then a positive relationship between temperature and growth efficiency would also be expected. Newell and Lucas (1981) and Roland and Cole (in press), for example, found higher BGE in summer than in winter. Some studies have reported that BGE tends to decline with increasing temperature in natural aquatic systems (Iturriaga and Hoppe 1977; Griffiths et al. 1984; Chin-Leo and Benner 1992; Daneri et al. 1994), even though growth rates tend to increase. In all these cases, however, the effect of temperature was very weak. Others have found no effect of temperature on BGE in the 8–25°C range (Barillier and Garnier 1993). All we can conclude from these data is that temperature is not an overriding regulating factor of BGE in natural aquatic systems. There has been a surge in studies of the effect of light, and particularly UV, on bacterial metabolism, but studies have focused on bacterial production and enzymatic activity (Müller-Niklas et al. 1995). Pakulski et al. (1998) observed photoinhibition of both bacterial production and respiration under ambient solar radiation, and recovery of both during darkness, but there was no consistent pattern in BGE with exposure to light (but see also Chapter 7).

Few studies have specifically addressed the changes in BGE along gradients of salinity. Griffiths et al. (1984) found a weak negative relationship between BGE on glucose and glutamate and salinity in a large-scale study, but concluded that there was no direct effect of salinity on BGE. A gradient of increasing salinity may well have corresponded to a gradient of declining productivity from coastal to open waters, and as we have seen, a pattern of declining BGE would be expected. There is experimental evidence of an inverse relationship between salinity and BGE in yeasts (Watson 1970; Gustafsson and Larsson 1990). The only study to our knowledge that has assessed the effect of pressure on BGE is the one by Turley and Lochte (1990), who concluded that deep-sea bacteria are able to mineralize more organic carbon at 450 atm than at 1 atm, but BGE is lower under high pressure. Biddanda and Benner (1997) found that growth efficiencies decreased from 0.15 on the surface to 0.08 at 500 m, probably again a substrate effect. Bacterial production decreased ninefold over depth, but respiration decreased only fivefold, indicating that bacterial cells were processing larger amounts of carbon to produce a given quantity of cell material.

The Energetic Cost of Survival and Growth

Microbial ecologists have focused on how relative availability of nutrients and C, and to a lesser extent the nature of the C and N sources, affect BGE. Very little research has been done on the energetic costs associated with living in an extremely dilute and variable environment such as the bulk of the oceans. Although the low growth efficiencies that are observed in the least productive areas may be linked to changes in the supply and quality of carbon and nutrient substrates, it is also likely that generally low BGE values reflect increased maintenance energetic costs that cells must face in extremely dilute aquatic systems. From this point of view, the extreme oligotrophy that characterizes most marine systems can be considered to be an environmental challenge that disturbs the balance between maintenance and production at the expense of the latter function (Koch 1997, Morita 1997). Next we briefly explore how specific metabolic functions might influence the overall energy budget of cells, particularly in highly dilute systems. It is important to bear in mind, however, that the bioenergetics of none of these processes has ever been quantified in a natural system.

Cost of Uptake and Transport. The cost associated with the uptake and transport of substrates should not be counted as maintenance if growth is dependent on these two processes. This would be the case if only transport systems for the available substrates that actually support growth were maintained by the cells. But it is now known that bacteria maintain a wide variety of uptake and transport systems, with considerable cost of energy, even if the corresponding substrates are not present (Ferenci 1996). This allows cells to be able to capture nutrients when they fluctuate but nevertheless represents a maintenance cost because it is not directly linked to growth. For example, low concentrations of glucose (in the nanomolar range) also lead to induction of transport systems for other sugars, increasing the scavenging potential of nutrient-limited bacteria for other substrates (Ferenci 1996). This, however, is accomplished at a significant energetic cost (Henderson and Maiden 1987).

Bacteria are also able to adapt to growth at widely varying substrate concentrations by drastically adjusting their kinetic properties (in Monod terms, μ_{max} and K_s), and the strategies that have been reported include (1) a single-uptake system that exhibits different kinetic properties depending on the concentration of its substrate (Nissen et al. 1984), (2) the bacteria switch between two or more transport systems of different affinity (Tempest and Neijssel 1978; Azam and Hodson 1981; Henderson and Maiden 1987; Ferenci 1996), and (3) other less well- defined changes, such as variations in the catabolic and/or anabolic capacity (Kurlanzka et al. 1991). The energetic costs associated with these different strategies have never been explicitly assessed in natural bacterial assemblages, but uptake and transport are a significant fraction of the total energy budget of cells in culture (Russell and Cook 1995), and it is expected that these functions exert an even larger energetic toll in dilute systems (Dawes 1985; Marden et al. 1987; Egli et al. 1993; Morita 1997).

Cost of Enzymatic Breakdown. Besides the energetic costs of uptake and transport of solutes, bacteria must invest energy in the production and excretion of extracellular enzymes used to hydrolyze polymers. A large fraction of DOC in natural aquatic systems is composed of polymeric substances that cannot be incorporated directly into bacteria. Large molecules and colloids present in the DOC pool must be acted upon by exoenzymes if they are to be utilized by bacteria (Hoppe 1991), and the hydrolysis of polymers has been suggested as the rate-limiting process for bacterial production in aquatic systems (Chróst 1990). The synthesis and excretion of enzymes must be coupled to active transport systems that can capture the products of extracellular hydrolysis and of enzymatic systems capable of catabolizing these substrates; thus these processes may represent a major energy expenditure of bacteria in natural aquatic systems. For example, Middelboe and Søndergaard (1993) found an inverse relationship between lake BGE and β-glucosidase activity. Extracellular enzyme production increased toward the end of batch culture incubations of lake bacterioplankton, when most of the labile DOC had been consumed and the submicrometer and colloidal fractions were increasingly utilized (Middelboe and Søndergaard 1993). There have been no direct measures of the energetic cost associated with exoenzymatic hydrolysis in natural systems, but recently Vetter et al. (1998) attempted to model the bioenergetics of exoenzymatic hydrolysis in terms of cost/benefit in the context of optimal foraging theory.

Cost of Excretion. In addition to the excretion of enzymes, bacteria are also capable of producing large amounts of extracellular mucopolysaccharides (Decho 1990), which form mucilaginous capsules (Decho 1990; Heissenberger et al. 1996) and also loosely associated slimes and fibrils around the cells (Leppard 1995). The chemical nature of these extracellular compounds varies greatly, but uronic acids often form the bulk of these materials (Kennedy and

Sutherland 1987). It has been suggested that excretion of certain metabolites is a pathway of energy dissipation that may contribute to the maintenance of intracellular stoichiometry (Decho 1990; Linton 1990). But excretion of organic metabolites, including polysaccharides, lipids, proteins, and humiclike substances, has also been reported under conditions of extreme carbon and nutrient limitation in aquatic bacteria (Iturriaga and Zsolnay 1981; Goutx et al. 1990; Tranvik 1992; Jørgensen and Jensen 1994). It has been suggested that mucilage and other excreted substances might play a role in the scavenging of substrates in very dilute systems (Decho 1990; Heissenberger et al. 1996). Most of the excretion products are polymeric, and the biosynthesis of these substances typically exerts high energy requirements on the cell (Stouthamer 1973; Russell and Cook 1995). Metabolite excretion appears to be greater when the organic substrate is in excess of the growth requirement and depends also on the nature of the organic substrate (Tempest and Neijssel 1992); and not surprisingly, there is a general inverse relationship between the overproduction and excretion of metabolites and growth efficiency in bacterial cultures (Linton 1990). There is also evidence for the excretion of secondary metabolites, defined as those having a more restricted distribution (which is almost species specific) and no obvious function in central metabolism. Some authors have argued that secondary metabolites may also be waste or overflow products, or evolutionary leftovers of a former autophysiological function, lacking a modern function (Kell et al. 1995). Current bacterial production measurements, whether based on changes in bacterial biomass, or on the incorporation of leucine or thymidine, are unlikely to include the production of exopolymers, and this will result in severe underestimation of BGE (Decho 1990).

Regulation of BGE at the Community Level

The question addressed here is whether the BGE that we measure in aquatic systems represents the energetics of the majority of cells composing the assemblage, or whether there are populations within the assemblage characterized by widely different BGE. If the latter is true, then the bulk BGE that we measure is simply a reflection of the proportions of these different physiologic or taxonomic populations and is not itself representative of the energetics of natural bacteria. The coexistence of multiple bacterial species and of cells in multiple physiologic conditions is what determines the bulk metabolic processes we measure in bacterial assemblages. We barely begin to understand the multitude of factors that can influence the taxonomic and physiologic composition of bacterial assemblages (see Chapters 2 and 3). Among these, trophic interactions, particularly grazing and viral infection, probably play leading roles.

Viral Infection. The effect of viral infection on microbial metabolism in aquatic systems is not well understood but has potentially opposite impacts. On the one hand viral lysis of phytoplankton, particularly during blooms, results in large inputs of fresh DOC and most likely an increase in production and BGE (Bratbak et al. 1998). But viral infection and subsequent lysis of bacteria should result in a decline in BGE, for a number of reasons. Middelboe et al. (1996) observed a decline in BGE in infected cultures, which they attributed to the increased energetic cost of hydrolyzing polymeric P and C from dead bacteria. Mortality from viral infection has been shown to be size-selective (Weinbauer and Höfle 1998a), so that larger, presumably more active cells are removed, leaving cells that either are not growing or are doing so very slowly, at low BGE. Differential mortality due to viral infection among different species with distinct physiologic traits has also been shown (Weinbauer and Höfle 1998b). Although BGE was not measured for individual species within a complex assemblage, it is likely that species that differ substantially in the preferred substrate and in maximum growth rate will also differ in BGE, so that differential infection by viruses will alter the BGE measured for the entire assemblage. Regardless of whether viral lysis selectively removes certain components of assemblage, most of the products of viral lysis return to the dissolved pool and can be further utilized by bacteria. As Fuhrman points out in Chapter 11, viral lysis of bacteria produces a semiclosed loop that essentially burns carbon and regenerates nutrients. This process may be confused with high respiration and can result in low apparent growth efficiency, even when individual cells are converting DOC into biomass with higher efficiencies.

Predation. Grazing of bacteria can also have a profound impact on the structure of bacterial assemblages and on the distribution of taxa and of populations in different physiologic conditions. There is ample evidence that protozoan grazing is often highly selective; for example, heterotrophic nanoflagellates have been shown to preferentially remove dividing cells (Sherr et al. 1992), highly active cells (del Giorgio et al. 1996), and even different species of bacteria (Hahn and Höfle 1998), presumably based on size differences (see also Chapter 12). Small cells are usually less active than larger cells (Gasol et al. 1995; Pernthaler et al. 1996; Posch et al. 1997) and since more active cells presumably are faster growing and have a higher BGE than nongrowing or slowly growing cells, so the size- and taxon-dependent mortality through selective removal by protists and other bacterial grazers should have an impact on the average growth efficiency of the assemblage.

Phylogenetic Composition. In terms of the specific genetic complement, one can ask whether the measured BGE is the result of phenotypic adaptation or genotypic adaptation. We know little about the differences in energetics among species in natural assemblages, but there is ample evidence that there can be large genetically determined differences in energy handling depending on taxonomic composition (Heijnen and van Dijken 1992).The same combination of substrates and environmental conditions will result in different patterns of energy and C allocation depending on the species involved. The effects of this on competition, for example, have been discussed for yeasts in marine systems (Gustafsson and Larsson 1990). We have no information on how the taxonomic composition of the bacterioplankton assemblage may affect bacterial growth efficiency in natural aquatic systems. But the advent of a new generation of molecular techniques is rapidly opening the genetic black box of planktonic bacteria (reviewed in Chapter 3), and soon we may be able to link broad taxonomic composition to aspects of microbial energetics and thus to explain some of the variance in BGE not accounted for by resource regulation.

ECOLOGICAL CONSEQUENCES OF MICROBIAL ENERGETICS

According to Teixeira de Mattos and Neijssel (1997), two basic strategies are possible to sustain growth despite a lowering of the availability of a nutrient: (1) the rate of consumption can be kept constant by increasing the maximum specific consumption rate or by lowering the affinity constant, or (2) growth efficiency can be increased allowing the same growth rate at decreased consumption rate. The declining trend in BGE toward more oligotrophic sites, where presumably the supply of carbon and energy is also lower, suggests that bacterioplankton respond with the first strategy. The hypothesis that bacteria maximize utilization rather than efficiency is important in interpreting the numerous measurements of bacterial production in the oceans. Maximizing the rate of utilization would imply that carbon consumption by bacteria, which can be approximated by the sum of bacterial production and respiration, effectively reflects the total amount of organic matter available to bacteria. The total carbon consumption by bacteria is thus a function of the total bioavailable organic C in the water, but the partition of this organic matter into catabolism and anabolism is regulated by factors other than the total availability of C. It has often been assumed that bacterial production is a good index of the supply of organic matter to bacteria (Cole et al. 1988; Ducklow and Carlson 1992), but the foregoing hypothesis implies that bacterial production is not necessarily determined by the supply of organic matter alone. Although the sum of bacterial production and respiration (total C consumption by bacteria) is positively correlated to primary production, it has a much smaller range than bacterial production alone, suggesting that the total amount of organic matter available to bacteria is less variable than previously recognized. The fate of this C utilized by bacteria varies greatly, however, because of differences in BGE along gradients of productivity. In our data set, the average total carbon flux through bacterioplankton (production + respiration) differed by only threefold between open-ocean systems and more productive coastal areas and estuaries (Figure 6). Because bacterial respiration is, in relative terms, so high in most systems, bacterial production alone is not a good index of the C available for bacterial use.

We discussed earlier the hypothetical link between energetics and physiological state of individual cells. In particular, we emphasized that cells need to consume substrates to grow but the inverse is not true: that is, substrates can be consumed without growth. At a population level, the links between energetics, particularly growth efficiency and physiological state and single-cell activity, are keys to interpreting metabolic measurements of mixed natural bacterial assemblages. It is important to understand that the factors that regulate carbon consumption, the net production of bacterial biomass, and cell-specific growth rates may not be the same. This variance has a number of ecological implications. For example, many studies have used micro-autoradiography to assess single-cell activity in natural bacterioplankton (e.g., Karner and Fuhrman 1997). These studies have found that often a large fraction of the bacterial assemblage can take up single radioactive substrates; and because consumption is equated to growth, this has been taken as evidence that most bacterioplankton cells are metabolically active and growing (Karner and Fuhrman 1997), which is probably not the case (Sherr et al. 1999).

In studies of natural assemblages it is customary to determine the bulk rate of biomass production (e.g., from leucine or thymidine uptake), and scale this to the total density or biomass of bacteria to obtain a cell-specific production rate and growth rate. Rates of bacterial production are extremely low in most of the oceans, but because bacterial density remains relatively high (Gasol et al. 1997), the resulting growth rates are often exceedingly low, sometimes in the order of days or weeks, and even months or years (see Chapter 4 and 8). In this respect, one aspect of bacterial energetics that is often overlooked in studies of natural bacterioplankton assemblages is the existence of limits to how slowly, bacteria will grow. Under poor conditions, instead of growing more slowly, the cells may cease growth and form resistant forms, such as

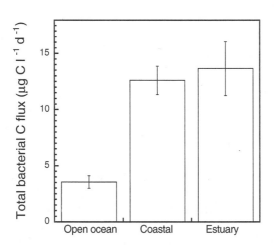

Figure 6. Total bacterial carbon consumption, calculated as the sum of bacterial production and respiration, from published measurements summarized in Table 1, average by system. Bars represent 1 standard error.

spores, or periodically enter and leave other quiescent states (Koch 1997). Cells need to complete a macromolecule once synthesis has begun, because there are negative consequences of not doing so; a common observation in slow-growing cultures is that not all cells are engaged in protein synthesis. The implication is that cells have a choice about their global synthesis of protein and energy expenditure and that in a highly carbon-stressed environment they sometimes do not attempt to synthesize protein if they do not have enough resources to finish making the macromolecule (Koch 1997). But cells may still consume carbon, albeit in small quantities, to maintain key cellular functions; and in spite of the presumably low average proportion of highly active and fast growing bacteria in most marine systems (Sherr et al. 1999) and the resulting low average bacterial growth rates, the microbial assemblage often maintains a relatively high adenylate energy charge (AEC) and ATP level (Morita 1997). The low growth rates reported for many marine systems could be an artifact of having a small subpopulation of bacteria that are actually growing and a large component of the bacteria that are not growing but nevertheless consuming DOM.

SUMMARY

1. Bacterial growth efficiencies in natural marine systems are generally low (< 0.3). The low BGE values are the combined result of the nature and supply of organic and inorganic substrates, and most likely also of the high energetic cost associated with life in very oligotrophic systems. There is a positive relationship between growth efficiency and primary production.

2. The energetic costs of the production of exoenzymes, active solute transport, and the excretion of a variety of polymers have not been investigated in natural bacterioplankton. In particular, the concept of maintenance energy requirements has never been explicitly investigated for bacterioplankton.

3. Bacterial respiration is much less variable than bacterial production across marine systems, and it is clear that BGE is driven by changes in bacterial production rather than in bacterial respiration. Bacterial production alone is not a good index of either organic carbon consumption or availability and grossly underestimates the carbon flow through bacterioplankton.

4. There are relatively few direct estimates of bacterioplankton respiration, and we are barely beginning to understand the regulation of microbial respiration, although this is possibly the largest single component of organic carbon flow in most marine systems.

5. The distinction between energy and organic carbon limitation of BGE should be further explored at both the conceptual and experimental levels. The growth efficiency measured in natural assemblages is an

integrated measure of the efficiency of utilization of a large number of organic compounds.

6. The community BGE that we measure represents the average efficiencies of different subpopulations of bacteria that coexist within the bacterioplankton assemblage. Understanding what controls the distribution of subpopulations of highly active versus dormant or slow-growing cells in bacterioplankton assemblages will no doubt advance our understanding of what controls BGE in natural aquatic systems.

7. One fundamental unanswered question is whether the low BGE measured in marine systems is the result of genotypic or phenotypic adaptation of the bacterial assemblage.

REFERENCES

Amon, R. M. W., and Benner, R. (1996) Bacterial utilization of different size classes of dissolved organic matter. *Limnol. Oceanogr.* 41:41–51.

Anderson, T. R. (1992) Modeling the influence of food C:N ratio, and respiration on growth and nitrogen excretion in marine zooplankton and bacteria. *J. Plankton Res.* 14:1645–1671.

Azam, F., and Hodson, R. E. (1981) Multiphasic kinetics for D-glucose uptake by assembles of natural marine bacteria. *Mar. Ecol. Prog. Ser.* 6:213–222.

Barillier, A., and Garnier, J. (1993) Influence of temperature and substrate concentration on bacterial growth yield in Seine River water batch cultures. *Appl. Environ. Microbiol.* 59:1678–1682.

Battley, E. H. (1987) *Energetics of Microbial Growth.* Wiley, New York.

Bauerfeind, S. (1985) Degradation of phytoplankton detritus by bacterial: estimation of bacterial consumption and respiration in an oxygen chamber. *Mar. Ecol. Prog. Ser.* 21:27–36.

Benner, R., Lay, J., K'nees, E., and Hodson, R. E. (1988) Carbon conversion efficiency for bacterial growth on lignocellulose:Implications for detritus-based food webs. *Limnol. Oceanogr.* 33:1514–1526.

Benner, R., Opsahl, S., Chin-Leo, G., Richey, J. E., and Forsberg, B. R. (1995) Bacterial carbon metabolism in the Amazon River system. *Limnol. Oceanogr.* 40:1262–1270.

Biddanda, B. (1988) Microbial aggregation and degradation of phytoplankton-derived detritus in seawater. 2. Microbial metabolism. *Mar. Ecol. Prog. Ser.* 42:89–95.

Biddanda, B. and Benner, R. (1997) Major contribution from mesopelagic plankton to heterotrophic metabolism in the upper ocean. *Deep-Sea Res.* 44:2069–2085.

Biddanda, B., Opsahl, S., and Benner, R. (1994) Plankton respiration and carbon flux through bacterioplankton. *Limnol. Oceanogr.* 39:1259–1275.

Billen, G. (1984) Heterotrophic utilization and regeneration of nitrogen. In J. E. Hobbie and P. J. le B. Williams, eds., *Heterotrophic Activity in the Sea.* Plenum Press, New York, pp. 313–355.

Billen, G., and Fontigny, A. (1987) Dynamics of *Phaeocystis*-dominated spring bloom in Belgian coastal waters. *Mar. Ecol. Prog. Ser.* 37:249–257.

Bjørnsen, P.K. (1986) Bacterioplankton growth yield in continuous seawater cultures. *Mar. Ecol. Prog. Ser.* 30:191–196.

Bjørnsen, P. K., and Kuparinen, J. (1991) Determination of bacterioplankton biomass, net production and growth efficiency in the Southern Ocean. *Mar Ecol. Prog. Ser.* 71:185–194.

Bratbak, G., Jacobsen, A., and Heldal, M. (1998) Viral lysis of *Phaeocystis pouchetii* and bacterial secondary production. *Aquat. Microb. Ecol.* 16:11–16.

Carlson, C. A., and Ducklow, H. W. (1996) Growth of bacterioplankton and consumption of dissolved organic carbon in the Sargasso Sea. *Aquat. Microb. Ecol.* 10:69–85.

Carlson, C. A., Bates, N. R., Ducklow, H. W., and Hansell, D. A. (1999) Estimation of bacterial respiration and growth efficiency in the Ross Sea, Antarctica. *Aquat. Microb. Ecol.* 27:229–244.

Cherrier, J., Bauer, J. E., and Druffel, E. R. M. (1996) Utilization and turnover of labile dissolved organic matter by bacterial heterotrophs in eastern North Pacific surface waters. *Mar. Ecol. Prog. Ser.* 139:267–279.

Chin-Leo, G., and Benner, R. (1992) Enhanced bacterioplankton production and respiration at intermediate salinities in the Mississippi River plume. *Mar. Ecol. Prog. Ser.* 87:87–103.

Chróst, R. H. (1990) Microbial ectoenzymes in aquatic environments. In J. Overbeck and R. J. Chróst, eds., *Aquatic Microbial Ecology: Biochemical and Molecular Approaches.* Springer-Verlag, Berlin, pp. 47–78.

Coffin, R. B., Connolly, J. P., and Harris, P. S. (1993) Availability of dissolved organic carbon to bacterioplankton examined by oxygen utilization. *Mar. Ecol. Prog. Ser.* 101:9–22.

Cole, J. J., Findlay, S., and Pace, M. L. (1988) Bacterial production in fresh and saltwater ecosystems: A cross-system overview. *Mar. Ecol. Prog. Ser.* 43:1–10.

Connolly J. P., Coffin, R. B., and Landeck, R. E. (1992) Modeling carbon utilization by bacteria in natural water systems. In C. Hurst, ed., *Modeling the Metabolic and Physiologic Activities of Microorganisms.* New York, Wiley, pp. 249–276.

Crawford, C. C., Hobbie, J. E., Webb, K. L. (1974) Utilization of dissolved free amino acids by estuarine microorganisms. *Ecology* 55:551–563

Daneri, G., Riemann, B., and Williams, P. J. le B. (1994) In situ bacterial production and growth yield measured by thymidine, leucine and fractionated dark oxygen uptake. *J. Plankton Res.* 16:105–113.

Dawes, E. A. (1985) Starvation, survival and energy reserves. In M. Fletcher, ed., *Bacteria in Their Natural Environment.* Academic Press, New York, pp. 43–79.

Decho, A.W. (1990) Microbial exopolymer secretions in oceanic environments. *Oceanogr. Mar. Biol. Annu. Rev.* 28:73–153.

del Giorgio, P. A., and Cole, J. J. (1988) Bacterioplankton growth efficiency in natural aquatic systems. *Annu. Rev. Ecol. Syst.* 29:503–541.

del Giorgio, P. A., Gasol, J. M., Mura, P., Vaqué, D., and Duarte, C. M. (1996) Protozoan control of the proportion of metabolically active bacteria in coastal marine plankton. *Limnol. Oceanogr.* 41:1169–1179.

del Giorgio P. A., Cole, J. J., and Cimbleris, A. (1997) Respiration rates in bacteria exceed phytoplankton production in unproductive aquatic systems. *Nature* 385:148–151.

Ducklow H. W., and Carlson, C. A. (1992) Oceanic bacterial production. *Adv. Microb. Ecol.* 12:113–181.

Egli, T., Lendenmann, U., and Snozzi, M. (1993) Kinetics of microbial growth with mixtures of carbon sources. *Antonie van Leeuwenhoek* 63:289–298.

Ferenci, T. (1996) Adaptation to life at micromolar nutrient levels: The regulation of *Escherichia coli* glucose transport by endoinduction and cAMP. *FEMS Microbiol. Rev.* 18:301–317.

Findlay, S., Pace, M. L., Lints, D., and Howe, K. (1992) Bacterial metabolism of organic carbon in the tidal freshwater Hudson Estuary. *Mar. Ecol. Prog. Ser.* 89:147–153.

Gasol, J. M., P. A. del Giorgio, R. Massana and Duarte, C. M. (1995) Active vs inactive bacteria:size-dependence in a coastal marine plankton community. *Mar. Ecol. Prog. Ser.* 128:91–97.

Gasol, J. M., del Giorgio, P. A., and Duarte, C. M. (1997) Biomass distribution in marine planktonic communities. *Limnol. Oceanogr.* 42:1353–1363.

Goldman, J. C., and Dennett, M. R. (1991) Ammonium regeneration and carbon utilization by marine bacteria grown on mixed substrates. *Mar. Biol.* 109:369–378.

Goldman, J. C., Caron, D. A., and Dennett, M. R. (1987) Regulation of gross growth efficiency and ammonium regeneration in bacteria by substrate C:N ratio. *Limnol. Oceanogr.* 32:1239–1252.

Goutx, M., Acquaviva, M., and Bertrand, J.-C. (1990) Cellular and extracellular carbohydrates and lipids from marine bacteria during growth on soluble substrates and hydrocarbons. *Mar. Ecol. Prog. Ser.* 61:291–296.

Griffith, P. C., Douglas, D. J., and Wainright, S. C. (1990) Metabolic activity of size-fractioned microbial plankton in estuarine, nearshore, and continental shelf waters of Georgia. *Mar. Ecol. Prog. Ser.* 59:263–270.

Griffiths, R. P., Caldwell, B. A., and Morita, R. Y. (1984) Observations on microbial percent respiration values in Arctic and subarctic marine waters and sediments. *Microb. Ecol.* 10:151–164.

Gustafsson, L., and Larsson, C. (1990) Energy budgeting in studying the effect of environmental factors on the energy metabolism of yeasts. *Thermochim. Acta* 172:95–104.

Hahn, M. W., and Höfle, M. G. (1998) Grazing pressure by a bacterivorous flagellate reverses the relative abundance of *Comomonas acidovorans* PX54 and *Vibrio* strain CB5 in chemostat cocultures. *Appl. Environ. Microbiol.* 64:1910–1918.

Haines, E. B., and Hanson, R. B. (1979) Experimental degradation of detritus made from the salt marsh plants *Spartina alterniflora* Loisel., *Salicornia virginica* L., and *Juncus roemerianus* Scheele. *J. Exp. Mar. Biol. Ecol.* 40:27–40.

Hansell, D. A., Bates, N. R., and Gundersen, K. (1995) Mineralization of dissolved organic carbon in the Sargasso Sea. *Mar. Chem.* 51:201–212.

Harder, J. (1997) Species-independent maintenance energy and natural populations sizes. *FEMS Microbiol. Ecol.* 23:39–44.

Heijnen, J. J., and van Dijken, J. P. (1992) In search of a thermodynamic description of biomass yields for the chemotrophic growth of microorganisms. *Biotechnol. Bioeng.* 39:833–858.

Heissenberger, A., Leppard, G. G., and Herndl, G. J. (1996) Relationship between the intracellular integrity and the morphology of the capsular envelope in attached and free-living marine bacteria. *Appl. Environ. Microbiol.* 62:4521–4529.

Henderson, P. J. F., and Maiden, C. J. (1987) Transport of carbohydrates by bacteria. In D. J. Stowell, A. J. Beardsmore, C. W. Keevil, and J. R. Woodward, eds., *Carbon Substrates in Biotechnology.* IRL Press, Oxford, pp. 67–92.

Hessen, D. O. (1992) Dissolved organic carbon in a humic lake: Effects on bacterial production and respiration. *Hydrobiologia* 229:115–123.

Hobbie, J. E., and Crawford, C. C. (1969) Respiration corrections for bacterial uptake of dissolved organic compounds in natural waters. *Limnol. Oceanogr.* 14:528–532.

Hoppe, H. G. (1991) Microbial extracellular enzyme activity: A new key parameter in aquatic ecology. In R.J. Chróst ed., *Microbial Enzymes in Aquatic Environments.* Springer-Verlag, New York, pp. 60–83.

Iturriaga, R., and Hoppe, H.-G. (1977) Observations of heterotrophic activity on photoassimilated matter. *Mar. Biol.* 40:101–108.

Iturriaga, R., and Zsolnay, A. (1981) Transformation of some dissolved organic compounds by a natural heterotrophic population. *Mar. Biol.* 62:125–129.

Jahnke, R. A., and Craven, D. B. (1995) Quantifying the role of heterotrophic bacteria in the carbon cycle: A need for respiration rate measurements. *Limnol. Oceanogr.* 40:436–441.

Johnson, M. D., and Ward, A. K. (1997) Influence of phagotrophic protistan bacterivory in determining the fate of dissolved organic matter in a wetland microbial food web. *Microb. Ecol.* 33:149–162.

Jørgensen, N. O. G., and Jensen, R. E. (1994) Microbial fluxes of free monosaccharides and total carbohydrates in freshwater determined by PAD-HPLC. *FEMS Microbiol. Ecol.* 14:79–94.

Jørgensen, N. O. G., Kroer, N., Coffin, R. B., Yang, X.-H., and Lee, C. (1993) Dissolved free amino acids, combined amino acids, and DNA as sources of carbon and nitrogen to marine bacteria. *Mar. Ecol. Prog. Ser.* 98:135–148.

Jørgensen, N. O. G., Kroer, N., Coffin, R. B., and Hoch, M. P. (1999) Relations between bacterial nitrogen metabolism and growth efficiency in an estuarine and an open-water ecosystem. *Aquat-Microb. Ecol.* 18:247–261.

Karner, M., and Furhman, J. A. (1997) Determination of active marine bacterioplankton: A comparison of universal 16S RNA probes, autoradiography, and nucleoid staining. *Appl. Environ. Microbiol.* 59:3187–3196.

Kell, D. B., Kapreyants, A. S., and Grafen, A. (1995) Pheromones, social behavior and the functions of secondary metabolism in bacteria. *TREE* 10:126–129.

Kennedy, A. F. D., and Sutherland, I. W. (1987) Analysis of bacterial exopolysaccharides. *Biotechnol. Appl. Biochem.* 9:12–19.

King, G. M., and Berman, T. (1985) Potential effects of isotopic dilution on apparent respiration in ^{14}C heterotrophy experiments. *Mar. Ecol. Prog. Ser.* 19:175–180.

Kirchman, D. L. (1990) Limitation of bacterial growth by dissolved organic matter in the subarctic Pacific. *Mar. Ecol. Prog. Ser.* 62:47–54.

Kirchman, D. L., and Rich, J. H. (1997) Regulation of bacterial growth rates by dissolved organic carbon and temperature in the equatorial pacific ocean. *Microb. Ecol.* 33:11–20.

Kirchman, D. L., Suzuki, Y., Garside, C., and Ducklow, H. W. (1991) High turnover rates of dissolved organic carbon during a spring phytoplankton bloom. *Nature* 352:612–614.

Koch, A. L. (1997) Microbial physiology and ecology of slow growth. *Microbiol. Mol. Biol. Rev.* 61:305–318.

Köhler, P., Bjørnsen, P. K., Lochte, K., and Anita, A. (1997) Dissolved organic matter and its utilization by bacteria during spring in the Southern Ocean. *Deep-Sea Res. II* 44:341–353.

Kovárová-Kovar, K., and Egli, T. (1998) Growth kinetics of suspended microbial cells: From single-substrate-controlled growth to mixed-substrate kinetics. *Microbiol. Mol. Biol. Rev.* 62:646–666.

Kristiansen, K., Nielsen, H., Riemann, B., and Fuhrman, J. A. (1992) Growth efficiencies of freshwater bacterioplankton. *Microb. Ecol.* 24:145–160.

Kroer, N. (1993) Bacterial growth efficiency on natural dissolved organic matter. *Limnol. Oceanogr.* 38:1282–1290.

Kurlandzka, A., Rosenzweig, R. F., and Adams, J. (1991) Identification of adaptive changes in an evolving population of *Escherichia coli:* The role of changes with regulatory highly pleiotrophic effects. *Mol. Biol. Evol.* 8:261–281.

Laanbroek, H. J., and Verplanke, J. C. (1986) Tidal variation in bacterial biomass, productivity and oxygen uptake rates in a shallow channel in the Oosterschelde Basin, the Netherlands. *Mar. Ecol. Prog. Ser.* 29:1–5.

Landwell, P., and Holme, T. (1979) Removal of inhibitors of bacterial growth by dialysis culture. *J. Gen. Microbiol.* 103:345–352.

Leppard, G. G. (1995) The characterization of algal and microbial mucilages and their aggregates in aquatic ecosystems. *Sci. Total Environ.* 165:103–131.

Linley, E. A. S., and Newell, R. C. (1984) Estimates of bacterial growth yields based on plant detritus. *Bull. Mar. Sci.* 35:409–425.

Linton, J. D. (1990) The relationship between metabolite production and the growth efficiency of the producing organism. *FEMS Microbiol. Rev.* 75:1–18.

Linton, J. D, and Stephenson, R. J. (1978) A preliminary study on growth yields in relation to the carbon and energy content of various organic growth substances. *FEMS Microbiol. Lett.* 3:95–98.

Lucas, M. I., Newell, R. C., and Velimirov, B. (1981) Heterotrophic utilization of kelp (*Ecklonia maxima* and *Laminaria pallida*). II. Differential utilization of dissolved organic components from kelp mucilage. *Mar. Ecol. Prog. Ser.* 4:43–55.

Marden, P., Nystrom, T., and Kjelleberg, S. (1987) Uptake of leucine by a marine gram-negative heterotrophic bacterium during exposure to starvation conditions. *FEMS Microbiol. Ecol.* 45:233–241.

Meyer, J. L., Edwards, R. T., and Risley, R. (1987) Bacterial growth on dissolved organic carbon from a blackwater river. *Microb. Ecol.* 13:13–29.

Middelboe, M., and Søndergaard, M. (1993) Bacterioplankton growth yield: A close coupling to substrate lability and β-glucosidase activity. *Appl. Environ. Microbiol.* 59:3916–3921.

Middelboe, M., and Søndergaard, M. (1995) Concentration and bacterial utilization of sub-micron particles and dissolved organic carbon in lakes and a coastal area. *Arch. Hydrobiol.* 133:129–147.

Middelboe, M., Nielsen, B., and Søndergaard, M. (1992) Bacterial utilization of dissolved organic carbon (DOC) in coastal waters — Determination of growth yield. *Arch. Hydrobiol. Ergebn. Limnol.* 37:51–61.

Middelboe, M. B, Jørgensen, N. O. J., and Kroer, N. (1996) Effects of viruses on nutrient turnover and growth efficiency of noninfected marine bacterioplankton. *Appl. Environ. Microbiol.* 62:1991–1997.

Monod, J. (1942) *Recherches sur la Criossance des Cultures Bactriennes.* Hermann, Paris.

Moran, M. A., and Hodson, R. E. (1989) Formation and bacterial utilization of dissolved organic carbon derived from detrital lignocellulose. *Limnol. Oceanogr.* 34:1034–1047.

Morita, R. Y. (ed.) (1997) *Bacteria in Oligotrophic Environments.* Chapman & Hall, New York.

Müller-Niklas, G., Heissenberger, A., Puskaric, S., and Herndl, G. J. (1995) Ultraviolet-B radiation and bacterial metabolism in coastal waters. *Aquat. Microb. Ecol.* 9:111–116.

Newell, R. C., and Lucas, M. (1981) The quantitative significance of dissolved and particulate organic matter released during fragmentation of kelp in coastal waters. *Kieler Meeresforsch.* 5:356–369.

Newell, R. C., Lucas, M., and Linley, E. A. S. (1981) Rate of degradation and efficiency of conversion of phytoplankton debris by marine microorganisms. *Mar. Ecol. Prog. Ser.* 6:123–136.

Nissen, H., Nissen, P., and Azam, F. (1984) Multiphasic uptake of D-glucose by an oligotrophic marine bacterium. *Mar. Ecol. Prog. Ser.* 16:155–160.

Pakulski, J. D., Benner, R., Amon, R., Eadie, B., and Whitledge, T. (1995) Community metabolism and nutrient cycling in the Mississippi River plume: Evidence for intense nitrification at intermediate salinities. *Mar. Ecol. Prog. Ser.* 117:207–218.

Pakulski, J. D., Aas, P., Jeffrey, W, Lyons, M., Von Waasenbergen, L., Mitchell, D., and Coffin, R. (1998) Influence of light on bacterioplankton production and respiration in a subtropical coral reef. *Aquat. Microb. Ecol.* 14:137–148.

Payne, W. J., and Wiebe, W. J. (1978) Growth yield and efficiency in chemosynthetic microorganisms. *Annu. Rev. Microbiol.* 32:155–183.

Pernthaler, J., Posch, T., Simek, K., Schwarzenbacher, A., and Psenner, R. (1996) Top-down effects on the size–biomass distribution of a freshwater bacterioplankton community. *Aquat. Microb. Ecol.* 10:255–263.

Pirt, S. J. (1965) The maintenance energy of bacteria in growing cultures. *Proc. R. Soc. London Ser.* B 163:224–231.

Pirt, S.J. (1982) Maintenance energy: A general model for energy-limited and energy-sufficient growth. *Arch. Microbiol.* 133:300–302.

Platpira, V. P., and Filmanovicha, R. S. (1993) Respiration rate of bacterioplankton in the Baltic Sea. *Hydrobiol. J.* 29:87–94.

Poindexter, J. S. (1987) Bacterial responses to nutrient limitation. In M. Fletcher, T. R. G. Gray, and J. G. Jones, eds., *Ecology of Microbial Communities.* Cambridge University Press, Cambridge, pp. 283–317.

Pomeroy, L. R., and Wiebe, W. J. (1993) Energy sources for microbial food webs. *Mar. Microb. Food Webs* 7:101–118.

Pomeroy, L. R., Hanson, R. B., McGillivary, P. A., Sherr, B. F., Kirchman, D. (1984) Microbiology and chemistry of fecal products of pelagic tunicates: Rates and fates. *Bull. Mar. Sci.* 35:426–439.

Pomeroy, L. R., Wiebe, W. J., Deibel, D., Thompson, R. J., and Rowe, G. T. (1991) Bacterial responses to temperature and substrate concentration during the Newfoundland spring bloom. *Mar. Ecol. Prog. Ser.* 75:143–159.

Pomeroy, L. R., Sheldon, J. E., Sheldon, W. M., and Peters. F. (1995) Limits to growth and respiration of bacterioplankton in the Gulf of Mexico. *Mar. Ecol. Prog. Ser.* 117:259–268.

Posch, T., Pernthaler, J., Alfreider, A., and Psenner, A. (1997) Cell-specific respiratory activity of aquatic bacteria studied with the tetrazolium reduction method, Cyto-clear slides, and image analysis. *Appl. Environ. Microbiol.* 63:867–873.

Robertson, M. L., Mills, A. L., and Zieman, J. C. (1982) Microbial synthesis of detritus-like particulates from dissolved organic carbon released by tropical sea-grasses. *Mar. Ecol. Prog. Ser.* 7:279–285.

Robinson, J. D., Mann, K. H., and Novitsky, J. A. (1982) Conversion of the particulate fraction of seaweed detritus to bacterial biomass. *Limnol. Oceanogr.* 27:1072–1079.

Roland, F. and Cole, J. J. Bacterial growth efficiency and its regulation in a tidal, freshwater river, the Hudson River Estuary. *Limnol. Oceanogr.* (in press).

Russell, J. B and Cook, G. M. (1995) Energetics of bacterial growth: Balance of anabolic and catabolic reactions. *Microbiol. Rev.* 59:48–62.

Sand-Jensen, K., Jensen, L. M., Marcher, S., and Hansen, M. (1990) Pelagic metabolism in eutrophic coastal waters during a late summer period. *Mar. Ecol. Prog. Ser.* 65:63–72.

Schwaerter, S., Søndergaard, M., Riemann, B., and Jensen, L. M. (1988) Respiration in eutrophic lakes: The contribution of bacterioplankton and bacterial growth yield. *J. Plankton Res.* 3:515–531.

Schweitzer, B., and Simon, M. (1995) Growth limitation of planktonic bacteria in a large mesotrophic lake. *Microb. Ecol.* 30:89–104.

Sepers, A. B. J. (1984) The uptake capacity for organic compounds of two heterotrophic bacterial strains at carbon-limited growth. *Z. Allg. Mikrobiol.* 24:261–267.

Sherr, E. B., and Sherr, B. F. (1996) Temporal offset in oceanic production and respiration process implied by seasonal changes in atmospheric oxygen: The role of heterotrophic microbes. *Aquat. Microb. Ecol.* 11:91–100.

Sherr, B. F., Sherr, E. B., and McDaniel, J. (1992) Effect of protistan grazing on the frequency of dividing cells in bacterioplankton assemblages. *Appl. Environ. Microbiol.* 58:2381–2385.

Sherr, B. F., del Giorgio, P. A., and Sherr, E. B. (1999) Estimating the abundance and single-cell characteristics of respiring bacteria via the redox dye, CTC. *Aquat. Microb. Ecol.* 18:117–131.

Søndergaard, M., and Theil-Nielsen, J. (1997) Bacterial growth efficiency in lakewater cultures. *Aquat. Microb. Ecol.* 12:115–122.

Sorokin, Y. I., Mameva T. I. (1980) Rate and efficiency of the utilization of labile organic matter by planktonic microflora in coastal Peruvian waters. *Pol. Arch. Hydrobiol.* 27:447–456

Stouthamer, A. H. (1973) A theoretical study on the amount of ATP required for synthesis of microbial cell material. *Antonie Van Leeuwenhoeck* 39:545–565.

Stuart, V., Newell, R. C., and Lucas, M. I. (1982) Conversion of kelp debris and faecal material from the mussel *Aulacomya ater* by marine micro-organisms. *Mar. Ecol. Prog. Ser.* 7:47–57.

Tempest, D. W., and Neijssel, O. M. (1978) Eco-physiological aspects of microbial growth in aerobic nutrient-limit environments. *Adv. Microb. Ecol.* 2:105–153.

Tempest, D. W, and Neijssel, O. M. (1992) Physiological and energetic aspects of bacterial metabolite overproduction. *FEMS Microbiol. Lett.* 100:169–176.

Teixeira de Mattos, M. J., and Neijssel, O. M. (1997) Bioenergetic consequences of microbial adaptation to low-nutrient environments. *J. Biotechnol.* 59:117–126.

Tortell, P. D., Maldonado, M. T., and Price, N. M. (1996) The role of heterotrophic bacteria in iron-limited ocean ecosystems. *Nature* 383:330–332.

Tranvik, L. J., (1988) Availability of dissolved organic carbon for planktonic bacteria in oligotrophic lakes of differing humic content. *Microb. Ecol.* 16:311–322.

Tranvik, L. J. (1990) Bacterioplankton growth on fractions of dissolved organic carbon of different molecular weights from humic and clear lakes. *Appl. Environ. Microbiol.* 56:1672–1677.

Tranvik, L. J. (1992) Rapid microbial production and degradation of humic-like substances in lake water. *Arch. Hydrobiol. Beih.* 37:43–50.

Tulonen, T., Salonen, K., and Arvola, L. (1992) Effects of different molecular weight fractions of dissolved organic matter on the growth of bacteria, algae and protozoa from a highly humic lake. *Hydrobiologia* 229:239–252.

Tupas, L., and Koike, I. (1990) Amino acid and ammonium utilization by heterotrophic marine bacteria grown in enriched seawater. *Limnol. Oceanogr.* 35:1145–1155.

Turley, C. M., and Lochte, K. (1990) Microbial response to the input of fresh detritus to the deep-sea bed. *Paleogeogr. Paleoclimatol. Paleoecol.* 89:3–23.

Vallino, J. J, Hopkinson, C. S., and Hobbie, J. E. (1996) Modeling bacterial utilization of dissolved organic matter: Optimization replaces Monod growth kinetics. *Limnol. Oceanogr.* 41:1591–1609.

Van der Kooij, D., Oranje, J. P., and Hijnen, W. A. M. (1982) Growth of *Pseudomonas aeruginosa* in tap water in relation to utilization of substrates at concentrations of few micrograms per liter. *Appl. Environ. Microbiol.* 44:1086–1095.

Vetter, Y. A., Deming, J. W., Jumars, P. A., and Krieger-Brockett, B. B. (1998) A predictive model of bacterial foraging by means of freely released extracellular enzymes. *Microb. Ecol.* 36:75–92.

Watson, T. G. (1970) Effects of sodium chloride on steady-state growth and metabolism of *Saccharomyces cerevisiae*. *J. Gen. Microbiol.* 64:91–99.

Weinbauer, M. G., and Höfle, M. G. (1998a) Size-specific mortality of lake bacterio-plankton by natural virus communities. *Aquat. Microb. Ecol.* 15:103–113.

Weinbauer, M. G., and Höfle, M. G. (1998b) Distribution and life strategies of two bacterial populations in a eutrophic lake. *Appl. Environ. Microbiol.* 64:3776–3783.

Westerhoff, H. V., Hellingwerf, K. J., and Van Dam, K. (1983) Thermodynamic efficiency of microbial growth is low but optimal for maximal growth rate. *Proc. Natl. Acad. Sci. USA* 80:305–309.

Wieser, W. (1994) Cost of growth in cells and organisms: General rules and comparative aspects. *Biol. Rev.* 68:1–33.

White, P. A, Kalff, J., Rasmussen, J. B., and Gasol, J. M. (1991) The effect of temperature and algal biomass on bacterial production and specific growth rate in freshwater and marine habitats. *Microb. Ecol.* 21:99–118.

Wright, R. T., and Hobbie, J. E. (1965) The uptake of organic solutes in lake water. *Limnol. Oceanogr.* 10:22–28.

Zweifel, U. L, Riemann, B., and Hagström, Å. (1993) Consumption of dissolved organic carbon by marine bacteria and demand for inorganic nutrients. *Mar. Ecol. Prog. Ser.* 101:23–32.

11

IMPACT OF VIRUSES ON BACTERIAL PROCESSES

Jed Fuhrman

Department of Biological Sciences,
University of Southern California,
Los Angeles, California

The microbial loop as a major part of marine ecosystems became a topic of wide general interest about 25 years ago. It started with the discovery of high bacterial abundance as learned by epifluorescence microscopy of stained cells, with counts typically 10^9 L^{-1} in the plankton (Francisco et al. 1973; Ferguson and Rublee 1976; Hobbie et al. 1977). With such high abundance, it became important to learn how fast they were dividing, and this was discovered by development and application of methods examining the frequency of dividing bacteria (Hagström et al., 1979) and bacterial DNA synthesis (Fuhrman and Azam 1980, 1982). These methods showed that bacterial doubling times were on the order of a day, and it quickly became apparent that bacteria were consuming a significant amount of dissolved organic matter, typically at a carbon uptake rate equivalent to about half the total primary production (Azam et al. 1983). Given that bacteria are too small to sink out of the water column and that their abundance stays relatively constant over the long term, there must be mechanisms within the water to remove bacteria at rates similar to the bacterial production rate. The original thinking on this considered protists to be the only "sink" for bacterial production. This conclusion followed from the observations that heterotrophic protists of the types known to

Microbial Ecology of the Oceans, Edited by David L. Kirchman.
ISBN 0-471-29993-6   Copyright © 2000 by Wiley-Liss, Inc.

consume bacteria are ubiquitous in seawater, that they are capable of growth at typical natural bacterial abundances, and that they are apparently capable of controlling bacterial abundances near natural levels (Azam et al. 1983).

While it is correct that grazing by protists has the potential for effective control of bacteria, this does not exclude the possibility that other agents are also involved in controlling bacterial processes. In the late 1980s, careful review of several studies showed that the best estimates of grazing by protists often fell short of balancing the best estimates of bacterial production, suggestive of additional loss processes (McManus and Fuhrman 1988). About that same time, evidence began to accumulate that viruses may also be important as a mechanism for removing bacteria. By now the evidence is fairly clear that this is so. This chapter briefly summarizes much of what we know about how viruses interact with marine bacteria, including general properties, abundance, distribution, infection of bacteria, mortality rate comparisons with protists, biogeochemical effects, effects on species compositions, and roles in genetic transfer and evolution.

WHAT ARE VIRUSES AND WHAT DO THEY DO?

Viruses are small particles, usually about 20–200 nm long, consisting of genetic material (DNA or RNA, single- or double-stranded) surrounded by a protein coat (some have lipid as well). They have no metabolism of their own and function only via the cellular machinery of a host organism. All cellular organisms appear to be susceptible to infection by some kind of virus. From culture studies, it has been learned that a given type of virus usually has a restricted host range, most often a single species or genus, although some viruses infect only certain subspecies, and a very few ($< 0.5\%$) may infect more than one genus (Ackermann and DuBow 1987). It has been suggested that some groups, such as cyanobacteria, may be infected by viruses that tend to have a wider host range (Safferman et al. 1983), but it is not clear whether this is due simply to differences in the taxonomy of the group or instead to some difference in the underlying biology. Viruses, which have no motility, contact the host cell by passive diffusion. They attach to the cell usually via some normal exposed cellular component, such as a transport protein or flagellum. There are three basic kinds of virus reproduction (Figure 1):

1. Lytic infection, where the virus attaches to a host cell, injects its nucleic acid, which causes the host to produce numerous progeny viruses, and then bursts the cell to release the progeny and begin the cycle again.

2. Chronic infection, where the progeny virus release from the host cell is nonlethal and occurs by extrusion or budding over several generations.

3. Lysogeny, where after injection, the viral genome becomes part of the genome of the host cell and reproduces as genetic material in the host cell line unless an "induction" event causes a switch to lytic infection. Induction is

VIRUS LIFE CYCLES

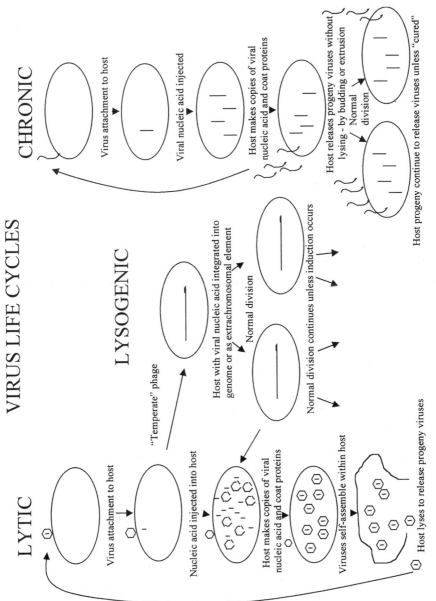

Figure 1. Virus life cycles.

329

commonly caused by DNA damage, such as from UV light or agents like mitomycin C. Viruses or virus like particles (VLP) may also be involved in killing cells by mechanisms that do not result in virus reproduction (Ackermann and DuBow 1987; also see Chiura, 1997).

DISTRIBUTION OF VIRUSES

Before one can determine how viruses are distributed, it is necessary to develop techniques to observe and enumerate them. Because of their small size, near or below the resolution limit of light microscopy (≈ 0.1 μm), the only way to observe any detail of viruses is to use electron microscopy. Sample preparation requires some means of concentrating the viruses from the water onto a flat electron microscopy grid (coated with a thin transparent organic film), and because viruses are denser than seawater, this can be done by ultracentrifugation, typically at forces of at least $100,000 \times g$ for a few hours (Sharp 1949; Bergh et al. 1989; Børsheim et al. 1990). Note that under ordinary gravity, forces like drag and Brownian motion prevent viruses from sinking.

To be observable, the viruses must be made electron-dense by staining with heavy metals like uranium salts. The viruses are recognized by their size, shape, and staining properties (usually electron-dense hexagons or ovals, sometimes with a tail; Figure 2), and counted. Typical counts are on the order of 10^{10} viruses per liter in surface waters, with abundance patterns similar to those of heterotrophic bacteria (see below). Recently it has been found that viruses can also be stained with nucleic acid stains like diamidino-2-phenylindole (DAPI), YoPro, or SYBR Green I, and observed and counted by epifluorescence microscopy (Hara et al. 1991; Hennes and Suttle 1995; Weinbauer and Suttle 1997; Noble and Fuhrman 1998b). Figure 3, an epifluorescence micrograph of SYBR Green–I stained bacteria and viruses, demonstrates this approach and illustrates the high relative virus abundance. Epifluorescence microscopy of viruses is possible even though the viruses are below the resolution limit of light because the stained viruses are a source of light and appear as bright spots against a dark background (just as stars are visible at night despite their small apparent size). Such epifluorescence counts are reported to be similar to or even higher than transmission electron microscopy (TEM) counts from seawater, especially at higher abundance levels (Weinbauer and Suttle 1997; Noble and Fuhrman 1998b); counts with SYBR Green tend to be about 30% higher than with TEM (Noble and Fuhrman 1998b).

Perhaps epifluorescence counts are higher because the TEM counts miss unexpectedly shaped viruses or ones obscured by other dark-stained material in the TEM preparations. Particular benefits of the epifluorescence methods are rapidity, ability to work in the field (e.g., onboard ship), and lower cost. Another important advantage is that with epifluorescence, larger sample sizes are much easier to prepare, thus affording increased statistical accuracy. However, such methods do not yield data on virus size or morphology, and

Figure 2. Electron micrograph of viruses and other microbes from Santa Monica Bay, California. The viruses are the small dark hexagonal or oval particles, some with tails. Scale bar = 1 μm.

there is a possibility that a small fraction of the objects counted as viruses are instead tiny nonviral particles containing condensed nucleic acids. The newest advance in the use of fluorescence for this work is the ability to count viruses stained with SYBR Green by flow cytometry (Marie et al. 1999), a method with great potential.

Electron or epifluorescence microscopy reveals the total (recognizable) virus community, and it is reasonable to ask what kinds of virus make up this community. It is commonly assumed that most of the total virus community is made up of bacteriophages. This assumption is made because viruses lack metabolism and have no means of actively swimming from host to host (and depend on random diffusion), and thus the most common viruses would be

Figure 3. Epifluorescence micrograph of prokaryotes and viruses stained with SYBR Green I. The viruses are the very numerous small dots, and the prokaryotes are the larger dots. (Method from Noble and Fuhrman 1998b). (A color version of this photomicrograph can be seen at http://www.wiley.com/products/subject/life/kirchman.)

expected to infect the most common organisms, and heterotrophic bacteria are generally the most abundant organisms in the plankton, often by far (Fuhrman et al. 1989). Data from field studies also show a robust and strong correlation between viral and bacterial abundances, while the correlations between viruses and chlorophyll are weaker and often not statistically significant (Boehme et al. 1993; Cochlan et al. 1993). This set of correlations has been cited as evidence for the majority of viruses being bacteriophages rather than infecting phytoplankton or other eukaryotes.

On the other hand, it is known that viruses infecting culturable cyanobacteria (*Synechococcus*) are also quite common and sometimes particularly abundant in seawater, exceeding 10^5 mL^{-1} in some cases (Suttle and Chan 1993, 1994; Waterbury and Valois 1993).

Similarly, viruses infecting some common culturable eukaryotic picoplankton, such as *Micromonas pusilla*, have been found to be sometimes quite abundant as well, occasionally near 10^5 mL^{-1} in coastal waters (Cottrell and Suttle 1995). These studies have been done with most probable number (MPN) methods, looking for agents from seawater that lyse pure cultures when tested at various dilutions. The MPN method uses several replicate serial dilutions to make a statistical estimate of the infectious viral abundance by seeing how much the original water can be diluted yet still cause infection.

Overall, the data suggest that most viruses from seawater infect non-photosynthetic bacteria, but viruses infecting prokaryotic and eukaryotic phytoplanton also occur regularly and can make up a significant fraction of the total. As a final note, we are only now learning about the distribution in seawater of Archaea, which may constitute a significant fraction of the total countable prokaryotes, especially in deeper waters (Fuhrman and Ouverney 1998) (see also Chapter 3). While most of this chapter discusses "bacteria," it must be considered that this practical, generic term includes the Archaea plus Bacteria for the purposes of general discussion. Thus, many of the viruses we observe may infect archaea microorganisms.

The distribution of virus abundance has so far been examined in many locations and habitats worldwide. Counts come from environments of all sorts —coastal, offshore, temperate, polar, tropical, and deep sea (Table 1,

Table 1. Typical counts of viruses from various marine planktonic environments; see also Figure 4

Location[a]	Viruses (10^8 L^{-1})	Method[b]	Ref.
North Atlantic, spring	150	TEM	Bergh et al. (1989)
Raunefjord (Norway)	100	TEM	Bergh et al. (1989)
Raunefjord, late winter	5	TEM	Bratbak et al. (1990)
Raunefjord, spring	20–100	TEM	Bratbak et al. (1990)
Southern California, nearshore	111–282	TEM	Cochlan et al. (1993)
Southern California, offshore	13–124	TEM	Cochlan et al. (1993)
Southern California offshore, 50 m	4–57	TEM	Cochlan et al. (1993)
Southern California offshore, 900 m	25	TEM	Cochlan et al. (1993)
Bering and Chukchi Seas	20–360	TEM	Steward et al. (1996)
Northern Adratic Sea	10–600	TEM	Weinbauer et al. (1995)
Gulf of Mexico, University of Texas pier	104	TEM	Weinbauer and Suttle (1997)
Gulf of Mexico, offshore	3–57	TEM	Weinbauer and Suttle (1997)
Gulf of Mexico, offshore	3–82	Yo-Pro	Weinbauer and Suttle (1997)
Southern California, 190 km offshore	135	TEM	Noble and Fuhrman (1998b)
Southern California, 190 km offshore	170	SYBR	Noble and Fuhrman (1998b)
Equatorial Pacific	53	FCM	Marie et al. (1999)
Mediterranean Sea	23	FCM	Marie et al. (1999)

[a]Near-surface and summer unless otherwise indicated.
[b]TEM is ultracentrifugation directly onto TEM grids for counting, without prior concentration steps. Yo-Pro and SYBR are stains used in epifluorescence direct counts of Anodisc-filtered samples. FCM is flow cytometry with SYBR Green stain.

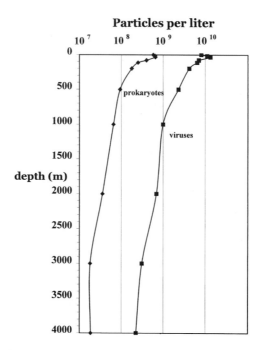

Figure 4. Depth profile of prokaryote (Bacteria + Archaea) and viral abundance from the Coral Sea (April 1998), as determined by epifluorescence microscopy of SYBR Green–stained samples; note the log scale. (Method from Noble and Fuhrman 1998b.)

Figure 4). Within the plankton, typical virus abundance is $1 - 5 \times 10^{10}\,\text{L}^{-1}$ in rich nearshore surface waters, dropping to about $0.1 - 1 \times 10^{10}\,\text{L}^{-1}$ in the euphotic zone of offshore low-nutrient areas, and also decreasing with depth, by about an order of magnitude. A typical deep-offshore profile is shown in Figure 4. Onshore–offshore gradients are also evident, similar to those of the bacteria (Boehme et al. 1993; Cochlan et al. 1993). As may be expected, seasonal changes are also common, with viruses following general changes in phytoplankton, bacteria, etc. (Bratbak et al. 1990) (e.g., see Raunefjord data in Table 1). One report indicated that virus abundance can sometimes be remarkably dynamic, changing drastically in time frames of minutes to hours (Bratbak et al. 1996), and this behavior has been interpreted as synchronized release from some hosts followed by rapid decay of many of the viruses. Given the relative youth of this part of the field, many parts of the world have not been investigated with regard to virus abundance, and the studies are patchy.

Counts are often also compared as ratios of virus to bacteria. Such ratios are typically 5–25, and commonly are close to 10, even as abundance drops to low levels in the deep sea. It is unknown what factors keep this ratio in such a relatively narrow range, but a link between these organisms, and also some reasonably tight regulatory mechanisms, are suggested. Correlations to chlorophyll and other parameters have been shown in some studies, apparently depending on the scale examined. Given that bacteria and chlorophyll are often correlated (Fuhrman et al. 1980), a strong possibility is that the relationship to chlorophyll is primarily a general relationship with the trophic status of the water rather than a specific relationship.

VIRAL ACTIVITIES

When the topic of viral activity is raised, the primary concern is usually lytic infection. While this is an important topic and serves as the main issue for the discussion that follows, it should also be realized that lysogeny is common. Evidence of lysogeny is usually from culture studies or direct induction with DNA-damaging agents (Jiang and Paul 1998b). Lysogens can readily be found and isolated from seawater, and lysogeny, which is linked to genetic transfer in a variety of bacteria, probably has important impacts in microbial population dynamics and evolution. However, the natural induction rate seems low under ordinary conditions, and lysogenic induction appears to be responsible for only a tiny fraction of total virus production in marine systems. Evidence for this conclusion comes from two kinds of study:

1. Growth of "seawater cultures" of natural marine bacteria inoculated into filtered seawater and grown under simulated in situ conditions shows rapid growth of bacteria but no appearance of viruses (as would be expected from lysogenic induction) even after several days, as long as the initial filtered seawater is cleared of viruses by filtration through a membrane with pore size of 0.02 μm. However, when the filtered seawater is prepared by filtration through a 0.2 μm filter and thus starts with many viruses, the virus abundance increases after a few days of bacterial growth, suggesting that infection from the viruses in the inoculum is occurring (Wilcox and Fuhrman 1994).

2. When lysogen abundance is estimated from artificial induction experiments (e.g., with mitomycin C), calculation of the maximum likely number of viruses released from lysogens under optimal conditions usually represents only a few percent of the total estimated virus production rate (Weinbauer and Suttle 1996; Jiang and Paul, 1998b).

Chronic infection in the ocean has not examined yet, but it is often presumed to be low. Release of filamentous (or other kinds of budding) viruses from native marine bacteria has not been noted in TEM studies, nor have significant numbers of free filamentous viruses been reported. However, these may be hard to recognize or differentiate from other filamentous objects (e.g., cilia, bits of cells), and the presence of filamentous or nondistinct viruses may partly explain why epifluorescence methods (looking for tiny particles containing densely packed nucleic acids) usually count more viruses than TEM methods. Therefore it may turn out that chronic infection is more common than currently thought.

One of the main issues regarding viral activity in seawater is the effect on bacterial mortality via lyic infection. Several recent studies on this topic tend to converge on the conclusion that viruses cause approximately 10–50% of total microbial mortality, depending on location, season, and so on. These estimates are fairly robust, having been determined several independent ways, as described in the subsections that follow.

Percentage of Infected Bacteria

Infection (i.e., assembled viruses within host cells) is visible by TEM only at the last step before lysis. Observations from marine waters ranging from the relatively rich coastal Long Island Sound to the oligotrophic Sargasso Sea showed that about 1–4% of the bacteria and cyanobacteria are visibly infected (Proctor and Fuhrman 1990), and subsequent measurements from other habitats have revealed similar results (Weinbauer and Peduzzi 1994). While this percentage sounds low, it represents only the final stage of infection, which apparently covers the last 10–20% of the infection cycle as observed with pure cultures (Proctor et al. 1993). The total infection rate has been calculated from the visibly infected fraction via a relatively simple model (Proctor and Fuhrman 1990; Proctor et al. 1993), and the final interpretation has been that the percentage of total mortality due to viruses is approximately 5–10 times the percent visibly infected, or about 5–40%.

Viral Decay

If virus production is stopped but viral decay continues, one may estimate the virus production rate that would be needed to maintain observed levels. Virus production is linked to mortality of hosts via lysis, and the burst size (number of viruses released per lysed host cell) is the conversion factor. Heldal and Bratbak (1991) used this concept by treating seawater with cyanide to stop production but allow virus destruction processes to proceed. Rate measurements from Norwegian coastal waters suggested virus turnover times on the order of a few hours. This method has sometimes implied rapid bacterial mortality considerably in excess of bacterial production (Bratbak et al. 1992), suggesting a substantial overestimate of mortality or underestimate of production. Decay has also been studied by adding marine viral cultures to seawater as tracers and observing the decline in infectivity as determined by plaque assays over time (Suttle and Chen 1992; Wommack et al. 1996; Noble and Fuhrman, 1997). However, recent work has shown that hosts may repair damaged viruses, especially in the light (Weinbauer et al. 1997), suggesting that some of the decay estimates may be too high and that a large fraction of the total viruses may be infective (Wilhelm et al. 1998b). A study that modeled the balance between decay, production, and repair of viruses concluded that about 40–80% of the sunlight-damaged viruses are repaired daily by hosts (Wilhelm et al. 1998a).

Viral DNA Synthesis

Steward et al. (1992b) adapted a method that had been used to measure bacterial production with tritiated thymidine and applied it to estimate viral production. The idea is that the appearance of nuclease-resistant label in viral size fraction after incubation with tritiated thymidine or $[^{33}P]O_4$ is indicative of viral DNA or RNA synthesis and can be used to calculate viral production. An empirically derived conversion factor is used to calculate the production rate. This approach has been applied in Southern California (Steward et al. 1992a; Fuhrman and Noble 1995) and also the Arctic. The results from these environments show a range of results, with viruses typically causing 5–50% of the total ascribed mortality of bacteria. Although some of the data may suggest higher percentages in richer coastal waters, there was no consistent pattern of the variation in this percentage with trophic status. The sensitivity level of this method was found to be most suitable for rich coastal environments.

Disappearance of Bacterial DNA in the Absence of Protists

An interesting approach to the investigation of bacterial mortality — namely, by means of measuring the decay of labeled DNA — was developed by Servais and colleagues (Servais et al. 1985, 1989). In this approach, cellular DNA in natural communities is pulse-labeled with $[^3H]$thymidine such that all the added tracer is taken up in a matter of hours. The subsequent decline of labeled DNA is thought to track bacterial mortality, on the presumption that DNA is not destroyed in healthy living cells. When protists are removed by size fractionation, the decline in labeled DNA has been considered to give an estimate of viral-caused mortality. Results from this approach in southern California coastal waters indicate that protist-free mortality is about half the total, and this has been interpreted as implying a significant viral impact (Fuhrman and Noble 1995).

Fluorescent Virus Tracers

Our next method is based on the same idea as so-called isotope dilution studies — namely, that one can measure both the production and the loss rate of a substance if some of the substance can be tagged and the tagged and untagged proportions can be monitored over time. In this case, fluorescently labeled viruses are made by concentrating native viruses from seawater, and staining with SYBR Green I. These are added back to seawater, which is incubated under simulated in situ conditions. Over time, the amount of added labeled viruses (no extra staining) and total viruses (stained just prior to counting with SYBR Green) are counted. Production of viruses adds unstained ones to the system, reducing the proportion of stained ones. However removal of viruses takes away both stained and unstained ones, a modification that should reduce the number of stained viruses but not change the relative proportions. By means of calculations analogous to those used for isotope dilution studies, one can use these results to calculate simultaneously the production and decay rates of viruses (Noble 1998). This method has measured virus turnover times in southern California nearshore and offshore waters of about 1–2 days, estimated to cause the majority of the total bacterial mortality (Noble 1998).

COMPARISON TO MORTALITY FROM PROTISTS

Given the earlier thinking that protists are the primary cause of bacterial mortality in marine planktonic systems, it is reasonable to ask how the impact of viruses on bacterial mortality compares to that of protists. A few studies have addressed this question directly, by different approaches. Weinbauer and Peduzzi (1995) used multiple correlation analysis of abundances of bacteria, viruses, and flagellates to conclude that virus-induced mortality of bacteria could occasionally prevail over flagellate grazing, especially at high bacterial abundances. However, a more direct approach is to use rate measurements. A few studies have compared virus-caused mortality with other causes directly and gone on to balance total mortality and loss rates with independent estimates of bacterial production. One study used three virus methods (frequency of infected bacteria, virus production estimated with tritiated thymidine, and size fractionated disappearance of labeled bacterial DNA) and two protist methods (removal of fluorescently labeled bacteria, size fractionated disappearance of labeled DNA) simultaneously with California coastal waters (Fuhrman and Noble 1995). The investigators found that the total mortality balanced production (thymidine and leucine incorporation methods) within 30% and that the other methods agreed remarkably well; it was concluded that viruses were responsible for about 40% of the total mortality.

Steward et al. (1996) examined viral processes by two methods: frequency of infected bacteria and viral incorporation of labeled phosphate, and separately estimated protist-caused mortality in the Bering and Chukchi Seas (Arctic). They found that viruses and protists were responsible for similar amounts of bacterial mortality, with protists dominating in some water samples and viruses in others. In the latter study, the total mortality estimates typically failed to balance production estimates, often by more than 50%. Those authors also concluded also that the viral effect is probably larger in eutrophic than in oligotrophic waters. Weinbauer and Höfle (1998) compared virus-caused and grazing mortality of bacteria in the epilimnion (aerobic), metalimnion (boundary), and hypolimnion (anaerobic) layers of Lake Plussee in Germany. They found that mortality from viruses was strongly dominant in the hypolimnion and metalimnion (where protists do poorly owing to oxygen deprivation) and that viruses were responsible for up to 30–50% of the mortality in the epilimnion, a result similar to that obtained for marine plankton.

Overall, the consensus is that viruses often are responsible for a significant fraction of bacterial mortality in marine planktonic systems, typically in the range of 10–40%. In some waters, viruses dominate bacterial mortality, while in others they have little impact on it. Some evidence suggests that the impact is usually higher in richer coastal systems than in relatively nutrient-poor ones. But there are not enough studies to back up broad generalizations or to address the reasons behind any patterns. An obvious factor would be the host abundance, since when hosts are rarer, the viruses are more likely to be inactivated before diffusing to a suitable host. Host abundance is a combination of total cell abundance (lower in oligotrophic systems) and species composition (see Chapter 3), and one must also factor in the appropriate host range for the viruses. Of these three, we really know only total bacterial abundance at this time. One may speculate that the bacteria are probably more diverse in oligotrophic systems than eutrophic systems (analogous to zooplankton and phytoplankton: Valiela 1984), and this system property, plus lower bacterial abundance, would tend to work against viruses in oligotrophic waters. Quantitative evidence on natural species diversity in most marine systems is not available, however, although it should be soon (see Chapter 3). There are also new methods to estimate natural virus diversity with pulsed field gel electrophoresis to show genome sizes (Steward and Azam 1999; Wommack et al. 1999), but this characteristic has not yet been compared between eutrophic and oligotrophic systems. Also, there is a possibility that unknown factors, such as broadened host range, may compensate for changes in host diversity.

ROLES IN FOOD WEB

Our ideas about the roles of microorganisms in marine food webs have been revised considerably following the initial discovery of high bacterial abundance and productivity, and it is now well established that a significant fraction of the total carbon and nutrient flux in marine systems passes through the heterotrophic bacteria via the dissolved organic matter (Azam et al. 1983; Fuhrman 1992). How do viruses fit into this picture? Given the accumulated data indicating that viruses can be major agents in the mortality of bacteria, one must ask how this mechanism may alter our view of matter and energy flux in the system. Focus has been given to three pertinent features of viruses: small size, composition, and mode of causing cell death, which is to release cell contents and progeny viruses to the surrounding seawater.

Viruses and the cellular debris produced when a host cell lyses consist of readily used protein and nucleic acid, plus all other cellular components, in a nonsinking form that is operationally defined as dissolved organic material. DOM is composed of dissolved molecules (monomers, oligomers, and polymers) plus colloids and cell fragments. Recent studies have shown that the marine DOM pool contains readily detectable remains of bacteria, such as membrane porins and peptidoglycan (Tanoue et al. 1995; McCarthy et al. 1998). While such fragments could conceivably be the result of grazing by protists, it seems probable that at least the more labile components (e.g., proteins) are free in the water as a result of viral lysis of bacteria because a protist may tend to digest them rather than release them.

What becomes of this material released by lysis? The most likely assumption about its fate is immediate or eventual availability to bacteria (Bratbak et al. 1990; Proctor and Fuhrman 1990; Fuhrman 1992). The release and availability of the lysis products to bacteria has been confirmed experimentally (Middelboe et al. 1996; Noble et al. 1998), although a few percent of the viruses may be grazed directly by heterotrophic flagellates (Gonzalez and Suttle 1993). Recent work indicates that the P in the released lysis products is selectively utilized much faster than the organic components, at least in the Mediterranean Sea, which is thought to be P-limited (Noble and Fuhrman 1998a, in press). A detailed study of the viral lysis of the "brown tide" alga, *Aureococcus anophagefferens*, has shown that lysis leads to extensive release of dissolved compounds (organic and inorganic), although much was initially in particulate form, and that most of this material is immediately or eventually available to bacteria and algae for growth (Gobler et al. 1997). Similarly, Bratbak et al. (1998) found that infection of another common bloom-forming alga, *Phaeocystis pouchetii*, effectively converts the entire cell biomass into dissolved organic carbon. If the cell lysed is a bacterium, then uptake by other bacteria represents a semiclosed trophic loop, whereby bacterial biomass is consumed mostly by other bacteria (see also Chapter 5). Because of respiratory losses and inorganic nutrient regeneration associated with utilization of dissolved organic substances, this loop has the net effect of oxidizing organic matter and regenerating inorganic nutrients (Figure 5) (Bratbak et al. 1990; Proctor and Fuhrman 1990; Fuhrman 1992). This bacterial–viral loop essentially "robs" production from protists that would otherwise consume the bacteria (McManus and Fuhrman 1988) and sequesters the biomass and activity into the dissolved and smallest particulate forms.

The net effect has been illustrated by a model showing that a food web with 50% bacterial mortality from viruses has 27% more bacterial respiration and production, and 37% less bacterial grazing by protists, culminating in a 7% reduction in macrozooplankton production compared with the same system with no viruses (Fuhrman 1992). That original steady state model had only bacteria being infected and all the viral matter being consumed by bacteria (Fuhrman 1992). A modification of that model, now including a small amount of viral infection of phytoplankton (7% loss) and also flagellate grazing of 3% of the virus production, has essentially the same net effect of increasing bacterial production and respiration (by 33%) and reducing protist and animal production (Fuhrman 1999).

Sequestration of materials in viruses, bacteria, and dissolved matter leads to better retention of nutrients in the euphotic zone in virus-infected systems

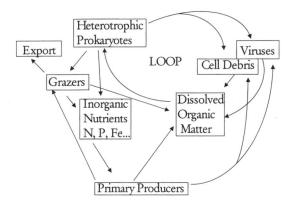

Figure 5. The bacterial–viral loop within the microbial food web; arrows represent transfer of matter.

because more material remains in these small nonsinking forms (Fuhrman 1992). In contrast, reduced viral activity leads to more material in larger organisms that sink, either themselves or as detritus, transporting carbon and inorganic nutrients to depth. This effect may be particularly important for potentially limiting nutrients (e.g., N, P, Fe) that are relatively concentrated in bacteria in comparison to eukaryotes (Kirchman 1996; see also Chapter 9). Thus, viral activity has the potential effect of helping to maintain higher levels of biomass and productivity in the system as a whole.

Lysis of organisms and release of their cell contents to the water have other potential geochemical effects because of the chemical and physical nature of the released materials and the location in the water column of the lysis itself. For example, polymers released from lysed cells contribute to small-scale viscosity of seawater that influences many biological and microscale physical–chemical processes and may facilitate aggregation and sinking of material from the euphotic zone (Proctor and Fuhrman 1991; Peduzzi and Weinbauer 1993; Azam 1998). On the other hand, viral lysis of microorganisms within sinking aggregates may effectively "dissolve" the particles, converting some sinking particulate matter into nonsinking dissolved material and colloids at whatever depth the lysis occurs (Proctor and Fuhrman 1991). This in turn contributes to the dissolution of sinking organic matter and its availability to free-living bacteria in the ocean's interior, as discussed by Cho and Azam (1988). Another important point is that viral lysis tends to cause the release of organic matter from cells, whereas grazing leads to more immediate release of inorganic matter.

The roles of viruses in producing long-lived DOM is not currently known. DOM with this property probably would come mostly from cell lysis products, since viruses themselves are protein and nucleic acid, which are thought to be relatively labile, while bacterial cell walls and some other components are probably more recalcitrant to degradation. The results of Gobler et al. (1997) indicate that some of the organic lysis products from infected algae are relatively refractory. If a large fraction of the lysis products turn out to be relatively nonlabile materials, quantitative aspects of conclusions above would have to be altered, but most of the qualitative aspects would be the same. A change may be that any long-lived DOM could remain in the system for a long time or be transported to deep or distant locations, to eventually fuel bacterial production.

EFFECTS ON SPECIES COMPOSITIONS

Viruses are density-dependent and generally species- or genus-specific. This combination means that dominant bacteria are most susceptible to infection, and rare ones least so. Lytic viruses can get ahead only when their average time to diffuse from host to host is shorter than the average time a burst remains infectious. Thus, when a species or strain becomes more densely populated, it is more susceptible to infection; conversely, when it is sparsely distributed it is much less susceptible. This makes viral infection work in opposition to competitive dominance (Fuhrman and Suttle 1993).

This relationship may have direct relevance to solving Hutchinson's "paradox of plankton," which asks how so many different kinds of phytoplankton can coexist on only a few potentially limiting resources, when competition theory predicts one or a few competitive winners (Hutchinson 1961). While there have been several possible explanations for this paradox (Siegel 1998; Tilman 1999), viral activity may also help solve it, because as stated earlier, competitive dominant organisms become particularly susceptible to infection, while rare species are relatively protected. It may be expected that the same principle applies to bacteria in addition to phytoplankton, although we know much less about bacterial diversity.

There is relatively little experimental evidence regarding viral control of species compositions. This is hardly surprising, given that we are now only starting to understand the species distributions and dynamics of marine bacteria (see Chapter 3). However, there are some data showing shifts in overall bacterioplankton community composition on scales of weeks to months (Rehnstam et al. 1993; Pinhassi et al. 1997). These reports indicate that the dominant members of the microbial community shift significantly over time. However, the cause of those shifts is not known. One report did correlate community composition with viruses. Waterbury and Valois (1993) examined over several seasons the relative abundance in temperate Atlantic waters (near Woods Hole, Massachusetts, and also offshore) of various culturable strains of *Synechococcus* cyanobacteria and viruses that infect them. They reported that although only a small percentage of *Synechococcus* were lysed daily (from < 1 to about 3% per day according to their calculations), the culturable cyanobacteria tended to be resistant to the co-occurring phage. They concluded that the cyanobacteria affected the species or strain composition of the community more than the total abundance.

Thingstad and Lignell (1997) have modeled the potential factors controlling aquatic microbial systems, regarding biomass and species compositions. Their models include growth limitation by organic carbon, inorganic phosphate, or nitrogen (inorganic or organic); cell losses include grazing by protists and viral lysis. Even when bacterial abundance is assumed to be controlled by protist grazing, the models have the robust result of showing that viruses control the steady state diversity of the bacterial community, whether bacterial growth rate limitation is by organic or inorganic nutrients. Thus, both empirical and theoretical analyses indicate a major role of viruses in regulating patterns of diversity.

RESISTANCE

What about development of resistance by the hosts? Such resistance, whereby bacteria mutate to resist the viral attack, is well known from nonmarine experiments, which usually involve highly simplified laboratory systems (Lenski 1988). In the natural marine world, this phenomenon was cited to explain the results of Waterbury and Valois (1993), discussed earlier.

Along the same lines, Olofsson and Kjellenberg (1991) have suggested that because of the development of resistance, significant mortality from marine viruses would be expected as a transient effect only as new virulent virus strains emerge. Suttle et al. (1991) responded that in dynamic, diverse, and sparse natural populations, other factors may exert more control over strain compositions than resistance. It makes sense that natural systems with many species and various trophic levels have far more interactions than the simple laboratory systems. Obviously, a species with a large fraction of mortality for one type of virus benefits from developing resistance, and this must occur in nature.

But resistance is not always a good thing. First, it often confers some competitive disadvantage via loss of some important receptor (Ackermann and DuBow 1987; Lenski 1988). Even complete resistance to viral attachment, without any receptor loss, if that were possible, would not necessarily be an advantage. For a bacterium in an oligotrophic environment whose growth may be limited by N, P, or organic carbon, unsuccessful infection by a virus (e.g., because the virus was stopped intracellularly by a restriction enzyme or encountered a genetic incompatibility) may be a significant nutritional benefit to the host organism, since the virus injection of DNA is a nutritious boost rich in C, N, and P (Proctor and Fuhrman 1990)! Even the protein coat, remaining outside the cell, is probably digestible by cell-surface-associated proteases (Hollibaugh and Azam 1983). In this situation, one may even imagine bacteria using "decoy" virus receptors to lure viral strains that cannot successfully infect them. Given favorable distributions of bacteria and viruses of various types, the odds could be tilted toward the bacteria; and if an infectious virus (i.e., one with a protected restriction site) occasionally gets through, the cell line as a whole may still benefit from this strategy.

Another consideration that works against resistance relates back to the system model results, which show that the heterotrophic bacteria as a group benefit substantially from viral infection, boosting their production significantly by essentially taking carbon and energy away from larger organisms. Recall also that viruses boost the entire system biomass and production by helping to maintain nutrients in the lighted surface waters. However, these arguments require us to invoke some sort of group selection theory to explain how individuals could benefit from not developing resistance (i.e., why not "cheat" by developing resistance and letting all the other organisms give the group benefits of infection?). Nevertheless, for whatever reasons, resistance of native communities to viral infection may be common but cannot be close to complete, given the continued ubiquitous existence of viruses roughly 10 times as abundant as bacteria and with turnover times on the order of a day. Simple mass balance calculations, typically reported in several papers on viral decay cited earlier, show that significant numbers of hosts must be infected and constantly releasing viruses. For example, with a typical lytic burst size of 50 and viral turnover time of one day, maintenance of a 10-fold excess of viruses over bacteria requires that 20% of the bacteria lyse daily. We have yet to learn whether lack of comprehensive resistance is due to frequent development of new virulent strains, rapid dynamics or patchiness in species compositions, or to a stable coexistence of viruses and their hosts. All these are possible, and they are not mutually exclusive.

GENETIC TRANSFER

Viruses also can be agents of genetic transfer between microorganisms, through two processes. In the more direct process, known as transduction, viruses package some of the host's own DNA into the phage head (sometimes with active phage genes and sometimes without, depending on the virus type) and then inject it into another potential host. Transduction in aquatic environments is frequent enough to measure and has been demonstrated (Saye et al. 1990; Jiang and Paul 1998a). Although transduction usually occurs within a restricted host range, a recent report by Chiura (1997) indicates that some marine bacteria and phage are capable of transfer across a wide host range. A second mechanism whereby viruses mediate genetic transfer is by causing the release of DNA from lysed host cells that may be taken up and used as genetic material by another microorganism. This process is called transformation.

While the extent of the foregoing mechanisms in natural systems is unknown at this time, they could have important roles in population genetics (by homogenization of genes within a potential host population) and also on evolution at relatively long time scales. Horizontal gene transfer is an integral component of microbial evolution, and the genomes of modern-day microbes contain numerous genes that have obviously been transferred from other species (Stephens et al. 1998). On shorter time scales, this process is of interest in studies of the possible dissemination of genes that may code for novel properties, whether introduced to native communities naturally or via genetic engineering.

SUMMARY

1. Viruses, which are obligate parasites of cellular organisms that are usually specific for certain hosts, are very abundant in marine plankton, typically 10^7 mL^{-1}, or about 10 times the abundance of marine bacteria.
2. Viruses infect bacteria, cyanobacteria, and protists, although it appears that most of the viruses present in seawater infect bacteria.
3. Several lines of evidence suggest that viruses are responsible for about 10–40% of the total bacterial mortality, and sometimes most of bacterial mortality is due to viral lysis.
4. The release of DOM during lysis of microbes is thought to stimulate bacterial activity at the expense of larger organisms and also to lead to increased retention of nutrients in the euphotic zone.
5. Other important roles of viruses include influence on species compositions and genetic transfer.

ACKNOWLEDGMENTS

I thank colleagues who have worked with me and discussed various aspects of virus research, including Lita Proctor, Rachel Noble, Robin Wilcox, John Griffith, Curtis Suttle, Farooq Azam, Frede Thingstad, Gunnar Bratbak, John Paul, Åke Hagström, and Doug Capone. David Kirchman and an anonymous reviewer provided useful comments on the manuscript. This work was supported by National Science Foundation grant OCE 9634028 and OCE 9906989.

REFERENCES

Ackermann, H.-W. and DuBow, M. S. (1987) *Viruses of prokaryotes*, Vol. 1: *General Properties of Bacteriophages*. CRC Press, Boca Raton, FL.

Azam, F. (1998) Microbial control of oceanic carbon flux: The plot thickens. *Science* 280(5364):694–696.

Azam, F., Fenchel, T., Gray, J. G., Meyer-Reil, L. A., and Thingstad, T. (1983) The ecological role of water-column microbes in the sea. *Mar. Ecol. Prog. Ser.* 10:257–263.

Bergh, O., Børsheim, K. Y., Bratbak, G., and Heldal, M. (1989) High abundance of viruses found in aquatic environments. *Nature*, 340:467–468.

Boehme, J., et al. (1993) Viruses, bacterioplankton, and phytoplankton in the southeastern Gulf of Mexico: Distribution and contribution to oceanic DNA pools. *Mar. Ecol. Prog. Ser.* 97:1–10.

Børsheim, K. Y., Bratbak, G., and Heldal, M. (1990) Enumeration and biomass estimation of planktonic bacteria and viruses by transmission electron microscopy. *Appl. Environ. Microbiol.* 56:352–356.

Bratbak, G., Heldal, M., Norland, S., and Thingstad, T. F. (1990) Viruses as partners in spring bloom microbial trophodynamics. *Appl. Environ. Microbiol.* 56:1400–1405.

Bratbak, G., Heldal, M., Thingstad, T. F., Riemann, B., and Haslund, O. H. (1992) Incorporation of viruses into the budget of microbial C-transfer. A first approach. *Mar. Ecol. Prog. Ser.* 83:273–280.

Bratbak, G., Heldal, M., Thingstad, T. F., and Tuomi, P. (1996) Dynamics of virus abundance in coastal seawater. *FEMS Microb. Ecol.* 19(4):263–269.

Bratbak, G., Jacobsen, A., and Heldal, M. (1998) Viral lysis of *Phaeocystis pouchetti* and bacterial secondary production. *Aquat. Microb. Ecol.* 16:11–16.

Chiura, H. X. (1997) Generalized gene transfer by virus-like particles from marine bacteria. *Aquat. Microb. Ecol.* 13(1):75–83.

Cho, B. C., and Azam, F. (1988) Major role of bacteria in biochemical fluxes in the ocean's interior. *Nature* 332:441–443.

Cochlan, W. P., Wikner, J., Steward, G. F., Smith, D. C., and Azam, F. (1993) Spatial distribution of viruses, bacteria and chlorophyll *a* in neritic, oceanic and estuarine environments. *Mar. Ecol. Prog. Ser.* 92:77–87.

Cottrell, M. T., and Suttle, C. A. (1995) Dynamics of a lytic virus infecting the photosynthetic marine picoflagellate *Micromonas pusilla*. *Limnol. Oceanogr.* 40(4):730–739.

Ferguson, R. L., and Rublee, P. (1976) Contribution of bacteria to standing crop of coastal plankton. *Limnol. Oceanogr.* 21:141–145.

Francisco, D. E., Mah, R. A., and Rabin, A. C. (1973) Acridine orange epifluorescence technique for counting bacteria in natural waters. *Trans. Am. Microsc. Soc.* 92:416–421.

Fuhrman, J. A. (1992) Bacterioplankton roles in cycling of organic matter: The microbial food web. In P. G. Falkowski and A. D. Woodhead eds., *Primary Productivity and Biogeochemical Cycles in the Sea*. Plenum Press, New York, pp. 361–383.

Fuhrman, J. A., (1999) Marine viruses and their biogeochemical and ecological effects. *Nature* 399:541–548.

Fuhrman, J. A., and Azam, F. (1980) Bacterioplankton secondary production estimates for coastal waters of British Columbia, Antarctica, and California. *Appl. Environ. Microbiol.* 39:1085–1095.

Fuhrman, J. A., and Azam, F. (1982) Thymidine incorporation as a measure of heterotrophic bacterioplankton production in marine surface waters: Evaluation and field results. *Mar. Biol.* 66:109–120.

Fuhrman, J. A., and Noble, R. T. (1995) Viruses and protists cause similar bacterial mortality in coastal seawater. *Limnol. Oceanogr.* 40(7):1236–1242.

Fuhrman, J. A., and Ouverney, C. C. (1998) Marine microbial diversity studied via 16S rRNA sequences: Cloning results from coastal waters and counting of native archaea with fluorescent single cell probes. *Aquat. Ecol.* 32:3–15.

Fuhrman, J. A., Ammerman, J. W., and Azam, F. (1980) Bacterioplankton in the coastal euphotic zone: Distribution, activity, and possible relationships with phytoplankton. *Mar. Biol.* 60:201–207.

Fuhrman, J. A., and Suttle, C. A. (1993) Viruses in marine planktonic systems. *Oceanography* 6:51–63.

Fuhrman, J. A., Sleeter, T. D., Carlson, C. A. and Proctor, L. M. (1989) Dominance of bacterial biomass in the Sargasso Sea and its ecological implications. *Mar. Ecol. Prog. Ser.* 57:207–217.

Gobler, C. J., Hutchins, D. A., Fisher, N. S., Cosper, E. M., and SanudoWilhelmy, S. A. (1997) Release and bioavailability of C, N, P, Se, and Fe following viral lysis of a marine chrysophyte. *Limnol. Oceanogr.* 42(7):1492–1504.

Gonzalez, J. M., and Suttle, C. A. (1993) Grazing by marine nanoflagellates on viruses and virus-sized particles — ingestion and digestion. *Mar. Ecol. Prog. Ser.* 94(1):1–10.

Hagström, Å., Larsson, U., Horstedt, P., and Normark, S. (1979) Frequency of dividing cells, a new approach to the determination of bacterial growth rates in aquatic environments. *Appl. Environ. Microbiol.* 37:805–812.

Hara, S., Terauchi, K., and Koike, I. (1991) Abundance of viruses in marine waters: Assessment by epifluorescence and transmission electron microscopy. *Appl. Environ. Microbiol.* 57(9):2731–2734.

Heldal, M., and Bratbak, G. (1991) Production and decay of viruses in aquatic environments. *Mar. Ecol. Prog. Ser.* 72:205–212.

Hennes, K. P., and Suttle, C. A. (1995) Direct counts of viruses in natural waters and laboratory cultures by epifluorescence microscopy. *Limnol. Oceanogr.* 40(6):1050–1055.

Hobbie, J. E., Daley, R. J., and Jasper, S. (1977) Use of Nuclepore filters for counting bacteria by fluorescence microscopy. *Appl. Environ. Microbiol.* 33:1225–1228.

Hollibaugh, J. T., and Azam, F. (1983) Microbial degradation of dissolved proteins in seawater. *Limnol. Oceanogr.* 28(6):1104–1116.

Hutchinson, G. E. (1961) The paradox of the plankton. *Am. Nat.* 45:137–145.

Jiang, S. C., and Paul, J. H. (1998a) Gene transfer by transduction in the marine environment. *Appl. Environ. Microbiol.* 64(8):2780–2787.

Jiang, S. C., and Paul, J. H. (1998b) Significance of lysogeny in the marine environment — Studies with isolates and a model of lysogenic phage production. *Microb. Ecol.* 35(3):235–243.

Kirchman, D. L. (1996) Oceanography — Microbial ferrous wheel. *Nature* 383(6598):303–304.

Lenski, R. E. (1988) Dynamics of interactions between bacteria and virulent bacteriophage. *Adv. Microb. Ecol.* 10:1–44.

Marie, D., Brussaard, C. P. D., Thyrhaug, R., Bratbak, G., and Vaulot, D. (1999) Enumeration of marine viruses in culture and natural samples by flow cytometry. *Appl. Environ. Microbiol.* 65(1):45–52.

McCarthy, M. D., Hedges, J. I., and Benner, R. (1998) Major bacterial contribution to marine dissolved organic nitrogen. *Science* 281(5374):231–234.

McManus, G. B., and Fuhrman, J. A. (1988) Control of marine bacterioplankton populations: Measurement and significance of grazing. *Hydrobiologia* 159:51–62.

Middelboe, M., Jørgensen, N. O. G., and Kroer, N. (1996) Effects of viruses on nutrient turnover and growth efficiency of non-infected marine bacterioplankton. *Appl. Environ. Microbiol.* 62:1991–1997.

Noble, R. T. (1998) The fates of viruses in the marine environment. Ph. D. thesis, University of Southern California, Los Angeles.

Noble, R. T. and Fuhrman, J. A. (1997) Virus decay and its causes in coastal waters. *Appl. Environ. Microbiol.* 63(1):77–83.

Noble, R. T., and Fuhrman, J. A. (1998a) Estimates of virus production and removal rates: Implications to bacterial mortality in the marine environment. *Eighth International Symposium on Microbial Ecology*, Halifax, Nova Scotia, Canada, Abstract.

Noble, R. T., and Fuhrman, J. A. (1998b) Use of SYBR Green I for rapid epifluorescence counts of marine viruses and bacteria. *Aquat. Microb. Ecol.* 14(2):113–118.

Noble, R. T., and J. A. Fuhrman. Breakdown and microbial uptake of marine viruses and other lysis products. *Aquat. Microb. Ecol.* (in press).

Noble, R. T., Middelboe, M., and Fuhrman, J. A. (1999) The effects of viral enrichment on the mortality and growth of heterotrophic bacterioplankton. *Aquat. Microb. Ecol.* 18:1–13.

Olofsson, S., and Kjelleberg, S. (1991) Virus ecology. *Nature* 351(6328):612–612.

Peduzzi, P., and Weinbauer, M. G. (1993) Effect of concentrating the virus-rich 2–200 nm size fraction of seawater on the formation of algal flocs (marine snow). *Limnol. Oceanogr.* 38:1562–1565.

Pinhassi, J., Zweifel, U., and Hagström, Å. (1997) Dominant marine bacterioplankton species found among colony-forming bacteria. *Appl. Environ. Microbiol.* 63(9):3359–3366.

Proctor, L. M., and Fuhrman, J. A. (1990) Viral mortality of marine bacteria and cyanobacteria. *Nature* 343:60–62.

Proctor, L. M., and Fuhrman, J. A. (1991) Roles of viral infection in organic particle flux. *Mar. Ecol. Prog. Ser.* 69:133–142.

Proctor, L. M., Okubo, A., and Fuhrman, J. A. (1993) Calibrating estimates of phage-induced mortality in marine bacteria:Ultrastructural studies of marine bacteriophage development from one-step growth experiments. *Microb. Ecol.* 25:161–182.

Rehnstam, A. S., Backman, S., Smith, D. C., Azam, F., and Hagström, Å. (1993) Bloom of sequence-specific culturable bacteria in the sea. *FEMS Microbiol. Ecol.* 102:161–166.

Safferman, R. S. et al. (1983) Classification and nomenclature of viruses of cyanobacteria. *Intervirology* 19:61–66.

Saye, D. J., Ogunsteitan, O. A., Slayer, G. S., and Miller, R. V. (1990) Transduction of linked chromosomal genes between *Pseudomonas aeruginosa. Appl. Environ. Microbiol.* 56:140–145.

Servais, P., Billen, G., and Rego, J. V. (1985) Rate of bacterial mortality in aquatic environments. *Appl. Environ. Microbiol.* 49:1448–1454.

Servais, P., Billen, G., Martinez, J., and Vives-Rego, J. (1989) Estimating bacterial mortality by the disappearance of ^{3}H-labeled intracellular DNA. *FEMS Microbiol. Ecol.* 62:119–126.

Sharp, G. D. (1949) Enumeration of virus particles by electron micrography. *Proc. Soc. Exp. Biol. Med.* 70:54–59.

Siegel, D. A. (1998) Resource competition in a discrete environment: Why are plankton distributions paradoxical? *Limnol. Oceanogr.* 43(6):1133–1146.

Stephens, R. S., et al. (1998) Genome sequence of an obligate intracellular pathogen of humans: *Chlamydia trachomatis. Science* 282(5389):754–759.

Steward, G. F., and Azam, F. (1999) Analysis of marine viral assemblages. In P. Johnson-Green, ed. *Proceedings of the Eighth International Symposium on Microbial Ecology*, Halifax, Nova Scotia, Canada.

Steward, G. F., Wikner, J., Cochlan, W. P., Smith, D. C., and Azam, F. (1992a) Estimation of virus production in the sea. II. Field results. *Mar. Microb. Food Webs.* 6(2):79–90.

Steward, G. F., Wikner, J., Smith, D. C., Cochlan, W. P., and Azam, F. (1992b) Estimation of virus production in the sea. I. Method development. *Mar. Microb. Food Webs* 6(2):57–78.

Steward, G. F., Smith, D. C., and Azam, F. (1996) Abundance and production of bacteria and viruses in the Bering and Chukchi Seas. *Mar. Ecol. Prog. Ser.* 131:287–300.

Suttle, C. A., and Chan, A. M. (1993) Marine cyanophages infecting oceanic and coastal strains of *Synechococcus:* Abundance, morphology, cross-infectivity, and growth characteristics. *Mar. Ecol. Prog. Ser.* 92:99–109.

Suttle, C. A., and Chan, A. M. (1994) Dynamics and distribution of cyanophages and their effect on marine *Synechococcus* spp. *Appl. Environ. Microbiol.* 60(9):3167–3174.

Suttle, C. A., Chan, A. M., and Cottrell, M. T. (1991) Virus ecology — Reply. *Nature* 351(6328):612–613.

Suttle, C. A., and Chen, F. (1992) Mechanisms and rates of decay of marine viruses in seawater. *Appl. Environ. Microbiol.* 58:3721–3729.

Tanoue, E., Nishiyama, S., Kamo, M., and Tsugita, A. (1995) Bacterial membranes — Possible source of a major dissolved protein in seawater. *Geochim. Cosmochim. Acta* 59(12):2643–2648.

Thingstad, T. F., and Lignell, R. (1997) Theoretical models for the control of bacterial growth rate, abundance, diversity and carbon demand. *Aquat. Microb. Ecol.* 13(1):19–27.

Tilman, D. (1999) Ecology — Diversity by default. *Science* 283(5401):495–496.

Valiela, I. (1984) *Marine Ecological Processes.* Springer-Verlag, New York.

Waterbury, J. B., and Valois, F. W. (1993) Resistance to co-occurring phages enables marine *Synechococcus* communities to coexist with cyanophages abundant in seawater. *Appl. Environ. Microbiol.* 59(10):3393–3399.

Weinbauer, M. G., and Höfle, M. G. (1998) Significance of viral lysis and flagellate grazing as factors controlling bacterioplankton production in a eutrophic lake. *Appl. Environ. Microbiol.* 64(2):431–438.

Weinbauer, M. G., and Peduzzi, P. (1994) Frequency, size, and distribution of bacteriophages in different marine bacterial morphotypes. *Mar. Ecol. Prog. Ser.* 108:11–20.

Weinbauer, M. G., and Peduzzi, P. (1995) Significance of viruses versus heterotrophic nanoflagellates for controlling bacterial abundance in the northern Adriatic Sea. *J. Plankton Res.* 17(9):1851–1856.

Weinbauer, M. G., and Suttle, C. A. (1996) Potential significance of lysogeny to bacteriophage production and bacterial mortality in coastal waters of the Gulf-of-Mexico. *Appl. Environ. Microbiol.* 62(12):4374–4380.

Weinbauer, M. G., and Suttle, C. A. (1997) Comparison of epifluorescence and transmission electron microscopy for counting viruses in natural marine waters. *Aquat. Microb. Ecol.* 13(3):225–232.

Weinbauer, M. G., Fuks, D., Puskaric, S., and Peduzzi, P. (1995) Diel, seasonal, and depth-related variability of viruses and dissolved DNA in the Northern Adriatic Sea. *Microb. Ecol.* 30:25–41.

Weinbauer, M. G., Wilhelm, S. W., Suttle, C. A., and Garza, D. R. (1997) Photoreactivation compensates for UV damage and restores infectivity to natural marine virus communities. *Appl. Environ. Microbiol* 63(6):2200–2205.

Wilcox, R. M., and Fuhrman, J. A. (1994) Bacterial viruses in coastal seawater: Lytic rather than lysogenic production. *Mar. Ecol. Prog. Ser.* 114:35–45.

Wilhelm, S. W., Weinbauer, M. G., Suttle, C. A., and Jeffrey, W. H. (1998a) The role of sunlight in the removal and repair of viruses in the sea. *Limnol. Oceanogr.* 43(4):586–592.

Wilhelm, S. W., Weinbauer, M. G., Suttle, C. A., Pledger, R. J., and Mitchell, D. L. (1998b) Measurements of DNA damage and photoreactivation imply that most viruses in marine surface waters are infective. *Aquat. Microb. Ecol.* 14(3):215–222.

Wommack, K. E., Hill, R. T., Muller, T. A., and Colwell, R. R. (1996) Effects of sunlight on bacteriophage viability and structure. *Appl. Environ. Microbiol.* 62(4):1336–1341.

Wommack, K. E., Ravel, J., Hill, R. T., Chun, J. S., and Colwell, R. R. (1999) Population dynamics of Chesapeake bay virioplankton: Total-community analysis by pulsed-field gel electrophoresis. *Appl. Environ. Microbiol.* 65(1):231–240.

12

BACTERIVORY: INTERACTIONS BETWEEN BACTERIA AND THEIR GRAZERS

Suzanne L. Strom

Shannon Point Marine Center,
Western Washington University,
Anacortes, Washington

Bacterivory is the process of consuming bacteria. Organisms that consume bacteria, particularly if they obtain a substantial portion of their nutrition this way, are termed bacterivores (or, equivalently, bactivores). This chapter focuses on bacterivory in the water column of oceans, although bacterivory in marine sediments is an equally fascinating and even more intractable issue. For planktonic bacterivores, the primary problem is that of obtaining minute particles from a dilute suspension in quantity sufficient to sustain life. For many years it was believed that oceanic bacterial concentrations were, except in unusual cases, too low to permit growth or even survival of bacterivores, and the relative constancy of planktonic bacterial biomass was due to sluggish, near-zero bacterial growth rates. The development of methods such as tritiated thymidine incorporation for the measurement of bacterial production, however, changed this concept radically (Chapter 4). As explicated in a groundbreaking paper by Pomeroy (1974), high bacterial growth rates and near-constant bacterial stocks mean that removal processes must be important in regulating stock size. Free-living bacteria are too small to be strongly affected by physical removal processes. Thus the top candidates for regulation of bacterial biomass are biological controls, including bacterivory and viral lysis (Chapter 11).

Microbial Ecology of the Oceans, Edited by David L. Kirchman.
ISBN 0-471-29993-6   Copyright © 2000 by Wiley-Liss, Inc.

Spurred by new observations of high bacterial production rates, ocean-ographers began to search for candidate bacterivores, as well as for methods to measure rates of bacterivory. Although early investigators such as Brandt (1901, cited in Mills 1989) and Johannes (1965) realized that cycles of production and recapture of dissolved substances must be important in marine waters, such cycles were not widely recognized until the early 1980s. At that time Azam et al. (1983) formalized the concept of the microbial loop. The microbial loop acknowledges explicitly that uptake of dissolved organic matter (DOM) by bacteria, followed by consumption of bacteria by bacterivores, is the dominant process by which the dissolved products of photosynthesis are returned to planktonic food webs.

In the years since the early 1980s, rate measurements have shown that bacterivory can be, and often is, equivalent to a substantial fraction of bacterial production. Like any prey community that is heavily regulated by consumers ("top-down" control), bacterial communities appear to be shaped in many ways by the behaviors and capabilities of bacterivores. Some influences of bacterivory — for example, on the size structure of bacterial communities — have been investigated extensively, while influences on community chemical composition, species diversity, and other features are as yet little known and represent important future research areas.

In this chapter, I first describe the most important bacterivores inhabiting the ocean's water column and discuss how they have solved the general problem of capturing small particles from a dilute suspension. Then we turn to the issue of rates. How is bacterivory measured? What do measured rates tell us about regulation of bacterial communities and links between bacteria and other planktonic organisms? Next we will look at bacterivory as a structuring force in planktonic ecosystems, considering the issue from two sides: the consequences of bacterivory for bacterivores, and the consequences for the bacterial communities themselves. Finally we will consider some thought-provoking examples of how bacterivory may play a role in the global cycling of carbon and nitrogen.

THE CAST OF CHARACTERS: WHO ARE THE BACTERIVORES?

Bacterivory in the ocean's water column is thought to be dominated by protists (see Chapter 2). Some of these are strict heterotrophs, which obtain all their nutrition from the consumption of other organisms; however, mixotrophs — photosynthetic protists that are also consumers — are increasingly seen as an important component of the bacterivore community (Chapter 16). The assignment of protists to the category "bacterivore" should not be taken to mean that bacteria are the sole prey of these grazers. Laboratory experiments, as well as field observations of protist food vacuole contents, indicate that bacteria-consuming protists also feed on (and may thrive on) phytoplankton, detritus, and heterotrophic protists (E. B. Sherr 1988; Caron et al. 1990a; Strom 1991;

Verity 1991b; Posch and Arndt 1996). In the dilute and compositionally variable prey environment of the ocean, grazers with narrow food preferences are unlikely to be important or persistent members of the community.

There are also metazoan bacterivores. In coastal waters, the planktonic larvae of benthic invertebrates such as asteroids can contribute to bacterivory (Rivkin et al. 1986). Many metazoans, especially the copepods and other common planktonic crustaceans, cannot capture individual bacterial cells effectively and thus are not quantitatively important bacterivores. Gelatinous zooplankton, such as salps, doliolids, and particularly larvaceans (appendicularians) that feed by means of fine-meshed mucous nets, can capture bacteria and even smaller particles efficiently, and may be important bacterial consumers (King et al. 1980; Mullin 1983; Deibel and Lee 1992; Bedo et al. 1993). Their occurrence is considerably more sporadic, and thus their impact more variable, than that of the bacterivorous protists. It is important to note that in environments where many bacteria are attached to larger particles (e.g., river mouths, salt marshes, other regions rich in marine snow and detritus: Mann 1988; Crump and Baross 1996), their effective size transformation makes them available to an entirely new cast of planktonic grazers.

THE PROTIST BACTERIVORES

Most bacterivorous activity seems to occur among the very smallest planktonic size classes (Figure 1). This has been shown both in studies that progressively

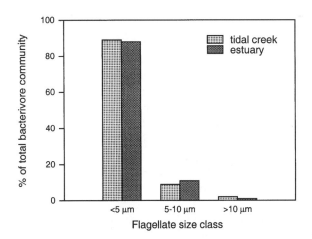

Figure 1. Percentage of total number of heterotrophic flagellates with ingested fluorescently labeled bacteria that fall into each of three size classes. In this study, conducted with water from a salt marsh tidal creek (lightly shaded bars) and from the open estuary (darkly shaded bars), both adjacent to Sapelo Island, Georgia, the smallest ($<5\,\mu$m) flagellates dominated the bacterivore community. (Redrawn from B. F. Sherr and E. B. Sherr 1991.)

removed grazer size classes by filter fractionation and in studies that directly observed the appearance of fluorescently labeled bacterial cells in consumers (Wikner and Hagström 1988; Wikner et al. 1990; B. F. Sherr and E. B. Sherr 1991). The organisms in these size classes are predominantly small ($1-10\,\mu$m) flagellates, including bodonids, choanoflagellates, and chrysomonads (see Chapter 2 for illustrations of bacterivorous taxa). Flagellates that are typically somewhat larger ($\sim 5-50\,\mu$m) and can consume bacteria include euglenoids, cryptomonads, and dinoflagellates. The special volume edited by Patterson and Larsen (1991) provides a taxonomic overview of free-living heterotrophic flagellates with a useful ecological slant. Small ($<20\,\mu$m) ciliates can also consume bacteria at high rates, and holotrichous forms in particular may be important components of the bacterivore community in highly productive coastal waters (E. B. Sherr and B. F. Sherr 1987). Less is known in general about the grazing impact of sarcodines such as foraminiferans and radiolarians: their "spider web–like" feeding mode, in which prey cells are entrapped by sticky pseudopodia, should allow them to feed effectively on bacteria. Both radiolarians and naked amoebae have been observed to ingest, and the latter to grow on, bacteria (Gowing 1989; Mayes et al. 1997) . There is even a report of bacterivory by the amoeboid stage of two thraustochytrid species, funguslike organisms found associated with decaying organic matter and other surfaces in the sea (Raghukumar 1992). The typically low abundance of sarcodine and other amoeboid consumers, however, means that their contribution to total bacterivory is likely small.

METAZOAN BACTERIVORES

Two recent studies (Turner and Tester 1992; Roff et al. 1995) have shown that the larval (naupliar) and juvenile (copepodite) stages of small marine copepod species can ingest bacteria. Feeding rates are not well known, although individuals have been observed with guts packed with fluorescently labeled bacteria (FLB), suggesting effective uptake. On the other hand, the FLB used in these studies were prepared from *Escherichia coli*, which are considerably larger on average than planktonic bacteria. Thus the potential for bacterivory by these zooplankters may have been overestimated. The smallest marine copepod species, which might be expected to have the greatest potential for bacterivory, are those that live in the tropics, and these species are poorly studied in general.

Larvaceans, salps, and doliolids are gelatinous zooplankters in the phylum Chordata; their distribution is worldwide. Larvaceans in particular have been implicated as potentially important bacterivores (King et al. 1980; Deibel and Lee 1992). Their mode of living is to construct a gelatinous house rigged with fine-meshed ($<1\,\mu$m) filtering structures, through which they propel a current of water [see Flood (1991) for diagrams]. Although these grazers are large (millimeters to centimeters) relative to those discussed previously, their filtering mesh is sufficiently fine and its sticky, retentive properties sufficiently great that they can clear bacteria and even smaller, colloidal particles from the water (Bedo et al. 1993). When the filters clog, the larvacean discards its house and constructs another, an event that can happen many times per day in productive waters. The discarded house with attached bacteria and other particles tends to sink, and can be an important source of "marine snow" in coastal regions (Alldredge 1976). Thus bacterivory by larvaceans short-circuits the microbial food web, transferring bacterial biomass directly to relatively large planktonic organisms and to sub–euphotic zone ocean waters (Michaels and Silver 1988; Deibel and Lee 1992).

FEEDING MECHANISMS: HOW DO THEY DO IT?

Average bacterial concentrations in the surface ocean are often about 10^6 cells mL^{-1} (Chapter 4). A million of anything sounds like a great deal, but in fact this constitutes an extremely dilute suspension. Given a representative oceanic bacterial cell diameter of 0.6 μm and thus a cell volume of 0.11 μm^3, bacteria are occupying only 1.1×10^{-7}th portion (0.1 ppm) of a milliliter. For comparison, assume that a person occupies a 1 m^2 area of the earth's surface. At a population density comparable to bacteria in seawater, everyone would have a personal space of 10 km^2 (3.6 mi^2). This is a population density about half that of the state of Alaska, ensuring plenty of solitude!

This analysis demonstrates that one of the fundamental problems faced by bacterivores is the harvest of small, widely dispersed cells from seawater. The calculations assume that bacteria are uniformly distributed throughout the volume. As mentioned earlier, a significant fraction of marine bacteria may be associated with larger particles in some environments (Caron et al. 1982; Caron 1991), and some bacterivores may be capable of, or even adapted to, feeding on particle-attached bacteria (Silver et al. 1984; Caron 1987). Recent evidence indicates that particularly in highly productive waters, many bacteria can be attached to "transparent exopolymeric substances" (TEP), sheets or strands of polymerized polysaccharides (Alldredge et al. 1993). In the presence of mucus-producing phytoplankton taxa such as *Phaeocystis* spp., polysaccharide gels may give structure to the water (Chin et al. 1998). It is not known how TEP or gels affect feeding by bacterivorous protists.

Bacterivory—indeed, suspension feeding in general—can be thought of as comprising a two-part process: encountering the bacterial cell, then entrapping, retaining, and ingesting it. Bacterivorous ciliates and flagellates encounter bacteria by creating water currents with cilia or flagella. (For free-living protists, feeding currents are identical to swimming currents; no distinction is made between water moving past the protist and the protist moving through the water.) The fluid environment of these microbes is a strong determinant of prey encounter and capture mechanisms. Because of the size scales and flow rates involved, water movement is dominated by viscous forces; that is, it is governed by laminar flow, perfect reversibility, and a complete lack of inertially driven phenomena such as glide. Our intuition about such environments works best if we consider ourselves swimming through cold molasses, or hot tar. The Reynolds number, the ratio of inertial to viscous forces, is the scaling function used to predict flow regimes under different conditions; Reynolds numbers much less than 1 characterize protist bacterivory. Cogent and entertaining descriptions of life at low Reynolds number are provided by Purcell (1976) and Vogel (1994).

The low Reynolds number world of protist bacterivores has profound consequences for the evolutionary design and function of their feeding organelles. For example, to create unidirectional flow, flagellae and cilia must move through the water asymmetrically. Many ciliates accomplish this by "feathering their oars": moving cilia through the water in a rowing fashion on the propulsion stroke, then keeping cilia close to the cell surface on the return stroke. Flagellates trace a spiral path through the water in a manner analogous to a corkscrew, using whiplike undulations or other motions of the flagellae (Sleigh 1973; Anderson 1988).

The extent to which protist bacterivores can "process" water is astonishing. Maximum clearance rates that can be obtained by these organisms exceed 10^5 body volumes per grazer per hour (Hansen et al. 1997); that is, they can remove all prey cells from a volume exceeding 100,000 times that of their own body each hour. To remove prey cells from the flow, bacteria may be entrapped by finely spaced structures such as oral cilia or the pseudopodial collar of choanoflagellates (Andersen 1988/89). Alternatively, most flagellates simply rely on contact with their cell surface, though prymnesiophytes can capture particles with their haptonema (Kawachi et al. 1991). Because encounter is a critical phase of the feeding process, factors that increase encounter rates, including larger predator or prey size and higher swimming speeds, will tend to increase feeding rates (Figure 2; see also Gerritsen and Strickler 1977; Shimeta 1993). Turbulence, because it can increase shear (hence relative motion and particle encounter rates) at the size scales of bacterivores, has been hypothesized to increase rates of protist bacterivory. So far the experimental evidence is equivocal, perhaps because shear interferes with the feeding currents of strong protist swimmers (Peters and Gross 1994; Shimeta et al. 1995).

Several investigators (Rubenstein and Koehl 1977; Jørgensen 1983; Shimeta and Jumars 1991) have applied aerosol filtration theory to suspension feeding in an effort to understand the dominant forces influencing particle capture and retention. Others, primarily Fenchel (e.g., 1980b, 1986), have adopted an approach based on protist functional morphology. An overall conclusion from these efforts to date is that protist bacterivores must be exceedingly good at what they do to sustain measured feeding rates. Monger and Landry (1990) provide an instructive example. They attempted to model the feeding rates of zooflagellates from a first-principles approach, examining the balance of forces involved in prey encounter and retention. They concluded that attractive (e.g.,

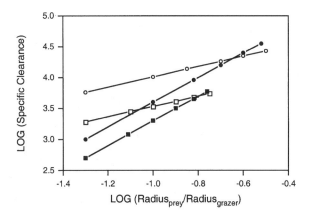

Figure 2. Combined effects of grazer and prey size on hourly volume-specific clearance rate, as predicted by two different encounter models: open symbols: force–balance model; solid symbols, geometric model. Predicted effect of relative size shown for grazer diameters of 4 μm (circles) and 8 μm (squares). The geometric model predicts a larger increase in clearance rate for a given increase in the ratio of prey radius to grazer radius. (Data replotted from Monger and Landry 1990.)

London–van der Waals) and repulsive (e.g., hydrophobic) interactions between predator and prey cell surfaces were critical for predicting flagellate clearance rates; even accounting for these surface interactions, the model tended to underestimate, sometimes substantially, real rates of bacterivory. Cell surface properties and their role in predator–prey interaction are areas rife with possibility for future study.

RATES OF BACTERIVORY: HOW MUCH DO THEY EAT?

As mentioned earlier, the combined observations of high bacterial production rates coupled with slight (or no) accumulation of new bacterial biomass were key to the recognition that bacterial loss processes, including grazing, must be important in marine waters. Indeed, careful temporal sampling of coastal waters has revealed cycles of bacterial and nanoflagellate biomass (Figure 3) that, in some cases, look remarkably similar to those that can be obtained in simple, controlled laboratory cultures (Sorokin 1977; Fenchel 1982; Kivi et al. 1993; Tanaka et al. 1997). Such predator–prey cycles are compelling evidence for the importance of bacterivory. Furthermore, the out-of-phase oscillations of bacteria and nanoflagellates demonstrate that bacterial production and bacterivory are not likely to be equivalent on any given sampling date. To assess the importance of bacterivory in removing bacterial production, it is essential to obtain rate measurements over a time span long enough to encompass such predator–prey oscillations.

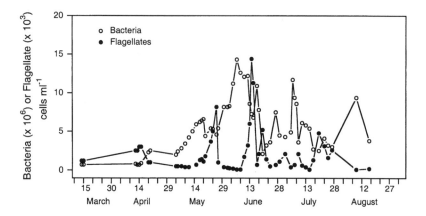

Figure 3. Abundance of bacteria (open symbols) and heterotrophic nanoflagellates (solid symbols) from March through August 1983 in a shallow marine fjord (Limfjorden, Denmark). Abundances shown are means of samples collected from 1 and 2 m depths. Note cyclic behavior of grazer and prey abundances. (Data replotted from Andersen and Sorensen 1986.)

MEASUREMENT TECHNIQUES

Finding methods to accurately estimate rates of production and consumption in the plankton has been a challenge of the first order for microbial ecologists. Bacterivory has proven to be one of the more intractable processes to measure. There are two major reasons for this: first, the organisms involved are some of the smallest in the ocean, so that visual methods (e.g., detection of fluorescent particles in bacterivore food vacuoles) tend to be tedious and imprecise, involving as they do the lower limits of microscopy and the upper limits of human capabilities. Second, bacteria are largely dependent on the products of other biological activity (e.g., dissolved organic carbon, regenerated nutrients), and the coupling between bacteria and the remainder of the planktonic food web is readily (sometimes deliberately) disrupted by the measurement techniques themselves, with unknown but potentially profound effects on rates (e.g., Snyder and Hoch 1996).

Owing to these overriding challenges, methods for measuring bacterivory have been described, reviewed, and critiqued at length in the literature. This section summarizes briefly the methods in current use and presents a general overview of the applications and problems associated with the two main classes of methods. The reader is directed to the appropriate literature for more in-depth investigation.

Methods for the measurement of bacterivory fall into two main classes: those that use a tracer of some sort to follow bacteria into bacterivore food vacuoles or cell cytoplasm (class I methods) and those that manipulate the community in such a way as to alter encounter rates between grazers and bacteria (class II methods) (Table 1). Methods in the former class typically

Table 1. Methods for the measurement of bacterivory in marine waters

Method	Key References	Brief Description
Class I[a]		
Ingestion of fluorescently labeled particles (FLP) or bacteria (FLB)	Børsheim (1984), B. F. Sherr et al. (1987), E. B. Sherr and B. F. Sherr (1993)	Fluorescent particles are counted inside of grazer food vacuoles after short incubation periods.
Radio-isotope labeling	Hollibaugh et al. (1980), Lessard and Swift (1985), Nygaard and Hessen (1990)	Bacteria labeled with radioisotopes are followed into individual grazer cells (picked from sample) or into grazer communities (isolated by postincubation size fractionation).
Minicell removal	Wikner et al. (1986), Wikner (1993)	Genetically marked, nondividing minicells derived from cultured bacteria are added to seawater and their disappearance rate monitored, giving a whole-community estimate of bacterivory.
Class II[b]		
Seawater dilution	Landry and Hassett (1982), Tremaine and Mills (1987), Landry (1993a)	Whole microplankton communities are progressively diluted with particle-free seawater to create a gradient in predator–prey encounter rates; rates of change in prey density are monitored.
Size fractionation	Wright and Coffin (1984), Weisse and Scheffel-Moser (1991)	Bacteria and grazers are separated by filtration; rates of change of bacterial abundance are monitored in grazer-free and grazer-containing incubations.
Metabolic inhibitors	Newell et al. (1983), Fuhrman and McManus (1984), B. F. Sherr et al. (1986)	Prokaryotic and eukaryotic inhibitors are used to quench the effects of bacterial growth and bacterivore grazing on bacterial abundance changes

[a]Methods involving bacterivore uptake of an added tracer.
[b]Methods in which the microplankton community is manipulated to uncouple bacterial growth and bacterivory.

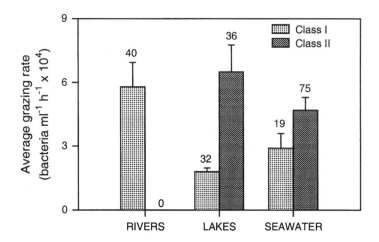

Figure 4. Average estimates of bacterivory ($\pm$1SE) based on two different types of measurement method, and sorted by environment; number of observations is indicated on top of each bar. Class I methods (lighter bars): bacterivory measured through the uptake of fluorescent particles, including fluorescently labeled bacteria. Class II methods (darker bars): whole-community bacterivory measured using dilution, size fractionation, inhibition, or minicell removal techniques. (Data replotted from Table 2 of Vaqué et al. 1994.)

provide feeding rates for individual grazer taxa or "morphotypes" (though note exceptions in Table 1); the latter exclusively provide whole-community rates. A third class of methods, those that assess bacterivory noninvasively by looking at some instantaneous property of the bacterivore community (e.g., digestive enzyme activity, food vacuole fullness) are being eagerly investigated but remain elusive at present. Comprehensive overviews of methodology are provided by McManus and Fuhrman (1988b) and Landry (1994). There is growing evidence that some bacterivory measurement techniques consistently provide higher or lower rate estimates than others; Vaqué et al. (1994), in a detailed analysis of methodological bias, found that methods involving tracer particle uptake systematically delivered lower rate estimates than those involving whole-community manipulation (Figure 4). In addition, rates estimated using tracer particle uptake were positively correlated with temperature, bacterial abundance, and bacterial production, while almost no such correlations were found for community manipulation techniques. It is crucial to remember that our understanding of marine microbial interactions is colored by the methods chosen to study them.

Class I Methods

Of methods that use tracers to assess bacterivory, uptake of fluorescently labeled bacteria (FLB, Table 1) is probably most widely employed. Visualization of surrogate food particles in food vacuoles gives specific information about which members of the community are bacterivores. Short-term incubations and minimal manipulation of seawater should obviate some containment artifacts. Potential problems with the method include egestion of tracer particles when samples are fixed (M. E. Sieracki et al. 1987) and, most worrisome, feeding selectivity against the tracer particles (Landry et al. 1991; Gonzalez et al. 1993b). Such selectivity can be substantial but is neither readily measurable nor consistent from one environment to the next. A useful review of methodological issues surrounding the use of surrogate food particles is provided by McManus and Okubo (1991).

Radiolabeling of bacteria to measure bacterivory can be done in two ways: isotope, often [^{3}H]thymidine, can be added to natural water samples, or bacteria from the environment can be prelabeled to achieve a higher specific activity, then added back to natural samples. The real strength of the isotope approach is that naturally occurring bacteria are the prey, eliminating concerns about selection against surrogate particles. The method can give rates of bacterivory for taxa large enough to be picked as individuals from the seawater sample. Whole-community rates are obtained by size-fractionating the sample after incubation to retain only organisms larger than bacteria. The primary concern with the method is uncertainty regarding the fate of the radiolabel within the microbial food web. For example, thymidine may be released during digestion of bacteria (Zubkov and Sleigh 1995) or as a consequence of cell breakage during postincubation size fractionation. Conversely, label may be transferred from bacteria to nanoflagellates, which are then eaten by larger protists, negating the ability to distinguish between bacterivory and other grazing pathways. For these reasons, short (<2 h) incubations were recommended in a review of the method (Caron et al. 1993).

Minicell removal is a third tracer-type technique, but with a difference. Minicells are nondividing bacteria produced from a laboratory-cultured strain such as *Escherichia coli* (see Wikner 1993 for an overview). They are "mini" relative to the parent cultured strain but are close to the size of naturally occurring marine bacteria. The minicells harbor plasmids that express high levels of specific proteins (e.g., β-lactamase, chloramphenicol acetyltransferase) that can be radiolabeled; recovery and quantification of the protein using gel electrophoresis and scintillation counting provides a sensitive biomarker of minicell biomass. Besides the technical sophistication, the difference between minicell removal and other class I methods is that minicells are not followed into bacterivores; rather, their disappearance from seawater is monitored over time. However, appearance of minicells in different planktonic size fractions can be used to indicate the size structure of the bacterivore community (Wikner et al. 1990). The technique has not been widely adopted by biological oceanographers, probably because of the complex molecular biological equipment and the degree of expertise required.

Class II Methods

The seawater dilution technique, while more widely used to estimate micro-zooplankton herbivory, has also been used to assess bacterivory (e.g., Ducklow and Hill 1985; Tremaine and Mills 1987). The technique involves setting up a dilution series in which natural seawater is progressively diluted with particle-free seawater. The rate of change in prey biomass is then monitored over an incubation period typically lasting 12–48 hours. The idea is that dilution proportionally reduces predator–prey encounter rates, hence grazing, while leaving the intrinsic growth rate of the prey cells unaltered (Landry and Hassett 1982). Thus prey biomass changes due to growth become progressively uncoupled from biomass changes due to grazing across the dilution series. A key assumption of the technique is that dilution does not alter the growth rate of the prey cells. For bacteria, whose resource supply rate is largely dependent on the activities of other organisms, this assumption may well be violated as the planktonic community is diluted. Conversely, preparation of the particle-free seawater by filtration may cause cell breakage (Fuhrman and Bell 1985), increasing the availability of dissolved compounds to the bacteria, while the reduction in bacterial density via dilution may reduce competition for scarce resources. Landry (1993a) and Vaqué et al. (1994) review these issues, and users of this and other class II techniques are well advised to check the effects of community manipulation on bacterial cell–specific production rates (e.g., Tremaine and Mills 1987).

Size fractionation is conceptually the simplest of the class II techniques. Bacterivores are separated from their prey by filtration through a filter having the appropriate pore size, and changes in prey biomass are monitored during incubations of size-fractionated and intact seawater samples. Bacterial growth in the absence of grazers is factored out of bacterial biomass changes in the presence of grazers to yield estimates of bacterivory (Wright and Coffin 1984; Weisse and Scheffel-Moser 1991). The primary virtue of the technique is its simplicity. On the negative side, all the caveats discussed for the dilution technique apply and, as well, the choice of filter pore size is crucial. Bacteria and bacterivores vary in size depending on environment, and there may well be a size overlap between the two communities (Cynar et al. 1985).

Metabolic inhibitors have been used fairly extensively to estimate rates of bacterivory. The technique is conceptually similar to that of size fractionation in that grazers and prey are separated by suppressing the activity of one or the other using chemical inhibitors (Newell et al. 1983). Again, microbial community function is disrupted by the technique, with unknown effects on bacterial activity. Also, the action of the inhibitors is not always complete (e.g., grazing may be incompletely suppressed by eukaryotic inhibitors: Caron et al. 1991) or may be, in a sense, too complete (e.g., prokaryotic inhibitors may also depress protist grazing rates; B. F. Sherr et al. 1986). The latter problems appear to be environment-specific and may be addressed by the use of careful control experiments.

Method Summary: Where Do We Go From Here?

Development of new methods and improvement of existing ones will be key to advancing our understanding of bacterivory. If past history is any guide, progress is likely to be incremental and tedious. Careful selection of methods to fit both the environment and the question being asked is, of course, critical, and the use of methods in tandem to shore up weaknesses has been recommended (Landry 1994). A holy grail of sorts is the development of a technique that can assess bacterivory without sample incubation by measuring some property of the naturally occurring bacterivore community. A concerted effort has been made to use digestive enzyme activity—specifically, the activity of enzymes that hydrolyze compounds unique to bacteria—as a quantitative measure of bacterivory (Gonzalez et al. 1993a; Vrba et al. 1993). Various problems have stymied application of this method, including a lack of specificity in some enzymes for uniquely bacterial compounds (Jenkins et al. 1998). Furthermore, the relationship between lysozyme activity and other measures of bacterivory varies widely among different aquatic environments (Gonzalez et al. 1993a; Simek et al. 1994; Vrba et al. 1996).

Meanwhile, there have been other exciting recent developments, particularly in the use of FLBs. Flow cytometry, particularly in conjunction with new bacterial stains, is a promising avenue for the automated analysis of FLB uptake during grazing experiments (Cucci et al. 1989; Monger and Landry 1992). Newly developed imaging-in-flow systems (C. K. Sieracki et al. 1998) should improve counting precision for grazers in dilute samples such as open-ocean seawater. Fluorescent compounds that can be used to label live bacteria (Epstein and Rossel 1995) promise fewer problems with grazer selectivity, and tagging of bacteria with fluorescent antibodies (immunofluorescence: Christoffersen et al. 1997) or identification of bacterial genetic markers within food vacuoles (16S rRNA: Cynar et al. 1985) may eventually be useful for measuring ingestion of bacteria with specific metabolic capabilities (e.g., denitrification).

THE BALANCE OF PROCESSES: BACTERIVORY COMPARED WITH BACTERIAL PRODUCTION

A key question for those seeking to understand microbial food webs is, What happens to bacterial production? A food web in which bacterial production is recaptured by bacterivores with a high efficiency will function very differently from one in which most bacterial mortality is due to viral lysis and, consequently, most bacterial production is returned to the DOM pool. Over a decade ago Pace (1988) reviewed this issue and concluded that while modest accumulation of bacterial biomass in the face of high production seemed to demand high removal rates, in fact measured bacterivory was consistently and often substantially less than bacterial production. Pace called this mismatch "an enigma" that required a stern look at both methodology and alternate sources of bacterial mortality. In the intervening years research on both bacterivory and viral infection has burgeoned. Figure 5 plots results from 17 studies that directly measured both bacterial production and bacterivory. Note that while the studies are evenly divided between low and high productivity environments, only one (the Arctic Ocean work of E. B. Sherr et al. 1997) is from a truly oceanic region; the rest were conducted in estuarine or coastal waters.

On viewing the data, the first impression is of many studies in which production exceeded grazing, as indicated by the many data points falling below the diagonal line that denotes a one-to-one relationship. However, as shown by the inset, this imbalance disappears when only data from low productivity waters ($< 1.1 \times 10^6$ bacteria produced $mL^{-1} d^{-1}$) are considered. This compilation indicates that bacterivory is the fate of nearly all bacterial production in low productivity marine waters. In higher productivity environments, bacterivory can be significant, but there must be other major sources of mortality. Many of these latter data were collected in shallow estuaries, where removal by benthic filter feeders could be important. In addition, the higher bacterial abundance typical of these environments (Chapter 4) should aid the spread of viral or bacterial infection, promoting mortality through bacterial lysis (e.g., Weinbauer et al. 1993; Steward et al. 1996). A few recent studies (Fuhrman and Noble 1995; Steward et al. 1996) have suggested that viral lysis can account for substantial bacterial mortality in low productivity coastal marine waters as well (see also Chapter 12). It is difficult to reconcile these findings with the close balance between bacterial production and grazing shown in Figure 5, inasmuch as a population cannot consistently sustain losses that exceed production. We seem to have moved from Pace's bacterivory enigma to a viral lysis enigma. Additional studies that address both processes directly and simultaneously should do much to clarify the situation.

A similar compilation of production and grazing (Figure 6) was made based on eight studies of the cyanobacterial genera *Synechococcus* and *Prochlorococcus*. It is important to realize that since these organisms can dominate phytoplankton biomass and primary productivity in oligotrophic ocean regions (Campbell et al. 1994; Liu et al. 1998), consumption of bacteria represents substantial removal of primary production as well. In Figure 6 the data are presented as biomass-specific growth and grazing rates: because some of the experiments measured changes in accessory pigment concentration rather than bacterial cell density, cell-based production and grazing estimates could not always be calculated. Although there are fewer data, a similar pattern results: nearly all the data points showing a large excess of growth over grazing are from shallow, high productivity environments (e.g., Kaneohe Bay, Hawaii; Chesapeake Bay; San Francisco Bay). The observations of Campbell and Carpenter (1986) from a warm-core Gulf Stream eddy are the one exception.

In summary, methodological issues aside, the data collected to date strongly indicate that bacterivory plays different roles in low versus high productivity

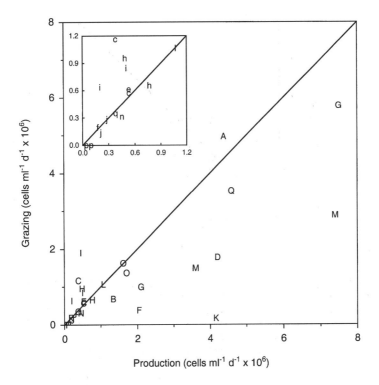

Figure 5. Bacterial production versus bacterivory for 17 marine studies. Only studies that estimated production and grazing independently were used. Production estimates based on uptake of tritiated thymidine unless otherwise indicated. Location, number of experiments, and grazing methodology given for points A–Q as follows: A, Long Island Sound, inhibitors, $n = 12$ (Fuhrman and McManus 1984); B, Kaneohe Bay, Hawaii, dilution (for production and bacterivory), $n = 2$ (Landry et al. 1984); C, Mediterranean Sea ($n = 2$), Baltic ($n = 4$), minicells (Wikner et al. 1986); D, Duplin River estuary, Georgia, inhibitors (production not measured at the same time), $n = 8$ (B. F. Sherr et al. 1986); E, Mediterranean Sea, minicells, $n = 6$ (Hagström et al. 1988); F, Chesapeake Bay, FLP uptake, $n = 7$ (February), $n = 8$ (June) (McManus and Fuhrman 1988a); G, Georgia coast, FLB uptake, $n = 9$ (open sound), $n = 10$ (tidal creek) (B. F. Sherr et al. 1989); H, Red Sea ($n = 10$), Gulf of Aden ($n = 6$), inhibitors (for production and grazing) (Weisse 1989); I, Bothnian Sea ($n = 11$), Finnish Bay ($n = 5$), Skagerrak ($n = 40$), minicells (Wikner et al. 1990); J, Gulf of Bothnia estuary, minicells, $n = 17$ (4 m depth), $n = 17$ (18 m depth) (Wikner and Hagström 1991); K, Atlantic coast of Spain, FLB uptake (two different conversion factors used to estimate bacterial production from thymidine uptake), $n = 6$ (Barcina et al. 1992); L, Adriatic coast, FLB uptake, $n = 12$ (Solic and Krstulovic 1994); M, California coast, FLB uptake, $n = 2$ for each of two mesocosms (Fuhrman and Noble 1995); N, Mediterranean coast of Spain, minicells (production estimated as net bacterial growth–grazing), $n = 6$ (del Giorgio et al. 1996); O, St. Lawrence estuary, dilution (for production and bacterivory), $n = 10$ each for upper and lower water column (Painchaud et al. 1996); P, Arctic Ocean, FLB (separate thymidine- and leucine-based production estimates), $n = 7$ (E. B. Sherr et al. 1997); Q, San Francisco Bay, dilution (for production and bacterivory), $n = 6$ (Suisun Bay), $n = 2$ (Tomales Bay) (Murrell and Hollibaugh 1998).

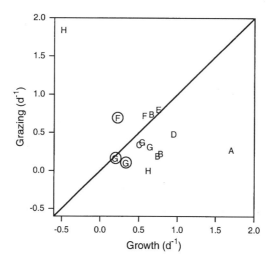

Figure 6. Specific growth rates of photosynthetic bacteria *Synechococcus* sp. (Syn) and *Prochlorococcus* sp. (Pro, circled symbols) versus specific grazing rates. All bacterial production values estimated with the same techniques used for grazing estimates. Letter references arranged as for Figure 6. A, Kaneohe Bay, Hawaii, dilution, $n = 2$ (Landry et al. 1984); B, Gulf of Maine ($n = 3$), North Atlantic warm core eddy ($n = 4$), Long Island Sound ($n = 1$), dilution and inhibitors (Campbell and Carpenter 1986); C, Open subarctic Pacific, dilution, $n = 6$ (Strom and Welschmeyer 1991); D, Chesapeake Bay, dilution, $n = 6$ (McManus and Ederington-Cantrell 1992); E, Northwest Indian Ocean, dilution, $n = 3$ (Burkill et al. 1993); F, central equatorial Pacific, dilution, $n = 5$ each for Syn and Pro (Landry et al. 1995); G, Kaneohe Bay, Hawaii ($n = 2$ for Syn, $n = 1$ for Pro), Station ALOHA, central North Pacific ($n = 8$ for Syn, $n = 12$ for Pro), inhibitors (Liu et al. 1995); H, San Francisco Bay, dilution, $n = 4$ (Suisun Bay), $n = 2$ (South Bay) (Murrell and Hollibaugh 1998).

systems. Low productivity systems overall exhibit an essentially 1:1 relationship between bacterial production and bacterivory. How other sources of bacterial mortality, in particular viral lysis, fit into this picture is not yet clear. Furthermore, studies of open-ocean waters, which cover most of the planet's surface, are needed to determine whether a bacterial production — bacterivory balance pertains there as well. In high productivity waters a substantial fraction of bacterial production is, in many cases, unaccounted for by bacterivory, indicating that additional sources of mortality (consumption by benthic organisms, viral lysis, and perhaps others) must be important.

WHAT ARE THE CONSEQUENCES OF BACTERIVORY FOR GRAZERS?

Consumption of bacteria has profound consequences for planktonic communities, enabling the transfer of bacterial biomass to larger consumers that would otherwise be unable to access this food source. On the level of individual grazers, the unusual biochemical composition of bacteria relative to eukaryotes almost certainly influences digestion, growth efficiency, nutrient regeneration, and other biological processing by bacterivores.

THE PLANKTONIC FLYWHEEL: DOM RECAPTURE BY THE MICROBIAL FOOD WEB

A fundamental role of bacterivory in marine food webs is the repackaging of bacterial biomass into protist cells, DOM, and inorganic wastes. Indeed, it is these various transformations that complete the "microbial loop," now viewed as a much more important pathway for the flow of energy and materials than the large-organism-dominated "classic food chain" in much of the world's ocean (see Chapters 5 and 6). "Repackaging" of bacterial biomass occurs when a bacterial cell is eaten by a larger consumer. As discussed earlier, most bacteria are though to be consumed by nanoflagellates; these, in turn, are known to be eaten by larger protists (ciliates, dinoflagellates), copepods, and other planktonic grazers (Stoecker and Capuzzo 1990; Gifford and Dagg 1991; Verity 1991b; Deibel and Lee 1992). Thus an important function of bacterivory is to make production by very small bacterial cells available to larger organisms in planktonic environments.

The growth efficiencies of protist grazers do not appear to be high, in general (reviewed by Straile 1997). Growth efficiencies of 30% or less mean that much ingested bacterial biomass is released by grazers as wastes. Protist grazers have been found to release 25% or more of ingested prey carbon as dissolved organic carbon (Taylor et al. 1985; Strom et al. 1997); some of this appears to be quite labile and is readily taken up again by bacteria, while other material is relatively refractory and may contribute to longer-lived oceanic DOM pools (Nagata and Kirchman 1992; Tanoue et al. 1995, see Chapter 5). There are very few studies in which the biochemical nature of DOM resulting from bacterivory has been determined. In terms of bulk DOC, however, high levels of DOC production during bacterivory and subsequent uptake by bacteria mean that the organic carbon ultimately derived from photosynthesis can cycle through the microbial food web many times (Strayer 1988).

Ingested bacterial prey may also be transformed into and released as inorganic waste (CO_2, NH_4^+, PO_4^{3-}, etc.). The relatively high respiration and excretion rates of protist grazers, especially the smallest forms (reviewed by Caron and Goldman 1990; Caron et al. 1990b), lead to inefficient transfer of carbon and nutrients from bacteria through bacterivores to higher trophic levels. This is demonstrated quantitatively in food web models presented elsewhere in this volume (Chapters 2 and 5). Even though the "microbial loop" recaptures dissolved products of photosynthesis, losses to respiration and excretion mean that transfer of these products to higher trophic levels is ultimately inefficient, becoming more so in food webs with many microbial interactions.

COMPOSITIONALLY UNIQUE: THE INFLUENCE OF BACTERIAL CHEMICAL COMPOSITION ON BACTERIVORES

The elemental composition of bacterial cells differs substantially from that of most eukaryotes. Because they are so small, and because, as prokaryotes, bacteria lack the organelles and other internal structures of eukaryotes, bacteria have a high protein and nucleic acid content and relatively little carbohydrate-rich structural material (Simon and Azam 1989; Raven 1994). Thus their C:N and C:P ratios are lower than those of most protists (Table 2, although see Chapter 9 for a slightly different view on this topic). Bacteria may also harbor relatively more iron than planktonic eukaryotes (Tortell et al. 1996). To a first approximation, regeneration of nutrients depends on the degree to which a given nutrient is present in excess in the diet. A bacterivore with a C:N ratio of 6.8 feeding on bacteria with a C:N ratio of 5.0 is receiving excess N relative to its C needs (assuming equal digestive assimilation and biological turnover time of the two elements). In contrast, phytoplankton tend to have C:N and C:P ratios similar to or higher than those of protists (average of 7.2 for 13 species according to Verity et al. 1992; average of 11.3 for 3 species grown under a range of nutrient regimes according to Goldman et al. 1979), and detrital elemental ratios are typically even higher. These observations suggest that a diet of bacteria ought to support higher rates of regeneration of nutrients (ammonium, phosphate, and possibly iron) than other diets of

Table 2. Carbon : nitrogen ratios (by atoms) for heterotrophic protists (nanoflagellates and bacterivorous ciliates) and bacteria

Protists	C:N[a]	Ref.[b]	Bacteria	C:N[a]	Ref.[b]
Strombidium sp.	7.1 (5.2–9.0)	1	*Pseudomonas putida*	5.1 (4.5–5.6)	4
Uronema sp.	4.4	1	Natural assemblage	5.9 (4.8–6.7)	4
Euplotes sp.	4.7	3	Natural assemblage	6.3 (3.8–8.7)	5
Cafeteria sp.	8.3	2	Natural assemblage	5.1 (4.1–9.8)	6
Bicoeca cf *maris*	8.0	2	*Vibrio* sp.	4.5 (4.1–5.0)	1
Paraphysomonas sp.	7.8	2	*Vibrio* sp.	4.4	3
Pseudobodo sp.	7.3	2			
Average	6.8			5.2	

[a] Ratios in parentheses show the range of values for the respective studies.
[b] 1, Ohman and Snyder (1991); 2, Verity (1991b); 3, Zubkov and Sleigh (1995); 4, Bratbak (1985); 5, Nagata (1986); 6, Goldman et al. (1987).

planktonic protists (Caron and Goldman 1990; Landry 1993b; Chase and Price 1998; Chapter 9). Indeed, one postulated role of bacterivory for mixotrophs is as a means of acquiring scarce nutrients in oligotrophic environments (Jones 1994; Riemann et al. 1995; Maranger et al. 1998; Chapter 16). Although protist grazers are believed to be significant and possibly dominant sources of planktonic nutrient regeneration (reviewed by Caron and Goldman 1990; Chapter 9), the influence of different diets on this process remains to be examined empirically.

Bacteria also have unique lipid profiles. Unlike marine algae, they contain branched chain fatty acids; there is also a wide variety of bacterial sterols (Saliot et al. 1991). The significance of this is twofold. First, lipids are a key component of the nutritional value or "food quality" of a prey organism; for example, certain polyunsaturated fatty acids have been shown to appear consistently in diets supporting high growth rates of protists, rotifers, larval fish, and other small marine organisms (e.g., Fraser et al. 1987; Volkman et al. 1989) . The aforementioned fatty acids are of algal origin, yet bacterial lipids can also be assimilated without structural alteration by bacterivores (Ederington et al. 1995). Might these bacterial lipids play a key role in the nutrition of bacterivores? Such a role could be either positive, contributing essential nutrients to the diet, or negative, substituting for more nutritionally desirable algal lipids or actively impeding metabolic or biosynthetic activity in the consumer. A second significant role of these unique bacterial lipid profiles, at least from the viewpoint of the oceanographer, is their utility as biomarkers for bacteria, hence bacterivory (e.g., Skerratt et al. 1995; Canuel et al. 1995). This approach has not been employed widely in planktonic food web studies, though it shows much promise.

WORKING RELATIONSHIPS: FORMATION OF SYMBIOSES THROUGH BACTERIVORY

That bacteria can form mutualistic relationships with protists is indubitable: consider the harboring of methanogens (Finlay and Fenchel 1992) or photosynthetic purple bacteria (Bernard and Fenchel 1994) by anaerobic ciliates. That many planktonic protists commonly contain bacteria has also been established (Lee et al. 1985). Some time ago Laval-Peuto (1991) made a plea for further study of these exciting relationships. Yet today we still know little about the ecological significance of bacterial endosymbioses for planktonic protists in general. On evolutionary time scales, Margulis (1981) has convincingly argued that eukaryote organelles such as mitochondria and chloroplasts arose when ingested bacteria and some of their metabolic capabilities were permanently retained within their bacterivorous "hosts." Perhaps the widespread retention of the capacity for phagocytosis in photosynthetic protists represents evolutionary pressure to retain the capacity for organelle acquisition (Cavalier-Smith and Lee 1985).

WHAT ARE THE CONSEQUENCES OF BACTERIVORY FOR BACTERIAL COMMUNITIES?

Bacterivores remove a substantial portion of the ocean's bacterial production, particularly in areas of low productivity (Figure 5). This means that bacterivory has tremendous potential to determine both biomass and structure of the bacterial community that remains. Bacterial abundance in low productivity waters is typically $0.5–1.0 \times 10^6$ cells mL^{-1} (Ducklow, Chapter 4). The feeding behaviors of bacterivores at low prey density must play a large role in regulating this abundance (e.g., Strom et al. in press). Why is the abundance not 1×10^3, or 1×10^7? Such a question has profound biogeochemical importance, for bacterial cells provide the largest reservoir of cell surface area and, potentially, living organic carbon, in open-ocean waters (Cho and Azam 1988, 1990). Because we know so little about regulation of bacterivore feeding behavior at low prey densities, we must leave this fascinating question for the present. We can, however, say more about the role of bacterivory in shaping bacterial community structure.

SAFETY IN SMALLNESS: HOW DOES BACTERIVORY INFLUENCE BACTERIAL SIZE STRUCTURE?

Planktonic suspension feeders are constrained by both hydrodynamic and mechanical forces. The net effect of these forces on prey encounter and retention determines the particle size range that a grazer can utilize. For planktonic protists, such size ranges tend to have well-defined lower limits (Figure 7); below these limits, particles may be captured and ingested, but with a substantially lower efficiency. This has led to the suggestion that small size is a refuge from grazing for planktonic bacteria (see the comprehensive review of Jürgens and Güde 1994, and references therein). Very large or morphologically complex shapes may similarly confer grazing resistance, although such forms appear to be more common in freshwater systems than in the ocean (Jürgens and Güde 1994).

From the viewpoint of marine bacterial communities, grazing pressure promotes an assemblage dominated by small bacterial cells. This has been shown in a range of studies. The removal of bacterivores from natural seawater samples generally results in growth of bacteria larger than those typical of the ocean's plankton (Ammerman et al. 1984; Kuuppo-Leinikki 1990; Gasol et al. 1995). Similarly, adding flagellate grazers to bacterial assemblages can shift the size distribution markedly to one dominated by small cells (Figure 8). Finally, size selectivity can be measured directly by adding tracer particles of different sizes to natural or cultured bacterivore assemblages (e.g., Gonzalez et al. 1990; Epstein and Shiaris 1992; Monger and Landry 1992). Particle uptake rates uniformly indicate a preference for prey larger than the smallest marine bacteria, although the strength of that preference is a matter of some debate (Monger and Landry 1991; Gonzalez 1996).

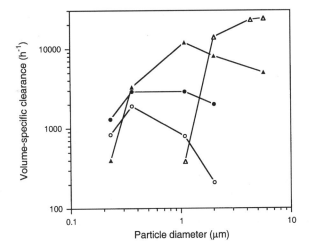

Figure 7. Prey size dependency of hourly volume-specific clearance rate for four species of bacterivorous ciliates. Prey particles were latex beads or baker's yeast (4.35 μm only). Ciliate species: *Glaucoma scintillans* (solid circles), *Colpodium campylum* (open circles), *Paramecium trichium* (solid triangles), and *Cyclidium glaucoma* (open triangles). (Data replotted from Fenchel 1980a.)

Size structuring of bacterial communities by grazers is especially interesting when combined with observations that the smallest marine bacteria may be metabolically inactive or, indeed, dead (Gasol et al. 1995; Choi et al. 1996). Marine bacteria may need to reach a certain size before they contain enough material for cell division. At the same time, bacterivores may preferentially

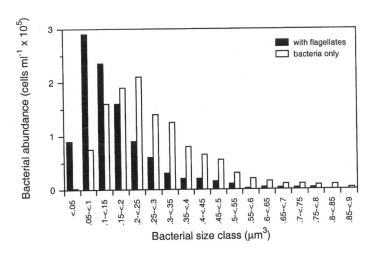

Figure 8. Size distribution of bacterial cells from a marine assemblage in the presence (solid bars) and absence (open bars) of the bacterivorous flagellate *Ochromonas* sp. (Data replotted from Andersson et al. 1986.)

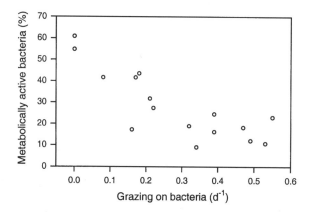

Figure 9. Metabolically active cells (as determined by CTC reduction) as a percentage of the whole bacterial community in samples of Mediterranean seawater, plotted as a function of bacterivory rates. Protist density was manipulated by size fractionation and bacterivory measured by minicell removal. (Data replotted from del Giorgio et al. 1996.)

graze this actively growing portion of the bacterial community (Figure 9), leaving behind, as the bulk of the ocean's standing stock, bacteria that are growing slowly or not at all.

A further twist in this story of bacteria–grazer interaction is the role of nutrient regeneration by grazers in supplying nutrients to bacteria. Protists are thought to dominate nutrient regeneration in much of the ocean, with bacteria primarily taking up, rather than releasing, regenerated ammonium and phosphate (Chapter 9). Experiments with bacterivores have shown that bacterivory can stimulate bacterial production, presumably by transforming ingested prey cells into regenerated nutrients and DOC (Goldman et al. 1985; Caron and Goldman 1990; Nagata and Kirchman 1991; Snyder and Hoch 1996). These are the sorts of couplings that can cause problems for class II bacterivory methods (see above). At the same time that bacterivores are cropping the larger, more rapidly growing bacterial cells in a community, they may be stimulating the growth of the remainder by excreting nutrients and DOC (Sieburth 1984; Jumars et al. 1989; Strom et al. 1997). In this view, production, consumption, and regeneration are tightly coupled processes that promote high cell-specific rates of bacterial production even in oligotrophic ocean waters (Azam et al. 1983; Goldman 1984).

PROTIST FINICALITY: EVIDENCE FOR OTHER TYPES OF FEEDING SELECTIVITY

In addition to discriminating against bacteria on the basis of size, protist bacterivores may feed preferentially on motile bacterial cells (Figure 10). This may be the source of bias against fluorescently labeled (hence heat-killed and

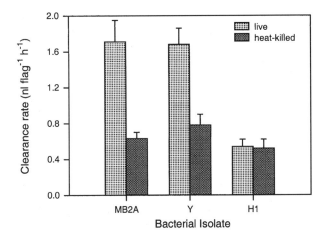

Figure 10. Clearance rates (mean ± 1SD) by a natural (Oregon coast) flagellate assemblage on live-stained bacteria (lighter shaded bars) and heat-killed, fluorescently stained bacteria (darker shaded bars). Each of three bacterial isolates was prepared using the two treatments. Isolates MB2A and Y were motile; isolate H1 was nonmotile. A significant decrease in clearance upon heat killing was detected only for the motile strains. (Data from Table 2 of Gonzalez et al. 1993b.)

nonmotile) bacteria found by some investigators (Monger and Landry 1992; Gonzalez et al. 1993b). Preferential feeding based on size or motility could be a passive selection (sensu Verity 1991a) in that both larger size and prey motility will increase encounter rates between grazer and prey cells (see above). In an ocean undersaturated with bacterial prey, more frequent encounters will increase bacterivore ingestion rates, and more complex predator–prey interactions need not be invoked to explain such selectivity.

The extent to which active selection occurs in bacterivory is unknown. Are bacterivores able to sense prey properties before capture, "deciding" to eat some cells and not others? Can prey cells be assessed after capture and rejected before ingestion? There is mounting evidence that such behaviors are possible for larger protists feeding on phytoplankton (Stoecker 1988; Taniguchi and Takeda 1988). Two species of bacterivorous flagellates showed feeding selectivity that was stronger at higher bacterial abundance, indicating behavioral plasticity that responds to environmental cues (Figure 11, Jürgens and DeMott 1995). The heterotrophic dinoflagellate *Oxyrrhis marina*, which can ingest bacteria, was found to discriminate between identically sized but compositionally differing strains of the prymnesiophyte alga *Emiliania huxleyi*, suggesting the possibility of chemical defenses among the microplankton (Wolfe et al. 1997). Coupling feeding studies with newly developed probes for bacterial taxonomy (Cynar et al. 1985; Simek et al. 1997; Chapter 3) would be a fascinating way to investigate the power of bacterivory in shaping taxonomic and biochemical aspects of bacterial communities.

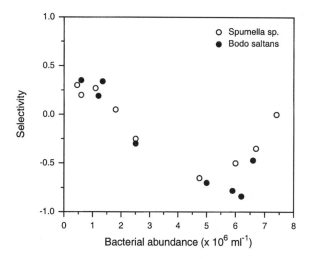

Figure 11. Selective feeding by two species of heterotrophic flagellates (*Spumella* sp. and *Bodo saltans*) as a function of live bacterial abundance. Latex beads (0.88 μm) and live-stained (RITC) bacteria were offered in equal concentrations against a background of varying concentrations of unstained bacteria. Selectivity was calculated as $(F_B - F_S)/(F_B + F_S)$ where F_B and F_S are clearance rates for latex beads and stained bacteria, respectively. Strong selection against low quality latex beads (large negative selection values) was associated with high overall prey abundances for both flagellates. (Data replotted from Jürgens and DeMott 1995.)

BIOGEOCHEMICAL CONSEQUENCES OF BACTERIVORY

Clearly bacterivory is important in regulating bacterial biomass and community structure in the sea. Just as clearly, marine bacteria, through their metabolic activities, regulate both stocks and transformation rates of numerous globally important compounds (Chapters 6, 13, and 14). Rarely have these two observations been combined in an attempt to determine how grazing regulation of bacterial communities affects bacterially mediated elemental cycling. For example, denitrification in low oxygen waters results in a major ocean–atmosphere transfer of fixed nitrogen, reducing the amount of nitrogen available to support primary production in the sea. Investigation of controls on bacterial denitrification has so far focused entirely on physical and chemical factors, including temperature, oxygen level, and DOC supply. While protists have not been studied in the vast oceanic low-oxygen water masses, distinctive species of hypoxic and anoxic ciliates are known to inhabit coastal regions such as seasonally anoxic embayments (Fenchel et al. 1990; Dolan and Coats 1991). Many such ciliates are bacterivorous and harbor bacterial endosymbionts. Might these and related hypoxic protists consume denitrifying bacteria and so aid in the regulation of global denitrification rates? Similar questions could be asked about grazing control of bacteria that mediate nitrification (e.g., Lavrentyev et al. 1997), nitrogen fixation, and other important elemental transformations in the sea.

As with nitrogen, global carbon cycles are critical in regulating earth's climate and level of biological activity. Dissolved organic matter in the sea constitutes one of the largest organic carbon reservoirs on the earth's surface (Hedges 1992) and, in both coastal and oceanic waters, DOM concentrations often exhibit an annual cycle, reaching local maxima in the summer (Hansell et al. 1995; Williams 1995; Chapter 6). Thingstad et al. (1997) have hypothesized that such accumulations might be the result of a "malfunctioning" microbial loop. In other words, grazers may prevent bacterial communities from achieving the biomass necessary to take up and metabolize all the DOM released by other organisms, especially during summer months of high phytoplankton and zooplankton productivity. In turn, such DOM accumulations may "buffer" seasonality in planktonic food webs, providing an alternative energy source during winter low productivity months. An exciting future for the study of bacterivory lies in the coupling of molecular marker techniques (e.g., to identify the sources and reactivity of accumulated DOM) with field experimentation to identify biological controls on these important bacterially driven cycles.

SUMMARY

1. While a variety of marine organisms can ingest bacteria, very small ($< 5 \mu$m) phagotrophic nanoflagellates probably account for most of the bacterivory in the ocean's water column.

2. Measurement of bacterivory is problematic. Prey surrogates may not be ingested to the same extent as natural bacteria, while other methods disrupt the close production–consumption–regeneration cycles on which microbial rates depend. Less invasive methods are needed.

3. In low productivity waters, bacterivory is generally equivalent to bacterial production, while production often exceeds grazing removal where productivity is high. Bacterivores consume a substantial fraction of primary production in the vast ocean regions dominated by *Synechococcus* and *Prochlorococcus*.

4. The transformation of bacterial biomass into protist biomass via bacterivory makes bacterial production available to larger consumers. This transformation is associated with relatively high rates of DOM production, nutrient regeneration, and CO_2 release, so that overall, complex microbial communities are relatively inefficient links to higher trophic levels.

5. Bacterivory has a profound influence on the size structure, biomass, taxonomic composition, and metabolic activity of bacterial communities in the sea. "Top-down" regulation of specific bacterial groups almost certainly influences global biogeochemical cycles.

REFERENCES

Alldredge, A. L. (1976) Discarded appendicularian houses as sources of food, surface habitats, and particulate organic matter in planktonic environments. *Limnol. Oceanogr.* 21:14–23.

Alldredge, A. L., Passow, U., and Logan, B. E. (1993) The abundance and significance of a class of large, transparent organic particles in the ocean. *Deep-Sea Res.* 40:1131–1140.

Ammerman, J. W., Fuhrman, J. A., Hagström, Å., and Azam, F. (1984) Bacterioplankton growth in seawater. I. Growth kinetics and cellular characteristics in seawater culture. *Mar. Ecol. Prog. Ser.* 18:31–39.

Andersen, P. (1988/89) Functional biology of the choanoflagellate *Diaphanoeca grandis* Ellis. *Mar. Microb. Food Webs* 3:35–50.

Andersen, P., and Sorensen, H. M. (1986) Population dynamics and trophic coupling in pelagic microorganisms in eutrophic coastal waters. *Mar. Ecol. Prog. Ser.* 33:99–109.

Anderson, O. R. (1988) *Comparative Protozoology.* Springer-Verlag, New York,

Andersson, A., Larsson, U., and Hagström, Å. (1986) Size-selective grazing by a microflagellate on pelagic bacteria. *Mar. Ecol. Prog. Ser.* 33:51–57.

Azam, F., Fenchel, T., Field, J. G., Gray, J. S., Meyer-Reil, L. A., and Thingstad, F. (1983) The ecological role of water-column microbes in the sea. *Mar. Ecol. Prog. Ser.* 10:257–263.

Barcina, I., Ayo, B., Unanue, M., Egea, L., and Iriberri, J. (1992) Comparison of rates of flagellate bacterivory and bacterial production in a marine coastal system. *Appl. Environ. Microbiol.* 58:3850–3856.

Bedo, A. W., Acuna, J. L., Robins, D., and Harris, R. P. (1993) Grazing in the micron and submicron particle size range: The case of *Oikopleura dioica* (Appendicularia). *Bull. Mar. Sci.* 53:2–14.

Bernard, C., and Fenchel, T. (1994) Chemsensory behaviour of *Strombidium purpureum*, an anaerobic oligotrich with endosymbiotic purple non-sulphur bacteria. *J. Eukaryotic Microbiol.* 41:391–396.

Børsheim, K. Y. (1984) Clearance rates of bacteria-sized particles by freshwater ciliates, measured with monodisperse fluorescent latex beads. *Oecologia* 63:286–288.

Bratbak, G. (1985) Bacterial biovolume and biomass estimations. *Appl. Environ. Microbiol.* 49:1488–1493.

Burkill, P. H., Edwards, E. S., John, A. W. G., and Sleigh, M. A. (1993) Microzooplankton and their herbivorous activity in the northeastern Atlantic Ocean. *Deep-Sea Res. II* 40:479–493.

Campbell, L., and Carpenter, E. J. (1986) Estimating the grazing pressure of heterotrophic nanoplankton on *Synechococcus* spp. using the seawater dilution and selective inhibitor techniques. *Mar. Ecol. Prog. Ser.* 33:121–129.

Campbell, L., Nolla, H. A., and Vaulot, D. (1994) The importance of *Prochlorococcus* to community structure in the central North Pacific Ocean. *Limnol. Oceanogr.* 39:954–961.

Canuel, E. A., Cloern, J. E., Ringelberg, D. B., Guckert, J. B., and Rau, G. H. (1995)

Molecular and isotopic tracers used to examine sources of organic matter and its incorporation into the food webs of San Francisco Bay. *Limnol. Oceanogr.* 40:67–81.

Caron, D. A. (1987) Grazing of attached bacteria by heterotrophic microflagellates. *Microb. Ecol.* 13:203–218.

Caron, D. A. (1991) Heterotrophic flagellates associated with sedimenting detritus. In D. J. Patterson and J. Larsen, eds., *The Biology of Free-Living Heterotrophic Flagellates.* Clarendon Press, Oxford, pp. 77–92.

Caron, D. A., and Goldman, J. C. (1990) Protozoan nutrient regeneration. In G. M. Capriulo, ed., *Ecology of Marine Protozoa.* Oxford University Press, New York, pp. 283–306.

Caron, D. A., Davis, P. G., Madin, L. P., and Sieburth, J. M. (1982) Heterotrophic bacteria and bacterivorous protozoa in oceanic macroaggregates. *Science* 218:795–797.

Caron, D. A., Goldman, J. C., and Dennett, M. R. (1990a) Carbon utilization by the omnivorous flagellate *Paraphysomonas imperforata. Limnol. Oceanogr.* 35:192–201.

Caron, D. A., Goldman, J. C., and Fenchel, T. (1990b) Protozoan respiration and metabolism. In G. M. Capriulo, ed., *Ecology of Marine Protozoa.* Oxford University Press, New York, pp. 307–322.

Caron, D. A., Lim, E. L., Miceli, G., Waterbury, J. B., and Valois, F. W. (1991) Grazing and utilization of chroococcoid cyanobacteria and heterotrophic bacteria by protozoa in laboratory cultures and a coastal plankton community. *Mar. Ecol. Prog. Ser.* 76:205–217.

Caron, D. A., Lessard, E. J., Voytek, M., and Dennett, M. R. (1993) Use of tritiated thymidine (TdR) to estimate rates of bacterivory: Implications of label retention and release by bacterivores. *Mar. Microb. Food Webs* 7:177–196.

Cavalier-Smith, T., and Lee, J. J. (1985) Protozoa as hosts for endosymbioses and the conversion of symbionts in organelles. *J. Protozool.* 32:376–379.

Chase, Z., and Price, (1998) Metabolic consequences of iron deficiency in heterotrophic marine protozoa. *Limnol. Oceanogr.* 42:1673–1684.

Chin, W.-C., Orellana, M. V., and Verdugo, P. (1998) Spontaneous assembly of marine dissolved organic matter into polymer gels. *Nature* 391:568–572.

Cho, B. C., and Azam, F. (1988) Major role of bacteria in biogeochemical fluxes in the ocean's interior. *Nature* 332:441–443.

Cho, B. C., and Azam, F. (1990) Biogeochemical significance of bacterial biomass in the ocean's euphotic zone. *Mar. Ecol. Prog. Ser.* 63:253–259.

Choi, J. W., Sherr, E. B., and Sherr, B. F. (1996) Relation between presence–absence of a visible nucleoid and metabolic activity in bacterioplankton cells. *Limnol. Oceanogr.* 41:1161–1168.

Christoffersen, K., Nybroe, O., Jürgens, K., and Hansen, M. (1997) Measurement of bacterivory by heterotrophic nanoflagellates using immunofluorescence labelling of ingested cells. *Aquat. Microb. Ecol.* 13:127–134.

Crump, B. C., and Baross, (1996) Particle-attached bacteria and heterotrophic plankton associated with the Columbia River estuarine turbidity maxima. *Mar. Ecol. Prog. Ser.* 138:265–273.

Cucci, T. L., Shumway, S. E., Brown, W. S., and Newell, C. R. (1989) Using phytoplankton and flow cytometry to analyze grazing by marine organisms. *Cytometry* 10:659–669.

Cynar, F. J., Estep, K. W., and Sieburth, J. M. (1985) The detection and characterization of bacteria-sized protists in "protist-free" filtrates and their potential impact on experimental marine ecology. *Microb. Ecol.* 11:281–288.

Deibel, D., and Lee, S. H. (1992) Retention efficiency of sub-micrometer particles by the pharyngeal filter of the pelagic tunicate *Oikopleura vanhoeffeni. Mar. Ecol. Prog. Ser.* 81:25–30.

Del Giorgio, P. A., Gasol, J. M., Vaqué, D., Mura, P., Agustí, S., and Duarte, C. M. (1996) Bacterioplankton community structure: Protists control net production and the proportion of active bacteria in a coastal marine community. *Limnol. Oceanogr.* 41:1169–1179.

Dolan, J. R., and Coats, D. W. (1991) Changes in fine-scale vertical distributions of ciliate microzooplankton related to anoxia in Chesapeake Bay waters. *Mar. Microb. Food Webs* 5:81–93.

Ducklow, H. W., and Hill, S. M. (1985) The growth of heterotrophic bacteria in the surface waters of warm core rings. *Limnol. Oceanogr.* 30:239–259.

Ederington, M. C., McManus, G. B., and Harvey, H. R. (1995) Trophic transfer of fatty acids, sterols, and a triterpenoid alcohol between bacteria, a ciliate, and the copepod *Acartia tonsa. Limnol. Oceanogr.* 40:860–867.

Epstein, S. S., and Rossel, J. (1995) Methodology of in situ grazing experiments: Evaluation of a new vital dye for preparation of fluorescently labeled bacteria. *Mar. Ecol. Prog. Ser.* 128:143–150.

Epstein, S. S., and Shiaris, M. P. (1992) Size-selective grazing of coastal bacterioplankton by natural assemblages of pigmented flagellates, colorless flagellates, and ciliates. *Microb. Ecol.* 23:211–225.

Fenchel, T. (1980a) Suspension feeding in ciliated protozoa: Functional response and particle size selection. *Microb. Ecol.* 6:1–11.

Fenchel, T. (1980b). Suspension feeding in ciliated protozoa: Structure and function of feeding organelles. *Arch. Protistenkol.* 123:239–260.

Fenchel, T. (1982) Ecology of heterotrophic microflagellates. IV. Quantitative occurrence and importance as bacterial consumers. *Mar. Ecol. Prog. Ser.* 9:35–42.

Fenchel, T. (1986) Protozoan filter feeding. *Prog. Protistol.* 1:65–113.

Fenchel, T., Kristensen, L. D., and Rasmussen, L. (1990) Water column anoxia: Vertical zonation of planktonic protozoa. *Mar. Ecol. Prog. Ser.* 62:1–10.

Finlay, B. J., and Fenchel, T. (1992) Methanogens and other bacteria as symbionts of free-living anaerobic ciliates. *Symbiosis* 14:375–390.

Flood, P. R. (1991) Architecture of, and water circulation and flow rate in, the house of the planktonic tunicate *Oikopleura labradoriensis. Mar. Biol.* 111:95–111.

Fraser, A. J., Sargent, J. R., Gamble, J. C., and MacLachlan, P. (1987) Lipid class and fatty acid composition as indicators of the nutritional condition of larval Atlantic herring. *Am. Fish. Soc. Symp.* 2:129–143.

Fuhrman, J. A., and Bell, T. M. (1985) Biological considerations in the measurement of dissolved free amino acids in seawater and implications for chemical and microbiological studies. *Mar. Ecol. Prog. Ser.* 25:13–21.

Fuhrman, J. A., and McManus, G. B. (1984) Do bacteria-sized marine eukaryotes consume significant bacterial production? *Science* 224:1257–1260.

Fuhrman, J. A., and Noble, R. T. (1995) Viruses and protists cause similar bacterial mortality in coastal seawater. *Limnol. Oceanogr.* 40:1236–1242.

Gasol, J. M., del Giorgio, P. A., Massana, R., and Duarte, C.M. (1995) Active versus inactive bacteria: Size dependence in a coastal marine plankton community. *Mar. Ecol. Prog. Ser.* 128:91–97.

Geider, R. J. (1992) Respiration: Taxation without representation? In P. G. Falkowski and A. D. Woodhead, eds., *Primary Productivity and Biogeochemical Cycles in the Sea.* Plenum Press, New York, pp. 333–360.

Gerritsen, J., and Strickler, J. R. (1977) Encounter probabilities and community structure in zooplankton: A mathematical model. *J. Fish. Res. Bard. Can.* 34:73–82.

Gifford, D. J., and Dagg, M. J. (1991) The microzooplankton–mesozooplankton link: Consumption of planktonic protozoa by the calanoid copepods *Acartia tonsa* Dana and *Neocalanus plumchrus* Murukawa. *Mar. Microb. Food Webs* 5:161–177.

Goldman, J. C. (1984) Oceanic nutrient cycles. In M. J. Fasham, ed., *Flow of Energy and Materials in Marine Ecosystems: Theory and Practice.* Plenum Press, New York, pp. 137–170.

Goldman, J. C., McCarthy, J. J., and Peavey, D. G. (1979) Growth rate influence on the chemical composition of phytoplankton in oceanic waters. *Nature* 279:210–215.

Goldman, J. C., Caron, D. A., Andersen, O. K., and Dennett, M. R. (1985) Nutrient cycling in a microflagellate food chain. I. Nitrogen dynamics. *Mar. Ecol. Prog. Ser.* 24:231–242.

Goldman, J. C., Caron, D. A., and Dennett, M. R. (1987) Regulation of gross growth efficiency and ammonium regeneration in bacteria by substrate C:N ratio. *Limnol. Oceanogr.* 32:1239–1252.

Gonzalez, J. M. (1996) Efficient size-selective bacterivory by phagotrophic nanoflagellates in aquatic systems. *Mar. Biol.* 126:785–789.

Gonzalez, J. M., Sherr, E. B., and Sherr, B. F. (1990) Size-selective grazing on bacteria by natural assemblages of estuarine flagellates and ciliates. *Appl. Environ. Microbiol.* 56:583–589.

Gonzalez, J. M., Sherr, B. F., and Sherr, E. B. (1993a). Digestive enzyme activity as a quantitative measure of protistan grazing: The acid lysosyme assay for bacterivory. *Mar. Ecol. Prog. Ser.* 100:197–206.

Gonzalez, J. M., Sherr, E. B., and Sherr, B. F. (1993b) Differential feeding by marine flagellates on growing vs. starving, and on motile vs. non-motile, bacterial prey. *Mar. Ecol. Prog. Ser.* 102:257–267.

Gowing, M. M. (1989) Abundance and feeding ecology of Antarctic phaeodarian radiolarians. *Mar. Biol.* 103:107–118.

Hagström, Å., Azam, F., Andersson, A., Wikner, J., and Rassoulzadegan, F. (1988) Microbial loop in an oligotrophic pelagic marine ecosystem: Possible roles of cyanobacteria and nanoflagellates in the organic fluxes. *Mar. Ecol. Prog. Ser.* 49:171–178.

Hansell, D. A., Bates, N. R., and Gundersen, K. (1995) Mineralization of dissolved organic carbon in the Sargasso Sea. *Mar. Chem.* 51:201–212.

Hansen, P. J., Bjørnsen, P. K., and Hansen, B. W. (1997) Zooplankton grazing and growth: Scaling within the $2-2000\,\mu$m body size range. *Limnol. Oceanogr.* 42:687–704.

Hedges, J. I. (1992) Global biogeochemical cycles: Progress and problems. *Mar. Chem.* 39:67–93.

Hollibaugh, J. T., Fuhrman, J. A., and Azam, F. (1980) Radioactive labelling of natural assemblages of bacterioplankton for use in trophic studies. *Limnol. Oceanogr.* 25:172–181.

Jenkins, D. G., Atkinson, C. F., and Garland, J. L. (1998) A cautionary note on measuring protistan bacterivory by acid lysozyme. *Invert. Biol.* 117:181–185.

Johannes, R. E. (1965) Influence of marine protozoa on nutrient regeneration. *Limnol. Oceanogr.* 10:434–442.

Jones, R. I. (1994) Mixotrophy in planktonic protists as a spectrum of nutritional strategies. *Mar. Microb. Food Webs* 8:87–96.

Jorgensen, C. B. (1983) Fluid mechanical aspects of suspension feeding. *Mar. Ecol. Prog. Ser.* 11:89–103.

Jumars, P. A., Penry, D. L., Baross, J. A., Perry, M. J., and Frost, B. W. (1989) Closing the microbial loop: Dissolved carbon pathway to heterotrophic bacteria from incomplete ingestion, digestion and absorption in animals. *Deep-Sea Res.* 36:483–495.

Jürgens, K., and DeMott, W. R. (1995) Behavioral flexibility in prey selection by bacterivorous nanoflagellates. *Limnol. Oceanogr.* 40:1503–1507.

Jürgens, K., and Güde, H. (1994) The potential importance of grazing-resistant bacteria in planktonic systems. *Mar. Ecol. Prog. Ser.* 112:169–188.

Kawachi, M., Inouye, I., Maeda, O., and Chihara, M. (1991) The haptonema as a food-capturing device: Observations on *Chrysochromulina hirta* (Prymnesiophyceae). *Phycologia* 30:563–573.

King, K. R., Hollibaugh, J. T., and Azam, F. (1980) Predator–prey interactions between the larvacean *Oikopleura dioica* and bacterioplankton in enclosed water columns. *Mar. Biol.* 56:49–57.

Kivi, K., Kaitala, S., Kuosa, H., Kuparinen, J., Leskinen, E., Lignell, R., Marcussen, B., and Tamminen, T. (1993) Nutrient limitation and grazing control of the Baltic plankton community during annual succession. *Limnol. Oceanogr.* 38:893–905.

Kuuppo-Leinikki, P. (1990) Protozoan grazing on planktonic bacteria and its impact on bacterial populations. *Mar. Ecol. Prog. Ser.* 63:227–238.

Landry, M. R. (1993a) Estimating rates of growth and grazing mortality of phytoplankton by the dilution method. In P. F. Kemp, B. F. Sherr, E. B. Sherr, and J. J. Cole, eds., *Current Methods in Aquatic Microbial Ecology.* Lewis Publishers, Boca Raton, FL, pp. 715–722.

Landry, M. R. (1993b) Predicting excretion rates of microzooplankton from carbon metabolism and elemental ratios. *Limnol. Oceanogr.* 38:468–472.

Landry, M. R. (1994) Methods and controls for measuring the grazing impact of planktonic protists. *Mar. Microb. Food Webs* 8:37–57.

Landry, M. R., and Hassett, R. P. (1982) Estimating the grazing impact of marine microzooplankton. *Mar. Biol.* 67:283–288.

Landry, M. R., Haas, L. W., and Fagerness, V. L. (1984) Dynamics of microbial plankton communities: Experiments in Kaneohe Bay, Hawaii. *Mar. Ecol. Prog. Ser.* 16:127–133.

Landry, M. R., Lehner-Fournier, J. M., Sundstrom, J. A., Fagerness, V. L., and Selph, K. E. (1991) Discrimination between living and heat-killed prey by a marine zooflagellate, *Paraphysomonas vestita* (Stokes). *J. Exp. Mar. Biol. Ecol.* 146:139–151.

Landry, M. R., Constantinou, J., and Kirshtein, J. (1995) Microzooplankton grazing in the central equatorial Pacific during February and August, 1992. *Deep-Sea Res. II* 42:657–671.

Laval-Peuto, M. (1991) Endosymbiosis in the protozoa. In P. C. Reid, C. M. Turley, and P. H. Burkill, eds., *Protozoa and Their role in Marine Processes.* Springer-Verlag, Berlin, pp. 143–160.

Lavrentyev, P. J., Gardner, W. S., and Johnson, J. R. (1997) Cascading trophic effects on aquatic nitrification: Experimental evidence and potential implications. *Aquat. Microb. Ecol.* 13:161–175.

Lee, J. J., Soldo, A. T., Reisser, W., Lee, M. J., Jeon, K. W. and Gošrtz, H.-D. (1985) The extent of algal and bacterial endosymbioses in Protozoa. *J. Protozool.* 32:391–403.

Lessard, E. J., and Swift, E. (1985) Species-specific grazing rates of heterotrophic dinoflagellates in oceanic waters, measured with a dual-label radioisotope technique. *Mar. Biol.* 87:289–296.

Liu, H., Campbell, L., and Landry, M. R. (1995) Growth and mortality rates of *Prochlorococcus* and *Synechococcus* measured with a selective inhibitor technique. *Mar. Ecol. Prog. Ser.* 116:277–287.

Liu, H., Campbell, L., Landry, M. R., Nolla, H. A., Brown, S. L., and Constantinou, J. (1998) *Prochlorococcus* and *Synechococcus* growth rates and contributions to production in the Arabian Sea during the 1995 Southwest and Northeast Monsoons. *Deep-Sea Res. II* 45:2327–2352.

Mann, K. H. (1988) Production and use of detritus in various freshwater, estuarine, and coastal marine ecosystems. *Limnol. Oceanogr.* 33:910–930.

Maranger, R., Bird, D. F., and Price, N. M. (1998) Iron acquisition by photosynthetic marine phytoplankton from ingested bacteria. *Nature* 396:248–251.

Margulis, L. (1981) *Symbiosis in Cell Evolution.* W. H. Freeman, San Francisco.

Mayes, D. F., Rogerson, A., Marchant, H., and Laybourn-Parry, J. (1997) Growth and consumption rates of bacterivorous Antarctic naked marine amoebae. *Mar. Ecol. Prog. Ser.* 160:101–108.

McManus, G. B., and Ederington-Cantrell, M. C. (1992) Phytoplankton pigments and growth rates, and microzooplankton grazing in a large temperate estuary. *Mar. Ecol. Prog. Ser.* 87:77–85.

McManus, G. B., and Fuhrman, J. A. (1988a) Clearance of bacteria-sized particles by natural populations of nanoplankton in the Chesapeake Bay outflow plume. *Mar. Ecol. Prog. Ser.* 42:199–206.

McManus, G. B., and Fuhrman, J. A. (1988b) Control of marine bacterioplankton populations: Measurement and significance of grazing. *Hydrobiologia* 159:51–62.

McManus, G. B., and Okubo, A. (1991) On the use of surrogate food particles to measure protistan ingestion. *Limnol. Oceanogr.* 36:613–617.

Michaels, A. F., and Silver, M. W. (1988) Primary production, sinking fluxes and the microbial food web. *Deep-Sea Res.* 35:473–490.

Monger, B. C., and Landry, M. R. (1990) Direct-interception feeding by marine zooflagellates: The importance of surface and hydrodynamic forces. *Mar. Ecol. Prog. Ser.* 65:123–140.

Monger, B. C., and Landry, M. R. (1991) Prey-size dependency of grazing by free-living marine flagellates. *Mar. Ecol. Prog. Ser.* 74:239–248.

Monger, B. C., and Landry, M. R. (1992) Size-selective grazing by heterotrophic nanoflagellates: An analysis using live-stained bacteria and dual-beam flow cytometry. *Arch. Hydrobiol.* 37:173–185.

Mullin, M. M. (1983) In situ measurement of filtering rates of the salp, *Thalia democratica*, on phytoplankton and bacteria. *J. Plankton Res.* 5:279–288.

Murrell, M. C., and Hollibaugh, J. T. (1998) Microzooplankton grazing in northern San Francisco Bay measured by the dilution method. *Aquat. Microb. Ecol.* 15:53–63.

Nagata, T. (1986) Carbon and nitrogen content of natural planktonic bacteria. *Appl. Environ. Microbiol.* 52:28–32.

Nagata, T., and Kirchman, D. L. (1991) Release of dissolved free and combined amino acids by bactivorous marine flagellates. *Limnol. Oceanogr.* 36:433–443.

Nagata, T., and Kirchman, D. L. (1992) Release of macromolecular organic complexes by heterotrophic marine flagellates. *Mar. Ecol. Prog. Ser.* 83:233–240.

Newell, S. Y., Sherr, B. F., Sherr, E. B., and Fallon, R. D. (1983) Bacterial response to presence of eukaryote inhibitors in water from a coastal marine environment. *Mar. Environ. Res.* 10:147–157.

Nygaard, K., and Hessen D. O. (1990) Use of ^{14}C-protein-labelled bacteria for estimating clearance rates by heterotrophic and mixotrophic flagellates. *Mar. Ecol. Prog. Ser.* 68:7–14.

Ohman, M. D., and Snyder, R. A. (1991) Growth kinetics of the omnivorous oligotrich ciliate *Strombidium* sp. *Limnol. Oceanogr.* 36:922-935.

Pace, M. L. (1988) Bacterial mortality and the fate of bacterial production. *Hydrobiologia* 159:41–49.

Painchaud, J., Lefaivre, D., Therriault, J.-C., and Legendre, L. (1996) Bacterial dynamics in the upper St. Lawrence estuary. *Limnol. Oceanogr.* 41:1610–1618.

Patterson, D. J., and Larsen, J., eds. (1991) *The Biology of Free-Living Heterotrophic Flagellates.* Systematics Association special volume, Clarendon Press, Oxford.

Peters, F., and Gross, T. (1994) Increased grazing rates of microplankton in response to small-scale turbulence. *Mar. Ecol. Prog. Ser.* 115:299–307.

Pomeroy, L. R. (1974) The ocean's food web, a changing paradigm. *BioScience* 24:499–504.

Posch, T., and Arndt, H. (1996) Uptake of sub-micrometre- and micrometre-sized detrital particles by bacterivorous and omnivorous ciliates. *Aquat. Microb. Ecol.* 10:45–53.

Purcell, E. M. (1976) Life at low Reynolds number. *Am. J. Phys.* 45:3–11.

Raghukumar, S. (1992) Bacterivory: A novel dual role for thraustochytrids in the sea. *Mar. Biol.* 113:165–169.

Raven, J. A. (1994) Why are there no picoplanktonic O_2 evolvers with volumes less than 10^{-19} m³? *J. Plankton Res.* 16:565–580.

Riemann, B., Havskum, H., Thingstad, F., and Bernard, C. (1995) The role of mixotrophy in pelagic environments. *NATO ASI Ser. Mol. Ecol. Aquat. Microbes* 38:89–114.

Rivkin, R. B., Bosch, I., Pearse, J. S., and Lessard, E. J. (1986) Bacterivory: A novel feeding mode for asteroid larvae. *Science* 233:1311–1314.

Roff, J. C., Turner, J. T., Webber, M. K.. and Hopcroft, R. R. (1995) Bacterivory by tropical copepod nauplii: Extent and possible significance. *Aquat. Microb. Ecol.* 9:165–175.

Rubenstein, D. I., and Koehl, M. A. R. (1977) The mechanisms of filter feeding: Some theoretical considerations. *Am. Sci.* 111:981–994.

Saliot, A., Laureillard, J., Scribe, P., and Sicre, M. A. (1991) Evolutionary trends in the lipid biomarker approach for investigating the biogeochemistry of organic matter in the marine environment. *Mar. Chem.* 36:233–248.

Sherr, B. F., and Sherr, E. B. (1991) Proportional distribution of total numbers, biovolume, and bacterivory among size classes of 2–20 μm nonpigmented marine flagellates. *Mar. Microb. Food Webs* 5:227–237.

Sherr, B. F., Sherr, E. B., Andrew, T. L., Fallon, R. D., and Newell, S. Y. (1986) Trophic interactions between heterotrophic protozoa and bacterioplankton in estuarine water analyzed with selective metabolic inhibitors. *Mar. Ecol. Prog. Ser.* 32:169–179.

Sherr, B. F., Sherr, E. B., and Fallon, R. D. (1987) Use of monodispersed, fluorescently labeled bacteria to estimate in situ protozoan bacterivory. *Appl. Environ. Microbiol.* 53:958–965.

Sherr, B. F., Sherr, E. B., and Pedrós-Alió, C. (1989) Simultaneous measurement of bacterioplankton production and protozoan bacterivory in estuarine water. *Mar. Ecol. Prog. Ser.* 54:209–219.

Sherr, E. B. (1988) Direct use of high molecular weight polysaccharide by heterotrophic flagellates. *Nature* 335:348–351.

Sherr, E. B., and Sherr, B. F. (1987) High rates of consumption of bacteria by pelagic ciliates. *Nature* 325:710–711.

Sherr, E. B., and Sherr, B. F. (1993) Protistan grazing rates via uptake of fluorescently labeled prey. In P. F. Kemp, B. F. Sherr, E. B. Sherr, and J. J. Cole, eds., *Current Methods in Aquatic Microbial Ecology.* Lewis Publishers, Boca Raton, FL, pp. 695–701.

Sherr, E. B., Sherr, B. F., and Fessenden, L. (1997) Heterotrophic protists in the Central Arctic Ocean. *Deep-Sea Res. II* 44:1665–1682.

Shimeta, J. (1993) Diffusional encounter of submicrometer particles and small cells by suspension feeders. *Limnol. Oceanogr.* 38:456–465.

Shimeta, J., and Jumars, P. A. (1991) Physical mechanisms and rates of particle capture by suspension feeders. *Oceanogr. Mar. Biol. Annu. Rev.* 29:191–257.

Shimeta, J., Jumars, P. A., and Lessard, E. J. (1995) Influences of turbulence on suspension feeding by planktonic protozoa: Experiments in laminar shear fields. *Limnol. Oceanogr.* 40:845–859.

Sieburth, J. M. (1984) Protozoan bacterivory in pelagic marine waters. In J. Hobbie and P. le B. Williams, eds., *Heterotrophic Activity in the Sea.* Plenum Press, New York, pp. 405–444.

Sieracki, C. K., Sieracki, M. E., and Yentch, C. S. (1998) An imaging-in-flow system for automated analysis of marine microplankton. *Mar. Ecol. Prog. Ser.* 168:285–296.

Sieracki, M. E., Haas, L. W., Caron, D. A., and Lessard, E. J. (1987) Effect of fixation on particle retention by microflagellates: Underestimation of grazing rates. *Mar. Ecol. Prog. Ser.* 38:251–258.

Silver, M. W., Gowing, M. M., Brownlee, D. C., and Corliss, J. O. (1984) Ciliated protozoa associated with oceanic sinking detritus. *Nature* 309:246–248.

Simek, K., Vrba, J., and Lavrentyev, P. (1994) Estimates of protozoan bacterivory: From microscopy to ectoenzyme assay? *Mar. Microb. Food Webs* 8:71–85.

Simek, K., Vrba, J., Posch, T., Hartman, P., Nedoma, J., and Psenner, R. (1997) Morphological and compositional shifts in an experimental bacterial community influenced by protists with contrasting feeding modes. *Appl. Environ. Microbiol.* 63:587–595.

Simon, M., and Azam, F. (1989) Protein content and protein synthesis rates of planktonic marine bacteria. *Mar. Ecol. Prog. Ser.* 51:201–213.

Skerratt, J. H., Nichols, P. D., McMeekin, T. A., and Burton, H. (1995) Seasonal and inter-annual changes in planktonic biomass and community structure in eastern Antarctica using signature lipids. *Mar. Chem.* 51:93–113.

Sleigh, M. (1973) *The Biology of Protozoa.* American Elsevier, New York.

Snyder, R. A., and Hoch, M. P. (1996) Consequences of protist-stimulated bacterial production for estimating protist growth efficiencies. *Hydrobiologia* 341:113–123.

Solic, M., and Krstulovic, N. (1994) Role of predation in controlling bacterial and heterotrophic nanoflagellate standing stocks in the coastal Adriatic Sea: Seasonal patterns. *Mar. Ecol. Prog. Ser.* 114:219–235.

Sorokin, Y. (1977) The heterotrophic phase of plankton succesion in the Japan Sea. *Mar. Biol.* 41:107–117.

Steward, G. F., Smith, D. C., and Azam, F. (1996) Abundance and production of bacteria and viruses in the Bering and Chukchi Seas. *Mar. Ecol. Prog. Ser.* 131:287–300.

Stoecker, D. K. (1988) Are marine planktonic ciliates suspension-feeders? *J. Protozool.* 35:252–255.

Stoecker, D. K., and Capuzzo, J. M. (1990) Predation on protozoa: Its importance to zooplankton. *J. Plankton Res.* 12:891–908.

Straile, D. (1997) Gross growth efficiencies of protozoan and metazoan zooplankton and their dependence on food concentration, predator–prey weight ratio, and taxonomic group. *Limnol. Oceanogr.* 42:1375–1385.

Strayer, D. (1988) On the limits to secondary production. *Limnol. Oceanogr.* 33:1217–1220.

Strom, S. L. (1991) Growth and grazing rates of the herbivorous dinoflagellate *Gymnodinium* sp. from the open subarctic Pacific Ocean. *Mar. Ecol. Prog. Ser.* 78:103–113.

Strom, S. L., and Welschmeyer, N. A. (1991) Pigment-specific rates of phytoplankton growth and microzooplankton grazing in the open subarctic Pacific Ocean. *Limnol. Oceanogr.* 36:50–63.

Strom, S. L., Benner, R., Ziegler, S., and Dagg, M. J. (1997) Planktonic grazers are a potentially important source of marine dissolved organic carbon. *Limnol. Oceanogr.* 42(6):1364–1374.

Strom, S. L., Miller, C. B., and Frost, B. F. What sets the lower limits to phytoplankton biomass in high nitrate, low chlorophyll ocean regions? *Mar. Ecol. Prog. Ser.* (in press).

Tanaka, T., Fujita, N., and Taniguchi, A. (1997) Predator–prey eddy in heterotrophic nanoflagellate–bacteria relationships in a coastal marie environment: A new scheme for predator–prey associations. *Aquat. Microb. Ecol.* 13:249–256.

Taniguchi, A., and Takeda, Y. (1988) Feeding rate and behavior of the tintinnid ciliate *Favella taraikaensis*, observed with a high speed VTR system. *Mar. Microb. Food Webs* 3:21–34.

Tanoue, E., Nishiyama, S., Kamo, M., and Tsugita, A. (1995) Bacterial membranes:Possible source of a major dissolved protein in seawater. *Geochim. Cosmochim. Acta* 59:2643–2648.

Taylor, G. T., Iturriaga, R., and Sullivan, C. W. (1985) Interactions of bactivorous grazers and heterotrophic bacteria with dissolved organic matter. *Mar. Ecol. Prog. Ser.* 23:129–141.

Thingstad, T. F., Hagström, Å., and Rassoulzadegan, F. (1997) Accumulation of degradable DOC in surface waters: Is it caused by a malfunctioning microbial loop? *Limnol. Oceanogr.* 42:398–404.

Tortell, P. D., Maldonado, M. T., and Price, N. M. (1996) The role of heterotrophic bacteria in iron-limited ocean ecosystems. *Nature* 383:330–332.

Tremaine, S. C., and Mills, A. L. (1987) Tests of critical assumptions of the dilution method for estimating bacterivory by microeucaryotes. *Appl. Environ. Microbiol.* 53:2914–2921.

Turner, J. T., and Tester, P. A. (1992) Zooplankton feeding ecology: Bacterivory by metazoan microzooplankton. *J. Exp. Mar. Biol. Ecol.* 160:149–167.

Vaqué, D., Gasol, J. M., and Marrase, C. (1994) Grazing rates on bacteria: The significance of methodology and ecological factors. *Mar. Ecol. Prog. Ser.* 109:263–274.

Verity, P. G. (1991a) Feeding in planktonic protozoans: Evidence for non-random acquisition of prey. *J. Protozool.* 38:69–76.

Verity, P. G. (1991b) Measurement and simulation of prey uptake by marine planktonic ciliates fed plastidic and aplastidic nanoplankton. *Limnol. Oceanogr.* 36:729–750.

Verity, P. G., Robertson, C. Y., Tronzo, C. R., Andrews, M. G., Nelson, J. R., and Sieracki, M. E. (1992) Relationships between cell volume and the carbon and nitrogen content of marine photosynthetic nanoplankton. *Limnol. Oceanogr.* 37:1434–1446.

Vogel, S. (1994) *Life in Moving Fluids.* Princeton University Press, Princeton, NJ.

Volkman, J. K., Jeffrey, S. W., Nichols, P. D., Rogers, G. I., and Garland, C. D. (1989) Fatty acid and lipid composition of 10 species of microalgae used in mariculture. *J. Exp. Mar. Biol. Ecol.* 128:219–240.

Vrba, J., Simek, K., Nedoma, J., and Hartman, P. (1993) 4-Methylumbelliferyl-β-N-acetylglucosaminide hydrolysis by a high-affinity enzyme, a putative marker of protozoan bacterivory. *Appl. Environ. Microbiol.* 59:3091–3101.

Vrba, J., Simek, K., Pernthaler, J., and Psenner, R. (1996) Evaluation of extracellular, high-affinity β-N-acetylglucosaminidase measurements from freshwater lakes: An enzyme assay to estimate protistan grazing on bacteria and picocyanobacteria. *Microb. Ecol.* 32:81–99.

Weinbauer, M. G., Fuks, D., and Peduzzi, P. (1993) Distribution of viruses and dissolved DNA along a coastal trophic gradient in the northern Adriatic Sea. *Appl. Environ. Microbiol.* 59:4074–4082.

Weisse, T. (1989) The microbial loop in the Red Sea: Dynamics of pelagic bacteria and heterotrophic nanoflagellates. *Mar. Ecol. Prog. Ser.* 55:241–250.

Weisse, T., and Scheffel-Moser, U. (1991) Uncoupling the microbial loop: Growth and grazing loss rates of bacteria and heterotrophic nanoflagellates in the North Atlantic. *Mar. Ecol. Prog. Ser.* 71:195–205.

Wikner, J. (1993) Grazing rates of bacterioplankton via turnover of genetically marked miinicells. In P. F. Kemp, B. F. Sherr, E. B. Sherr, and J. J. Cole, eds., *Current Methods in Aquatic Microbial Ecology.* Lewis Publishers, Boca Raton, FL, pp. 703–714.

Wikner, J., and Hagström, Å (1988) Evidence for a tightly coupled nanoplanktonic predator–prey link regulating the bacterivores in the marine environment. *Mar. Ecol. Prog. Ser.* 50:137–145.

Wikner, J., and Hagström, Å. (1991) Annual study of bacterioplankton community dynamics. *Limnol. Oceanogr.* 36:1313–1324.

Wikner, J., Andersson, A., Normark, S., and Hagström, Å. (1986) Use of genetically marked minicells as a probe in measurement of predation on bacteria in aquatic environments. *Appl. Environ. Microbiol.* 52:4–8.

Wikner, J., Rassoulzadegan, F., and Hagström, Å. (1990) Periodic bacterivore activity balances bacterial growth in the marine environment. *Limnol. Oceanogr.* 35:313–324.

Williams, P. J. le B. (1995) Evidence for the seasonal accumulation of carbon-rich dissolved organic matter, its scale in comparison with changes in particulate material and the consequential effect on net C/N assimilation ratios. *Mar. Chem.* 51:17–29.

Wolfe, G. V., Steinke, M., and Kirst, G. O. (1997) Grazing-activated chemical defense in a unicellular marine alga. *Nature* 387:894–897.

Wright, R. T., and Coffin, R. B. (1984) Measuring microzooplankton grazing on planktonic marine bacteria by its impact on bacterial production. *Microb. Ecol.* 10:137–149.

Zubkov, M. V., and Sleigh, M. A. (1995) Ingestion and assimilation by marine protists fed on bacteria labeled with radioactive thymidine and leucine estimated without separating predator and prey. *Microb. Ecol.* 30:157–170.

13

MARINE NITROGEN FIXATION

Hans W. Paerl

Institute of Marine Sciences,
University of North Carolina at Chapel Hill,
Morehead City, North Carolina

Jonathan P. Zehr

Ocean Sciences Department,
University of California, Santa Cruz,
Santa Cruz, California

INTRODUCTION

Nitrogen is a major element of many biological metabolites and structural molecules (amino acids, proteins, nucleic acids). In the marine environment, the availability of N is often believed to be key in limiting growth and productivity (Dugdale 1967; Ryther and Dunstan 1971; Carpenter and Capone 1983). Nitrogen can be lost from ecosystems through the activities of microorganisms that use oxidized forms of nitrogen (nitrate and nitrite) as electron acceptors in a respiration that results in the release of nitrous oxide and N_2 gas, and nitrogen fixation balances this loss (see Chapter 15). Thus, the availability of N is of critical importance in sustaining primary and secondary productivity, the balance of biogeochemical cycles, and possibly genetic diversity in the

Microbial Ecology of the Oceans, Edited by David L. Kirchman.
ISBN 0-471-29993-6   Copyright © 2000 by Wiley-Liss, Inc.

ocean. Biological demands for N exceed availability in a range of estuarine, coastal, and pelagic waters, creating intense physiological, ecological, and evolutionary pressures to circumvent or compensate for N limitation. Geochemists historically have argued that N is not the nutrient that ultimately limits production in the oceans, since the presence of the genetic potential for obtaining N from the large atmospheric reservoir should alleviate nitrogen deficiencies over long time scales (Redfield 1958; S. V. Smith 1984). Biologists have pointed out, however, that N_2 fixation rates in the ocean are controlled by a myriad of interacting environmental factors. These include oxygen tension, turbulence, and availability of nutrients and energy needed to sustain this process on short- and long-time scales (Howarth et al. 1988; Paerl 1990; Falkowski 1997).

Ironically, N is the most common element in the atmosphere as gaseous N_2. Even though N_2 is not highly soluble, the high concentration of atmospheric N_2 results in a concentration of N_2 in seawater that is plentiful relative to other biologically available forms of N. To render N_2 biologically available, however, the triple bond of the N_2 ($N \equiv N$) molecule must be broken, which requires a high activation energy. Chemically, N_2 can be "fixed" to ammonium (NH_4^+) in the Haber process, which requires high temperature and pressure, and a catalyst. Biologically, N_2 can be fixed through an enzymatically mediated process known as biological nitrogen fixation, having the following stoichiometry:

$$N_2 + 8H^+ + 8e^- + 16MgATP \rightarrow 2NH_3 + H_2 + 16MgADP + 16P_i$$

While biological N_2 fixation would appear to provide a relatively inexhaustible supply of N for biological productivity (both terrestrial and aquatic), nitrogen fixation rates in various environments are constrained by a number of environmental factors. Nitrogen fixation is inhibited by oxygen, and requires large stoichiometric amounts of ATP and reductant. In addition, the N_2-fixing apparatus and energy-supplying processes require specific nutrients as cofactors and structural components; these include phosphorus, iron, and trace metals (Mo, Co, V). These requirements alone or jointly constrain when and where N_2 fixation can occur. Phylogenetically and metabolically diverse microorganisms have developed an array of biochemical and physiological strategies to exploit N_2 fixation as a means of obtaining N. Different types of microorganism are involved in N_2 fixation over a range of times and places, constrained by complex ecophysiological determinants. In this chapter we outline how these factors determine the types of organism mediating N_2 fixation in different marine habitats, explore how and why N_2 fixation is regulated, and discuss how N_2 fixation rates, as well as the potential for N_2 fixation, can be investigated.

THE NITROGEN FIXATION APPARATUS: ENERGETIC, MOLECULAR, BIOCHEMICAL, AND PHYSIOLOGICAL FEATURES

Nitrogenase, the enzyme complex that catalyzes the reduction of atmospheric N_2 to ammonium, is composed of two proteins called component I and component II. Component I is often called the molybdenum iron (MoFe) protein or dinitrogenase, and component II the iron (Fe) protein or dinitrogenase reductase (Figure 1). The function of the Fe protein is to reduce the MoFe protein through a series of single electron transfers that each require docking and undocking of the two proteins and hydrolysis of ATP (Howard and Rees 1996). The substrate-binding site for nitrogen reduction is in the MoFe protein (Howard and Rees 1996). The MoFe protein (~ 200 kDa) is composed of two sets of heterodimers, composed of alpha and beta subunits encoded by the *nifD* and *nifK* genes. The multisubunit structure coordinates FeS clusters, called the P clusters, and the MoFe cofactor, which is presumably the site for substrate (N_2) binding (Howard and Rees 1996). This basic structure differs slightly in some microorganisms, which have what are termed "alternative" nitrogenases (Bishop and Premakumar 1992). The alternative nitrogenases are often present as second and/or third copies of the *nif* operon that encode non-Mo-containing component I proteins. One of these (the "first

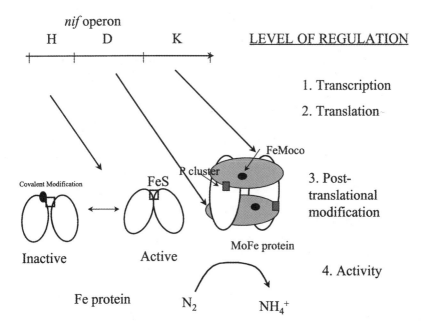

Figure 1. The structural arrangement and genetic regulation of nitrogenase, the enzyme complex mediating N_2 fixation.

alternative") contains vanadium (V) in place of Mo in the cofactor; the other (the "second alternative") contains only Fe. At the genetic level, the alternative nitrogenases differ from the Mo-containing nitrogenase in that they contain a third subunit in component I encoded by *nifG* (*nifDGK*, Bishop and Premakumar 1992). The *vnfDGK* operon can be separated from *vnfH*, compared to the *nifHDK* operon, which normally is transcribed as a single transcript. The ecological significance of the alternative nitrogenase genes is not known, but there are substantial differences in substrate specificity and activity at low temperatures (Bishop and Premakumar 1992).

The Fe protein ($\sim 60\,$kDa) is composed of a pair of identical subunits encoded by the *nifH* gene. The Fe protein is highly conserved among organisms, and even among the alternative nitrogenases (*nifH*, *vnfH*, *anfH*). The high degree of similarity of *nifH* genes among diverse taxa has made it an attractive target for molecular studies in the environment (Zehr and McReynolds 1989; Zehr and Paerl 1998).

Nitrogen fixation activity results from expression (transcription and translation) of the *nifHDK* structural genes. The nitrogen fixation apparatus is complex, involving a suite of other *nif* genes, involved in regulation and cofactor synthesis and insertion (Dean and Jacobsen 1992). At least 20 nitrogenase (*nif*) genes have been characterized in *Klebsiella pneumoniae*, one of the first organisms in which nitrogen fixation genes were described by genetic manipulation. The other *nif* genes have a variety of functions, including controlling synthesis of the FeMo cofactor (*nifEN*), facilitating molybdenum transport (*nifQ*), or regulating transcription (*nifLA*) of the *nif* genes in response to external factors such as inorganic nitrogen availability or oxygen.

The nitrogen fixation apparatus exhibits a striking degree of genetic and physiological conservation among prokaryotes thus far examined (Postgate 1982), although the details of regulation are well understood in only a few model microorganisms (see Merrick 1992 for review). Nitrogen fixation is an expensive process, requiring large expenditures of ATP, and reductant (NADH or NADPH). One of the features of nitrogenase is that it also reduces H^+ to H_2, which consumes ATP. Cellular costs include maintaining the multiple genes in the chromosome and synthesizing multiple structural and regulatory gene products; in addition, there are stoichiometric costs required for N_2 fixation by nitrogenase.

The large metabolic costs of N_2 fixation activity explain why nitrogenase is highly regulated. Two major factors that regulate nitrogenase synthesis are the availability of fixed inorganic N (primarily ammonium, the product of the nitrogen fixation reaction) and the availability of oxygen. Clearly, the presence of fixed N precludes the necessity for N_2 fixation as a means of obtaining nitrogen. Oxygen regulates the N_2-fixing apparatus in most organisms for a different reason: the extreme oxygen sensitivity of the nitrogenase proteins.

Nitrogen fixation is regulated by a number of genetic mechanisms and physiological adaptations. The expression of the structural genes for nitrogenase (*nifHDK*) is regulated by a cascade of regulatory genes. Gene regulation is important in regulating N_2 fixation activity, and the regulatory system is complex and coordinates control in response to oxygen, as well as nitrogen availability. Once the nitrogenase protein has been synthesized, posttranslational mechanisms can also control activity (Kanemoto and Ludden 1984; Pope et al. 1985). These mechanisms are characterized in only a few model microorganisms. An uncharacterized but possibly similar mechanism appears to be present in cyanobacteria, including marine species (Villbrandt et al. 1992; Zehr et al. 1993). In the planktonic diazotroph *Trichodesmium*, the posttranslational modification may help to provide protection against a transient elevation in oxygen concentrations (Zehr et al. 1993).

Finally, nitrogen fixation activity can also be regulated by physiological or behavioral mechanisms to avoid oxygen inactivation (Fay 1992; Gallon 1992). First, the nitrogen fixation apparatus, at the transcriptional and posttranslational levels, is regulated by the presence of oxygen (Merrick 1992). Many aerobic heterotrophic or facultative anaerobic heterotrophs express nitrogenase only under anoxic or microaerophilic conditions (e.g., *Klebsiella*) or maintain a high respiration rate to reduce the intracellular concentration of oxygen (e.g., *Azotobacter*) (Postgate 1982). The "oxygen problem" is exacerbated in cyanobacteria because of the oxygen evolved during oxygenic photosynthesis. This group of microorganisms demonstrates a number of strategies for maintaining N_2 fixation (Fay 1992; Gallon and Stal 1992). These include (1) forming biochemically and structurally differentiated O_2-devoid cells, termed heterocysts, that harbor nitrogenase (Volk 1982), (2) among undifferentiated taxa (e.g., *Lyngbya, Trichodesmium*), spatially separating these processes, (3) temporally separating oxygenic photosynthesis from O_2-inhibited N_2 fixation by confining the former to daylight and the latter to nighttime, (4) forming consortial mutualistic and symbiotic associations with other microbes and higher organisms, where the diazotroph provides fixed N, while the nondiazotrophic partner offers respiratory protection from ambient O_2 (Paerl and Kellar 1978), (5) modifying the nitrogenase enzyme complex so that the enzyme remains inactive but not destroyed by O_2 (Kanemoto and Ludden 1984; Zehr et al. 1993; Zehr and Paerl 1998), and possibly (6) autoprotection, where the Fe protein itself reduces O_2 (Thorneley and Ashby 1989; Bergman et al. 1997).

The multiple physiological and molecular mechanisms regulating N_2 fixation provide the potential for diverse strategies to adapt to different habitats, microenvironments, and fluctuations in oxygen concentrations. For example, in laminated benthic mats, some cyanobacteria fix nitrogen primarily during the dark phase to avoid oxygen inactivation (Stal and Krumbein 1985; Villbrandt et al. 1990; Bebout et al. 1993; Figure 2). Other cyanobacterial strains fix N_2 during the day (Figure 2), which is advantageous from an energetics perspective (Stewart 1973). These two types of strategy can co-occur in the same environment.

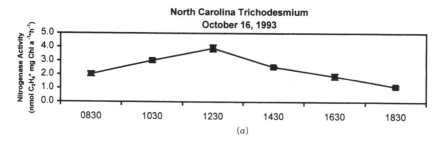

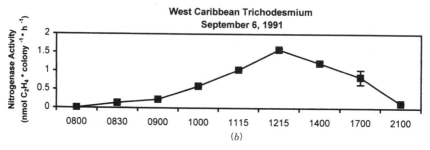

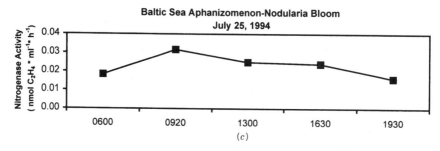

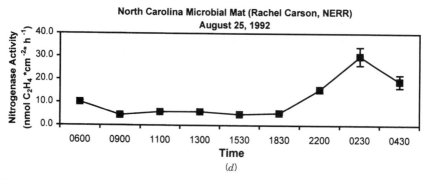

Figure 2. Four patterns of N₂ fixation (nitrogenase activity as measured by the acetylene reduction technique) in different marine habitats supporting morphologically and physiologically distinct cyanobacterial taxa for (a) a bloom of the planktonic nonheterocystous filamentous genus *Trichodesmium* in the Gulf Stream, western Atlantic ocean, off the coast of North Carolina, (b) *Trichodesmium* populations in the western Caribbean Sea (data from D. G. Capone), (c) a bloom of the heterocystous cyanobacteria *Aphanizomenon* and *Nodularia*, in the Baltic Sea (data from D. G. Capone), and (d) a laminated intertidal microbial mat (Atlantic coastal waters near Beaufort, North Carolina); this mat was dominated by nonheterocystous filamentous and coccoid cyanobacteria.

Trichodesmium, which fixes N_2 only during the day (Figure 2) (even though it does not form heterocysts), regulates nitrogen fixation by transcription controlled by a circadian rhythm (Chen et al. 1998). Transcription of the N_2 fixation apparatus is initiated prior to the light phase (sun-up), presumably to coordinate N_2 fixation with photosynthesis during the day. It may be that circadian rhythms are important for timing nitrogen fixation (and other metabolic activities) in many diazotrophic cyanobacteria, since it has been shown that circadian rhythms control many aspects of cyanobacterial metabolism (Golden et al. 1997).

It can be concluded that different microorganisms have developed very different strategies for protecting the nitrogen fixation apparatus from oxygen inactivation. This helps explain why certain (and in some cases unique) nitrogen-fixing (i.e., diazotrophic) microbial taxa are found in specific habitats and microenvironments (Paerl 1996; Paerl and Pinckney 1996).

EVOLUTION OF NITROGEN FIXATION: BIOLOGICAL VERSUS GEOLOGICAL TIME SCALES

Nitrogenase appears to be of ancient origin. First, nitrogenase is present in diverse prokaryotic organisms, including representatives of the Archaea, as well as the Bacteria (Figure 3). Second, the enzyme is irreversibly inactivated in the presence of O_2, which is indicative of an ancient process, dating back to the evolution of prokaryotic life in the O_2-deprived Precambrian era up to 3.9 billion years ago (bya) (Knoll 1979; Schopf and Walter 1982). The formation of the ocean basins following the separation of the ancient landmass Pangea 200 million years ago (mya) (Ehrlich 1996) was very recent relative to the evolution of the N_2 fixation genes (Figure 3). There are indications that cyanobacteria-like microorganisms were present up to 3.5 bya, but the oxygen atmosphere did not develop until nearly 2.3 bya (Figure 3). Thus, the nitrogenase enzyme probably first evolved in an anoxic atmosphere, and the evolution of N_2-fixing microorganisms during the development of the oxygen-containing atmosphere had to coincide with evolution of biochemical and physiological strategies to avoid inactivation of nitrogenase by oxygen. Furthermore, the presence of nitrogenase so early in evolution coincides with a reducing atmosphere, in which it is not clear that N_2 fixation would be advantageous (because ammonia may have been present). It is, therefore, possible that the original function of nitrogenase was not to fix N_2, but perhaps to reduce other substrates (Postgate and Eady 1988). Nitrogenase catalyzes the reduction of multiple triple-bonded substrates, including toxic compounds such as cyanide and carbon monoxide, which may have been present in the early atmosphere.

Phylogenetic analysis of the nitrogenase gene, and of the 16S rRNA genes of organisms containing nitrogenase indicates that nitrogenase is an ancient enzyme (Young 1992; Hirsch et al. 1995). Nitrogenase is found in microorganisms

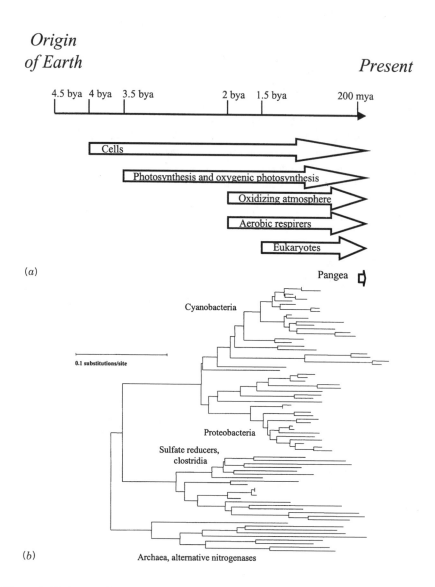

Figure 3. Phylogeny of nitrogenase genes relative to evolution of the biosphere. The phylogeny of nitrogenase (b) accompanies the timeline (a) of major developments of Earth biosphere (per Ehrlich 1996), showing diversification of nitrogenase genes among extant microorganisms. The phylogenetic tree is an unrooted distance tree, and so there is not a direct relationship between the timeline and phylogenetic radiations.

ranging from strict anaerobes to oxygen-evolving phototrophs (cyanobacteria) (Figure 4). A particularly interesting aspect of nitrogenase phylogeny is the clustering of *nifH* genes from anaerobes, which does not reflect the phylogenetic relationship of these organisms (on the basis of molecular sequences of ribosomal RNA). In contrast, the phylogenetic affiliation of *nifH* from aerobes is largely consistent with ribosomal RNA-based phylogeny (Zehr and Capone 1996). Cyanobacterial phylogeny based on *nifH* is also largely consistent with 16S rRNA phylogeny, at least in the tight clustering of the heterocystous cyanobacteria (Zehr et al. 1996). The evolution of the heterocyst was a response to the development of an oxygen atmosphere, if not to the evolution of oxygenic photosynthesis.

The physiology and molecular evolution of nitrogenase have apparently changed little since the transition from anoxic to oxic conditions on Earth (≈ 2 bya), one of the planet's major biogeochemical and evolutionary events. During this transition, previously dominant anaerobic microbes gradually witnessed both reductions in and disappearances of their niches, especially in well-mixed aquatic environments exchanging O_2 with the atmosphere. For the N_2-fixing taxa, ever-increasing levels of biospheric O_2 over geological time scales (i.e., millions of years) translated into a need to molecularly, physiologically, and ecologically adapt to a set of changing environmental "ground rules." While increasing O_2 levels represented a serious environmental constraint, an oxic biosphere also influenced the availability of metal cofactors, specifically iron, which is essential for functional nitrogenase. Under anaerobic conditions, Fe mainly exists in its soluble reduced Fe^{2+} form. Under aerobic conditions, the far less soluble Fe^{3+} ion dominates. Recent work has demonstrated Fe limitation of both photosynthetic production and N_2 fixation in euphotic N-limited ocean waters (Martin et al. 1994; Paerl et al. 1994).

NITROGEN-FIXING MICROBES: DIVERSITY AND HABITATS

The capability for N_2 fixation has thus far only been demonstrated in prokaryotes (*Bacteria* and *Archaea*). Nitrogen-fixing prokaryotes are diverse, and N_2 fixation appears to be widely distributed throughout diverse microbial groups (Figure 4) but is scattered such that even closely related microorganisms do not necessarily have the capacity to fix N_2. Diazotrophic groups include representatives of (1) anoxygenic phototrophs (photosynthetic bacteria: *Chlorobium*, *Chromatium*, *Rhodospirillum*), (2) oxygenic phototrophs (cyanobacteria), including all heterocystous filamentous (*Aphanizomenon*, *Calothrix*, *Nodularia*), some nonheterocystous filamentous (*Oscillatoria*, *Lyngbya*, *Trichodesmium*), and nonfilamentous (*Gloeothece*, *Synechococcus*) genera, (3) anaerobic heterotrophic bacterial genera (*Clostridium*, *Desulfovibrio*), (4) numerous microaerophilic heterotrophic bacterial genera (*Klebsiella*, *Vibrio*), (5)

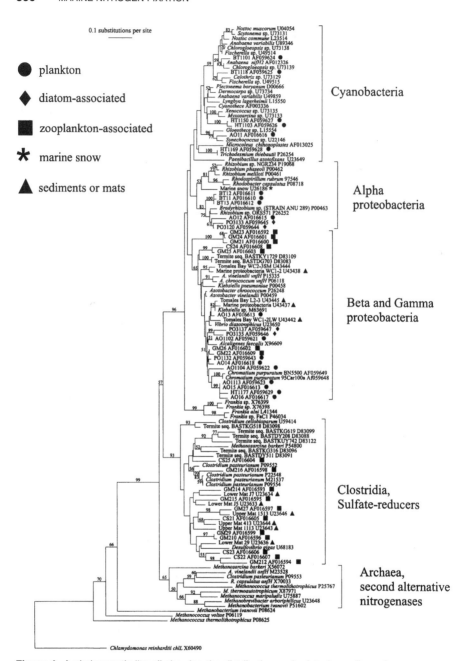

Figure 4. A phylogenetic "tree" showing the distribution and relatedness (based on sequence analysis of the structural gene encoding for the Fe protein subunit of nitrogenase, *nifH*) of N_2-fixing microorganisms. Symbols denoted by specific habitats and biotic associations in which marine N_2 fixers have been found and characterized.

a few aerobic heterotrophic genera, most notably *Azotobacter*, (6) chemolitho-autotrophic bacterial genera (*Thiobacillus*), and (7) some Archaea (methano-gens). All these groups are found in the marine environment, although different groups are most abundant in specific habitats consistent with requirements for oxygen, energy (light, organic matter), and macro- and micronutrients (Potts 1980; Capone 1983; Paerl 1990). Plant and animal symbiotic associations with diazotrophs are numerous and widespread in the marine environment (Paerl in press). Representative marine diazotrophs available in culture collections are shown in Table 1.

The marine environment offers a range of diazotrophic microbial habitats in the water column, deep sediments, shallow sediments or intertidal regions, and estuarine regions exhibiting wide fluctuations in salinity and other envi-ronmental characteristics (Table 2). The microbial assemblages that inhabit these different habitats reflect differences in physical and chemical characteris-tics as well as biota capable of forming associations with diazotrophs.

ENVIRONMENTAL RATES AND MEASUREMENTS

Nitrogen fixation can be measured either indirectly, by assessing the activity of the nitrogenase enzyme complex using the acetylene reduction assay, or directly by $^{15}N_2$ assimilation or the accumulation of fixed N into microbial biomass. Because of its ease of operation (amenability to field and laboratory studies), high sensitivity, and good reproducibility, the most commonly used technique is the acetylene reduction assay (ARA) (Stewart et al. 1967). This technique is based on the ability of nitrogenase to reduce triple-bonded substrates other than N_2. In particular, the triple-bonded acetylene gas can be reduced to ethylene in an approximate 3:1 molar ratio compared with N_2 fixation (i.e., reduction of N_2 to NH_3). The exact ratio depends partially on the rate that nitrogenase reduces H^+ to H_2 (Hardy et al. 1968), a reaction that accompanies the N_2 to NH_3 reaction (Postgate 1982). Both acetylene and ethylene are easily quantified by flame ionization gas chromatography. A shortcoming of this and other bottle-based incubation techniques is that delicate N_2-fixing associations and aggregates may be disrupted, leading to potential underestimates of rates.

When N_2 fixation rates are high (i.e., high sensitivity is not needed), direct assessment of N_2 fixation by means of $^{15}N_2$ tracer enrichment and natural abundance techniques may be preferred. The ^{15}N technique has been used extensively in culture studies (Bergerson 1980), and to a lesser extent under field conditions (Montoya et al. 1992). The fixation of $^{15}N_2$ and its incorpor-ation into particulate matter is measured either by mass spectrometry, which provides highest sensitivity and precision, or by emission spectrophotometry, which offers a simple and rapid means of sample preparation and analysis. Periodic ^{15}N calibration checks of the ARA are required to establish and confirm the molar relationship between acetylene reduction and N_2 fixation.

Table 1. Representative nitrogen-fixing isolates from the marine environment, available in culture collections

Name	Type of Organism	Metabolism	Strain and/or reference Reference	Environment
Chromatium purpuratum	Purple sulfur, gamma proteobacteria	Anaerobe, phototroph	ATCC 700430	Planktonic crustacean invertebrate
Chromatium vinosum	Gamma proteobacterium	Anaerobe, phototroph	ATCC 35206	Organic matter, marine basin, Western Australia
Rhodobacter sp.	Alpha proteobacterium	Aerobic heterotroph (dark) or faculative anaerobic phototroph (light)	ATCC 700304	Marine coastal water, Izu Peninsula, Japan
Rhodovulum adriaticum	Alpha proteobacterium	Aerobic heterotroph (dark) or faculative anaerobic phototroph (light)	ATCC 35885	Marine mud and water, Winogradsky column
Vibrio diazotrophicus	Gamma proteobacterium	Aerobe, faculative anaerobe	ATCC 33466	Sea urchin gastrointestinal tract
Richelia sp.	Group IV cyanobacterium	Aerobe, photoautotroph	Uncultivated	Open ocean, symbiont
Oscillatoria sp.	Group III cyanobacterium	Aerobe, photoautotroph	PCC 6401	Marine mud, California
Chroococcus sp.	Cyanobacterium	Aerobe, photoautotroph	PCC 9106	Limestone, intertidal, Gulf of Dlat, Israel
Myxosarcina sp.	Group II cyanobacterium	Aerobe, photoautotroph	PCC 7312, ATCC 29377	Snail shell, intertidal, Puerto Penasco, Mexico
Lyngbya sp.	Group III cyanobacterium	Aerobe, photoautotroph	PCC 8106	Cyanobacterial mat, intertidal, North Sea, Germany
Lyngbya sp.	Group III cyanobacterium	Aerobe, photoautotroph	PCC 8992	Salt marsh
Trichodesmium sp.	Group III cyanobacterium	Aerobe, photoautotroph	CCMP 1765, IMS 101	Coastal North Carolina
Cyanothece sp.	Group I cyanobacterium	Aerobe, photoautotroph	ATCC 51472, ATCC 51142	Seawater and intertidal, Port Aransas, Texas
Anabaena sp. CA	Group IV cyanobacterium	Aerobe, photoautotroph	ATCC 33047	Marine mat, Texas coast
Symploca sp. (*Microcoleus chthonoplastes*)	Group III cyanobacterium	Aerobe, photoautotroph	PCC 8002	Intertidal mud, Menai Strait, North Wales

Organism	Group	Metabolism	Strain number	Source
Leptolyngbya sp.	Group III cyanobacterium	Aerobe, photoautotroph	PCC 7004	Mangrove root, Port Aransas, Texas
Leptolyngbya sp.	Group III cyanobacterium	Aerobe, photoautotroph	PCC 7375, ATCC 29409	Plankton, Woods Hole, Massachusetts
Lyngbya sp.	Group III cyanobacterium	Aerobe, photoautotroph	PCC 7104, ATCC 29117	Rock, Montauk Point, Long Island, New York
Xenococcus sp. (*Dermocarpa*)	Group II cyanobacterium	Aerobe, photoautotroph	PCC 7305, ATCC 29373	Marine aquarium
Pleurocapsa sp.	Group II cyanobacterium	Aerobe, photoautotroph	PCC 7516	Rock, Marseilles, France
Clostridium oceanicum	Gram positive	Anaerobe, heterotroph	ATCC 25648	Marine sediments
Clostridium litorale	Gram positive	Anaerobe, heterotroph	ATCC 49638	Marine sediments, North Sea, Germany
Bacillus marinus	Gram positive	Aerobe/faculative anaerobe, heterotroph	ATCC 29840	Marine sediment, Northeast Atlantic
Bacillus sphaericus	Gram positive	Aerobe/faculative anaerobe, heterotroph	ATCC 19005	Marine sediment, Grand Bahama Banks
Deleya marina (deposited as *Arthrobacter marinus*)	Gram positive, high G + C group	Aerobe/faculative anaerobe	ATCC 25374	Seawater, Woods Hole, Massachusetts
Listonella (Beneckea) pelagia	Gamma proteobacterium	Aerobe	ATCC 25916	
Methylophaga thalassica	Gamma proteobacterium	Aerobic chemotroph	ATCC 33146	Coastal seawater
Desulfobacter curvatus W	Delta proteobacterium	Anaerobe	ATCC 43919, DSM 3379	Marine mud, Italy
Alcaligenes faecalis	Gamma proteobacterium	Heterotroph	ATCC 15554	Feces, marine aquarium
Roseobacter denitrificans		Heterotroph	ATCC 33942	Seaweed, *Enteromorpha*
Methanococcus maripaludis	Archaea	Anaerobic chemoautotroph	ATCC 43000, DSM 2067	Marine marsh sediment
Desulfovibrio africanus	Delta proteobacterium	Anaerobic heterotroph	ATCC 19997	Marine mud

Table 2. Summary of marine environments supporting bacterial and cyanobacterial nitrogen fixing microorganisms

Environment	Habitat or Associated Organism	Organisms Identified	Ref.
Water column	Free-living heterotrophic bacteria	*Vibrio*	Guerinot and Colwell (1985), Tibbles and Rawlings (1994)
	Free-living cyanobacteria	*Trichodesmium, Aphanizomenon, Nodularia, Nostoc*	Fogg (1982), Capone et al. (1997)
	Diatom symbiotic cyanobacteria	*Richelia*	Villareal (1992)
	Diatom-associated bacteria		Martinez et al. (1983)
	Sargassum-associated	*Dichothrix*	Carpenter (1972)
Benthic	Sea urchin–associated		Guerinot and Patriquin (1981)
	Shipworm-associated		Carpenter and Culliney (1975)
	Ascidian-associated	*Prochloron[a]*	Paerl (1984)
	Codium fragile		Head and Carpenter (1975)
	Sponge-associated (e.g., *Halichondria*)		Shieh and Lin (1992)
	Mangrove-associated		Hicks and Silvester (1985)
	Sea grasses and marsh grasses (*Zostera marina* and *Spartina alterniflora*)		Shieh et al. (1989)
	Sediment	*Azotobacter, Azospirillum, Campylobacter, Beggiatoa, Enterobacter, Klebsiella, Vibrio, Desulfovibrio, Clostridium, Rhodopseudomonas*	See Capone (1983) for references
	Marine mats	*Anabaena, Calothrix, Lyngbya, Nostoc, Microcoleus,[a] Oscillatoria, Phormidium, Gloeocapsa, Synechococcus*	Stal and Krumbein (1985), Stal and Caumette (1994), Steppe et al. (1996)

[a]Not conclusively shown to be a nitrogen fixer.

N_2 fixation can also be measured as accumulation (over time) of N into particulate, dissolved, and total organic matter using CHN analyses, Kjeldahl digestion/oxidation, and high temperature catalytic combustion techniques. This method is best suited for waters having very high rates of N_2 fixation and when a significant portion of the particulate N is microbial biomass.

THE BIOGEOCHEMICAL AND ECOLOGICAL IMPORTANCE OF N$_2$ FIXATION IN THE MARINE ENVIRONMENT

From habitat and ecosystem N budget perspectives, N$_2$ fixation has been shown to be a significant, and, at times, dominant source of "new" N supporting marine primary and secondary production (Table 3). Field surveys of N$_2$-fixing activity (nitrogenase activity) have increased dramatically during the past three decades, enabling us to generalize where, when, and how much N$_2$ is fixed in planktonic and benthic habitats in geographically diverse habitats (Table 3; Chapter 15). Because they are often enriched with organic matter and N-limited, nearshore intertidal and subtidal benthic environments can be rich repositories of cyanobacterial and bacterial diazotrophs exhibiting relatively high rates of N$_2$ fixation (Paerl et al. 1981; Capone 1983; Paerl 1990). These include estuarine and coastal salt marshes, mudflats, where diazotrophy is dominated by epiphytic and laminated mat microbial communities (Stal and

Table 3. Estimates of rates of N$_2$ fixation and its contribution to nitrogen demands for sustaining primary productivity of various marine ecosystems

System	N$_2$ Fixation		% of N Demand	Ref.
	$mgN\,m^{-2}\,d^{-1}$	$gN\,m^{-2}\,y^{-1}$		
Coral reefs				
Shark Bay	0.9–1.6	0.3–0.6	52–104	Smith (1984)
Christmas Island	2.4–8.2	0.9–3.0	37–127	
Canton Atoll	6.7–26	2.4–9.5	31–121	
Salt marshes				
Sippewisset March		6.8	82	Valiela (1983)
Gulf of Mexico Coast		15.4	73	DeLaune and Patrick (1990)
Sea grass beds				
Thalassia	82		>100	Patriquin and Knowles (1972)
Thalassia	21		30–50	
Zostera	5.5		3–28	Capone (1983)
Planktonic ecosystems				
Baltic Sea	0.6		3	
Baltic Sea			17–29	*Ambio* (1990), Sörensson and Sahlsten (1987)
Oligotrophic open ocean				
Western Atlantic	0.001–0.3		?	Capone et al. (1997)
Pacific near Hawaii	0.3–0.7		>20	Karl et al. (1997), Letellier and Karl (1997)

Caumette 1994), macrophyte- (e.g., seagrass) dominated communities (Capone 1983), mangroves (Gotto and Taylor 1976), coralline and carbonate reefs (Webb et al. 1975; Capone 1983), and lithifying stromatolitic cyanobacterial mats (Bauld 1983; Paerl and Pinckney 1996). In these habitats, N_2 fixation can contribute from 20% to over 100% of community N needs (Capone 1983; see Chapter 15). Significant amounts of benthic fixed N inputs may be exported to the water column as detrital aggregates, invertebrate and fish grazing, and weathering (i.e., storm and tidal scouring and erosion).

For many years, the contribution of nitrogen fixation to the nitrogen dynamics of the open ocean was assumed to be negligible. The discovery of nitrogen fixation in association with the colony-forming filamentous cyanobacterium *Trichodesmium* (Figure 5) in the Sargasso Sea (Dugdale et al. 1961) was the first indication that nitrogen fixers were present in these oligotrophic waters. A few other nitrogen-fixing genera have been reported, the most common being heterocystous cyanobacterial endosymbionts of diatoms (such as *Richelia* in the diatom genera *Rhizosolenia* and *Hemiaulus*) (Kimor et al. 1978; Carpenter 1983; Mague et al. 1977; Villareal 1992). Epiphytic heterocystous cyanobacteria were also found on the planktonic macroalga *Sargassum* (Carpenter and Capone 1983). A few other nitrogen-fixing cyanobacterial genera have been noted (Carpenter 1983), but *Trichodesmium* and *Richelia* are usually believed to be the dominant diazotrophs in the open ocean. These genera frequently exhibit high rates of N_2 fixation in ultraoligotrophic waters. Recent examinations of shipboard-based N_2 fixation measurements in various pelagic oceanic regions (i.e., western Atlantic, Caribbean, central and southwestern Atlantic, northern Pacific), indicate that this process contributes significantly to "new" N inputs (Karl et al. 1997, Capone et al. 1997; Chapter 15). The current intensification and geographic expansion of surveys will yield a more accurate and comprehensive set of measurements.

Estimates of the contribution of nitrogen fixation in certain ultraoligotrophic waters (Sargasso Sea, North-Central Pacific gyre) in recent years have suggested that it may be more important from a budget perspective than previously believed and may rival upwelling and atmospheric deposition as dominant sources of new N (Michaels et al. 1996; Karl et 1997). Some of the evidence for the importance of nitrogen fixation comes from indirect evidence from biogeochemical estimates of nitrogen budgets and fluxes (Michaels et al. 1996) and suggests a rate of nitrogen fixation that is much larger than the amount that could be contributed by our current knowledge of *Trichodesmium* distribution and abundance. Thus, it is possible that other less obvious, less well-known diazotrophs could play a role in open-ocean planktonic nitrogen fixation (Lipschultz and Owens 1996).

Using a molecular approach, it was shown that there are diverse potentially nitrogen-fixing microorganisms in the planktonic environment. N_2-fixing genes derived from picoplanktonic cyanobacteria (related to *Cyanothece* or *Gloeothece nifH*) and heterotrophic eubacteria have been found (Zehr et al. 1998). Large unicellular nitrogen-fixing cyanobacterial genera had been

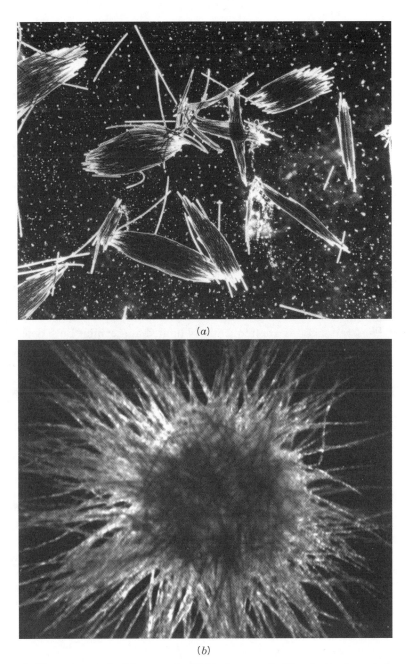

(a)

(b)

Figure 5. Photomicrographs (phase contrast light microscopy) of the planktonic, nonheterocystous, colony-forming, N_2-fixing cyanobacterium *Trichodesmium*. This diazotroph is widely distributed in tropical and subtropical oligotrophic open-ocean waters, where it is considered to be a major contributor to "new" nitrogen input (Carpenter and Romans 1991; Capone et al. 1997). Both the fusiform "tuft" (a) and spherical "puff" (b) forms are shown. These samples were obtained (1995 and 1998) from the Gulf Stream (western North Atlantic Ocean) approximately 100 km off the coast of North Carolina. (Color versions of this and the other photomicrographs can be seen at http://www.wiley.com/products/subject/life/kirchman.)

reported from tropical and subtropical waters (Waterbury et al. 1988), but their distribution and contribution to nitrogen fixation in the environment remain unknown. Zehr et al. (1998) detected heterotrophic bacterial diazotrophs belonging to the alpha and gamma proteobacterial classes. These bacteria could potentially fix nitrogen if they had adequate sources of carbon and a strategy for avoiding oxygen inactivation from the dissolved oxygen present in aerobic surface waters. Fixation in association with particles with anoxic microzones may be one such strategy. Finally, diverse diazotrophic bacterial types were identified in association with invertebrate planktonic zooplankton. The types of organism identified by phylogenetic analysis of the *nifH* gene indicate that many of the diazotrophic microorganisms associated with zooplankton are anaerobes, suggesting that they may be inhabitants of the gut. Diazotrophic anaerobes have been isolated from invertebrate guts (Proctor 1997). Thus, there are diverse potential contributors to nitrogen fixation in the open-ocean environment, but it is not clear whether (and if so how much) in situ N_2 fixation can be attributed to the picoplankton.

Estuaries and lagoons are frequently N-limited and exhibit both detectable rates and genetic potentials (based on *nif* gene detection and characterization) reflecting diverse N_2 fixing cyanobacterial and bacterial populations (Zehr et al. 1995; Steppe et al. 1996). In some highly nutrient-enriched coastal and estuarine systems, N_2-fixing cyanobacteria assemblages can be a dominant fraction of phytoplankton biomass, and can contribute significantly to N and C budgets. The filamentous heterocystous genera *Nodularia* and *Aphanizomenon* form large blooms in the eastern and central Baltic Sea (Figure 6). Here, N_2 fixation contributes approximately 25% of "new" N inputs (Sörensson and Sahlsten 1987). *Nodularia* and *Anabaena* can also form substantial fractions of phytoplankton biomass in nutrient-enriched estuaries (Australia, New Zealand, Colombia, Venezuela) and embayments (Huber 1986; Paerl in press). Here, P and organic matter enrichment are most commonly linked to growth and bloom potentials of nuisance (toxic, hypoxia-inducing, food web disrupting) genera (Paerl in press). Recent studies using enrichment, isolation, and molecular and immunological techniques have shown the presence of planktonic heterotrophic bacteria, potentially able to fix N_2 (i.e., *nif* genes are present). It is unlikely that these bacteria are active in a freely suspended state, since microaerophilic conditions are needed for nitrogenase expression (Merrick 1992). However, it has been shown that detrital aggregates and other surface-associated habitats (e.g., biofilms) can serve as reduced (O_2 deplete) microenvironments supporting N_2 fixation (Paerl et al. 1987; Paerl and Carlton 1988). In coastal regions where macrophyte (e.g., *Spartina, Zostera*) detritus can be exported from nearshore environments, the conditions for such N_2 fixation exist.

(a)

Figure 6. Examples of bloom-forming cyanobacteria responsible for planktonic N_2 fixation. (a) Surface view of the Baltic Sea (Gulf of Finland, approximately 50 km south of the Finnish coastline), during a summer bloom of the filamentous N_2 fixing cyanobacteria *Nodularia* and *Alphanizomenon* (photograph courtesy of Pia Moisander). (b) Filamentous, heterocystous N_2 fixing *Anabaena* sp. during mid-summer in the stratified surface waters of the eutrophying Neuse River estuary, North Carolina. (c) and (d) Populations of *Nodularia* spp. obtained from the Gulf of Finland.

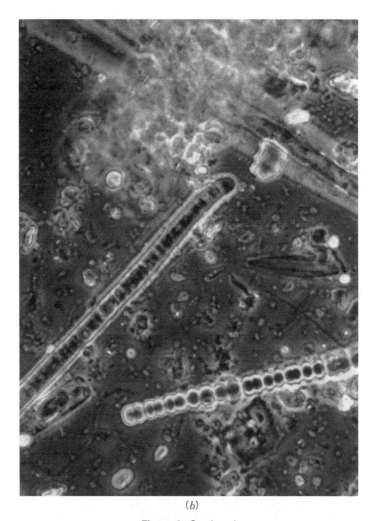

(*b*)

Figure 6. Continued

IDENTIFICATION OF NITROGEN-FIXING MICROBES WITH MOLECULAR AND IMMUNOLOGICAL PROBES

Historically, N_2 fixers have been identified by classical microbiological enrichment techniques (Guerinot and Colwell 1985). Cultivation techniques are extremely important and have provided us with models for many biochemical processes, including N_2 fixation. Cultivation has its limitations, though, owing to its inherent selectivity and the difficulty in relating findings with isolates back to the complex environment. Cultivated isolates may or may not be representative taxonomically or physiologically of organisms responsible for in situ N_2 fixation. In some cases, it has been difficult to ascertain whether

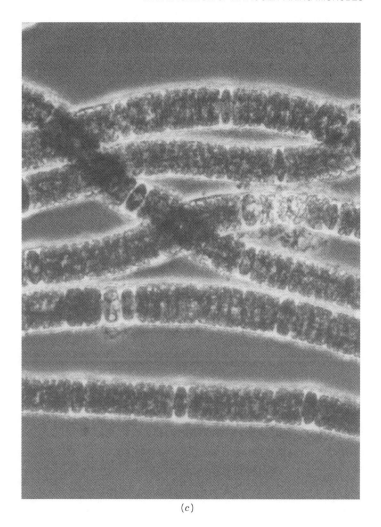

(c)

Figure 6. Continued

N_2-fixing organisms are present, or when rates are detected, to determine which organisms are responsible.

In an ecological context, N_2 fixation activity in a given habitat occurs at four levels: presence of organisms with nitrogenase genes, conditions appropriate for transcription, appropriate conditions for maintenance of protein synthesis, and protein activity at translational and posttranslational levels (Figure 1). Cultivation techniques can provide a tool for evaluating the presence of nitrogen-fixing microorganisms if selectivity for specific microbes or the issue of culturability can be avoided. Molecular and immunological tools provide means to detect and identify nitrogen-fixing

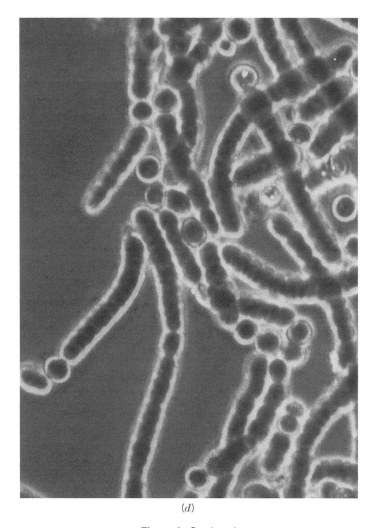

(*d*)

Figure 6. Continued

microorganisms that are present (including uncultivable strains) (Paerl et al. 1989; Zehr and McReynolds 1989; Currin et al. 1990; Kirshtein et al. 1993) and to assess whether the N_2-fixing genes are being transcribed and protein being synthesized (Figure 1).

By amplifying, cloning, and sequencing of nitrogenase genes, Zehr et al. (1995) and Steppe et al. (1996) showed that marine intertidal mats, which are dominated by cyanobacteria, harbor diverse microorganisms that contain nitrogenase genes. These studies indicated the presence in these environments of numerous nitrogenase genes that were phylogenetically related to anaerobic microorganisms such as sulfate reducers and clostridia.

Subsequent studies in the open ocean and in estuarine systems have shown that there are distinctly different potential N_2-fixing microbial populations in the different habitats.

The *nifH* sequence types found in different habitats are consistent with organisms of the types that should be selected from these environments. Thus, anaerobe-type *nifH* genes are found in benthic habitats (Zehr et al. 1995), which may exhibit anoxic conditions with depth or over diurnal cycles (Bebout et al. 1993). In the open ocean, *nifH* genes that cluster with those of typical aerobic or facultative heterotrophic microorganisms and cyanobacteria are found (Figure 4; Zehr et al. 1995). The diversity of *nifH* genes found in the benthic environments and mats is much higher than was found in the plankton (Zehr et al. 1995; Zehr and Paerl 1998), which is consistent with the relative rates of nitrogen fixation between the benthos and the water column. Planktonic invertebrates tend to also have large numbers of *nifH* sequences representing these anaerobic types of organisms, which may be associated with the invertebrate gut (Figure 4, Zehr et al. 1995; Proctor 1997; Braun et al. 1999). Some proteobacterial *nifH* genes are also found in these types of anoxic environments.

It remains to be demonstrated whether the diverse types of anaerobic microorganism with nitrogenase genes indeed do fix nitrogen. Assays directed toward nitrogenase messenger RNA (e.g., *nifH* mRNA) (Zehr et al. 1996) can potentially address this question. Molecular and immunological assays can provide insight at the cellular level. Assays for DNA or mRNA can be coupled to amplification techniques using fluorescent tags that allow investigators to visualize microscopically individual cells containing nitrogenase genes or *nif* mRNA (Hodson et al. 1995), or the nitrogenase protein itself (Currin et al. 1990; Bergman et al. 1997). These approaches are particularly useful for investigating complex microbial communities or consortia, where diversity of N_2-fixing microorganisms is high and there is a high degree of spatial complexity, such as in benthic sediments or intertidal mats (Figure 7).

Immunoassays directed toward the nitrogenase protein can potentially demonstrate not only that nitrogenase is expressed, but whether it is in an active or inactive form. The Fe protein of nitrogenase has been shown to undergo a covalent modification that is correlated with N_2 fixation activity (Kanemoto and Ludden 1984). This posttranslational modification of the Fe protein can be observed with proteins separated on polyacrylamide gels. Since the modification state can be observed by Western blotting of these gels, an indication of the activity of the nitrogenase protein can be obtained. The modification–demodification switch, which is catalyzed enzymatically by a pair of enzymes, responds quickly to environmental conditions (ammonium, oxygen, light: Kanemoto and Ludden 1984), and thus, the assay of the state of the Fe protein can provide information on whether nitrogenase is active or inactive.

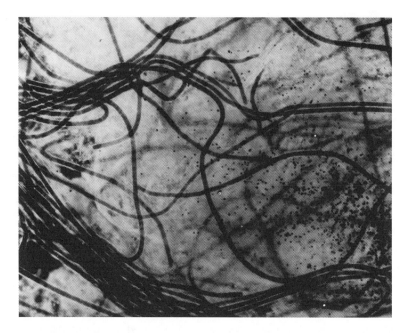

Figure 7. Photomicrograph illustrating the spatial heterogeneity and microbial diversity of N_2-fixing, cyanobacteria-dominated, laminated marine microbial mat. This intertidal mat was obtained from Bird Shoal, near Beaufort, North Carolina (cf. Bebout et al. 1993; Steppe et al. 1996). The "fabric" of this mat is comprised of the nonheterocystous filamentous species *Lyngbya aestuarii* and *Microcoleus chthonoplastes*. This barrier island is on the southern fringe of North Carolina's Outer Banks. It is periodically inundated with N-deplete coastal Atlantic Ocean water.

ENVIRONMENTAL AND ECOLOGICAL FACTORS CONTROLLING MARINE NITROGEN FIXATION

Field observations indicate that different controls can dominate in spatially and geographically distinct regions. For example, in the open ocean, turbulence, the availability of specific essential nutrients (e.g., Fe), and grazing may be key controls on diazotrophy (Paerl 1990; Rueter et al. 1992; Paerl et al. 1994), while in benthic habitats, the rate of accumulation of organic matter, P availability, and diel patterns of $[O_2]$ (due to intense localized daytime productivity and nighttime respiration) may regulate N_2 fixation potentials (Capone 1983; Paerl 1990). Chemical and physical constraints on this process are discussed in more detail below.

Marine N_2 fixation has multiple environmental requirements and controls.

1. N_2 fixation is an energy-demanding process (12–16 ATPs required to reduce N_2 to $2NH_3$). As such, adequate radiant or chemical energy supplies

are needed. These supplies may be restricted or unavailable under certain circumstances (e.g., light limitation, organic matter depletion).

2. Nitrogenase structure and function require certain elements, including P, Fe and a suite of trace metals (Mo, Co, Vn) (Stewart 1973; Merrick 1992). P and Fe, limitations of N_2 fixation have been demonstrated in natural communities (Paerl et al. 1987, 1994), while Mo, Co, and V limitations have been shown in culture only (Ter Steeg et al. 1986; Paulsen et al. 1991; Cole et al. 1993).

3. Oxygen represents a potent inhibitor of N_2 fixation (Fay 1992; Gallon 1992). Certain physical–chemical conditions, including high ambient O_2 saturation, excessive small-scale (shear) and large-scale (mixing) turbulence, can alter microscale (cellular) O_2 diffusive conditions, exacerbating the potential for nitrogenase inhibition by enhanced inward cellular diffusion of O_2 (Paerl et al. 1995; Kucera 1996).

4. High levels of salinity, including and exceeding average seawater concentrations (32–35 ppt), (Dubois and Kapustka 1981; LeRudilier et al. 1984; J. S. Thomas et al. 1988), desiccation (Potts 1994), excessive irradiance (especially UV) (Garcia-Pichel and Castenholz in press), extreme pH and temperature values (Stal and Caumette 1994), antibiotics, and anthropogenic pollutants (petrochemicals, heavy metals, herbicides, and pesticides) may also mediate activities and growth of diazotrophs.

5. Biological controls, including selective grazing by invertebrates, bacterial and viral lysis, as well as positive (consortial) nutrient exchange interactions with associated bacteria, may affect diazotrophic community dynamics (Paerl and Kellar 1978).

Specific nutrient deficiencies may restrict the activities and geographic extent of N_2 fixation. Among geochemists, phosphorus has been mentioned as the inorganic nutrient (excluding N) most likely, potentially to control oceanic phytoplankton primary productivity and growth (S. V. Smith 1984). The geochemical argument for P limitation is based on the assumption that biological N_2 fixation is capable of satisfying oceanic N requirements over geologic time scales (i.e., millions of years). On biologically relevant time scales (i.e., days to years), there is little experimental evidence to support this assumption and on the whole, the oceans remain N-limited (Carpenter and Capone 1983; Falkowski 1997). Most likely, oceanic N_2 fixation is controlled by a complex interplay of the physical–chemical and biotic factors mentioned above, going far beyond P supply alone (Paerl 1990; Falkowski 1997). Phosphorus supply rates have, however, been shown to control production and N_2 fixation in some localized, highly productive benthic habitats, including reefs, sea grass beds, microbial mats, carbonate sediments, and mangrove communities (Capone 1983).

Molybdenum is a cofactor of nitrogenase and as such has been investigated as a nutrient potentially limiting N_2 fixation (Howarth and Cole 1985; Cole et al. 1993). Howarth and Cole (1985) proposed that the relatively high (>20 mM) concentrations of sulfate (SO_4^{2-}), a structural analog of the most common form of molybdenum found in seawater, molybdate (MoO_4^{2-}), could competitively (via the uptake process) inhibit N_2 fixation, thereby limiting this process. Competitive inhibition of MoO_4^{2-} uptake by high SO_4^{2-} concentrations was shown by Cole et al. (1993). However, MoO_4^{2-} is highly soluble in seawater with concentrations on the order of 100 μM. Ter Steeg et al. (1986) and Paulsen et al. (1991) showed that despite the potential for SO_4^{2-} competition, Mo availability was ensured at concentrations much lower than 100 μM. In both coastal and pelagic ocean (western Atlantic) waters, N_2-fixing potentials of marine diazotrophs appear to be unaffected by this competition (Paulsen et al. 1991). Most likely, the small cellular Mo requirements for N_2 fixation are met through reduced but sufficient uptake and storage. In addition, recent work has shown the presence of "alternative" non-Mo-requiring nitrogenases in bacterial and cyanobacterial diazotrophs (Bishop and Premakumar 1992). If such microbes are broadly distributed in nature, a mechanism would exist by which Mo limitation could be circumvented.

Fe is another cofactor of nitrogenase, and diazotrophic microorganisms exhibit a relatively (compared to Mo) high demand for this metal. Constraints on Fe availability in N-limited waters may restrict the distributions and activities of pelagic diazotrophs, including *Trichodesmium* and the *Richelia–Rhizosolenia* symbiosis (Rueter et al. 1992). This possibility was evaluated in chronically N-deficient western Atlantic (Gulf Stream, Sargasso Sea) waters known to be frequented by these assemblages (Paerl et al. 1994). Using bioassays of naturally occurring and cultured (on Sargasso Sea water) *Trichodesmium* populations, it was shown that both chelated (EDTA) and nonchelated forms of Fe (as $FeCl_3$) stimulated N_2 fixation and growth (as chlorophyll *a* increase) relative to untreated controls. Maximum stimulation occurred in response to FeEDTA.

A freshwater *Anabaena* species was shown to produce potent chelators capable of sequestering Fe from ambient waters at exceedingly low levels, providing a competitive advantage over eukaryotic, combined N-requiring phytoplankton (Murphy et al. 1976). This mechanism may play a similar role in the marine environment, which exhibits both chronic N-limitation and restricted Fe inputs.

The atmosphere is considered to be the main route by which Fe can be resupplied to the open ocean. Rueter et al. (1992) suggested that by forming large buoyant aggregates near the water's surface, *Trichodesmium* was able to intercept Fe derived from aeolian sources (volcanic emissions, dust from desertification, air pollution from industrial and automotive emissions) (Duce and Tindale 1991), by trapping particulate Fe in the weblike matrix of trichomes.

Dissolved organic matter (DOM) content has additionally been mentioned as a possible modulator of N_2-fixing cyanobacterial growth and bloom potential. Early studies (Pearsall 1932) cited DOM as a factor potentially controlling cyanobacterial blooms in lakes. To account for DOM-stimulated cyanobacterial growth it was hypothesized that DOM "conditions" the water for cyanobacteria, possibly by inducing nutrient assimilatory enzymes and heterotrophy (Antia et al. 1991) and/or by providing a source of energy and nutrition for associated heterotrophic bacteria, which are known to optimize the growth of "host" cyanobacteria (Paerl and Pinckney 1996).

Turbulence exerts a strong impact on phytoplankton growth and structural integrity (W. Thomas and Gibson 1990). Among N_2-fixing cyanobacteria, relatively low rates of turbulence (e.g., gentle stirring) promote localized "phycosphere" nutrient cycling, alleviating certain forms of nutrient limitation (inorganic C, P, trace metals), and enhancing growth (Fogg et al. 1973). Increased levels of turbulence such as vigorous shaking can, however, inhibit growth, with excessive turbulence causing "crashes" through disaggregation, cell and filament damage, and rapid death among bloom-forming colonial genera in culture and in nature (Fogg 1982). Excessive small-scale ($< 100 \, \mu m$) turbulence, or shear, can negatively impact N_2 fixation and growth in dominant bloom-forming heterocystous cyanobacterial genera (e.g., *Anabaena*, *Nodularia*) (Kucera 1996; Moisander in preparation). In laboratory experiments where specific shear rates were applied to these taxa, Kucera (1996) and Moisander (in preparation) showed that turbulent conditions characterizing wind-exposed highly mixed estuarine and coastal surface waters might control N_2 fixation in these genera and therefore may represent a potential barrier to their expansion. Shear is an environmental impediment to N_2 fixation and thus negatively affects the competitive advantage that these diazotrophs might otherwise enjoy in N-limited waters experiencing excess shear stress. The negative impacts could be due to breakage of cyanobacterial filaments, specifically at the delicate heterocyst–vegetative cell junction, causing O_2 inactivation of nitrogenase in heterocysts (Fogg 1969), as well as to disruption of phycosphere consortial bacterial–cyanobacterial associations.

On the ecosystem level, physical–chemical and biotic regulatory variables often co-occur and may interact synergistically and antagonistically to control N_2 fixation (Paerl and Carlton 1988), precluding simple interpretations of environmental control of this process.

HUMAN IMPACTS ON N₂ FIXATION IN ESTUARINE AND COASTAL WATERS. EUTROPHICATION AND EXPANDING CYANOBACTERIAL BLOOMS

Man's activities may play an additional regulatory role on regional and global scales. These include urban, agricultural, and industrial nutrient discharges, which impact loading and N:P concentration ratios, known to control the abundance of N_2-fixing cyanobacteria in freshwater, estuarine, and coastal waters (Niemi 1979; V. H. Smith 1990). Cyanobacteria have exploited recent anthropogenic alterations of aquatic environments, most notably accelerating nutrient enrichment, or eutrophication (Fogg 1969; Paerl 1988; Paerl and Tucker 1995). A particularly problematic aspect of cyanobacterial opportunism is the development and proliferation of harmful (i.e., toxic, hypoxia/anoxia-inducing, food-web-altering) blooms in nutrified waters (Fogg 1969; Sellner 1997). In addition, nutrient fluxes and trophodynamics can be altered by the substantial (and sometimes dominant) contribution of cyanobacteria to phytoplankton biomass (Horne 1977; Porter and Orcutt 1980).

Cyanobacterial blooms are, to varying degrees, symptomatic of eutrophication in geographically diverse, nutrient-enriched marine waters, including estuaries, embayments, brackish coastal and pelagic seas (e.g., Baltic) and hypersaline lagoons (Peele-Harvey, Australia) (Niemi 1979; Huber 1986). Historically, the most notorious (i.e., toxin-producing, hypoxia-generating) nuisance diazotrophic genera, *Anabaena, Aphanizomenon, Lyngbya, Nodularia,* and *Oscillatoria,* have been confined to heavily polluted freshwater impoundments (Francis 1878; Fogg 1969; Paerl and Tucker 1995); however, regional and global expansion into more incipient eutrophying waters is under way. Examples include the appearance, persistence, and expansion of toxic (to wildlife, cattle, domestic animals, and humans), heterocystous, N_2-fixing genera (*Anabaena, Aphanizomenon, Nodularia*) in brackish fjords in Norway and Sweden and in estuaries and coastal embayments in South Africa, Australia and New Zealand, Brazil, Colombia, Canada, and the United States (e.g., Lake Ponchartrain, Louisiana; Florida Bay, Florida; the Albemarle–Pamlico Sound System, North Carolina; Puget Sound, Washington), all under the influence of increasing agricultural runoff, as well as groundwater and atmospheric loading of nutrients (Carmichael 1997; Paerl 1997). N_2-fixing taxa (*Anabaena, Aphanizomenon Nodularia, Cylindrospermopsis*) that produce toxins and bad tastes/odors are becoming increasingly prevalent and problematic in U.S. and Canadian aquaculture operations (Paerl and Tucker 1995; Carmichael 1997).

The Baltic Sea exemplifies the impacts of long-term (several centuries) eutrophication on cyanobacterial bloom potentials (*Ambio* 1990) (Figure 6a). Incipient, yet growing, invasions and outbreaks appear to be taking place in more recently impacted systems. Recently, we (Piehler et al. in preparation) observed actively N_2-fixing *Anabaena* strains in previously cyanobacteria-free mesohaline (5–15 ppt salinity) segments of eutrophying Neuse River estuary, North Carolina (Figure 6b). In a parallel laboratory study (Moisander et al. in preparation), it was shown that the two dominant, toxic Baltic Sea *Nodularia* strains (Sivonen et al. 1989) (Figures 6c, 6d) were capable of growth and bloom formation in Neuse River Estuary water over a wide range of salinities (0–15 ppt). A species of *Nodularia* sp. has recently been discovered in the plankton of Lake Michigan (McGregor et al. unpublished), possibly an indication of advancing eutrophication of this large lake.

These examples are testimony that increasingly, waters downstream of expanding urban and agricultural regions are prone to invasion by N_2-fixing cyanobacterial nuisance genera (*Anabaena, Aphanizomenon, Nodularia*) (Paerl 1988, 1996; Sellner 1997). This trend is of concern, because these genera are capable of growth in chronically N-deficient waters, typical of many of our estuarine and coastal ecosystems. These heterocystous cyanobacteria and some nonheterocystous diazotrophic genera (e.g., *Lyngbya*, *Oscillatoria*) should enjoy an obvious competitive advantage in N-deficient waters, and there are numerous examples of diazotrophic "cockroach" bloom genera exploiting N-deficient freshwater ecosystems replete with phosphorus and essential nutrients (Fe and other trace metals). Using an extensive data set from a range of freshwater lakes and reservoirs, V. H. Smith (1983) showed a strong relationship between total N:P ratios (by weight) and the prevalence of cyanobacterial bloom genera in freshwater environments. N:P ratios below 20 were conducive to the development and periodic persistence of N_2-fixing genera. This stoichiometric predictor of cyanobacterial dominance has received surprisingly little attention and scrutiny in the marine environment, even though many estuarine and coastal waters exhibit N:P ratios well below 20, are N-limited (D'Elia et al. 1986; Nixon 1986), and are undergoing various symptoms and stages of cyanobacterial expansion (Niemi 1979; Paerl 1996; Paerl and Millie 1996).

While diazotrophic cyanobacteria enjoy the advantage of being able to subsist on atmospheric N_2, they can also proliferate on combined N sources (including both inorganic and organic forms) (Paerl 1988). This nutritional flexibility may provide a competitive advantage in response to N-loading events currently characterizing estuarine and coastal eutrophication. Large pulses of non-point-source N loading, especially from atmospheric deposition and runoff, have increased markedly in these waters and are suspected of being key "drivers" of the eutrophication process (Nixon 1995; Paerl 1997). These pulses may be followed by blooms. In the N-limited Neuse River estuary of North Carolina, cyanobacterial growth responses closely track (in time and space) such events (Pinckney et al. 1998). In particular, organic N- and ammonium-enriched conditions may favor cyanobacterial dominance in these waters (Pinckney et al. 1998). Interestingly, earlier observations of such correlations in nature (Pearsall 1932; Fogg 1969) have been largely overlooked.

Many N_2-fixing cyanobacterial nuisance bloom genera are capable of regulating buoyancy, hence vertical orientation in the water, by gas means of vacuolation (Klemer and Konopka 1980; Romans et al. 1994). This mechanism can be highly advantageous under either thermally or salinity-stratified conditions, where the nutrient-rich hypolimnetic waters can be accessed periodically by vertical excursions throughout the water column. Buoyancy compensation promotes cyanobacterial dominance in stratified (and stagnant) waters by allowing surface bloom-formers access to nutrient-rich bottom waters within a matter of minutes to hours (vertical migration speeds can exceed meters per hour: c.f. Reynolds and Walsby 1975).

Growing population pressures and use of coastal watersheds will play an increasingly important role in determining community structure and function of marine diazotrophs and their impacts on N budgets. By the turn of the century, approximately 70% of the human population of North America and

Europe will live within 50 km of the coast (Paerl 1997). Nutrient-enriched (specifically non-point-N-enriched) estuaries may be the next frontier of cyanobacterial bloom expansion. Research efforts in the past two decades indicate that, a priori, salinity does not represent a barrier to either the establishment or expansion of diverse diazotrophic cyanobacterial genera (Paerl 1990). In addition, phytoplankton growth in these waters is generally P and trace element sufficient. Accordingly, supply rates of these nutrients do not seem to play a dominant role in explaining either the distribution or proliferation of N_2-fixing cyanobacterial bloom taxa (Paerl 1990; Pinckney et al. 1998).

Clearly, the freshwater-based paradigm that cyanobacterial bloom expansion can largely be controlled by P loading (Vollenweider and Kerekes 1982) requires reexamination and possible modification with regard to estuarine and coastal waters currently experiencing bloom expansion. While these waters exhibit favorable conditions for cyanobacterial expansion based on N:P ratios (V. H. Smith 1983), they do not necessarily conform to the freshwater P-limitation paradigm, since most estuarine and coastal waters currently supporting diazotrophic genera are limited with respect to N rather than P. This indicates that nutrient loading interacts with other environmental factors (mixing, turbulence, light, grazing, etc.) in the regulation of eutrophication.

Because they frequently "track" nutrient enrichment and its consequences on marine fertility, planktonic and benthic N_2-fixing cyanobacteria appear to be good indicators of eutrophication associated with human expansion into and use of the coastal zone. Ironically, nitrogen enrichment is often the key nutritional "driver" of coastal eutrophication, yet N_2 fixers appear to benefit in terms of increasing prevalence and dominance. How can this be?

Even though coastal waters are undergoing N enrichment, the amounts and rates of N enrichment do not exceed the ability of these waters to effectively assimilate nitrogenous compounds. This can be attributed to efficient N uptake by various microbes, which operate at submicromolar levels, thereby maintaining low ambient inorganic and organic N concentrations. In addition, denitrification, the microbial conversion of nitrate to N_2 gas, occurs at relatively high rates in estuarine and coastal ecosystems (Seitzinger and Giblin 1996). Both processes effectively strip biologically available forms of N from the water column, especially in the face of high N loading. Therefore, while N enrichment increases algal biomass, hence organic matter content (which is known to stimulate N_2 fixation), impacted waters generally exhibit N concentrations low enough to obviate either ammonium or nitrate inhibition of nitrogenase activity. As long as fertility is enhanced without parallel enhancement of ambient inorganic N concentrations, N_2 fixation will be favored according to this eutrophication scenario.

Because it is accompanied by increased rates of organic matter production, eutrophication is frequently associated with growing frequencies and magnitudes of excessive oxygen consumption (respiration of organic matter) in stratified bottom waters of estuaries and coastal ecosystems (e.g., Baltic Sea,

Chesapeake Bay, the Albemarle–Pamlico Sound system, the Mississippi River delta in the Gulf of Mexico). These events, termed hypoxia (O_2 concentrations $<4\,mg\,L^{-1}$) and anoxia (no detectable O_2), lead to severe water quality problems, including defaunation (i.e., finfish and shellfish kills) (Turner and Rabalais 1994; Paerl et al. 1998). Moreover, hypoxic and anoxic sediments can release large amounts of nutrients (especially P and Fe), which may further stimulate phytoplankton growth and bloom potentials. Such events have been shown to be particularly favorable for cyanobacterial bloom development in lakes and reservoirs.

Therefore, as accelerating eutrophication becomes a more common and prevalent feature of our coastal waters, the role of cyanobacteria (including diazotrophic bloom-forming genera) in the structure and function of phyto-plankton and benthic microalgal communities is expected to increase. As such, estuarine, coastal, and possibly oceanic N_2 fixation could play an increasingly important biogeochemical and trophic role in response to human perturbation and development of coastal watersheds.

SUMMARY

1. Nitrogen fixation is a critical "bottleneck" process in marine nitrogen cycling and production dynamics because it is the sole way by which biologically synthesized N can enter marine ecosystems. Depending on habitats and environmental conditions, the rate of nitrogen fixation can be high compared to other N inputs and cycling rates. In the subtropical and tropical oligotrophic oceans, N_2 fixation may be a dominant input of "new" N.

2. Recent microbiological and molecular studies point to a remarkably high diversity in the marine environment of heterotrophic, chemolithotrophic, and photosynthetic bacterial as well as cyanobacterial N_2-fixing taxa. In planktonic environments, relatively few cyanobacterial genera appear to dominate this process, while benthic N_2 fixation is distributed among a more diverse microbial assemblage.

3. Molecular and physiological studies have proven essential for character-izing diazotrophic community composition and activities. Cellular- and population-level transcriptional and translational (RNA) studies in con-junction with DNA-based and immunological studies will enable us to identify and quantify taxa actively engaged in N_2 fixation in a variety of marine habitats.

4. Depending on physical, chemical, and biotic characteristics of specific marine habitats, N_2 fixation may be controlled by nutrient (e.g., C, P, Fe) availability, turbulence, light, and synergistic and antagonistic biotic interactions. On the ecosystem level, these regulatory factors tend to interact in time and space.

5. By virtue of enhanced nutrient loading (eutrophication) and hydrological modifications of estuarine and coastal waters, man is altering diazotrophic community composition and activity, with potential impacts on regional and possibly global marine productivity and nutrient cycling.

ACKNOWLEDGMENTS

The authors thank the National Science Foundation (projects DEB 94-08471, DEB94-10325, OCE 92-02106, OCE 94-15985, OCE 95-03539, IBN 96-15772, IBN 96-29314), the U.S. Department of Agriculture (NRICRP project 9600509), and the state of North Carolina (Sea Grant: (project RMER/30) for support of their research. In addition, we are grateful to D. G. Capone, E. J. Carpenter, J. W. Waterbury, and other collaborators for sharing their data and perspectives on the topic of marine N_2 fixation dynamics.

REFERENCES

Ambio (1990) Special issue: Marine eutrophication. *Ambio* 19:101–141.

Antia, N. J., Harrison, P. J., and Oliveira, L. (1991) The role of dissolved organic nitrogen in phytoplankton nutrition, cell biology and ecology. *Phycologia* 30:1–89.

Bauld, J. (1983) Microbial mats in marginal marine environments: Shark Bay, Western Australia and Spencer Gulf, South Australia. In Y. Cohen, R. W. Castenholz, and H. O. Halvorson, eds., *Microbial Mats: Stromatolites*. Alan R. Liss, New York, pp. 39–58.

Bebout, B. M., Paerl, H. W, Crocker, K. M., and Prufert, L. E. (1987) Diel interactions of oxygenic photosynthesis and N_2 fixation (acetylene reduction) in a marine microbial mat community. *Appl. Environ. Microbiol.* 53:2353–2362.

Bebout, B. M., Fitzpatrick, M. W., and Paerl, H. W. (1993) Identification of the sources of energy for nitrogen fixation and physiological characterization of nitrogen-fixing members of a marine microbial mat community. *Appl. Environ. Microbiol.* 59:1495–1503.

Bergerson, F. J. (1980) Measurement of nitrogen fixation by direct means. In F. J. Bergerson, ed., *Methods for Evaluating Biological Nitrogen Fixation*. Wiley, Chichester, pp. 65–75.

Bergman, B. B., Gallon, J. R., Rai, A. N., and Stal, L. J. (1997) N_2 fixation by non-heterocystous cyanobacteria. *FEMS Microbiol. Rev.* 19:139–185.

Bishop, P. E., and Premakumar, R. (1992) Alternative nitrogen fixation systems. In G. Stacey, R. H. Burris, and H. J. Evans, eds., *Biological Nitrogen Fixation*. Routledge, Chapman and Hall, New York, pp. 736–762.

Braun, S., Proctor, L., Zani, S., Mellon, M. T., and Zehr, J. P. (1999) Molecular evidence for zooplankton-associated nitrogen-fixing anaerobes based on amplification of the *nifH* gene. *FEMS Microbial Ecol.* 28:273–279.

Capone, D. G. (1983) Benthic nitrogen fixation. In E. J. Carpenter and D. G. Capone, eds., *Nitrogen in the Marine Environment*. Academic Press, New York, pp. 105–137.

Capone, D. G., Zehr, J. P., Paerl, H. W., Bergman, B., and Carpenter, E. J. (1997) *Trichodesmium*, a globally significant marine cyanobacterium. *Science* 276:1221–1229.

Carmichael, W. W. (1997) The cyanotoxins. *Adv. Bot. Res.* 27:211–256.

Carpenter, E. J. (1972) Nitrogen fixation by a blue-green epiphyte on pelagic *Sargassum*. *Science* 178:1207–1208.

Carpenter, E. J. (1983) Nitrogen fixation by marine *Oscillatoria* (*Trichodesmium*) in the world's oceans. In E. J. Carpenter and D. G. Capone, eds., *Nitrogen in the Marine Environment*. Academic Press, New York, pp. 65–104.

Carpenter, E. J., and Capone, D. G., eds. (1983) *Nitrogen in the Marine Environment*. Academic Press, New York.

Carpenter, E. J., and Culliney, J. L. (1975) Nitrogen fixation in marine shipworms. *Science* 187:551–552.

Carpenter, E. J., and Romans, K. (1991) Major role of the cyanobacterium *Trichodesmium* in nutrient cycling in the North Atlantic Ocean. *Science* 254:1356–1358.

Chen, Y.-B., Dominic, B., Mellon, M. T., Zehr, J. P. (1998) Circadian rhythm of nitrogenase gene expression in the diazotrophic filamentous nonheterocystous cyanobacterium *Trichodesmium* sp. strain IMS 101. *J. Bacteriol.* 180:3598–3605.

Cole, J. J., Lane, J. M., Marino, R., and Howarth, R. W. (1993) Molybdenum assimilation by cyanobacteria and phytoplankton in freshwater and salt water. *Limnol. Oceanogr.* 38:25–35.

Currin, C. A., Paerl, H. W., Suba, G. K., and Alberte, R. S. (1990) Immunofluorescence detection and characterization of N_2-fixing microorganisms from aquatic environments. *Limnol. Oceanogr.* 35:59–71.

D'Elia, C. F., Sanders, J. G., and Boynton, W. R. (1986) Nutrient enrichment studies in a coastal plain estuary: Phytoplankton growth in large scale, continuous cultures. *Can. J. Fish. Aquat. Sci.* 43:397–406.

Dean, D. R., and Jacobsen, M. R. (1992) Biochemical genetics of nitrogenase. In G. Stacey, R. H. Burris, and H. J. Evans, eds., *Biological Nitrogen Fixation*. Routledge, Chapman and Hall, New York, pp. 763–834.

DeLaune, R. D., and Patrick, W. H. (1990) Nitrogen cycling in Louisiana Gulf Coast brackish marshes. *Hydrobiologia* 199:73–79.

Dubois, J. D., and Kapustka, L. A. (1981) Osmotic stress effects on the $N_2(C_2H_2)ASE$ activity of aquatic cyanobacteria. *Aquat.Bot.* 11:11–20.

Duce, R. A., and Tindale, N. W. (1991) Atmospheric transport of iron and its deposition in the ocean. *Limnol. Oceanogr.* 36:1715–1726.

Dugdale, R. C., Menzel, D. W., and Ryther, J. H. (1961) Nitrogen fixation in the Sargasso Sea. *Deep-Sea Res.* 7:298–300.

Dugdale, R. C. (1967) Nutrient limitation in the sea: Dynamics, identification and significance. *Limnol. Oceanogr.* 12:685–695.

Ehrlich, H. L. (1996) *Geomicrobiology*, 3rd ed. Marcel Dekker, New York.

Falkowski, P. G. (1997) Evolution of the nitrogen cycle and its influence on the biological sequestration of CO_2 in the ocean. *Nature* 387:272–275.

Fay, P. (1992) Oxygen relations of nitrogen fixation in cyanobacteria. *Microbiol. Rev.* 56:340–373.

Fogg, G. E. (1969) The physiology of an algal nuisance. *Proc. R. Soc. London B* 173:175–189.

Fogg, G. E. (1982) Marine plankton. In N. G. Carr and B. A. Whitton, eds., *The Biology of Cyanobacteria.* Blackwell Scientific, Oxford, pp. 491–514.

Fogg, G. E., Stewart, W. D. P., Fay, P., and Walsby, A. E. (1973) *The Blue-Green Algae.* Academic Press, London.

Francis, G. (1878) Poisonous Australian lake. *Nature* 18:11–12.

Gallon, J. R. (1992) Tansley Review No. 44/Reconciling the incompatible: N_2 fixation and O_2. *New Phytol.* 122:571–609.

Gallon, J. R., and Stal, L. J. (1992) N_2 fixation in non-heterocystous cyanobacteria: An overview. In E. J. Carpenter, D. G. Capone, and J. G. Rueter, eds., *Marine Pelagic Cyanobacteria: Trichodesmium and Other Diazotrophs.* Kluwer, Dordrecht, pp. 115–139.

Garcia-Pichel, F., and Castenholz, R. W. Cyanobacterial responses to UV-Irradiation. In B. A. Whitton and M. Potts, eds., *Ecology of Cyanobacteria: Their Diversity in Time and Space.* Kluwer, Dordrecht. In press.

Golden, S. S., Ishiura, M., Johnson, C. H., and Kondo, T. (1997) Cyanobacterial circadian rhythms. *Annu. Rev. Plant Physiol. Plant Mol. Biol.* 48:327–354.

Gotto, J. W., and Taylor, B. F. (1976) N_2 fixation associated with decaying leaves of the red mangrove *Rhizophora mangle. Appl. Environ. Microbiol.* 31:781–783.

Guerinot, M. L., and Colwell, R. R. (1985) Enumeration, isolation, and characterization of N_2-fixing bacteria from seawater. *Appl. Environ. Microbiol.* 50:350–355.

Guerinot, M. L., and Patriquin, D. G. (1981) N_2-fixing vibrios isolated from the gastrointestinal tract of sea urchins. *Can. J. Microb.* 27:311–347.

Hardy, R. W. F., Holsten, R. D., Jackson, E. K., Burns, R. C. (1968) The acetylene–ethylene assay for N_2 fixation: Laboratory and field evaluation. *Plant Physiol.* 43:1185–1207.

Head, W. D., and Carpenter, E. J. (1975) Nitrogen fixation associated with the marine macroalgae *Codium fragile. Limnol. Oceanogr.* 20:815–823.

Hicks, B. J., and Silvester, W. B. (1985) Nitrogen fixation associated with the New Zealand mangrove *Avicennia marina var. resinfera. Appl. Environ. Microbiol.* 49:955–959.

Hirsch, A. M., McKhann, H. I., Reddy, A., Liao, J., Fang, Y., and Marshall, C. R. (1995) Assessing horizontal transfer of *nif HDK* genes in eubacteria: Nucleotide sequence of *nif K* from *Frankia* strain HFPCcI3. *Mol. Biol. Evol.* 12:16–27.

Hodson, R. E., Dustman, W. A., Garg, R. M., and Moran, M. A. (1995) In situ PCR for visualization of microscale distribution of specific genes and gene products in prokaryotic communities. *Appl. Environ. Microbiol.* 61:4074–4082.

Horne, A. J. (1977) Nitrogen fixation: A review of this phenomenon as a polluting process. *Prog. Water Technol.* 8:359–372.

Howard, J. B., and Rees, D. C. (1996) Structural basis of biological nitrogen fixation. *Chem. Rev.* 96:2965–2982.

Howarth, R. W., and Cole, J. J. (1985) Molybdenum availability, nitrogen limitation, and phytoplankton growth in natural waters. *Science* 229:653–655.

Howarth, R. W., Marino, R., Lane, J., and Cole, J. J. (1988) Nitrogen fixation in freshwater, estuarine and marine ecosystems. 1. Rates and importance. *Limnol. Oceanogr.* 33:(2)619–687.

Huber, A. L. (1986) Nitrogen fixation by *Nodularia spumigena* Mertens (Cyanobacteria). I. Field studies on the contribution of blooms to the nitrogen budget of the Peel–Harvey Estuary, Western Australia. *Hydrobiologia* 131:193–203.

Kanemoto, R. H., and Ludden, P. W. (1984) Effect of ammonia, darkness, and phenazine methosulfate on whole-cell nitrogenase activity and Fe protein modification in *Rhodospirillum rubrum. J. Bacteriol.* 158:713–720.

Karl, D., Letelier, R., Tupas, L., Dore, J., and Christian, J. D. H. (1997) The role of nitrogen fixation in biogeochemical cycling in the subtropical North Pacific Ocean. *Nature* 386:533–538.

Kimor, B., Reid, F. M. H., and Jordan, J. B. (1978) An unusual occurrence of *Hemiaulus membranaceus* Cleve (Bacillariophyceae) with *Richelia intracellularis* (Cyanophyceae) off the coast of southern California in October 1976. *Phycologia* 17:162–166.

Kirshtein, J. D., Zehr, J. P., and Paerl, H. W. (1993) Determination of N_2 fixation potential in the marine environment: Application of the polymerase chain reaction. *Mar. Ecol. Prog. Ser.* 95:305–309.

Klemer, A. R., and Konopka, A. E. (1980) Causes and consequences of blue-green algal (cyanobacterial) blooms. *Lake Reservoir Manage.* 5(1):9–19.

Knoll, A. (1979) Archean photoautotrophy: Some alternatives and limits. *Origins Life* 9:313–327.

Kucera, S. (1996) The influence of small-scale turbulence on N_2 fixation and growth in heterocystous cyanobacteria. MS thesis, University of North Carolina, Chapel Hill.

LeRudilier, D. T., Bernard, T., Goas, G., and Hamelin, J. (1984) Osmoregulation in *Klebsiella pneumoniae*: Enhancement of anaerobic growth and nitrogen fixation under stress by proline betaine, γ-butyrobetaine, and other related compounds. *Can. J. Microbiol.* 30:299–305.

Letelier, R. M., and Karl, D. M. (1997) Role of *Trichodesmium* in the productivity of the subtropical North Pacific Ocean. *Mar. Ecol. Prog. Ser.* 133:263–273.

Lipschultz, F., and Owens, N. (1996) An assessment of nitrogen fixation as a source of nitrogen to the North Atlantic Ocean. *Biogeochemistry* 35:261–274.

Mague, T. H., Mague, F. C., and Holm-Hansen, O. (1977) Physiology and chemical composition of nitrogen-fixing phytoplankton in the central North Pacific Ocean. *Mar. Biol.* 41:213–227.

Martin, J. H. et al. (1994) Testing the iron hypothesis in ecosystems of the equatorial Pacific Ocean. *Nature* 28:159–182.

Martinez, L. A., Silver, M. W., King, J. M., and Alldredge, A. L. (1983) Nitrogen fixation by floating diatom mats: a source of new nitrogen to oligotrophic ocean waters. *Science* 221:152–154.

Merrick, M. J. (1992) Regulation of nitrogen fixation genes in free-living and symbiotic bacteria. In G. Stacey, H. J. Evans, and R. H. Burris, eds., *Biological Nitrogen Fixation.* Chapman & Hall, New York, pp. 835–877.

Michaels, A. F., Olson, D., Sarmiento, J. L., Ammerman, J. W., Fanning, K., Jahnke, R., Knap, A. H., Lipschultz, F., and Prospero, J. M. (1996) Inputs, losses and transformations of nitrogen and phophorus in the pelagic North Atlantic Ocean. *Biogeochemistry* 35:181–226.

Montoya, J. P., Wiebe, P. H., and McCarthy, J. J. (1992) Natural abundance of ^{15}N in particulate nitrogen and zooplankton in the Gulf Stream region and Warm Core Ring 86A. *Deep-Sea Res.* 39 (suppl. 1):S363–S392.

Murphy, T. O., Lean, D. R. S., and Nalewajko, C. (1976) Blue-green algae: Their excretion of selective chelators enables them to dominate other algae. *Science* 221:152–154.

Niemi, A. (1979) Blue-green algal blooms and N:P ratio in the Baltic Sea. *Acta Bot. Fenn.* 110:57–61.

Nixon, S. W. (1986) Nutrient dynamics and the productivity of marine coastal waters. In R. Halwagy, D. Clayton, and M. Behbehani, eds., *Marine Environment and Pollution.* Alden Press, Oxford, pp. 97–115.

Nixon, S. W. (1995) Coastal marine eutrophication: A definition, social causes, and future concerns. *Ophelia* 41:199–219.

Paerl, H. W. (1984) N$_2$ fixation (nitrogenase activity) attributable to a specific *Prochloron* (Prochlorophyta)-ascidian association in Palau, W. Micronesia. *Mar. Biol.* 81:251–254.

Paerl, H. W. (1988) Nuisance phytoplankton blooms in coastal, estuarine, and inland waters. *Limnol. Oceanogr.* 33:895–905.

Paerl, H. W. (1990) Physiological ecology and regulation of N$_2$ fixation in natural waters. *Adv. Microb. Ecol.* 11:305–344.

Paerl, H. W. (1996) Microscale physiological and ecological studies of aquatic cyanobacteria: Macroscale implications. *Microsc. Res. Tech.* 33:47–72.

Paerl, H. W. (1997) Coastal eutrophication and harmful algal blooms: Importance of atmospheric deposition and groundwater as "new" nitrogen and other nutrient sources. *Limnol. Oceanogr.* 42:1154–1165.

Paerl, H. W. Marine plankton. In M. Potts and B. A. Whitton, eds., *The Ecology of Cyanobacteria.* Kluwer, Dordrecht. In press.

Paerl, H. W., and Bebout, B. M. (1988) Direct measurement of O$_2$-depleted microzones in marine *Oscillatoria*: Relation to N$_2$ fixation. *Science* 242:441–445.

Paerl, H. W., and Carlton, R. G. (1988) Control of nitrogen fixation by surface-associated microzones. *Nature* 332:260–262.

Paerl, H. W., and Kellar, P. E. (1978) Significance of bacterial–*Anabaena* (Cyanophyceae) associations with respect to N$_2$ fixation in freshwater. *J. Phycol.* 14:254–260. 14:254–260.

Paerl, H. W., and Millie, D. F. (1996) Physiological ecology of toxic cyanobacteria. *Phycologia* 35(6):160–167.

Paerl, H. W., and Pinckney, J. L. (1996) Microbial consortia: Their role in aquatic production and biogeochemical cycling. *Microb. Ecol.* 31:225–247.

Paerl, H. W., and Tucker, C. (1995) Ecology of blue-green algae in aquaculture ponds. *J. World Aquacult. Soc.* 26:1–53.

Paerl, H. W., Webb, K. L., and Wiebe, W. (1981) Nitrogen fixation in waters. In W. J. Broughton, *Nitrogen Fixation*, Vol. 1: *Ecology.* Oxford Publications, Oxford, pp. 193–240.

Paerl, H. W., Crocker, K. M., and Prufert, L. E. (1987) Limitation of N$_2$ fixation in coastal marine waters: Relative importance of molybdenum, iron, phosphorus and organic matter availability. *Limnol. Oceanogr.* 32:525–536.

Paerl, H. W., Priscu, J. C., and Brawner, D. L. (1989) Immunochemical localization of nitrogenase in marine *Trichodesmium* aggregates: Relationship to N_2 fixation potential. *Appl. Environ. Microbiol.* 55:2965-2975.

Paerl, H. W., Prufert-Bebout, L. E., and Guo, C. (1994) Iron-stimulated N_2 fixation and growth in natural and cultured populations of the planktonic marine cyanobacterium *Trichodesmium*. *Appl. Environ. Microbiol.* 60:1044-1047.

Paerl, H. W., Pinckney, J. L., and Kucera, S. A. (1995) Clarification of the structural and functional roles of heterocysts and anoxic microzones in the control of pelagic nitrogen fixation. *Limnol. Oceanogr.* 40:634-638.

Paerl, H. W., Pinckney, J. L., Fear, J. M., and Peierls, B. L. (1998) Ecosystem responses to internal and watershed organic matter loading: Consequences for hypoxia in the eutrophying Neuse River estuary, North Carolina, USA. *Mar. Ecol. Prog. Ser.* 166:17-25.

Patriquin. D., and Knowles, R. (1972) Nitrogen fixation in the rhizosphere of marine angiosperms. *Mar. Biol.* 16:49-58.

Paulsen, D. M., Paerl, H. W., and Bishop, P. E. (1991) Evidence that molybdenum-dependent nitrogen fixation is not limited by high sulfate in marine environments. *Limnol. Oceanogr.* 36:1325-1334.

Pearsall, W. (1932) Phytoplankton in the English Lakes. 2. The composition of the phytoplankton in relation to dissolved substances. *J. Ecol* 20:241-262.

Pinckney, J. L., Paerl, H. W., Harrington, M. B., and Howe, K. H. (1998) Annual cycles of phytoplankton community structure and bloom dynamics in the Neuse River estuary, NC (USA). *Mar. Biol.* 131:371-381.

Pope, M. R., Murrell, S. A., and Ludden, P. W. (1985) Covalent modification of the iron protein of nitrogenase from *Rhodospirillum rubrum* by adenosine diphosphoribosylation of a specific arginine residue. *Proc. Natl. Acad. Scie. USA* 82:3173-3177.

Porter, K. G., and Orcutt, J. D. (1980) Nutritional adequacy, manageability, and toxicity as factors that determine the food quality of green and blue-green algae for *Daphnia*. *American Society of Limnology and Oceanography Special Symposium*, Vol. 3, pp. 268-281.

Postgate, J. R. (1982) *The Fundamentals of Nitrogen Fixation*. Cambridge University Press, London.

Postgate, J. R., and Eady, R. R. (1988) The evolution of biological nitrogen fixation. In H. Bothe, F. J. de Bruijn, and W. E. Newton, eds., *Nitrogen Fixation: Hundred Years After*. Gustav Fischer, Stuttgart, pp. 31-40.

Potts, M. (1980) Blue-green algae (cyanophyta) in marine coastal environment of the Sinai Peninsula: Distribution, zonation, stratification and taxonomic diversity. *Phycologia* 19:60-73.

Potts, M. (1994) Desiccation tolerance of prokaryotes. *Microbiol. Rev.* 58:755-805.

Proctor, L. M. (1997) Nitrogen-fixing, photosynthetic, anaerobic bacteria associated with pelagic copepods. *Aqua. Microb. Ecol.* 12:105-113.

Redfield, A. C. (1958) The biological control of chemical factors in the environment. *Am. Sci.* 46:205-221.

Reynolds, C. S., and Walsby, A. E. (1975) Water blooms. *Biol. Rev.* 50:437-481.

Romans, K. M., Carpenter, E. J., and Bergman, B. (1994) Buoyancy regulation in the colonial diazotrophic cyanobacterium *Trichodesmium tenue*: Ultrastructure and storage of carbohydrate, polyphosphate and nitrogen. *J. Phycol.* 30:935–942.

Rueter, J. G., Hutchins, D. A., Smith, R. W., and Unsworth, N. L. (1992) Iron nutrition in *Trichodesmium*. In E. J. Carpenter, D. G. Capone, and J. G. Rueter, eds., *Marine Pelagic Cyanobacteria: Trichodesmium and Other Diazotrophs*. Kluwer, Dordrecht, pp. 289–306.

Ryther, J. H., and Dunstan, W. M. (1971) Nitrogen, phosphorus and eutrophication in the coastal marine environment. *Science* 171:1008–1112.

Schopf, J. W., and Walter, M. R. (1982) Origin and early evolution of cyanobacteria: The geological evidence. In N. G. Carr and B. A. Whitton, *The Biology of Cyanobacteria*. Blackwell Scientific, Oxford, pp. 543–564.

Seitzinger, S. P., and Giblin, A. E. (1996) Estimating denitrification in North Atlantic continental shelf sediments. *Biogeochemistry* 35:235–260.

Sellner, K. G. (1997) Physiology, ecology, and toxic properties of marine cyanobacterial blooms. *Limnol. Oceanogr.* 42:1089–1104.

Shieh, W. Y., Simidu, U., and Maruyama, Y. (1989) Enumeration and characterization of nitrogen fixing bacteria in an eelgrass *Zostera marina* bed. *Microb. Ecol.* 18:249–260.

Shieh, W. Y., and Lin, Y. M. (1992) Nitrogen fixation (acetylene reduction) associated with the zoanthid *Palythoa tuberculosa* Esper. *J. Exp. Mar. Biol. Ecol.* 163:31–41.

Sivonen, K. et al. (1989) Occurrence of the hepatotoxic cyanobacterium *Nodularia spumigena* in the Baltic Sea and the structure of the toxin. *Appl. Environ. Microbiol.* 55:1990–1995.

Smith, S. V. (1984) Phosphorus vs. nitrogen limitation in the marine environment. *Limnol. Oceanog.* 29:1149–1160.

Smith, V. H. (1983) Low nitrogen to phosphorus ratios favor dominance by blue-green algae in lake phytoplankton. *Science* 221:669–671.

Smith, V. H. (1990) Nitrogen, phosphorus, and nitrogen fixation in lacustrine and estuarine ecosystems. *Limnol. Oceanogr.* 35:1852–1859.

Sörensson, F., and Sahlsten, E. (1987) Nitrogen dynamics of a cyanobacterial bloom in the Baltic Sea: New versus regenerated production. *Mar. Ecol. Prog. Ser.* 37:277–284.

Stal, L. J., and Caumette, P. (1994) *Microbial Mats: Structure, Development and Environmental Significance*, NATO ASI Series G: Ecological Sciences Vol. 35. Springer-Verlag, Berlin.

Stal, L. J., and Krumbein, W. E. (1985) Nitrogenase activity in the non-heterocystous cyanobacterium *Oscillatoria* sp. grown under alternating light–dark cycles. *Arch. Microbiol.* 143:67–71.

Steppe, T. F., Olson, J. B., Paerl, H. W., Litaker, R. W., and Belnap, J. (1996) Consortial N_2 fixation: A strategy for meeting nitrogen requirements of marine and terrestrial cyanobacterial mats. *FEMS Microbiol. Ecol.* 21:149–154.

Stewart W. D. P., Fitzgerald, G. P., and Burris, R. H. (1967) *In situ* studies on N_2 fixation, using the acetylene reduction technique. *Proc. Natl. Acad. Sci. USA* 58:2071–2078.

Stewart, W. D. P. (1973) Nitrogen fixation: In N. G. Carr and B. A. Whitton, eds., *The Biology of Blue-Green Algae*. Blackwell Scientific, Oxford, pp. 260–278.

Ter Steeg, P., Hanson, P. J., and Paerl, H. W. (1986) Growth-limiting quantities and accumulation of molybdenum in *Anabaena oscillarioides* (Cyanobacteria). *Hydrobiologia* 140:143–147.

Thomas, J. S., Apte, K., and Reddy, B. R. (1988) Sodium metabolism in cyanobacterial nitrogen fixation and salt tolerance. In H. Bothe, F. J. de Bruyn, and W. E. Newton, eds., *Nitrogen Fixation: Hundred Years After*. Gustav Fischer, Stuttgart, pp. 195–201.

Thomas, W., and Gibson, C. (1990) Effects of small-scale turbulence on microalgae. *J. Appl. Phycol.* 2:71–77.

Thorneley, R. N. F., and Ashby, G. A. (1989) Oxidation of nitrogenase iron protein by dioxygen without inactivation could contribute to high respiration rates of *Azotobacter* species and facilitate nitrogen fixation in other aerobic environments. *Biochem. J.* 261:181–187.

Tibbles, B. J., and Rawlings, D. E. (1994) Characterization of N_2 fixing organisms from a temperate salt marsh lagoon including isolates that produce ethane from acetylene. *Microb. Ecol.* 27:65–80.

Turner, R. E., and Rabalais, N. M. (1994) Coastal eutrophication near the Mississippi delta. *Nature* 368:619–621.

Valiela, I. (1983) Nitrogen in salt marsh ecosystems. In E. J. Carpenter and D. G. Capone, eds., *Nitrogen in the Marine Environment*. Academic Press, New York, pp. 649–678.

Villareal, T. A. (1992) Marine nitrogen fixing diatom–cyanobacterial symbioses. In E. J. Carpenter, D. G. Capone, J. G. Rueter, eds., *Marine Pelagic Cyanobacteria: Trichodesmium and other Diazotrophs*. Kluwer, Dordrecht, pp. 163–175.

Villbrandt, M., Stal, L. J., and Krumbein, W. E. (1990) Interactions between nitrogen fixation and oxygenic photosynthesis in a marine cyanobacterial mat. *FEMS Microbiol. Ecol.* 74:59–72.

Villbrandt, M., Stal, L. J., Bergman, B., and Krumbein, W. E. (1992) Immunolocalization and Western blot analysis of nitrogenase in *Oscillatoria limosa* during a light–dark cycle. *Bot. Acta* 105:90–96.

Vollenweider, R. A., and Kerekes, J. J. (1982) Eutrophication of waters: Monitoring, assessment and control. Organization for Economic Cooperation and Development, Paris.

Waterbury, J. B., Watson, S. W., and Valois, F. W. (1988) Temporal separation of photosynthesis and dinitrogen fixation in the marine unicellular cyanobacterium *Erythrospira marina*. *Eos* 69:1089.

Webb, K. L., Dupaul, W. D., Wiebe, W. J., Sottile, W., and Johannes, R. E. (1975) Enewetak Atoll: Aspects of the nitrogen cycle on a coral reef. *Limnol. Oceanogr.* 20:198–210.

Wolk, C. P. (1982) Heterocysts. In: N. G. Carr and B. A. Whitton, eds., *The Biology of Cyanobacteria*. Blackwell Scientific, Oxford.

Young, J. P. W. (1992) Phylogenetic classification of nitrogen-fixing organisms. In G. Stacey, H. J. Evans, and R. H. Burris, eds., *Biological Nitrogen Fixation*. Chapman & Hall, New York, pp. 43–86.

Zehr, J. P., and Capone, D. G. (1996) Problems and promises of assaying the genetic potential for nitrogen fixation in the marine environment. *Microb. Ecol.* 32:263–281.

Zehr, J. P., and McReynolds, L. A. (1989) Use of degenerate oligonucleotides for amplification of the *nifH* gene from the marine cyanobacterium *Trichodesmium thiebautii*. *Appl. Environ. Microbiol.* 55:2522–2526.

Zehr, J. P., Mellon, M. T., and Zani, S. (1998) New nitrogen fixing microorganisms detected in oligotrophic oceans by the amplification of nitrogenase (*nifH*) genes. *Appl. Environ. Microbiol.* 64:3444–3450.

Zehr, J. P., and Paerl, H. W. (1998) Nitrogen fixation in the marine environment: Genetic potential and nitrogenase expression. In K. E. Cooksey, ed., *Molecular Approaches to the Study of the Ocean*. Chapman & Hall, London.

Zehr, J. P., Wyman, M., Miller, V., Duguay, L., and Capone, D. G. (1993) Modification of the Fe protein of nitrogenase in natural populations of *Trichodesmium thiebautii*. *Appl. Environ. Microbiol.* 59:669–676.

Zehr, J. P., Mellon, M., Braun, S., Litaker, W., Steppe, T. F., and Paerl, H. W. (1995) Diversity of heterotrophic nitrogen fixation genes in a marine cyanobacterial mat. *Appl. Environ. Microbiol.* 61: 2527–2532.

Zehr, J. P., Braun, S., Chen, Y.-B., and Mellon, M. (1996) Nitrogen fixation in the marine environment: Relating genetic potential to nitrogenase activity. *J. Exp. Mar. Biol. Ecol.* 203:61–73.

14

NITRIFICATION AND THE MARINE NITROGEN CYCLE

Bess B. Ward

Department of Geosciences,
Princeton University,
Princeton, New Jersey

Nitrification is the process whereby ammonium is oxidized to nitrite and then to nitrate. It thus links the most oxidized and most reduced forms of nitrogen and helps determine their overall distributions. Ammonium rarely occurs at significant concentrations in oxygenated seawater. It is recycled rapidly between heterotrophic organisms (which excrete ammonium directly or release organic nitrogen, which is microbially degraded to ammonium) and photosynthetic phytoplankton (which utilize ammonium as a nitrogen source) in the surface ocean. Ammonium can accumulate in anoxic sediments and in seawater, where oxygen concentrations are very low.

Similarly, nitrite rarely accumulates in oxygenated seawater (see below for the exception of the primary nitrite maximum), although nitrite is an essential intermediate in several oxidation and reduction processes in the nitrogen cycle. Nitrate, the end product of nitrification, however, accumulates in the deep ocean where there is no demand for inorganic nitrogen by phytoplankton. This accumulation is largely due to nitrifying bacteria, and it represents the major fixed nitrogen pool in the ocean.

As far as is known, the oxidation of ammonia to nitrite and of nitrite to nitrate, although thermodynamically favorable when linked to reduction of oxygen, does not occur in the absence of biological activity. The only exception is the oxidation of ammonia and amino-level nitrogen in organic compounds to nitrogen gas via inorganic catalysis by manganese oxide, which has been demonstrated in sediments (Luther et al. 1997). Some biological process may

Microbial Ecology of the Oceans, Edited by David L. Kirchman.
ISBN 0-471-29993-6   Copyright © 2000 by Wiley-Liss, Inc.

yet be identified with this process, and this pathway may have significant ramifications for the nitrogen cycle of sediments. The production of nitrate from ammonium, however, appears to be entirely biologically driven.

The most important organisms in nitrification are the so-called nitrifying bacteria. This group includes several genera of bacteria, all within the proteobacterial phylum. Not all of these genera are closely related to each other, but they appear to have arisen from a common ancestor, diverging relatively soon after the ability to nitrify was developed (Teske et al. 1994). There are two functionally distinct groups of nitrifiers: those that oxidize ammonium to nitrite and those that oxidize nitrite to nitrate. No organism is known to carry out both reactions. These unique metabolic traits are not without costs; the nitrifiers are chemolithoautotrophic for the most part, a lifestyle that enables them to exploit a unique niche in natural systems but seems to confer a constraint of slow growth and inflexible nutritional requirements.

Nitrification is closely coupled with other important steps in the nitrogen cycle, such as nitrogen assimilation by phytoplankton and loss of fixed nitrogen by denitrification. Thus nitrification is part of the linkage between the carbon and nitrogen cycles at several levels. These and other aspects of nitrification and nitrifying bacteria in marine systems are explored in this chapter.

CHARACTERISTICS AND PHYLOGENY OF BACTERIA INVOLVED IN NITRIFICATION

Physiology and Phylogeny of Autotrophic Nitrifiers

The best-known nitrifying bacteria are chemolithoautotrophic bacteria. They are capable of obtaining all their carbon requirements via the fixation of carbon dioxide (using the Calvin cycle), using the reducing power they obtain from the oxidation of ammonium or nitrite as their only energy source. While several strains of nitrite-oxidizing bacteria have been shown to be able to augment this chemolithoautotrophic lifestyle with heterotrophic metabolism of simple carbon substrates, the known ammonia oxidizers are obligate chemolithoautotrophs. This metabolic habit has been described as a hard way to make a living. It constrains the nitrifiers to use only inorganic nutrient and energy sources and is probably responsible for the relatively slow growth rates observed for nitrifiers in culture.

The number and diversity of bacterial strains that are identified as autotrophic nitrifying bacteria is rather limited, in comparison, say, to the number and diversity of organisms that are capable of denitrification or of heterotrophic growth. The phylogeny of nitrifiers (Teske et al. 1994) shows them all to be descendents of a common ancestor, which was photosynthetic rather than descending from a common ancestral nitrifier. The ammonia oxidizers are found in the beta and gamma subdivisions of the Proteobacteria (Figure 1). The nitrite oxidizers are found in the alpha, delta, and gamma subdivisions.

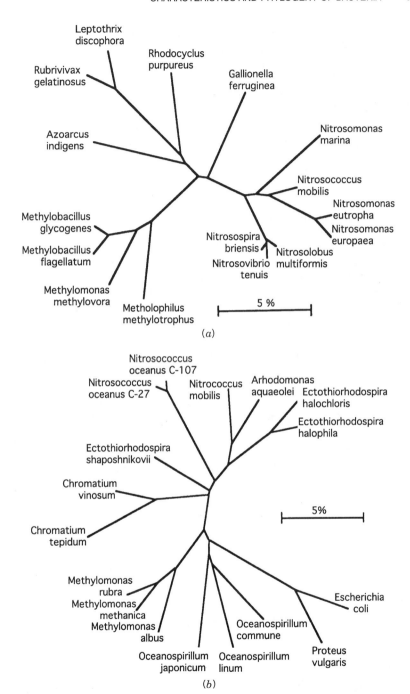

Figure 1. Phylogenetic relationships among selected representatives of the Proteobacteria. Both the beta (a) and gamma (b) subdivisions of the Proteobacteria contain closely related ammonia- and methane-oxidizing bacteria. (From Teske et al. 1994.)

The ammonia oxidizers show significant metabolic and morphological similarities with another group of autotrophic bacteria, the methane oxidizers. They are also closely related phylogenetically to the methane oxidizers, in both the gamma and beta subdivisions (Figure 1). Prior to the availability of ribosomal RNA sequence data for determination of phylogenetic relationships, it had been reported that ammonia-oxidizing nitrifiers were capable of methane oxidation and vice versa (see below). Thus the report that the central genes in the two pathways, genes encoding ammonia monooxygenase in the nitrifiers and methane monooxygenase in the methanotrophs, were evolutionarily related was an interesting verification of the metabolic studies (Holmes et al. 1995).

Heterotrophic Nitrification

Many denitrifying bacteria, other heterotrophic bacteria, and fungi have been reported to perform heterotrophic nitrification. "Heterotrophic nitrification" refers to several different oxidation reactions that parallel parts of the autotrophic nitrification process, including reactions that release nitrate or nitrite from breakdown of organic nitrogen compounds. It has been argued that heterotrophic nitrification involves quite different enzyme systems (Wehrfritz et al. 1993) and that it cannot serve as an energy generating mechanism (Castignetti 1990), as does the autotrophic process. Heterotrophic nitrification has been studied mostly in terrestrial systems, especially acid forest soils, where it has been difficult to document autotrophic nitrification. Experiments using isotopes to differentiate production of nitrate from inorganic and organic substrates in a forest system found that heterotrophic nitrification accounted for less than 10% of the total nitrification rate (Barraclough and Puri 1995). No information is available on the occurrence or significance of heterotrophic nitrification in marine systems. However, bacteria with metabolism that is homologous to that of the kinds of bacteria most often implicated in these processes are probably present in seawater. This is a process that should be investigated in aquatic systems.

DISTRIBUTION AND ABUNDANCE OF NITRIFIERS IN SEAWATER

Discovery and Description

Autotrophic nitrifiers were first isolated and characterized in the late 1800s by Sergei Winogradsky (1890), the "father of chemolithoautotrophy". Winogradsky's organisms were derived from soils; the first reported isolation and characterization of marine nitrifying bacteria was by Stanley Watson in the 1960s (Watson 1965). Nitrification had been detected in seawater prior to that time and nitrifiers had been implicated in the production of nitrite in seawater (reviewed by Vaccaro 1962; see also Rakestraw 1936). The first marine ammonia oxidizer was originally called *Nitrosocystis oceanus* and is now referred to as *Nitrosococcus oceanus*, the lone ammonia-oxidizing species in the gamma subdivision of the Proteobacteria. Watson also isolated marine strains of the best-known terrestrial genus of ammonia-oxidizing bacteria, *Nitrosomonas*, and he also reported the first isolations and characterizations of marine nitrite-oxidizing bacteria (Watson and Waterbury 1971; Watson et al. 1986). His culture collections have formed the basis of much of our modern knowledge of the biochemistry, physiology, phylogeny, distribution, and activity of nitrifiers in seawater and terrestrial systems.

Abundance

Since the original isolations, there have been several more reports of isolation of similar marine nitrifying bacteria from many marine environments. The first estimates of their abundance were obtained by cultivation methods (Watson 1965), and the most probable number (MPN) approach continues to be used in both soil and aquatic environments. Although time-consuming (many months), this approach provides, in addition to abundance estimate, isolates for further study. Although most strains have characteristic intracytoplasmic membrane structures that can be seen by electron microsocopy, it is not possible to distinguish the otherwise nondescript cells from other bacteria in seawater (by, e.g., epifluorescence microscopy with DNA fluorochromes). Immunological methods for detection and enumeration were used to study the distribution of several species of nitrifiers in seawater (Ward and Carlucci 1985), leading to the conclusion that individual species of nitrifiers represent on the order of 0.1% of the total bacterial assemblage in the water column. Maximum abundances have been reported near the bottom of the photic zone in the vicinity of the primary nitrite maximum in some environments (Ward et al. 1989a), but such characteristic patterns are not always detected. Relative abundances of beta and gamma subdivision ammonia-oxidizing bacteria have been estimated using semiquantitative PCR in Antarctic lakes (Voytek et al. 1998); the results compare favorably with those obtained by immunofluorescence in replicate samples. While nitrification is recognized as an essential process in sediments, there is less information on the abundance of nitrifying bacteria in sediment environments.

ROLE OF NITRIFICATION IN THE OCEAN'S NITROGEN CYCLE

The role of nitrification in the nitrogen cycle, whether on land or in aquatic systems, is to convert ammonium, a common nitrogenous waste product of heterotrophic metabolism, to nitrate. Many forms of organic and inorganic nitrogen can be utilized by plants, and the transformation of ammonium into nitrate does not really change the absolute availability of nitrogen for plant nutrition. In soils, the different ionic properties of ammonium and nitrate are important in determining how long the inorganic nitrogen stays available in the soil solution. Seitzinger et al. (1991) argued that the generally lower net nitrification/denitrification rates of marine versus freshwater sediments could be partially explained by the smaller size of the exchangeable ammonium pool in marine pore waters. In pelagic systems, however, the properties of these ions are less important to their distributions. Unlike nitrogen fixation, which introduces new fixed nitrogen into the system, and denitrification, which removes it from the system, nitrification simply changes the chemical nature and oxidation state of inorganic nitrogen. Because of the different chemical properties and varying preferences, abilities, and metabolic costs of utilizing ammonium versus nitrate, however, this transformation is very important in both agricultural and natural systems.

In the ocean, ammonium is often the first metabolic product of the breakdown of organic nitrogen; it is an excretory product of zooplankton and protozoans, and it is released from the microbial degradation of complex organic compounds. Because ammonium contains nitrogen at the oxidation level of proteins, it is readily assimilated by both phytoplankton and bacteria and is a preferred nitrogen source. Therefore, it rarely accumulates in surface ocean waters; it is assimilated as rapidly as it is produced by various members of the microbial food web. In fact, ammonia-oxidizing bacteria may be in competition for ammonium with other planktonic organisms. The different physiological requirements of phytoplankton and nitrifying bacteria probably play a role in determining exactly where in the water column ammonium assimilation and ammonium oxidation occur. As explained later in this chapter, highest nitrification rates occur near the base of the euphotic zone in the upper hundred or so meters of the ocean. There is usually very little nitrate in the surface ocean, however, because of its use by phytoplankton. Nitrate has accumulated in the deep ocean from nitrification in the absence of phytoplankton assimilation. It is because of nitrifying bacteria that the deep ocean is full of nitrate, rather than ammonium. The deep nitrate reservoir can be made available to phytoplankton by mixing and upwelling. These physical processes bring cold, deep, nitrate-rich water up to the surface where, in the presence of light, phytoplankton can assimilate the nitrate. Thus, although nitrate is not usually abundant in surface waters, it is a very important nitrogen source for phytoplankton.

The nitrate that is produced by nitrification, especially that which accumulates below the euphotic zone, may have a fate other than assimilation by phytoplankton. Denitrification (Chapter 15) is another step in the nitrogen cycle, which is also controlled by bacteria. Denitrifying bacteria are quite different from nitrifying bacteria in phylogeny as well as metabolism (but, see below for exceptions). Denitrifiers utilize nitrate and nitrite as respiratory substrates; they can respire using these oxides of nitrogen as electron acceptors in place of oxygen. In the process, nitrate is reduced sequentially to nitrite, then to nitric oxide and nitrous oxide, and finally to dinitrogen gas. The last three products are all gases. Most organisms that require nitrogen for nutrition are unable to assimilate the gases; thus denitrification results in a net loss from the system of fixed, (i.e., biologically available) nitrogen. Although denitrification involves several semi-independent steps that need not function together all the time, it is common for denitrifiers to begin the sequence with nitrate and to produce varying amounts of the other products depending on the environmental conditions. Thus, although denitrifiers have little in common with nitrifiers, the former are in fact dependent on the latter — other than lightning and fertilizers, nitrifiers are the only source of nitrate. The links between nitrification and denitrification are discussed further in the context of the environments in which they occur (see below). The link is mentioned here to emphasize that although nitrifiers by themselves do not change the fixed nitrogen inventory of the ocean, their activity makes possible the activity of denitrifiers, which are responsible for the major loss term in the global nitrogen budget.

Thus the role of nitrifiers in the nitrogen cycle of the ocean is to link the oxidizing and reducing processes of the nitrogen cycle by converting ammonium to nitrate. It is because of this conversion that the major fixed nitrogen pool in the ocean is in the form of nitrate. The deep oceanic nitrate reservoir is a huge pool of nitrogen whose availability to phytoplankton is controlled largely by physical processes. The nitrification link also makes possible the loss of fixed nitrogen via denitrification, by converting nitrogen released as a waste product of animal metabolism into a form that can be respired by denitrifiers.

DISTRIBUTION OF NITRIFICATION IN SEAWATER AND SEDIMENTS

It was long ago recognized that nitrification must be a process of some consequence in the ocean; the major nitrogen product of organic matter decomposition (the net result of biological processes in the deep sea away from the euphotic zone) is ammonium, but the huge volume of the deep ocean contains nitrate at concentrations of $40\,\mu M$ and ammonium at trace or undetectable levels. Therefore, it seemed obvious that nitrification must be occurring in the deep ocean (see Chapter 15 for background on stoichiometry of organic matter mineralization). Nitrate concentrations in the surface ocean are usually maintained at low levels because phytoplankton assimilate nitrate more rapidly than it can be supplied by mixing or diffusion from the deep nitrate reservoir. Ammonium, which is produced in the photic zone by heterotrophic processes, is also usually immediately assimilated by phytoplankton before it can be nitrified. The recognition that ammonium and nitrate have different sources, and that the differences are important both physically and biologically, led to the application of the new production paradigm (Dugdale and Goering 1967; Eppley and Peterson 1979) as a way to understand phytoplankton nitrogen demand and growth in the surface ocean, and the subsequent flux of nitrogen to the deep sea and ocean floor (Figure 2). Ammonium is considered to be a "regenerated" source of nitrogen because it is produced largely in situ, in the same water where phytoplankton live and

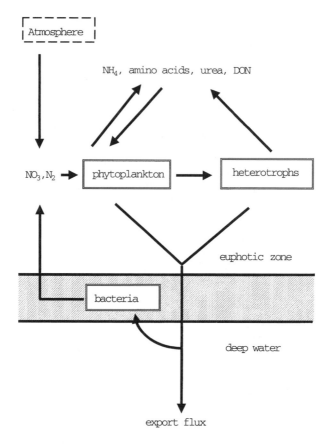

Figure 2. Schematic representation of the new production paradigm (after Eppley and Peterson 1979). If the shaded area is below the euphotic zone, then nitrate produced there will enter the surface layer as a new nutrient (via physical processes). If the shaded area is within the euphotic zone, then nitrification occurring there will supply regenerated nitrate.

utilize it. The nitrogen in ammonium is recycled rapidly and repeatedly among phytoplankton, the zooplankton that graze on them, and the inorganic nutrient that is released from heterotrophic metabolism and grazing.

Nitrate, on the other hand, is considered to be "new" nitrogen because it is not present in the euphotic zone most of the time. For phytoplankton to utilize it, nitrate must be transported into the system by physical means — mixing or upwelling from deep waters or falling in rain — and the rate of nitrate supply can be equated with the steady state rate of primary production based on nitrate as a nitrogen source (Eppley and Petersen 1979). This equality makes it possible to measure "new production," and by inference, the sinking flux of nitrogen, by measuring the assimilation of nitrate in incubated samples. This perception of nitrate as a new nutrient is consistent with the idea that nitrification, leading to the accumulation of nitrate, occurs in deep water, not in the euphotic zone itself.

Nitrification Rates: General Distribution

Actual measurements of the rate and its depth distribution, however, showed that this assumption that most nitrification occurred in the deep ocean was an oversimplification. The highest nitrification rates, both ammonium oxidation and nitrite oxidation, occur not in the deep ocean, but in a region near the bottom of the euphotic zone. In this depth interval, where the light intensity is very reduced, phytoplankton are light-limited, and their rates of nutrient assimilation are therefore reduced. It is in this interval that nitrifying bacteria can compete with phytoplankton for ammonium; one often observes a sharp peak in nitrification rate at a depth in the water column where light intensity has been reduced to 5–10% of surface light intensity (Ward 1987b).

The rate of nitrification in deep ocean water is minimal, because of the decreasing flux of ammonium from organic matter decomposition with increasing depth. The process has been detected at depths up to a few thousand meters (Ward and Zafiriou 1988).

Methods for Measuring Nitrification in Water and Sediments

Direct measurement of the rate of nitrification is problematic for several reasons having to do with the sensitivity of the methods and potential artifacts introduced by incubation methods. The easiest experimental design might be simply to incubate samples and measure the concentrations of nitrite or nitrate over time (Pakulski et al. 1995). Accumulation of nitrite or nitrate would indicate net nitrification; even a decrease in the concentration, however, would not mean that no nitrification had occurred, but rather that consumption had exceeded production in the incubation bottle. Given the limiting nature and low concentration of fixed nitrogen in the surface ocean, large fluxes can be obscured by tight coupling between production and consumption terms.

The addition of specific inhibitors has been used as a modification of the simple nutrient measurement approach just described. In this approach, chemicals that specifically inhibit either ammonium oxidation (e.g., allyl-thiourea, methyl fluoride, or N-serve) or nitrite oxidation (chlorate) are added to replicate incubation bottles (Bianchi et al. 1997). The method assumes that the nitrite concentration is at steady state in the sample and that nitrification is the only process that produces or consumes nitrite. Clearly, the bottles must be incubated in the dark to prevent consumption by phytoplankton. One need only measure the concentration of nitrite over time in the bottles in which ammonium oxidation was inhibited to estimate the nitrite oxidation rate (equal to the rate of nitrite decrease). The rate of nitrite increase in the bottles to which nitrite oxidation inhibitor was added equals the rate of ammonium oxidation. There are still problems with this approach:

1. Preventing production by phytoplankton probably has cascading effects on the activities of other microbes in the bottle, such that the rate of ammonium mineralization is reduced, therefore changing the source term for the nitrification substrate.

2. Incubating in the dark may release the nitrifying bacteria from light inhibition such that the measured rate exceeds the in situ rate.

3. Incubations typically last 48 hours (Bianchi et al. 1994a,b, 1997), which is sufficient to overcome the lag induced by light inhibition but is also long enough to create quite unnatural conditions (bottle effects of wall growth, perturbation of nutrient fluxes, etc.).

An alternative approach using specific inhibitors has the advantage of the increased sensitivity of radiotracers: being chemolithoautotrophs, nitrifiers fix CO_2 while oxidizing nitrogen. The amount of CO_2 fixation due to nitrifiers can be computed by taking the difference between incubations with and without addition of an inhibitor that specifically removes the contribution of nitrifiers (Billen 1976; Somville 1978; Dore and Karl 1996b). Then a conversion factor is used to translate the CO_2 fixation into ammonium and nitrite oxidation rates.

The other main approach to measuring nitrification rates directly is to use the stable isotope ^{15}N as a tracer (Olson 1981a; Ward et al. 1984). This approach is not without its problems, mainly because ^{15}N has a significant natural abundance and must be measured by means of a mass spectrometer or emission spectrometer, both more expensive and difficult processes than using a scintillation counter for radioisotopes. Somewhat shorter incubations are possible (a few hours to 24 h are commonly used, which is short enough to minimize bottle effects and avoid equilibration of the isotopic signal). The signal of transfer of the tracer from substrate to product pool (e.g., $[^{15}N]H_4$ to $[^{15}N]O_2$) can be detected regardless of what other processes are occurring in the incubation (so in situ light conditions can be used), and no assumptions of steady state need be made. The major drawback of this method is the necessity to add tracer, sometimes in excess of the natural concentration of substrate. This problem has been largely overcome with the advent of more sensitive mass spectrometers, however, and estimates obtained under conditions approaching in situ may be possible.

The ^{15}N approach is most useful in water samples because complete mixing of the tracer is possible. In sediments and soils, rate measurements are constrained by the inhomogeneous nature of the sample and the dependence of rates on the structure of the environment. Inhibitor approaches similar to those described above for water samples have been used in sediments (Henriksen et al. 1981; Sloth et al. 1992; Miller et al. 1993). The methyl fluoride method of Miller et al. (1993) seems particularly promising because the gas can diffuse thoroughly into the core with minimal disturbance of microzones and gradients. The ammonium oxidation inhibitor is added to cores, and the accumulation of NH_4 over time is assumed to represent the net rate of nitrification.

Other processes that consume ammonium would lead to an underestimate of the rate.

To overcome the biasing resulting from uneven dispersal of tracer or inhibitor, sediment rate measurements are often made in slurries, which destroy the sediments' gradient structure, which in turn is essential to the in situ fluxes. Even if rate measurements in sediments are made using whole-core incubations—for example, when the inhibitor is a gas—it is still difficult to obtain a depth distribution of the rate (usually, an areal rate is obtained). A sophisticated system based on measurements and a model that avoids direct rate measurements has been used to overcome this problem. Microelectrodes having very high vertical resolution are used to measure the fine-scale distribution of oxygen and nitrate in freshwater sediments. By assuming that the observed vertical gradients represent a steady state condition, reaction–diffusion models can then be used to estimate the rates of nitrification, denitrification, and aerobic respiration and to compute the location of the rate processes in relation to the chemical profiles (e.g., Jensen et al. 1994). The major drawback to this approach is the lack of microelectrodes for nitrate that can be used in seawater (although recent developments in biosensors may overcome the carbonate interference that is problematic in seawater).

Nitrification Rates in the Ocean

Rates of nitrification reported for the open ocean are in the range of a few to a few hundred nanomolar per day. Where profiles extending to a depth of several hundred to a few thousand meters are available, the main pattern that emerges is the association of highest rates of ammonia oxidation with the lower region of the photic zone (Figure 3). In the eastern tropical North Pacific, ammonia oxidation rates at the maximum were no more than 20 nM d^{-1} (Ward and Zafiriou 1988). In the Peru upwelling region, a maximum rate of 747 nM d^{-1} was reported (Lipschultz et al. 1990). In the temperate eastern Pacific Ocean off western North America, maximum rates of 45 nM d^{-1} were reported (Ward 1987b). Nitrite oxidation shows a less predictable distribution with depth; in the Peru upwelling system, maximum rates of 600 nM d^{-1} were observed near the lower boundary of the euphotic zone, but high rates (e.g., nearly 300 nM d^{-1}) were observed within the oxygen minimum zone (Lipschultz et al. 1990).

Several studies have focused on the primary nitrite maximum, rather than attempting to obtain complete depth profiles. Dore and Karl (1996b) reported a few rate measurements based on inhibitor experiments from the central Pacific Ocean. Ammonium and nitrite oxidation rates were usually comparable, and were maximal (about 135 nM d^{-1}) just below the primary nitrite maximum. Bianchi et al. (1994a) reported nitrification rates up to 1–2 μM d^{-1} in the Rhone River plume, with rates decreasing to the usual oceanic levels with increasing distance from shore in the Mediterranean Sea.

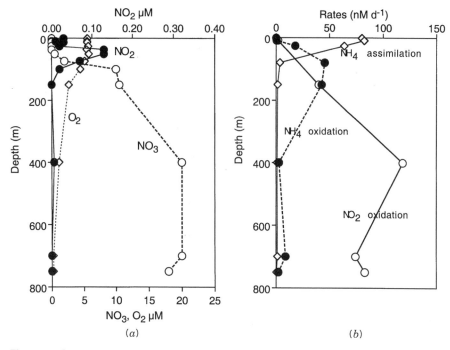

Figure 3. Depth distribution of measured nitrification rates in the ocean. Data from the Southern California Bight (33°18.7′N, 118°09.6′W) in about 900 m water depth (redrawn from Ward 1987). (a) Hydrographic data for NO$_3$ (open circles), NO$_2$ (solid circles) and O$_2$ (diamonds). (b) Rates in nanomoles per liter per hour for ammonium oxidation (solid circles), nitrite oxidation rate (open circles), and ammonium assimilation rate (diamonds). Rates measured using ^{15}N tracer techniques. (From Ward 1987b.)

Ammonia and nitrite oxidation rates are often reported to be highly coupled and to occur at about the same rate. This coupling would be consistent with the observation that neither ammonium nor nitrite accumulates to high levels in most of the world ocean. The exceptions to this rule are the primary nitrite maximum of near surface waters and the secondary nitrite maximum characteristic of oxygen minimum zones. The surface water feature has been attributed to two possible processes. Because the primary nitrite maximum is usually associated with low light intensities at the bottom of the euphotic zone, some interaction between biological processes and light is suspected. In one scenario, phytoplankton are responsible (Keifer et al. 1976): assimilation of nitrate (via reduction to nitrite and then to ammonium) requires energy, especially the step at which nitrite is reduced. Under low light intensity, phytoplankton might not have enough energy to reduce the nitrate completely, and some of the intermediate nitrite is allowed to leak out of the cell. In the alternative scenario, nitrifying bacteria are responsible (Olson 1981b): nitrite oxidizers are more sensitive to light inhibition than are ammonia oxidizers, so ammonia oxidizers are able to be active at slightly shallower depths in the water column than are nitrite oxidizers. This leads to an accumulation of nitrite in the interval between the depths at which ammonia- and nitrite-oxidizers are released from light inhibition. It may be that in many environments, both scenarios contribute to the net accumulation of the primary nitrite maximum. In the North Pacific Gyre, Dore and Karl (1996a) attribute two distinct features within the primary nitrite maximum to phytoplankton (the upper primary nitrite maximum) and nitrifiers (the lower primary nitrite maximum).

The magnitude of nitrification rates in sediments can be much higher, and it is certainly more variable than data reported from marine water column measurements. The variability arises not only from the small-scale heterogeneity inherent in sediments (partly due also to bioturbation and association of nitrification with the walls of faunal tubes in the sediments), but from the wide range in the level of organic matter input to sediments in shallow water. In both deep and shallow sediments, nitrification can be one of the main sinks, if not the main sink, for oxygen in sediments (Grundmanis and Murray 1977; Blackburn and Blackburn 1993).

ENVIRONMENTAL VARIABLES THAT AFFECT NITRIFICATION RATES AND DISTRIBUTION

As mentioned in the general discussion of the depth distribution of nitrification rates, variables such as light intensity and substrate concentration are important determinants of the magnitude and location of nitrification rates. These are the kinds of variable that determine much of the biogeochemical cycling of the ocean, and their influence on nitrification is not surprising. Their effects have been studied both in laboratory culture experiments and in the field samples, using incubations and measurements of natural assemblages. A very interesting historical overview of research into the ecological and environmental factors that influence nitrification was presented by Kaplan (1983).

Temperature

The effect of temperature, while of potential importance in wastewater systems where nitrifying bacteria are cultured under artificial conditions, is not generally considered to be an important environmental variable for nitrification because bacterial populations are generally adapted to the temperature of their environments. Thus, at least in sediments, one can demonstrate a classical dependence of the rate of nitrification on temperature in any particular environment (e.g., Thamdrup and Fleischer 1998), but temperature is not generally the limiting factor. The maximal rate is usually attained at temperatures exceeding ambient temperature, but the observed rate of nitrification indicates that the in situ population is adapted to its ambient temperature. In other words, nitrifiers adapted to low temperature can nitrify at rates comparable to the rates attained by high-temperature-adapted nitrifiers at those higher temperatures. While temperature is an important master variable for biological processes, nitrification is if anything less sensitive to regulation by temperature than other processes and is usually regulated in the environment by some other variable.

Inhibitory Compounds

Nitrifying bacteria, both ammonia oxidizers and nitrite oxidizers, but especially the former, are susceptible to inhibition by a wide range of compounds, and several different modes of action have been suggested (Bedard and Knowles 1989). The two most common modes of action are as follows: (1) interference with the active site of the primary enzyme (i.e., ammonia monoxygenase in ammonia oxidizers) by compounds that share structural homology with ammonia (e.g., methane and a large number of larger organic compounds) and (2) metal binding compounds that interfere with the action or availability of copper in the ammonia-oxidizing enzymes. In both ammonia- and nitrite-oxidizers, the susceptibility of key enzymes in the nitrification pathways forms the basis of methods used to measure the rate of nitrification (see above).

In terrestrial systems, the presence of certain organic compounds (e.g., monoterpenes produced by plants) has been proposed to limit the rate of nitrification, and the inhibition of nitrification in acid soils has long been of concern. The potential of naturally occurring organic compounds to inhibit nitrification in seawater has not been rigorously investigated. Inhibition by organosulfur compounds has been demonstrated in cultured marine ammonia oxidizers (Ly et al. 1993) and the inhibitory effect of sulfide on nitrification is thought to limit nitrification and coupled nitrification/denitrification in marine sediments (see below; Joye and Hollibaugh 1995). While naturally occurring organic compounds have not been investigated as potential inhibitors in seawater, the product of their photodecomposition, carbon monoxide, has been implicated (Jones and Morita 1984). While CO, like methane, acts as a substrate analog or a suicide inhibitor for ammonia oxidizers, the direct inhibitory effect of light on nitrifiers (see below) is considered to outweigh the potential effect of CO inhibition in surface waters.

Light

The inhibitory effect of light was first reported by German researchers in the 1960s (Muller-Nugluck and Engel 1961; Schon and Engel 1962), verified in enrichment cultures in seawater by Horrigan et al. (1981), and later described in more detail (e.g., Vanzella et al. 1989; Guerrero and Jones 1996a, b). Horrigan et al. (1981) showed that even in enrichment cultures of nitrifiers derived from the sea surface film, nitrification was severely inhibited by light, such that periods of darkness lasting more than 12 hours were necessary to allow net nitrification to occur over a 24-hour period. Vanzella et al. (1989)

found evidence that nitrite oxidizers were more sensitive to sunlight than were ammonia oxidizers, based on single culture studies, but Guerrero and Jones (1996a) showed that species-specific responses may obscure any generalizations among major groups. Horrigan and Springer (1990) found that nitrifying bacterial isolates from estuarine habitats were less sensitive to light inhibition than were those isolated from seawater, and they suggested that the importance of light in regulating nitrification might thus differ in the two systems.

Olson (1981a, b) used his observations of ammonia and nitrite oxidation in the Southern California Bight, as well as historical reports of light inhibition, to formulate a hypothesis that light inhibition of nitrifiers is responsible for the position of the primary nitrite maximum in near surface seawater. Several studies of nitrification rates in surface seawaters from various geographical region show profiles that are consistent with light inhibition (Ward et al. 1984; Ward 1985, 1987b; Lipschultz et al. 1990; Ward and Kilpatrick 1991). Simulated in situ rate measurements (i.e., measurements done under simulated in situ light conditions) show a clear negative relationship with ambient light intensity (Figure 4). Nitrifiers may be somewhat protected from light inhibition in surface waters by the presence of absorbant organic compounds in seawater. However, the data from simulated in situ rate measurements are consistent with light inhibition in surface waters, and the well-verified sensitivity of nitrifiers in culture strongly suggests an influence of light in the environment. Dore and Karl (1996b) did not directly assess the effect of light on the nitrification rates they measured (all rates were measured in the dark). Rather, they concluded that phytoplankton are responsible for the primary nitrite maximum in surface waters of the central gyre. The relative contribution of phytoplankton versus nitrifying bacteria to the maintenance of this feature is apparently still unresolved.

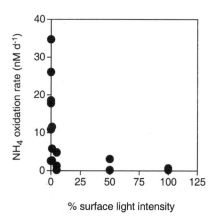

Figure 4. Ammonium assimilation rate (nM d^{-1}) versus surface light intensity for combined data from three stations in the Southern California Bight (approximately 33°20′N, 118°30′W). Rates measured using ^{15}N tracer techniques. (From Ward 1985.)

Nitrification in the Euphotic Zone

Independent of their potential involvement in the primary nitrite maximum, it is significant for other reasons that nitrification rates are often maximal in the vicinity of the bottom of the euphotic zone. The important distinction made in the new production paradigm (see above) between ammonium, which is the "regenerated" nutrient, and nitrate, the "new" nutrient, is the basis of ^{15}N tracer methods to assess new and regenerated primary production. However, if nitrification occurs in the same depth interval where nitrate assimilation occurs, the nitrate too is a "regenerated" nutrient. Then, total nitrate assimilation would depend on a combination of nitrate supplied by nitrification, plus that supplied by mixing from the deep reservoir. Several studies have addressed the question of whether significant nitrification occurs in the euphotic zone and have concluded that in situ nitrification could supply 100% or more of the phytoplankton nitrate demand (Ward et al. 1989b; Dore and Karl 1996b; Bianchi et al. 1997). Thus nitrification and nitrate assimilation by phytoplankton can be closely coupled, even though the two processes are favored by quite different environmental conditions.

Substrate Concentration

The influence of light may be compounded by the necessity for ammonia oxidizers to utilize ammonium, which is in short supply in the depth intervals at which light is most intense. When ammonium assimilation and ammonium oxidation are measured in the same incubation (e.g., Ward 1987b), it is seen that assimilation occurs in the upper portion of the euphotic zone and nitrification in the lower portion (Figure 3). This pattern suggests that in the well-lit upper waters, phytoplankton are able to assimilate ammonium, but nitrifiers either are unable to compete for ammonium in the presence of phytoplankton or are prevented by light inhibition from utilizing ammonium in that environment. In the lower portion of the euphotic zone, phytoplankton may be light-limited and unable to assimilate ammonium, whereas nitrifiers are released from light inhibition and able to utilize the ammonium being released by heterotrophic decomposition in that interval.

The influence of substrate concentration on nitrification rates is expected to be a first-order dependence, and indeed, this response is usually observed in culture. The instantaneous rate of ammonium oxidation or nitrite oxidation increases predictably with increasing ammonium or nitrite concentration in culture experiments. However, researchers have been unable to demonstrate consistent substrate dependence for ammonia oxidizers in natural assemblages. A general relationship between measured ammonium oxidation rate and ambient ammonium concentration (e.g. Ward 1985) supports the importance of substrate concentration (Figure 5). But lack of response to substrate perturbation (e.g., Olson 1981a; Ward and Kilpatrick 1990) has led to speculation that the affinity for ammonium of ammonia oxidizers in natural

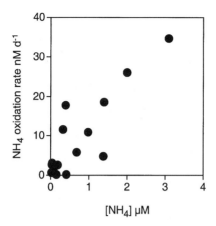

Figure 5. Ammonium assimilation rate (nM d^{-1}) versus ambient ammonium concentration (μM) for combined data from four stations in the northeastern Pacific Ocean off the coast of Washington State (approximately 47°07′N, and shoreward of 125°W). Rates measured using 15N tracer techniques. (From Ward 1985.)

communities may be so great (i.e., their half-saturation constants so low) that current experimental methods cannot detect a response to enhanced substrate concentrations. In contrast to ammonia oxidation, nitrite oxidation in natural samples usually shows a more conventional response to added substrate (e.g., Olson 1981a), but this second step in the nitrification process has received somewhat less attention.

Oxygen Concentration

Oxygen concentration is also a variable in determining nitrification rates; nitrifiers are reputed to be microaerophiles. While requiring molecular oxygen for reactions in the nitrogen oxidation pathways and for respiration, they are considered to thrive best under relatively low oxygen conditions. For example, Goreau et al. (1980) reported that an oxygen concentration of 1% in the headspace (vs. 20% in normal atmospheric air) supported the highest growth rate for *Nitrosomonas* sp. marine; this result has been extrapolated to nitrifiers in general.

Microaerophily may be important in interface environments such as the sediment–water interface and in the oxygen minimum zones of the ocean. In the sediment environment, for example, ammonium (derived from anaerobic decomposition in deeper layers) diffuses up toward the sediment–water interface and oxygen diffuses into the sediments from the overlying water. At the interface, nitrifiers find an optimal environment. In such an environment, nitrification can be directly linked to denitrification through their common intermediates (including nitrite, nitrate, and nitrous and nitric oxides). The nitrate produced during nitrification, along with the nitrate that diffuses into the sediments from the overlying water, helps support denitrification in subsurface sediments. In the sediment interface environment, the zone in which nitrification occurs may be less than a millimeter thick, or it may extend for several centimeters into the sediments, depending on the environment and the organic loading to the system.

The coupling between obligately aerobic nitrification and facultatively anaerobic denitrification is the classical way in which oceanographers view the linkage between the two processes in the environment. Vertical gradients of oxygen, nitrate, nitrite, and ammonium are consistent with the stratification of the processes imposed by the physiological constraints of the organisms in relation to the distribution of variables in the environment. The dependence of the depth distribution of nitrification and denitrification on oxygen distribution has been shown elegantly in freshwater sediments (e.g., Lorenzen et al. 1998). In this experiment, the oxygen penetration into microbial mat sediments varied on a diel light–dark cycle as photosynthesis proceeded at day and ceased at night. Oxygen and nitrate concentrations were measured by means of microelectrodes or biosensors, and nitrification and denitrification rates were modeled with a reaction–diffusion model. Nitrification occurred only in the daylight, when oxygen was available from photosynthesis. At night, photosynthesis ceased and oxygen penetration into the mat was insufficient to support nitrification. The counter-effect of light potentially inhibiting nitrification was apparently not a problem; light penetration in sediments even during the day was not sufficient to inhibit. The absence of nitrification in surface sediments at night was difficult to explain; it was attributed to inhibition of nitrifiers by unknown chemical components of the sediments and porewaters. Such direct demonstration of these interactions among photosynthesis and nitrification has not been possible in marine sediments to date. It seems likely that the same relationships pertain, however, except for the confounding effect of sulfide.

Nitrifying bacteria living in marine sediments that are periodically or persistently exposed to anoxic conditions would need to be able to survive periods of inactivity or even serious inhibition to be able to recover when conditions improved. The physiological basis of this survival or tolerance of anoxic conditions is unknown, and the degree to which nitrifiers can recover from serious anoxia is variable. Joye and Hollibaugh (1995) showed in microcosm studies that nitrification was almost completely inhibited by sulfide, the end product of bacterial sulfate reduction that occurs widely in anoxic marine sediments. A pulse of sulfide, which was detectable in the sediments for only a few hours, inhibited nitrification for at least 24 hours. Days to weeks were required for full recovery. Thus in marine sediments where sulfate reduction occurs, the ability of nitrifiers to respond to daily oxygen fluctuations may be impeded. Even when the oxygen–sulfide interface deepens during the day, nitrifiers may be unable to recover from the sulfide poisoning. In that case, both nitrification and denitrification (which is partially dependent on nitrate supply from nitrification) may occur at slower rates than would be predicted for similar environments in freshwater sediments (where sulfate, and therefore sulfide release from sulfate reduction, is much less prevalent).

This picture may be further complicated by the suggestion that nitrifiying bacteria are in fact not obligate aerobes. While net nitrification and growth at the expense of inorganic nitrogen occurs only under aerobic conditions, both ammonia- and nitrite-oxidizing nitrifiers are apparently capable of partial or even complete denitrification. Loss of fixed nitrogen has been observed in cultures of nitrifying bacteria growing on reduced oxygen tension, including the report by Goreau et al. (1980) cited above. Not only did the ammonia oxidizers grow best at 1% oxygen on a lithotrophic medium (no organic substrates), but they also produced the greatest amount of nitrous oxide relative to nitrite under those conditions. Production of nitrous oxide, nitric oxide, and dinitrogen was reported for *Nitrosomonas* growing in the presence of organic compounds in the absence of oxygen (Stuven et al. 1992). *Nitrosomonas* could also grow using hydrogen as a electron donor and nitrite as its electron acceptor (Bock et al. 1995). Nitrite oxidizers can grow via dissimilatory nitrate reduction in the presence of organic matter and the absence of oxygen (Freitag et al. 1987).

The potential ecological impact of this physiological versatility in nitrifying bacteria has not been widely investigated in natural systems. These microorganisms have received much more attention in connection with sewage and wastewater treatment, where there is economic incentive to enhance the conversion of ammonium to nitrogen gas under totally anaerobic or totally aerobic conditions (Muller et al. 1995; Strous et al. 1997). The conditions that are conducive to denitrification by nitrifying bacteria are the same ones that induce denitrification in classical denitrifying bacteria. It is conceivable that both metabolic types are involved in the process. Thus the net inorganic nitrogen distribution we observe might be a more complex function of multiple processes than is presently appreciated.

NITRIFICATION IN OXYGEN MINIMUM ZONES

The aspect of the nitrifiers' reductive metabolism that has received the most attention is the potential contribution of nitrifying bacteria to the production of trace gases such as nitrous and nitric oxides in the water column and sediments. These gases are intermediates in denitrification and could also result from nitrification under low or zero oxygen concentration. Both gases are involved in important atmospheric processes; they contribute to greenhouse warming and to catalytic destruction of stratospheric ozone. Thus, understanding which processes are responsible for their production could prove to be important for understanding or potentially regulating their fluxes.

In a few special places in the open ocean, oxygen concentration is depleted to a level low enough to allow denitrification to occur in the water column.

These regions, referred to as oxygen minimum zones, occur off the coast of Peru, in the Arabian Sea, and in the eastern tropical North Pacific Ocean (off the west coast of Mexico). The coupling between nitrification and denitrification has also been studied in these systems, which are essentially analogous to the sediment environments described earlier, except that the oxygen and nitrate gradients extend over tens to hundreds of meters. Suboxic and anoxic waters and sediments tend to have large fluxes, and sometimes large accumulations, of the gaseous intermediates of nitrification and denitrification. This is probably due to the sensitivity to oxygen concentration in the local environment of the microorganism of the various organisms and enzymes involved in the production and consumption of these intermediates.

In studies of nitrogen cycling in oxygen minimum zones, nitrification and denitrification appear to be linked, as might be expected from analogy with sediment systems. Figure 6 describes schematically the approximate locations

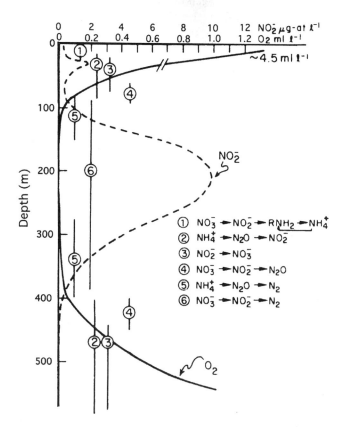

Figure 6. Schematic of expected chemical and activity distributions in the oxygen minimum zone off Peru (from Codispoti and Christensen 1985). Numbers in circles represent simplified pathways (shown in the figure) and the lines associated with each circled number denote the depth interval over which the activity is expected to occur.

of nitrifying and denitrifying activities relative to the chemical distributions typical of an oxygen minimum zone. Ammonium oxidation typically is maximal near the bottom of the euphotic zone close to the upper boundary of the oxygen minimum zone (Lipschultz et al. 1990), but nitrite oxidation is detected within the oxygen minimum zone itself. This finding is consistent with the ability of nitrite oxidizers to persist and metabolize at very low or zero oxygen conditions, but their metabolism under these conditions would be expected to be reductive rather than oxidative. The nitrous oxide that typically accumulates in the suboxic (low but not zero oxygen concentrations) regions of the oxygen minimum zones is thought to be due to nitrification, since nitrous oxide is depleted in the core of the oxygen minimum zone where denitrification rates are thought to be greatest. A direct tracer confirmation of this pathway in nature has not yet been accomplished. Stable isotope measurements of nitrous oxide from oxygen-depleted waters in the Arabian Sea imply that both nitrification and denitrification may contribute to the signal (Naqvi et al. 1998). Ward and Zafiriou (1988) attributed the nitric oxide production they observed in the oxygen minimum zone of the eastern tropical North Pacific to nitrification and found that it was equivalent to 18% of the total ammonia oxidation rate.

NITRIFICATION AND METHANE OXIDATION

As mentioned above in connection with the physiology and phylogeny of nitrifying bacteria, the ammonia oxidizers and the methanotrophs have important biochemical similarities. These similarities extend to the nature of the primary enzyme in the ammonia and methane oxidation pathways, the sensitivity of the enzymes to a wide range of metabolic inhibitors, the metabolic capabilities of the cell, and the ultrastructure of the cell (Bedard and Knowles 1989). Methanotrophs, like ammonia oxidizers, depend on two substrates with generally opposite sources. Methane, like ammonium, accumulates in anoxic habitats, where it is produced by strictly anaerobic methanogens. Oxygen diffuses into surface waters or surface sediments from the overlying oxygenated habitats. The classical environment for significant contributions by methanotrophs to system-wide carbon cycling is stratified lakes, in which a large fraction of the annual carbon fixation is cycled through methanogenesis and methane oxidation. The methanotrophic activity is highest at the interface where oxygen and methane coincide. As described above for sediments, such an interface would also be an interface for ammonium and oxygen and might be expected to harbor high nitrification activity as well.

Because of the presence of high sulfate concentrations in seawater, methanogenesis is not as important in marine sediments and seawater as it is in freshwater systems. Nevertheless, there are a few marine environments where methane is found. The situation in Scan Bay or Cape Lookout Bight sediments, or in the water column of the Black Sea or the Cariaco Basin, is largely analogous to that in stratified lakes (see above).

Biochemical and kinetic studies on pure cultures showed that methanotrophs are capable of oxidizing ammonia and that ammonia oxidizers are capable of oxidizing methane (Dalton 1977; Jones and Morita 1983; Ward 1987a; King and Schnell 1994). Both show a form of competitive inhibition that may be more realistically interpreted as suicide inhibition — occupation of the active site by the competitive molecule leads to inactivation at high concentrations of the competing substrate. These findings have led to uncertainty about which organisms are responsible for observed methane and ammonium fluxes in nature. Although both groups of microorganisms show similar regulation by environmental variables and similar sensitivities to a variety of inhibitors (Bedard and Knowles 1989), the possibility for differential regulation based on substrate affinity or competition suggested that "cross-oxidation" might have important implications for the rate of ammonium or methane oxidation in nature. Based on a combination of simulated in situ rate measurements, inhibitor studies and kinetic experiments with natural assemblages, it has been concluded largely that methanotrophs are mostly responsible for methane oxidation and nitrifiers for ammonium oxidation in both freshwater and marine environments (Bedard and Knowles 1997; Ward and Kilpatrick 1990). These conclusions do not rule out a role for cross-oxidation or participation by both groups in some environments, and whether the two processes can be separated entirely in the environment remains problematic.

CONCLUSIONS

The nitrogen cycle warrants study in the marine environment because of this element's role as the macronutrient most likely to be limiting for primary production. The diverse compounds and wide range of oxidation states in which nitrogen occurs play many different roles in the chemistry of the atmosphere and ocean, and in the biochemistry of organisms. Most of the transformations in the nitrogen cycle are solely the domain of microorganisms, and this is exemplified by classical autotrophic nitrifying bacteria. In recent years, our understanding of the rates and distribution of nitrification in the sea has converged on a model in which rates are greatest in the upper water column, within the euphotic zone, such that nitrification is linked to nitrogen assimilation by phytoplankton. The other major site for nitrification is the sediment–water interface and surficial sediments, where nitrification is tightly linked to denitrification, as it is in oxygen minimum zones of the ocean. The metabolic diversity of nitrifiers, while limited compared to many other microorganisms, is greater than usually admitted by our simple models of processes in relation to environmental variables such as oxygen. These aspects of environmental regulation of nitrification and the diversity of organisms involved in nitrification remain interesting areas of fruitful research.

SUMMARY

1. Nitrification is performed by chemolithoautotrophic bacteria that obtain metabolic energy for CO_2 fixation from the oxidation of inorganic nitrogen compounds, resulting in the net transformation of ammonia to nitrate.

2. Both nitrite- and ammonia-oxidizing nitrifiers are Proteobacteria. The ammonia oxidizers are closely related both phylogenetically and metabolically to methanotrophic bacteria.

3. Nitrification rates are highest near the base of the euphotic zone in the open ocean, and in sediments in the coastal region. High nitrate concentrations in deep water probably result from slow accumulation due to low nitrification rates coupled to the absence of a significant sink for nitrate in oxygenated deep water.

4. Environmental factors that mediate the rate of nitrification include light intensity, substrate concentration, and oxygen concentration. Nitrification is subject to inhibition by a number of different chemicals, which may influence its rate in the environment.

5. Nitrification has been implicated in the production of nitric and nitrous oxides in the ocean. Trace gas production is enhanced at low oxygen tensions. In oxygen minimum zones, nitrification is coupled to denitrification, leading to net loss of fixed nitrogen.

ACKNOWLEDGMENTS

The helpful reviews and suggestions for the improvement of this manuscript by Frederica Valois and Doug Capone are greatly appreciated.

REFERENCES

Barraclough, D., and Puri, G. (1995) The use of ^{15}N pool dilution and enrichment to separate the heterotrophic and autotrophic pathways of nitrification. *Soil Biol. Biochem.* 27:17–22.

Bedard, C., and Knowles, R. (1989) Physiology, biochemistry, and specific inhibitors of CH_4, NH_4^+ and CO oxidation by methanotrophs and nitrifiers. *Microbiol. Rev.* 53:68–84.

Bedard, C., and Knowles, R. (1997) Some properties of methane oxidation in a thermally stratified lake. *Can. J. Fish. Aquatic Sci.* 54:1639–1645.

Bianchi, M., Bonin, P., and Feliatraj, A. (1994a) Bacterial nitrification and denitrification rates in the Rhone River plume (northwestern Mediterranean Sea). *Mar. Ecol. Prog. Ser.* 103:197–202.

Bianchi, M., Morin., P., and Lecorre, P. (1994b) Nitrification rates, nitrite and nitrate distribution in the Almeria–Oran frontal systems (Eastern Alboran Sea). *J. Mar. Syst.* 5:327–342.

Bianchi, M., Feliatra, A., Treguer, P., Vincendeau, M. A., and Morvan, J. (1997) Nitrification rates, ammonium and nitrate distribution in upper layers of the water column and in sediments of the Indian sector of the Southern Ocean. *Deep-Sea Res.* 44:1017–1032.

Billen, G, (1976) Evaluation of nitrifying activity in sediments by dark ^{14}C-bicarbonate incorporation. *Water Res.* 10:51–57.

Blackburn, T. H., and Blackburn, N. D. (1993) Coupling of cycles and global significance of sediment diagenesis. *Mar Geol* 113:101–110.

Bock, E., Schmidt, I., Stuven, R., and Zart, D. (1995) Nitrogen loss caused by denitrifying *Nitrosomonas* cells using ammonium or hydrogen as electron donors and nitrite as electron acceptor. *Arch. Microbiol.* 163:16–20.

Castignetti, D. (1990) Bioenergetic examination of the heterotrophic nitrifier–denitrifier *Thiosphaera pantotropha*. *Antonie Van Leeuwenhoek Int. J. Gen. Mol. Microbiol.* 58:283–289.

Codispoti, L. A., and Christensen, J. P. (1985) Nitrification, denitrification and nitrous oxide cycling in the eastern tropical South Pacific Ocean. *Mar. Chem.* 16:277–300.

Dalton, L. (1977) Ammonia oxidation by the methane oxidising bacterium *Methylococcus capsulatus* strain bath. *Arch. Microbiol.* 114:273–279.

Dore, J. E., and Karl, D. M. (1996a) Nitrite distributions and dynamics at Station ALOHA. *Deep-Sea Res. II* 43:385–402.

Dore, J. E., and Karl, D. M. (1996b) Nitrification in the euphotic zone as a source for nitrite, nitrate, and nitrous oxide at Station ALOHA. *Limnol Oceanogr.* 41:1619–1628.

Dugdale, R. C., and Goering, J. J. (1967) Uptake of new and regenerated forms of nitrogen in marine production. *Limnol Oceanogr.* 12:196–206.

Eppley, R. W., and Peterson, B. J. (1979) Particulate organic matter flux and planktonic new production in the deep ocean. *Nature* 282:677–680.

Freitag, A., Rudert, M., and Bock, E. (1987) Growth of *Nitrobacter* by dissimilatoric nitrate reduction. *FEMS Microbiol Lett.* 48:105–109.

Goreau, T. J., Kaplan, W. A., Wofsy, S. C., McElroy, M. B., Valois F. W., and Watson, S. W. (1980) Production of NO_2^- and N_2O by nitrifying bacteria at reduced concentrations of oxygen. *Appl. Environ. Microbiol.* 40:526–532.

Grundmanis, V., and Murray, J. W. (1977) Nitrification and denitrification in marine sediments from Puget Sound. *Limnol Oceanogr.* 22:804–813.

Guerrero, M. A., and Jones, R. D. (1996a) Photoinhibition of marine nitrifying bacteria. 1. Wavelength-dependent response. *Mar. Ecol. Prog. Ser.* 141:183–192.

Guerrero, M. A., and Jones, R. D. (1996b) Photoinhibition of marine nitrifying bacteria. 2. Dark recovery after monochromatic or polychromatic irradiation. *Mar. Ecol. Prog. Ser.* 141:193–198.

Henriksen, K., Hansen, J. I., and Blackburn, T. H. (1981) Rates of nitrification, distribution of nitrifying bacteria, and nitrate fluxes in different types of sediment from Danish waters. *Mar. Biol.* 61:299–304.

Holmes, A. J., Costello, A., Lidstrom., M. E., and Murrell, J. C. (1995) Evidence that particulate methane monooxygenase and ammonia monooxygenase may be evolutionarily related. *FEMS Microbiol. Lett.* 132:203–208.

Horrigan S. G., and Springer, A. L. (1990) Oceanic and estuarine ammonium oxidation—Effects of light. *Limnol Oceanogr.* 35:479–482

Horrigan, S. G., Carlucci, A. F., and Williams, P. M. (1981) Light inhibition of nitrification in sea-surface films. *J. Mar. Res.* 39:557–565.

Jensen, K., Sloth., N. P., Risgaard-Petersen, N., Rysgaard, S., and Revsbech, N. P. (1994) Estimation of nitrification and denitrification from microprofiles of oxygen and nitrate in model sediment systems. *Appl. Environ. Microbiol.* 60:2094–2100.

Jones, R. D., and Morita, R. Y. (1983) Methane oxidation by *Nitrosococcus oceanus* and *Nitrosomonas europaea*. *Appl Environ. Microbiol.* 45:401–410

Jones, R. D., and Morita, R. Y. (1984) Effect of several nitrification inhibitors on carbon monoxide and methane oxidation by ammonium oxidizers. *Can. J. Microbiol.* 30:1276–1279.

Joye, S. B., and Hollibaugh, J. T. (1995) Influence of sulfide inhibition of nitrification on nitrogen regeneration in sediments. *Science* 270:623–625.

Kaplan, W. A. (1983) Nitrification. In E. G. Carpenter and D. G. Capone, eds. *Nitrogen in the Marine Environment*. Academic Press, New York, pp. 139–190.

Keifer, D. A., Olson, R. F., Holm-Hansen, O. (1976) Another look at the nitrite and chlorophyll maxima in the central North Pacific. *Deep-Sea Res.* 23:1199–1208.

King, G. M., and Schnell, S. (1994) Ammonium and nitrite inhibition of methane oxidation by *Methylobacter albus* BG8 and *Methylosinus trichosporium* OB3B at low methane concentrations. *Appl. Environ. Microbiol.* 60:3508–3513.

Lipschultz, F., Wofsy, S. C., Ward, B. B., Codispoti, L. A., Friederich, G., and Elkins, J. W. (1990) Bacterial transformations of inorganic nitrogen in the oxygen deficient waters of the eastern tropical south Pacific Ocean. *Deep-Sea Res.* 37:1513–1541.

Lorenzen, J., Larsen, L. H., Kjaer, T., and Revsbech, N. P. (1998) Biosensor determination of the microscale distribution of nitrate, nitrate assimilation, nitrification and denitrification in a diatom-inhabited freshwater sediment. *Appl. Environ. Microbiol.* 64:3264–3269.

Luther, G. W., Sundby, B., Lewis, G. L., Brendel, P. G., and Silverberg, N. (1997) Interactions of manganese with the nitrogen cycle: Alternative pathways to dinitrogen. *Geochim. Cosmochim. Acta.* 61:4043–4053.

Ly, J., Hyman, M. R., and Arp, D. J. (1993) Inhibition of ammonia oxidation in *Nitrosomonas europaea* by sulfur-compounds: Thioethers are oxidized to sulfoxides by ammonia monooxygenase. *Appl. Environ. Microbiol.* 59:3718–3727.

Miller, L. G., Coutlakis, M. D., Oremland, R. S., and Ward, B. B. (1993) Selective inhibition of nitrification (ammonium oxidation) by methylfluoride and dimethyl ether. *Appl. Environ. Microbiol.* 59:2457–2464.

Muller, E. B., Stouthamer, A. H., and van Verseveld, H. W. (1995) Simultaneous NH_3 oxidation and N_2 production at reduced O_2 tensions by sewage sludge subcultured with chemolithotrophic medium. *Biodegradation* 6:339–349.

Muller-Nugluck, M., and Engle, H. (1961) Photoinaktivierung von *Nitrobacter winogradskyi* Buch. *Arch. Mikrobiol.* 39:130–138.

Naqvi, S. W. A, Yoshinari, T., Jayakumar, D. A., Altabet, M. A., Narvekar, P. V., Devol, A. H., Brandes, J. A., and Codispoti, L. A. (1998) Budgetary and biogeochemical implications of N_2O isotope signatures in the Arabian Sea. *Nature* 391:462–464.

Olson, R. J. (1981a) ^{15}N tracer studies of the primary nitrite maximum. *J. Mar. Res.* 39:203–226.

Olson, R. J. (1981b) Differential photoinhibition of marine nitrifying bacteria: A possible mechanism for the formation of the primary nitrite maximum. *J. Mar. Res.* 39:227–238.

Pakulski, J. D., Benner, R., Amon, R., Eadie, B., and Whitledge, T. (1995) Community metabolism and nutrient cycling in the Mississippi River plume—Evidence for intense nitrification at intermediate salinities. *Mar. Ecol. Prog. Ser.* 117:207–218.

Rakestraw, N. W. (1936) The occurrence and significance of nitrite in the sea. *Biol. Bull.* 71:133–167.

Schon, G., and Engle, H. (1962) Den Einfluss des Lichtes auf *Nitrosomonas europaea* Win. *Arch. Mikrobiol.* 42:415–428.

Seitzinger, S. P., Gardner, W. S., and Spratt, A. K. (1991) The effect of salinity on ammonium sorption in aquatic sediments—Implication for benthic nutrient recycling. *Estuaries* 14:167–174.

Sloth, N. P., Nielsen, L. P., Blackburn, T. H. (1992) Nitrification in sediment cores measured with acetylene inhibition. *Limnol Oceanogr.* 37:1108–1112.

Somville, M. (1978) A method for the measurement of nitrification rates in water. *Water Res.* 12:843–838.

Strous, M., van Gerven, E., Kuenen, J. G., and Jetten, M. (1997) Effects of aerobic and microaerobic conditions on anaerobic ammonium-oxidizing (Anammox) sludge. *Appl. Environ. Microbiol.* 63:2446–2448.

Stuven, R., Vollmer, M., and Bock, E. (1992) The impact of organic matter on nitric oxide formation by *Nitrosomonas europaea*. *Arch. Microbiol.* 158:439–443.

Teske, A., Alm, E., Regan, J. M., Toze, S., Rittmann, B. E., and Stahl, D. A. (1994) Evolutionary relationships among ammonia- and nitrite-oxidizing bacteria. *J. Bacterial* 176:6623–6630.

Thamdrup, B., Fleischer. S. (1998) Temperature dependence of oxygen respiration, nitrogen mineralization, and nitrification in Arctic sediments. *Aquat Microb. Ecol.* 15:191–199.

Vaccaro, R. F. (1962) The oxidation of ammonia in sea water. *J. Cons. Perm. Int. Explor. Mer.* 27:3–14.

Vanzella, A., Guerrero, M. A., and Jones, R. D. (1989) Effect of CO and light on ammonium and nitrite oxidation by chemolithotrophic bacteria. *Mar. Ecol. Prog. Ser.* 57:69–76.

Voytek, M. A., Ward, B. B., and Priscu, J. C. (1998) The abundance of ammonia-oxidizing bacteria in Lake Bonney, Antarctica, determined by immunofluorescence, PCR and in situ hybridization. Antarctic Research Series, *The McMurdo Dry Valleys*, pp. 217–228.

Ward, B. B. (1985) Light and substrate concentration effects on marine ammonium assimilation and oxidation rates. *Mar. Chem.* 16:301–316.

Ward, B. B. (1987a) Kinetic studies on ammonia and methane oxidation by *Nitrosococcus oceanus*. *Arch. Microbiol.* 147:126–133.

Ward, B. B. (1987b) Nitrogen transformations in the Southern California Bight. *Deep-Sea Res.* 34:785–805.

Ward, B. B., and Carlucci, A. F. (1985) Marine ammonium- and nitrite-oxidizing bacteria: Serological diversity determined by immunofluorescence in culture and in the environment. *Appl. Environ. Microbiol.* 50:194–201.

Ward, B. B., and Kilpatrick, K. A. (1990) Relationship between substrate concentration and oxidation of ammonium and methane in a stratified water column. *Cont. Shelf Res.* 10:1193–1208.

Ward, B. B., and Kilpatrick, K. A. (1991) Nitrogen transformations in the oxic layer of permanent anoxic basins: The Black Sea and the Cariaco Trench. In E. Izdar and J. W. Murray, eds., *Black Sea Oceanography.* Kluwer, Dordrecht, pp. 111–124.

Ward, B. B., and Zafiriou, O. C. (1988) Nitrification and nitric oxide in the oxygen minimum of the eastern tropical North Pacific. *Deep-Sea Res.* 35:1127–1142.

Ward, B. B., Talbot, M. C., and Perry, M. J. (1984) Contributions of phytoplankton and nitrifying bacteria to ammonium and nitrite dynamics in coastal water. *Cont. Shelf Res.* 3:383–398.

Ward, B. B., Glover, H. E., and Lipschultz, F. (1989a) Chemoautotrophic activity and nitrification in the oxygen minimum zone off Peru. *Deep-Sea Res.* 36:1031–1051.

Ward, B. B., Kilpatrick, K. A., Renger, E., and Eppley, R. W. (1989b) Biological nitrogen cycling in the nitracline. *Limnol Oceanogr.* 34:493–513.

Watson, S. W. (1965) Characteristics of a marine nitrifying bacterium, *Nitrosocystis oceanus* sp. n. *Limnol Oceanogr.* 10 (suppl):R274–R289.

Watson, S. W., and Waterbury, J. B. (1971) Characteristics of two marine nitrite oxidizing bacteria, *Nitrospina gracilis* nov. gen. nov. sp. and *Nitrococcus mobilis* nov. gen. nov. sp. *Arch. Microbiol.* 77:203–230.

Watson, S. W., Bock, E., Valois, F. W., Waterbury, J. B., and Schlosser, U. (1986) *Nitrospira marina* gen. nov. sp. nov.: A chemolithotrophic nitrite-oxidizing bacterium. *Arch. Microbiol.* 144:1–7.

Wehrfritz, J. M., Reilly, A., Spiro, S., and Richardson, D. J. (1993) Purification of hydroxylamine oxidase from *Thiosphaera pantotropha:* Identification of electron acceptors that couple heterotrophic nitrification to aerobic denitrification. *FEBS Lett.* 335:246–250.

Winogradsky, S. (1890) Sur les organismes de la nitrofication. *Compt. Rend.* 110:1013–1016.

15

THE MARINE MICROBIAL NITROGEN CYCLE

Douglas G. Capone

Department of Biological Sciences and
Wrigley Institute for Environmental Studies,
University of Southern California,
Los Angeles, California

INTRODUCTION

Nitrogen (N) is a key constituent of life on Earth. It occurs in a complex array of different chemical pools and states in the biosphere. All organisms require it in stoichiometric proportions to carbon and other essential elements for balanced growth. For centuries, agriculturists and nutritionists have recognized the importance of N sources in food production and human health, respectively. Starting in the last century, chemists (e.g., Liebig) and microbiologists (e.g., Winogradsky, Beijerinck) began to unveil the primary features and complexities of the natural cycle of N, first in terrestrial environments and, not long thereafter, in marine ecosystems (e.g., Brandt, in Mills 1989, Waksman et al. 1933).

The pace of research on the marine N cycle has greatly accelerated in the last several decades, initially instigated by efforts to gain an understanding of the factors controlling the natural productivity of the sea (e.g., Thomas 1966; Ryther and Dunstan 1971). Indeed, N is thought to limit primary production through much of the world's oceans (McCarthy and Carpenter 1983; Howarth

Microbial Ecology of the Oceans, Edited by David L. Kirchman.
ISBN 0-471-29993-6   Copyright © 2000 by Wiley-Liss, Inc.

1988). A more recent impetus to better understand the relationship between N cycling and marine productivity has arisen from the observation of anthropogenic perturbations at regional and, possibly, global scales in the marine N cycle, a result of massive increases in fertilizer production.

N is a relevant factor in greenhouse warming and climate change (Vitousek et al. 1997). Nitrous oxide (N_2O), which can react with and destroy stratospheric ozone and is also a potent greenhouse gas, has been increasing in the atmosphere (Kahlil and Rasmussen 1992). More importantly, the availability of N in the oceans may control its capacity to fix and sequester atmospheric CO_2. The geological record indicates that key processes of the N cycle covary with marine primary production and with climate change over long time scales and may provide important feedbacks in the overall dynamics of global climate change.

FLUXES, POOLS, PROCESSES, AND PLAYERS

Relative to many of the biologically required elements, N has a complex cycle. The oceans, and most of their subsystems are open, thereby allowing free exchange of materials with adjacent systems. Nitrogen can enter the sea by several pathways. Once in the sea, N occurs in a variety of molecular forms, redox states, and phases. Specific pools of N may show various and distinct spatial and temporal patterns of abundance, thereby providing important clues to the underlying biogeochemical dynamics crucial to our understanding of this cycle. Physical, chemical, and biological processes within a system can affect the distribution of N among the various pools. Processes may also be spatially or temporally segregated from one another. Of particular interest to microbial ecology is the identity of the organisms carrying out the biologically catalyzed interconversions among N species, and the physiological and biogeochemical purpose these reactions serve.

Inputs and Outputs

Riverine delivery of inorganic and organic N, including some N in particulate form, is the most obvious source of combined nitrogen to the coastal seas (Meybeck 1993; Caraco and Cole 1999). Nitrogen also arrives into the sea by wet and dry atmospheric deposition of inorganic and organic N (Paerl 1993; Michaels et al. 1993; Cornell et al. 1995). Some gaseous forms of combined N (e.g., NH_3 and NO_x), may also represent an input in some areas of the coastal oceans (Paerl, personal communication). Soluble forms of nitrogen can also arrive through submarine groundwater discharge in coastal areas (Johannes 1980; Capone and Bautista 1985).

The relative importance of pathways of N delivery will vary with respect to location and climate: for example, the influence of the Amazon in the tropical Atlantic is profound (DeMaster and Pope 1996), while riverine delivery may represent only minor inputs in other locations (e.g., southwestern Pacific Ocean). Atmospheric inputs may be greater near landmasses, and the form of atmospheric input will vary as a function of land use and population density. The input into the ocean of elements other than N can have additional bearing on N cycling in the oceans (see below).

Nitrogen is removed from the sea by denitrification and gaseous evasion (Law and Owens 1990), sedimentation to the seafloor (Jahnke 1996), and biomass harvest.

Marine N Pools

Nitrogen occurs in a variety of inorganic and organic forms (Table 1). The largest reservoir of N in the sea is dissolved dinitrogen gas (N_2), which occurs in concentrations of about 1 mM and accounts for 23×10^6 Tg ($1 Tg = 10^{12}$ g). Concentrations are relatively uniform, and they vary largely as a function of temperature- and salinity-dependent solubility (Weiss 1970; Scranton 1983). On a localized basis, excesses of N_2 (saturation anomalies) may arise through microbial denitrification (Cline and Ben-Yaakov 1973) or through air injection, pressure, or temperature anomalies (Scranton 1983) — for instance, after upwelling and surface heating.

Two gases that occur as trace constituents in seawater are nitrous oxide (N_2O) and nitric oxide (NO). N_2O, which occurs at a concentration of about 310 ppb in the atmosphere and typically from 10 to 50 nM in seawater (Capone 1996a), arises from several microbial pathways (see below). In the sea, one may observe relatively high transients in N_2O concentrations in redox boundaries (Hahn 1981). NO, a highly reactive species, also occurs at trace concentrations in the atmosphere and in some oceanic environments (Chapter 14). Far less is known about the importance and cycling of NO in the sea.

The primary nongaseous inorganic combined forms of dissolved N are nitrate (NO_3^-), nitrite (NO_2^-) and ammonium (NH_4^+). Highly detailed work has been undertaken since the turn of the century to describe the distributions of these key plant nutrients. Much of what we know of the N transformations in the sea was first inferred from analysis of the spatial and temporal patterns of inorganic N distributions.

Nitrate is a relatively mobile species and can be the dominant form of N in runoff, riverine input, groundwater discharge, and atmospheric deposition to the ocean. In the sea, concentrations of N can vary widely in space and time. Relatively high concentrations (tens to hundreds micromolar) are characteristic of eutrophic coastal and upwelling environments, while lower concentrations occur in the surface waters of the tropical gyres (Sharp 1983) and in suboxic and anoxic zones (Table 1). Large reservoirs of nitrate exist below the permanent thermocline and account for about 677 Tg N (Capone 1991).

Nitrite generally occurs in the sea in much lower concentrations than nitrate (Table 1), although higher concentrations are associated with redox interfaces in both the water column and sediments. The primary nitrite maximum of the

Table 1. Primary pools and concentrations of nitrogen in the ocean

Formula	Form	State	Typical concentrations (μM)				Global Marine Pools (Pg)[b]	
			Oceanic	Ref.[a]	Coastal	Ref.[a]	Pools (Pg)[b]	Ref.[a]
GASEOUS								
N_2	Dinitrogen	Gas, dissolved	900–1100	1	900–1100	1	$22–23 \times 10^3$	2,3
N_2O	Nitrous oxide	Gas, dissolved	0.006–0.07	2	0–0.25	2	0.2–0.8	2,3
NO	Nitric oxide	Gas, dissolved	?		?		?	2,3
INORGANIC								
NO_3^-	Nitrate	Dissolved	<0.03–>40	1	<0.1–200	1	570–677	2,3
NO_2^-	Nitrite	Dissolved	<0.03–0.1	1	<0.03–10	1	?	3
NH_4^+	Ammonium	Dissolved, sorbed	<0.03–1	1	<0.03–>100	1	7–8	3,4
$(NH_2)_2CO$	Urea	Dissolved	<0.1–0.5	5	0–>2	5	0.17	3
ORGANIC								
DON	Dissolved organic	Dissolved	3–7	1	3–20	1	63–530	2,3
PON	Particulate organic	Particulate	0.07–0.5	1	0.1–30	1	3–24	3
Phyto biomass	Particulate organic	Particulate					0.15–0.3	3,4
Animal biomass	Particulate organic	Particulate					0.17	3
NH_2 | $R—CH—COOH$	Amino acids	Dissolved, free	0.05–0.5	1,5	0.04–2.2	1,5		
	Amino acids (Total)	Dissolved, combined	0.15–1.5	1	0.12–6.6	1	3–5	6
	Amino acids (Non-amino-acid amide)	Dissolved	0.04–0.12	5				
	Amino sugars	Dissolved	0.29–1.22	5				
Nonamide	Indole and pyrrole N: (e.g., purines, pyrimidines, porphyrins)	Dissolved	0.09–0.19	5				
Leu-Ala-Gly-(etc.)	Proteins	Dissolved, particulate	0.1–0.3	1				
ATCCTAGGG-(etc.)	DNA, RNA	Dissolved, particulate						
	Peptidoglycan	Dissolved, particulate						

[a] 1, Sharp (1983); 2, Capone (1991); 3, Sodulund and Svensson (1976); 4, Falkowski et al. (1998); 5, M. McCarthy et al. (1997, 1998); 6, Bada (1998).
[b] $1 \ Pg = 10^{15} \ g$.

open ocean occurs near the base of the euphotic zone and is generally associated with assimilatory NO_3^- reduction (and release of the NO_2^- intermediate), while a deeper, secondary NO_2^- maximum is associated with the differential inhibition of nitrification (Olson 1981; Chapter 14).

Ammonium concentrations also vary widely among environments. In oligotrophic regions, NH_4^+ concentrations are usually near the limit of detection by conventional colorimetric methods ($< 0.03\ \mu M$). In most open-ocean environments, there are no strong spatial trends in NH_4^+ concentration with depth. However, in suboxic and anoxic environments NH_4^+ concentrations increase. In highly reducing nearshore sediments, millimolar levels of NH_4^+ can be observed. High concentrations of NH_4^+ are also associated with nearshore waters subject to sewage and waste discharges (Ryther and Dunstan 1971; Garside et al. 1976).

Dissolved organic N (DON) can also represent a major N pool in marine systems (Table 1) with patterns of distribution emerging from recent research (Bronk et al. 1994). Analyses of organic N have focused on both specific compounds such as individual amino acids and nucleic acids (Sharp 1983), as well as on polymeric compounds (protein, DNA) and specific classes of dissolved organic N (DON < 0.3–$0.7\ \mu m$ fraction) (M. McCarthy et al. 1997) and particulate organic N (PON)(> 0.2–$0.7\ \mu m$) (Table 1). Amino and nucleic acids, while typically low in concentration, are highly dynamic pools in many marine ecosystems (Fuhrman and Ferguson 1986). The bulk of the organic N pool remains to be characterized. Recent evidence indicates that much of the uncharacterized fraction may consist largely of refractory biopolymers, possibly peptidoglycan remnants of bacterial cell walls (M. McCarthy et al. 1998).

Processes

Conversions among inorganic and organic species are generally biologically mediated. The principal transformations of the marine N cycle include the uptake and incorporation of inorganic forms of N into organic N; the regeneration and release of inorganic N, primarily as NH_4^+, from organic forms of N; the oxidation of ammonium and nitrite in nitrification; the reduction of nitrate or nitrite to the gaseous end products, N_2 and N_2O, in denitrification; and the reduction ("fixation") of N_2 to NH_4^+ (Figure 1). Bacteria are dominant in most of these transformations.

Inorganic N, as nitrate, nitrite, ammonium, and such organic forms of N as urea and other small molecular weight organic molecules (e.g., amino acids), often termed utilizable N, can be taken up and incorporated into organic matter by a variety of organisms. Very high rates of N uptake and assimilation into organic matter occur in photic environments in conjunction with photosynthetic C assimilation. These reactions are largely catalyzed by photoautotrophs including cyanobacteria, eukaryotic micro- and macroalgae, and sea grasses. Heterotrophic bacteria and archaea (as well as some fungi) can also assimilate inorganic N and may compete with photoautotrophs for these resources (Chapter 9).

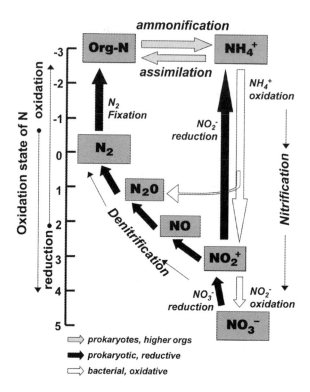

Figure 1. Marine nitrogen cycle emphasizing oxidation–reduction reactions and biological transformations.

The capacity for N_2 fixation, the reduction of N_2 gas to NH_4^+ for biosynthesis, occurs in a disparate array of prokaryotes, including representatives of both eubacteria and archaea. Organisms that can fix N_2 are at an advantage in environments with low levels of combined N. Through this process, N_2 fixers provide a source of utilizable N to the biosphere from the large pool of N_2 (Table 1) and balance losses of NO_3^- by denitrification. Marine N_2 fixers are found in greatest abundance in the tropical open ocean and in shallow tropical environments (Chapter 13).

Oxidized forms of N, such as NO_3^- and NO_2^-, that are taken up by cells must first be reduced to NH_4^+, at the -3 oxidation level, by assimilatory NO_3^- and NO_2^- reductase before they can be assimilated into cell biomass (Falkowski 1983) (Figures 1 and 2). In most marine organisms utilizing inorganic N, including N_2 fixers, the assimilation of NH_4^+ is catalyzed by the enzymes glutamine synthetase and glutamate synthase (GS/GOGAT pathway), with glutamate as the major end product (Figure 2). Glutamate is then utilized in the cell to synthesize other N-containing compounds. Alternative NH_4^+ pathways exist (e.g., the glutamate dehydrogenase pathway) but are less common in marine plankton growing in oligotrophic conditions (Falkowski 1983).

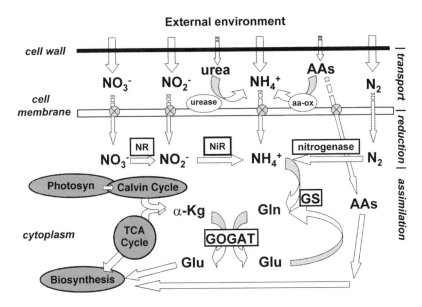

Figure 2. Pathways of N uptake and incorporation: NR, nitrate reductase; NiR, nitrite reductase; aa, amino acids; aa-ox, amino acid oxidase; GS, glutamine synthetase; GOGAT, glutamate synthase; αKg, α-ketoglutarate; Gln, glutamine; Glu, glutamate.

N is a constituent of a wide variety of simple and complex organic molecules, and the pathways of organic N breakdown largely remain to be detailed. The release of inorganic N from organic forms can occur through the metabolism of higher organisms by the urea cycle with release of various urea cycle end products (e.g., NH_4^+, urea) (Antia et al. 1991). The degradative catabolism of organic N by heterotrophic microorganisms including many bacteria and fungi, often termed ammonification, also results in the release of NH_4^+ (Fenchel et al. 1998). Labile DON that is released by excretion or cell lysis can also be further catabolized by heterotrophs or assimilated by heterotrophs or autotrophs (Paul 1983; Antia et al. 1991). A variety of algal taxa and bacteria possess cell surface oxidases (Palenik and Morel 1990). These may be important in liberating NH_4^+ from organic compounds, and the NH_4^+ released is subsequently available for uptake.

Nitrification is a two-step process involving first the conversion of NH_4^+ to NO_2^- followed by the oxidation of NO_2^- to NO_3^- (see Chapter 14). These reactions are catalyzed by two specialized groups of aerobic, autotrophic bacteria: the ammonium oxidizers and the nitrite oxidizers (Chapter 14). Pathways of heterotrophic nitrification have been described for soils but have not been well characterized in aquatic environments. Autotrophic nitrifiers have been implicated in the production of N_2O in many marine environments (Yoshida 1988; Capone 1991). The large reserves of NO_3^- maintained in the deep sea attest to the major role of nitrifiers in the oceans.

Biological denitrification is the conversion of oxidized forms of N to gaseous end products (Knowles 1982). Specifically, the denitrification pathway (Figure 1) is as follows:

$$NO_3^- \rightarrow NO_2^- \rightarrow NO \rightarrow N_2O \rightarrow N_2$$

Reductase	nitrate	nitrite	nitric oxide	nitrous oxide
Gene	*nar*	*nir*	*nor*	*nos*

In each reaction, the oxidized N substrates serve as terminal electron acceptors in respiration, generally coupled to the oxidation of organic compounds. The first step of the process, the reduction of NO_3^- to NO_2^-, is termed dissimilatory NO_3^- reduction (to distinguish it from the assimilatory pathway—see above) and occurs in a relatively wide variety of bacteria. The capacity to reduce NO_2^- to N_2O or N_2 occurs in a smaller subset of these bacteria (Knowles 1982). These reactions generally occur where O_2 is low. In some bacteria, NO_3^- reduction is followed by dissimilatory reduction of NO_2^- to NH_4^+, and the NH_4^+ released into the environment (Hattori 1983).

Pathways leading to N_2O or N_2 production represent a loss of utilizable (combined) N for marine ecosystems. When the first two reduction steps in either pathway are temporally or spatially uncoupled, NO_2^- may accumulate. Thus, NO_2^- is a key branch point in the N cycle as an intermediate in nitrification, and assimilatory and and dissimilatory NO_3^- reduction.

Denitrification occurs at substantial rates in a variety of benthic and pelagic marine environments, including midwaters areas of the Indian and Pacific Oceans (Hattori 1983), as well as estuarine, coastal, and shelf sediments (Christensen et al. 1987; Seitzinger 1988; Devol 1991).

HOW WE STUDY THE N CYCLE

The approaches to the study of the marine N cycle are many and varied, and new methods are constantly evolving. As in all analytical fields, we are ever seeking to improve the sensitivity for analytes and for specific transformations, to develop methods to detect new compounds and pathways, to minimize our perturbations to systems studied, to reduce the introduction of experimental artifacts, and to increase the efficiency, frequency, and coverage of our sampling (Christian and Capone 1996).

It is beyond the scope of this chapter to detail and critique the various analytical methods involved in assessing N cycling. Parsons et al. (1994a) serves as a starting point for methods involved in routine determination of major inorganic and organic pools. D'Elia (1983) has reviewed methods for N compounds.

Methodology for the various microbiological and biogeochemical assays involved in N cycling may be found in several recent compendia. See chapters in Kemp et al. (1993), including Capone (1993) and Seitzinger (1993); in Hurst et al. (1997) including Capone (1996b); and in Knowles and Blackburn (1993) including Glibert and Capone (1993).

At the broadest level, our approaches to studying the marine N cycle can be regarded as either observational or experimental. The earliest insights into the marine N cycle came from observations of the spatial and temporal distributions of key nutrients such as NO_3^- in marine waters and sediments. The application of the natural abundance of stable isotopes of biologically relevant elements has greatly enhanced the utility of observational approaches (Peterson and Howarth 1987). By applying a knowledge of source materials and the extent of fractionation by particular biological and chemical transformations, to the small differences in the ratio of the stable isotopes of nitrogen, ^{14}N and ^{15}N, one can glean substantially more information than just from the distributions of particular pools of the N cycle.

In parallel with observational studies, experimental studies using natural and cultured populations have expanded our understanding of microbially mediated N transformations. The isolation and characterization of nitrifying (Kaplan 1983), denitrifying (Hattori 1983), and N_2 fixing (Capone 1988) marine bacteria and archaea have allowed us to identify and characterize the microbiological basis of the major N transformations. Experimental approaches examining N transformations have allowed direct confirmation of inferences and hypotheses developed using results from observational and classical microbiological methods. Direct tracer methods are preferred to directly quantify specific pathways of biological N transformation (Harrison 1983). While N does have a radioisotope, ^{13}N, its very brief half-life and the sophisticated facilities necessary to generate it preclude its usefulness in field applications (Capone 1996b). In contrast, the stable isotope of N, ^{15}N, is available in a variety of forms useful for determining rates of specific pathways.

The enzymes mediating various N transformations (e.g., assimilatory NO_3^- reductase, glutamine synthetase) have been characterized and assays developed to estimate the activity in natural populations of marine microbes (Falkowski 1983). Immunological and molecular methods provide sensitive means of detecting proteins and genes involved in N cycle transformations (Ward 1995; Zehr and Capone 1995).

Providing mathematical formalisms that describe the cycling of N in the sea and test our conceptualizations of processes is an activity dating back to Redfield et al. (1963) (see below), Riley (1967), and Dugdale (1967). "Modeling" has become an indispensable tool with which to integrate our knowledge of the marine N cycle and to place this knowledge in a broader framework.

INTEGRATING CONCEPTS

The multiplicity of chemical species of N, their distribution among different phases, and the various biotic transformations among pools make the biogeochemistry of N relatively complex. In comparing the differences among systems, several key concepts regarding N distributions and their relationships to plant and animal production have emerged over the last several decades. These concepts are dynamically evolving and being refined as we learn more about marine systems.

Deep Nitrate, Redfield Ratios, and the Conveyor Belt

While surface waters are often depleted in inorganic nutrients such as NO_3^- and PO_4^{3-} throughout much of the open oceans, relatively high nutrient concentrations are found in deep waters beneath the thermocline (Figure 3). Assimilation of nutrients by phytoplankton accounts for the depletion of nutrients in the upper water column, where light is available for photosynthesis (i.e., the euphotic zone). The concentration of NO_3^- below the thermocline is greater in the Pacific Ocean than in the Atlantic Ocean (Figure 3). Observations made by geochemists and physical oceanographers have helped to identify the broad-scale circulation and mixing processes that result in these depth and basin-scale distributions (Broecker and Peng 1982).

Building on earlier observations of Harvey (1927), Redfield and colleagues (Redfield et al. 1963; Redfield 1934, 1958) pioneered and formalized a geochemical model relating the increase of nutrients at depth in the sea to the aerobic bacterial regeneration, coupled with nitrification, of planktonic organic matter sedimenting through the subeuphotic zone. They developed stoichiometric relationships, termed the Redfield equation, between the composition of generic plankton organic matter and the macronutrients released during aerobic decomposition:

$$(CH_2O)_{106}(NH_3)_{16}(H_3PO_4) + 138\ O_2 \rightleftarrows 106\ CO_2$$
$$+ 16\ HNO_3 + H_3PO_4 + 122\ H_2O \tag{1}$$

Modeling based on Redfield ratios provided important initial insights into key microbial processes operative in the deep sea, namely, ammonification, nitrification, and denitrification. The Redfield ratio (106:16:1 for the mole ratio of $CO_2:NO_3:PO_4$) can be used to predict the inorganic nutrient concentrations resulting from regeneration of plant material based on the relative degree of O_2 utilization in deep water. Briefly, the longer a parcel of water is at depth (i.e., removed from the euphotic zone), the greater the decomposition of organic matter, the O_2 consumed, and the concentration of inorganic nutrients in that parcel of water.

The relationships between deep-ocean circulation and global marine nutrient distributions were brought together by Broecker and Peng (1982) in the conveyor belt model of deep-ocean circulation (Figure 4). As the oceans are currently configured, deep waters largely form in the North Atlantic Ocean and flow south through to the South Atlantic. New deep water is entrained in the circumantarctic region. A portion of this water feeds the Indian Ocean deep-water flow before finally arriving in the Pacific Ocean. Upwelling occurs along the way, but waters entrained at depth the longest eventually upwell in

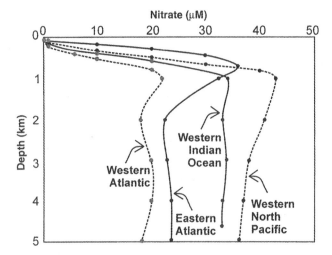

Figure 3. Representative depth profiles of NO_3^- from different ocean basins (Adapted from Wada and Hattori 1991.)

the Pacific. The oldest waters, in equilibrium with atmospheric O_2 when they first sink, can take upward of 2000 years to make this transit. Along the way, organic C sedimenting from the surface fuels O_2 consumption in these waters. Consequently, the highest concentrations of NO_3^- are found in midwaters of the Pacific Ocean, which have remained at depth the longest.

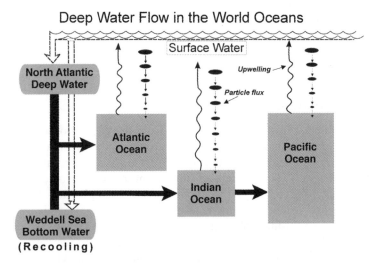

Figure 4. The conveyor belt model of deep ocean circulation after Broecker and Peng (1982). (Adapted from a representation from Wada and Hattori 1991.)

While Redfield's model adequately describes nutrient distributions through much of the deep sea, there are often significant departures from predicted nutrient distributions. Distinct provinces have been identified based on these anomalies and within which particular N cycle processes dominate (Fanning 1992; Longhurst 1998).

Waters found at middepths in the Indian Ocean and in the tropical Pacific Ocean have very low O_2 levels and exhibit substantial departures from Redfield model predictions (Hattori 1983). Lower than predicted concentrations of NO_3^- in these low O_2 waters are due to initiation of bacterial nitrate reduction and denitrification processes that consume NO_3^-. Large expanses of anoxic midwaters in the Indian Ocean, and in the eastern tropical North and South Pacific, are major global sites of denitrification (Hattori 1983).

Sambrotto et al. (1993) have noted dissolved inorganic C drawdown (i.e., net organic C production) substantially in excess of that predicted by NO_3^- uptake relative to Redfield stoichiometry. Michaels et al. (1994) have noted similar anomalies at the Bermuda Atlantic Time Series (BATS) site.

A new parameter, termed N^*, which considers departures from Redfield stoichiometry occurring during the decomposition of organic matter and regeneration of nutrients, has been derived (Michaels et al. 1996; Gruber and Sarmiento 1997). It is defined as follows:

$$N^* = (NO_3^- + 16PO_4^+ + 2.90) \times 0.87 \qquad (2)$$

The value of 2.90 adjusts N^* to 0 when one is considering global pools of N and P. The N^* parameter has been used to identify both negative and positive deviations from canonical Redfield values. For example, large areas of the tropical North Atlantic have been found to exhibit an excess of inorganic N relative to P at middepths (Michaels 1996, Gruber and Sarmiento 1997). It has been suggested that this excess of regenerated N, relative to P at these depths, results from the sedimentation of material enriched in N by N_2 fixation in near-surface waters.

N as a Limiting Nutrient

Much of the interest in the N cycle derives from the general observation that N is often the nutrient factor limiting plant growth and/or biomass accumulation in the ocean (Ryther and Dunstan 1971; J. J. McCarthy and Carpenter 1983; Howarth 1988). The concept of limiting nutrients, still actively debated today, originally derived from the application of Liebig's "law of the minimum." Liebig first articulated this tenet that held that the chemical factor in shortest supply would limit plant growth in agricultural systems. Oceanographers, observing the reciprocal relationship between the increase in plant biomass in the spring in the temperate North Atlantic Ocean with the decrease in NO_3^- (and PO_4^{3-}) concentrations, inferred N as the limiting nutrient factor resulting in the termination of the spring bloom (Parsons et al. 1984b). Biotic factors, such as grazing, also contribute to bloom termination, and this circumstance has promoted considerable discussion about the relative extent of "top-down" (i.e., herbivore) versus "bottom-up" (i.e., nutrient) control of plant biomass and production (e.g., Schindler et al. 1997). The observation that coastal eutrophication frequently results from N enrichment of systems (Ryther and Dunstan 1971; Nixon and Pilson 1983; see below) provides additional evidence for a controlling role of N in these systems.

When considering nutrient limitation, a number of factors need be considered. Various approaches have been used to assess nutrient limitation and can give widely divergent results, depending on the ecosystem component or process considered (e.g., biomass limitation vs net plant production vs net ecosystem production), the time scale of the observation, or the particular parameter assessed as an index of limitation and cellular capacity for growth (Smith 1984; Howarth 1988). Nutrient bioassays in which nutrients (N or P) are added to incubations of natural water are commonly used to assess nutrient limitation. Short-term (e.g., hours) nutrient bioassays often focus on physiological responses — for instance, in gross photosynthetic capacity as determined by $[^{14}C]O_2$ uptake — while assays over the somewhat longer term (days) are needed to monitor biomass (often as chlorophyll a) accumulation, which approximates net production. Biologists generally focus on nutrient limitation of growth on time scales of hours to days, while geochemists are typically more interested in net ecosystem production at larger spatial dimensions (e.g., basin scale) and often on much longer time scales (Smith 1984).

The simplistic view that N is the key limiting nutrient throughout most of the ocean is being rapidly replaced with a much more dynamic view of seasonal variation in limiting nutrients in estuaries (D'Elia et al. 1986; Fisher et al. 1992) and coastal waters (Dortch and Whitledge 1992). The oceanic regions with high nitrate but low chlorophyll (HNLC) concentration are primarily Fe-limited (Martin 1992). Phosphorus may be a key limiting nutrient in shallow tropical ecosystems (Smith 1984; Short et al. 1990; LaPointe et al. 1992). Moreover, we have also come to recognize that nutrient limitation may differ among contemporaneous species within a system. Examples include Si limitation of diatoms (Dortch and Whitledge 1992; Dugdale and Wilkerson 1998) and Fe (Rueter et al. 1992; Paerl et al. 1994) or P (Karl et al. 1992) limitation of N_2-fixing cyanobacteria in otherwise N-limited ecosystems.

New N and Oceanic Productivity

In 1967 Dugdale and Goering (1967) presented a conceptual model that has since provided an important framework for placing oceanic N cycle studies in the broader context of carbon productivity. They recognized two types of phytoplankton production:

1. "Recycled" production supported by N (termed "recycled N") regenerated from organic matter within the euphotic zone, and typified by NH_4^+.

2. "New production" that is supported by N imported from outside the euphotic zone and exemplified by NO_3^- diffusing or advecting up from deep pools, N_2 fixation, and atmospheric N deposition, as well as N derived from rivers and runoff in nearshore areas. Each represents a source of "new N" (Figure 5).

One very useful aspect of the concept is to provide a constraint on organic export from the euphotic zone, independent of more conventional (and controversial) procedures such as direct measurement of export flux (Platt et al. 1992). Interest in the capacity of the upper water column to draw down and export to depth atmospheric CO_2 reinvigorated interest in estimating new production (Eppley and Peterson 1979). Initial attention was focused on the role of deep NO_3^- in promoting atmospheric CO_2 drawdown, but the recognition that NO_3^- codiffuses or advects from depth with a stoichiometric amount of CO_2 (with respect to phytoplankton needs) mitigates the potential for deep NO_3^- to effect atmospheric CO_2 sequestration (Peng and Broecker 1991). On the other hand, new production dependent on N_2 fixation or atmospheric combined N deposition can support CO_2 sequestration (Karl et al. 1995; Capone et al. 1997).

Benthic–Pelagic Coupling

While the deep benthos of the open ocean receives organic debris from overlying waters and may exhibit seasonal responses to input (Altabet and Deuser 1985), only a small fraction of the export production reaches the

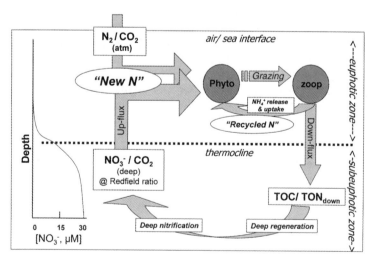

Figure 5. Conceptual representation of the new N model after Dugdale and Goering (1967): TOC, total organic carbon; TON, total organic nitrogen.

bottom (Suess 1980), and the benthos is sufficiently far removed that it has essentially no effect on processes on the upper, photic layers on short (annual–decadal) time scales. However, as one moves closer to the continents, the ability of the benthos to interact on biologically relevant time scales with the euphotic zone increases dramatically. In general, the higher levels of new production found in coastal surface waters (Eppley and Peterson 1979) result in a higher delivery of exported production to the sediments (Suess 1980) and higher sediment organic content (Jahnke 1996). Low O_2 and higher concentrations of NH_4^+ are often associated with nearshore and coastal sediments, both a result of the intense heterotrophic metabolism by the dense populations of bacteria in the sediments. Fluxes of inorganic nutrients from the benthos, in turn, can support a large fraction of the nutrient demand of water column primary productivity (Nixon and Pilson 1983). The high levels of spring production fueled by high nutrients in spring freshets in estuaries, or winter overturn in coastal waters, often result in organic accumulations in sediments that are metabolized through the year, providing a steady supply of inorganic nutrients to overlying waters (Boynton et al. 1995).

Nitrification–Denitrification Coupling

One particular aspect of shallow sediments with particular relevance to the N cycle is the phenomenon of nitrification–denitrification coupling (Koike and Sørensen 1988; Henriksen 1988). Nitrification is an aerobic, O_2-consuming process that produces NO_3^- from oxidation of NO_2^- and NH_4^+. Denitrification generally requires conditions of low O_2 and oxides of N, such as NO_3^-, and organic matter as its substrates. Denitrification can obtain its NO_3^- from influx of NO_3^- from the overlying waters or from sediment nitrification (Koike and Sørensen 1988). While the two processes would seem to be mutually exclusive with respect to their O_2 requirements, the sharp gradients of O_2, NH_4^+, and NO_3^- allow them to operate in close proximity to each other, and in many cases it appears that nitrification helps provide for both the conditions and the substrates for denitrification.

Coupling of Biogeochemical Cycles

In the sea as elsewhere, the nitrogen cycle does not operate in isolation but is directly coupled to the cycling of carbon and other biologically relevant elements. In addition to a carbon source, all organisms have a biosynthetic requirement for a suite of essential nutrients (e.g., N, P, S, Mg, Fe), needed in stoichiometric proportions for balanced growth (Fenchel et al. 1998). Nitrogen, which often constitutes several percent of cell biomass, is a macronutrient. Hence, all organisms regardless of their physiology devote portions of their overall energetic metabolism to N acquisition and incorporation. Furthermore, forms of inorganic N are directly involved in the energetic metabolism of certain heterotrophic (e.g., NO_3^- reducers and denitrifiers) and autotrophic (e.g., nitrifiers) bacteria.

Within particular biogeochemically significant species of bacteria and archaea, one can find the capacity to link SO_4^{2-} respiration and methanogenesis with N assimilation and ammonification, to couple sulfide, NH_4^+ or H_2 oxidation to denitrification (Zumft 1997), or methane oxidation, sulfate reduction, or denitrification to N_2 fixation (Chapter 13). Interestingly, bacteria and archaea of different presumed physiologies can exhibit considerable plasticity in metabolism: CH_4 oxidizers often have a capacity to nitrify, while many CH_4 oxidizers can nitrify (Bedard and Knowles 1989); some bacteria classified as denitrifiers can reduce metals (and vice versa) (Lovley 1991). Thus, complex interactions occur among various biogeochemical cycles.

Controls on N Cycle Pathways

Controls on N cycle processes range from molecular and physiological regulation (e.g., product feedback inhibition and substrate limitation of enzyme activity and gene transcription), to broad-scale physical phenomena (light, temperature, diffusion, convective mixing) (see Chapters 13 and 14). The mix of controlling factors determining the processes dominating various environments results in the characteristic N cycle of each particular system.

As alluded to earlier, the availability of elements other than N often is the controlling factor in a particular ecosystem, or of a subpopulation within an ecosystem. For instance, while the concentration of a particular substrate or product of a N transformation pathway may directly modulate a reaction rate in nature, the rate of transformation of specific forms of N can also be constrained by the availability of other required elements. Fe availability, which has been found to limit primary productivity in certain areas of the ocean (Martin 1992), may constrain the rates of NO_3^- assimilation or N_2 fixation, since the enzymes responsible for the utilization of these compounds are Fe containing (i.e., NO_3^- reductase and nitrogenase, respectively) (Falkowski 1983; Rueter et al. 1992). Atmospheric deposition of aeolian dust is thought to be a key pathway of Fe flux into the sea (Duce and Tindale 1991) and appears to stimulate primary productivity in some areas of the oceans (see below), possibly through enhancement of NO_3^- assimilation and N_2 fixation (Paerl et al. 1994).

The availability of other chemical constituents may also limit or determine the relative importance of particular N cycle pathways. Organic substrates can limit denitrification (Knowles 1982) or heterotrophic N_2 fixation (Capone 1988). O_2 concentration can profoundly influence the relative importance and rates of N cycle reactions such as nitrification, denitrification, and N_2 fixation. The source of P for planktonic N_2 fixers such as the cyanobacterium *Trichodesmium*, which proliferates in highly oligotrophic tropical oceanic waters, remains to be identified (Karl et al. 1992, 1997). P can be an important controlling factor in benthic N_2 fixation in shallow carbonate systems (Short et al. 1990).

Most recently, important interactions of Si, N, and Fe have been shown for oceanic systems dominated by diatoms (Dortch and Whitledge 1992; Dugdale and Wilkerson 1998; Hutchins and Bruland 1998).

Increased nutrient input into nearshore environments from anthropogenic sources can uncouple controls on natural biogeochemical processes. Such impacts can cause direct perturbations on rates of N cycling, with dramatic and profound effects on nearshore ecosystems. Examples of anthropogenic disturbance include the effect of duck excreta in stimulation of plankton blooms in Great South Bay, New York (Ryther and Dunstan 1971), and the overgrowth of coral reef ecosystems by macroalgae resultant from increased N loading (Smith et al. 1981; LaPointe 1997). More dramatic conditions that have been related to increased N fluxes to these systems include the apparent increases in the extent of anoxia over the last few decades in systems such as the Chesapeake Bay (Cooper and Brush 1991; Justic et al. 1997) or coastal waters of the Mississippi plume (R. E. Turner and Rabalais 1994)

CONTRASTING SYSTEMS

The details of the N cycle vary greatly among ecosystems with respect to the mode of N delivery and removal, the relative sizes of N pool and species, and the relative importance of particular N transformation pathways. Contrasts with respect to N distributions and processes are easily seen along the various spatial gradients that define the major oceanic provinces (Fanning 1992; Longhurst 1998) — for example, with respect to distance offshore, mixed-layer, and photic zone depth, and as a function of overall depth to the seafloor. Sharp contrasts in N distributions and transformation processes are evident along latitudinal gradients. Relatively nutrient-poor waters in the oceanic tropics grade into waters that are richer in N at temperate and boreal latitudes. In surface waters at higher latitudes, the processes of the N cycle exhibit strong seasonality. Substantial variations in N cycling and N distributions occur with depth in the water or sediment column. While it is far beyond the scope of this chapter to provide the details of differences among the various provinces of the sea, a few contrasting examples are provided.

Temperate Estuarine and Coastal Shelf Waters

In nearshore environments, such as estuaries and, to a lesser extent, coastal shelf waters, inorganic nutrient fluxes are often dominated by an array of external inputs arriving from riverine flow, runoff, atmospheric deposition, and groundwater discharge (Nixon and Pilson 1983) (Table 2). Input of organic N may be considerable, much of it derived from relatively refractory terrigenous sources. Large inorganic nutrient inputs and pools, combined with a high capacity for primary production, result in rapid incorporation of inorganic N

Table 2. Features of major ecosystems with particular respect to the N cycle

Biomes	Photic Zone Depth (m)[a]	Typical Biomass Chl a (mg m^{-2})[a]	Typical Biomass Dry wt (g m^{-2})[b,c]	Typical Primary Productivity [gC (m^{-2} y^{-1})][a,c,d]	Export Productivity (%)[e,f]	Major Pathways of combined N Delivery	Limiting Nutrient	N Cycle Reactions Nitrification	N Cycle Reactions Denitrification	N Cycle Reactions N$_2$ Fixation
PELAGIC/MICROPLANKTON										
Temperate estuary	5–20	10–200[g]		100–500	n/a	riv, atm, run, gw, sedf, ocn	N/P	+++[sed]	+++[sed]	0/+
Temperate coastal	20–40	5–50		100–200	25–30	riv, atm, sedf, curr, (uw, gw, run)	N	++[sed]	++[sed]	0
Temperate oceanic	40–60	1–10		50–100	10–15	vdif, conv. curr, (atm)	N	+	0	0/+
Polar oceanic	20–40	10–90		100–500	25–30	uw, diff, conv, curr, (atm)	N/Fe/(Si)	+	??	0
Tropical oceanic	60–80	1–5		10–100	~5	curr, (atm) vdi, atm	N	++	(+++)	++
Coastal upwelling	20–40	10–70		200–700	30–40	uw, riv, vdif atm	N/(Si)	+	0/++[sed]	0
Oceanic Upwelling	30–50	3–10		75–125	10–20	uw, vdif, atm	N/Fe/(Si)	++	0	0

BENTHIC/MACROPHYTE

Temperate salt marsh	+++	n/a	250–2250	n/a	tid, riv, atm run, gw	N	++	++	++
Kelp/macro-algae	+++	n/a	1000–5000	500–1500	curr, atm, run, (gw, riv)	N	?	?	?
Temperate sea grass	++	n/a	50–2500	50–500	curr, riv, run, atm, gw	N/P	++	++	+
Tropical sea grass	+++	n/a	1000–5000	500–1000	curr, riv, run, atm, gw	P/N	++/?	++/?	+++
Mangrove	+++	n/a	8000–50,000	500–2000	tid, riv, atm, run, gw	P/N?	++/?	++/?	+++/?
Coral reef	+++	n/a		500–5000	curr, gw, atm,	P/N	++	0/+?	+++

[a] Longhurst (1998).
[b] Woodwell et al. (1973).
[c] Mann (1982).
[d] De Vooys (1979).
[e] Eppley and Peterson (1979).
[f] Berger et al. (1987).
[g] Day et al. (1989).

Key to Ratings: O, absent or very minor; +, activity present but minor; ++, moderate activity; +++, high activity; parentheses, in some but not all cases; [se] activity in underlying sediments; ?, unknown contribution.

Key to Pathways: atm, atmospheric deposition; conv, convective overturn; curr, currents; lateral advection; gw, groundwater; riv, riverine input; run, runoff; sed sediment diffusion, tid, tidal input; uw, upwelling; vdif, vertical eddy diffusion.

into organic material. Subsequently, this dissolved and particulate organic material can be exported to the sediments or advected from the system.

Because of high plant biomass and suspended inorganic particles from runoff and resuspension, nearshore systems are typically characterized by a shallow euphotic zone, with relatively strong benthic–pelagic coupling (see above). Fluxes of nutrients and biological processes are strongly phased by season and along gradients from land to sea (i.e., estuarine Chl_{max} and rapid decreases in nutrients down the estuary). Primary productivity is low in the winter. Spring freshets bring large nutrient pulses to estuaries and nearshore waters (Boynton et al. 1995). Maxima in primary production, nutrient uptake and drawdown, and organic export often occur in the spring or early summer, with recycling becoming more prominent in the summer and fall.

In such organic rich systems, high NO_3^- concentrations can promote considerable levels of denitrification (Christensen et al. 1987; Seitzinger, 1988), which otherwise might be constrained by in situ nitrification rates (Henriksen and Kemp 1988; Koike and Sørensen 1988). In general, because relatively high supplies of externally loaded and recycled inorganic and organic N are available for plant growth, N_2 fixation is not generally of significance in temperate estuaries or shelf waters (Capone 1988; Howarth et al. 1988).

Temperate and Boreal Oceanic Waters

The open ocean receives substantially less continentally derived nutrient input (Michaels et al. 1996) and exhibits a far greater dependence on the seasonal injection of deep nutrients into the euphotic zone (Table 2). In temperate and boreal areas, plant production and nutrient drawdown are cyclic, seasonal phenomena (Parsons et al. 1984b). During the winter, deep convective mixing recharges the euphotic zone with nutrients from deeper waters. The classical spring diatom bloom of the North Atlantic is initiated when day length increases and a seasonal thermocline is established (Sverdrup 1953; Longhurst 1998). The phytoplankton are confined to the upper mixed layer, where there is abundant NO_3^-. As plant biomass increases, there is a corresponding drawdown of NO_3^- concentration. The bloom is terminated when NO_3^- is drawn down and, concurrently, active grazer populations have developed. In the summer populations shift to a dinoflagellate-dominated community, which depends largely on recycled N, vertical migration to deep NO_3^-, and the slow diffusion of NO_3^- across the thermocline (Parsons et al. 1984b; Longhurst 1998).

Coastal and Circumantarctic Upwelling Regions

Wind-driven circulation patterns in ocean gyres result in strong regional upwelling along the eastern flank of the major oceanic basins and in the circumantarctic region. Deep waters rich in NO_3^- and PO_4^{3-} are upwelled and create areas of intensive water column primary production (Ryther 1969).

Phytoplankton populations are often dominated by diatoms, and uptake of NO_3^- is a predominant process in the N cycle. However, Fe and Si may play key roles in controlling overall export production (Martin et al. 1990; Nelson and Treguer 1992; Dugdale and Wilkerson 1998) (see above). Much of this production can reach the sediments (Eppley and Peterson 1979). While these areas represent only a small fraction of the surface area of the global ocean, they account for a disproportionate fraction of global productivity and are often associated with high fish production (Ryther 1969; Walsh 1981). The Southern Ocean may play a critical role in the regulation of atmospheric CO_2 (François et al. 1997; Sarmiento et al. 1998). Predicted warming may increase surface stratification, which may in turn affect the specific populations and the dynamics of N cycling (Arrigo et al. 1999).

Equatorial Upwelling

In the equatorial Pacific, there are large areas of equatorial divergence that drive a weak upwelling system. These zones exhibit enhanced productivity relative to the surrounding oligotrophic gyres (Barber et al. 1996). Despite the upwelling of NO_3^- in these high-nitrate, low-chlorophyll (HNLC) areas, levels of phytoplankton biomass do not attain the concentrations expected, given the ambient nutrient concentrations. Among key nutrients, upwelled NO_3^- is not drawn down completely (Martin 1992; J. J. McCarthy et al. 1996). Martin (1994) posited that export production and N cycling in these areas far removed from continental sources might in fact be Fe-limited. Later experimental research confirmed that Fe is indeed a major factor in limiting phytoplankton populations in HNLC areas (Martin et al. 1994; Coale et al. 1996). Silicate also plays a key role in regulating the uptake of nitrate by diatoms and their export in these regions (Dugdale and Wilkerson 1998).

Subtropical/Tropical Oceanic Gyres

In the subtropical and tropical oligotrophic gyres, continental influences are at a minimum, surface water temperatures are high and relatively constant seasonally, the upper water column is permanently stratified and chronically depleted in nutrients and, consequently, primary productivity is also low. Primary production is largely dependent on recycled N (Eppley and Peterson 1979; Berger et al. 1987). The euphotic zone is typically very deep (≈ 80 m), and there is generally a chlorophyll a and algal biomass maximum toward the base of the euphotic zone (typically at 60–80 m). It is generally held that the primary source of "new" N for these vast systems is the slow vertical diffusion of NO_3^- across the thermocline (J. J. McCarthy and Carpenter 1983; Lewis et al. 1986; Ledwell et al, 1993) and, possibly, periodic eddy pumping (Falkowski et al. 1991; McGillicuddy et al. 1998). However, some NO_3^- transport from depth to the near surface is effected by vertically migrating diatom mats (*Rhizosolenia* sp.) (Villareal et al. 1999). Nonetheless, most of the NO_3^- moving upward is intercepted by the phytoplankton populations near the base of the euphotic zone. The upper mixed layers are generally typified by very low biomass and by nutrient concentrations well below the limits of detection by conventional colorimetric measurements (typically estimated to be about 30 nM for NO_3^- and NH_4^+).

N_2-fixing organisms such as the filamentous cyanobacteria *Trichodesmium* spp. inhabit the upper water column in these environments (Carpenter 1983; Letelier and Karl 1996; Capone et al. 1997). In the subtropical portions of these gyres, N_2 fixers may flourish for only a few months each year and are of limited significance in N cycling (Carpenter and J. J. McCarthy 1975). However, research has indicated a potentially significant role of N_2 fixers in tropical systems (Michaels et al. 1996; Gruber and Sarmiento 1997; Karl et al. 1997). In the tropical Atlantic Ocean, N_2 fixation can be of quantitative significance as a source of new N through the year (Capone et al. 1997). Moreover, blooms of *Trichodesmium* and the diatom endosymbiont *Richelia*, which can be observed by ocean color satellite (e.g., SeaWiFS), can bring large pulses of N into these systems as episodic blooms (Capone et al. 1998; Carpenter et al. 1999). Evidence also indicates that in addition to long-recognized inputs of DIN (Duce 1986), atmospheric deposition of DON can also provide a substantial input of new N to these very nutrient poor systems (Cornell et al. 1995).

In the tropical waters of the Indian Ocean, and in the eastern tropical North and South Pacific Ocean, anoxia develops in subsurface waters that have been out of contact with the atmosphere for long time periods (see above). Areas of globally significant denitrification occur at the oxic–anoxic interfaces of the intermediate waters of these areas (Hattori 1983). At these interfaces, substantial N_2O production can also occur (Capone 1991).

Our understanding of the subtropical oligotrophic ocean has greatly benefited over the last decade from the establishment of two time-series stations under the U.S. Joint Global Ocean Flux Study (JGOFS): the Hawaiian Ocean Time (HOT) series station north of Oahu in the subtropical Pacific and the BATS station near Bermuda have provided unmatched data sets with which to discern complex interactions among biogeochemical processes as well as to detect subtle changes and shifts in major processes over time. Karl and colleagues (Karl et al. 1995) have noted fundamental shifts in phytoplankton populations and predominant N cycle processes at HOT over the last decade, leading from a previously eukaryote-dominated, N-limited system to a prokaryote-dominated, P-limited system with a substantial input of N by N_2 fixers in the system.

Similarly, Michaels et al. (1994) have detected at the BATS site each summer large drawdowns of dissolved inorganic C that cannot be sustained by available pools of inorganic N or P. Along with large positive N^* anomalies at middepths, they have concluded that a substantial amount of N_2 fixation must be occurring at this station (Michaels et al. 1996, cf. Lipschultz and Owens 1996).

3.n,1Coral Reef, Algal Mats, and Large-Plant-Dominated Communities

In shallow marine systems, a variety of benthic plant communities proliferate. These systems exhibit particularly intense rates of primary production and N cycling and include temperate salt marshes (Valiela and Teal 1979), the rooted sea grasses of the tropics and temperate zones (Phillips and Helfferich 1980), tropical mangroves (Ewel et al. 1998), kelps (Chapman and Cragie 1977), and coral reefs (D'Elia and Wiebe 1990). Macroalgal communities (Hanisak 1983) and algal mats (see Chapter 13) often occur in association with each of these communities.

Rooted plant systems such as the sea grasses and salt marshes are very productive and, therefore, plant N demand is high (Table 2). Much of this demand is met by root uptake of N from sediment pore waters (Short 1987). Intensive N regeneration and N_2 fixation in the sediments provide ample dissolved N in these systems to meet demand (Capone 1983a and b). Algal mats are notable for their vertical compression of processes (Chapter 13). Coral reefs, which are largely defined by the coral–algal symbioses (but may also be dominated by macroalgae: Larkum, 1999) are generally thought to proliferate in nutrient-poor tropical waters because of high levels of N_2 fixation occurring in these environment (Capone 1983a and b; D'Elia and Wiebe 1990).

GLOBAL PERSPECTIVES

N Budgets and Balances

Processes and pools in the marine N cycle are often placed in the context of the terrestrial and global N cycles. Estimates of the gaseous N pools and some inorganic N (e.g., NO_3^-) pools have been refined based on results from diverse and far-ranging oceanographic studies such as the the Geochemical Ocean Sections Study (GEOSECS) and the World Ocean Circulation Experiment (WOCE), as well as intensive monitoring programs in many nearshore and coastal areas. Because of the difficulty in the analytical approaches involved in measuring organic DON and PON pools, far fewer measurements are available for these N compounds. Consequently, our projections of global pools of DON and PON are less robust. While we broadly understand the distribution and relative importance of N processes, we still lack quantitative estimates of the fluxes and internal transformations of N within many systems.

With increasing evidence of large-scale perturbations of the N cycle (see below), it has become more pressing to develop capabilities to predict changes that may occur in the near term. Estimating the overall current size of the various pools and magnitude of specific processes of the N cycle in the sea, improving our estimates as new data become available, and attempting to retrospectively infer changes that have occurred in earlier earth history will foster that ability.

Current estimates of the major fluxes of N in the ocean indicate that riverine inputs into the oceans are relatively well constrained, and these are thought to account for an input of between 21 and 70 Tg N y^{-1}. Atmospheric deposition to the sea surface accounts for an input of about 56 to 154 Tg N y^{-1} (Table 3). This value has recently been revised upward to include DON deposition (Cornell et al. 1995). Estimates for pelagic N$_2$ fixation range from 5 to over 100 Tg N y^{-1} (Capone and Carpenter 1999). Both biological and geochemical evidence indicates that this value may be revised upward (Michaels et al. 1996; Capone et al. 1997; Gruber and Sarmiento 1997). Direct studies of pelagic N$_2$ fixation are very limited both in their frequency and in their spatial extent. Benthic N$_2$ fixation has been estimated to account for an input of about 15 Tg N y^{-1} (Capone and Carpenter 1982).

The demand for N by primary producers can be inferred from extensive surveys of primary production in the world's oceans. Current estimates suggest that N demand for total oceanic primary productivity to be about 8000 Tg N y^{-1} (Falkowski et al. 1998). Of this total, export ("new") production is thought to accounts for about one-third, or 2400 Tg N y^{-1}. In theory, in a steady state ocean, the demand for new N should be met by the sum of external inputs, nitrification, and N$_2$ fixation (Capone 1991; 1996a). Even assuming maximum values for the other sources of N, nitrification would have to account for the bulk of new N demand. However, nitrification occurs at relatively low rates in the deep sea (Kaplan 1983; also see Chapter 14). Direct measurements of nitrification are among the most challenging to make and therefore the scarcest among the major processes involved in the marine N cycle.

Estimates of rates of N losses from the sea through sedimentation of PON and burial over the vast expanses of the seafloor are also difficult to make and are therefore also rare. However, given the great depth and relatively low productivity over much of the oceans, little detritus makes it to deep sediments (Suess 1980). Continental margins are thought to be key areas of organic burial in the current ocean (Walsh 1991; Hedges and Keil 1995). Current estimates of N burial range from 20 to 38 Tg N y^{-1} (Table 3).

Over the last few decades denitrification in the oceans has been examined more intensively, particularly in the Indian (Naqvi et al. 1982) and tropical Pacific Oceans (Hattori 1983; Codispoti and Christensen 1985), and on the continental shelves (Christensen et al. 1987; Devol 1991) (see above), and there have been several upward revisions of its quantitative significance (Codispoti 1995). Water column denitrification is currently thought to account for losses of from 64 to 290 Tg N y^{-1} while denitrification losses from coastal and shelf sediments have been estimated to be about 60 to 90 Tg N y^{-1} (Christensen et al. 1987; Seitzinger 1988). Other mechanisms whereby N may be lost from the system include gaseous evasion of N (e.g., Law and Owens 1990) and fisheries harvest.

Based on the current upper end values (the most recent revisions), total N inputs are about 100 Tg N y^{-1} less than estimates for its removal, suggesting

Table 3. Nitrogen fluxes in the ocean

Process	Flux (Tg N y^{-1})	Ref.
INPUTS		
Riverine and runoff	21–110	Walsh (1991), Meybeck (1993), Caraco and Cole (1999)
Atmospheric deposition	56–154	Cornell et al. (1995)
N$_2$ fixation		
Water column	10–110	Carpenter (1983), Capone et al. (1997), Gruber and Sarmiento (1997)
Benthic	15	Capone (1983a)
Total	102–389	
PRODUCTION DEMAND		
Planktonic		
Total	7200	Falkowski et al. (1998)
Export	2424	Falkowski et al. (1998)
Benthic — net annual	160[a]	DeVooys (1979)
OUTPUTS		
Sedimentation	16–38	Soderlund and Svensson (1976), McElroy (1983), Hedges and Keil (1995)[a]
Dentrification		
Water column	64–290	Hattori (1983), Codispoti (1989, 1995)
Benthic	60–94	Christensen et al. (1987), Seitzinger (1988), Hattori (1983)
Other	10	Liu (1979)
Total	150–432	

[a]Assuming a C:N ratio of 10.

an imbalanced N cycle. This may be due to underestimates of N inputs or overestimates of N losses. However, there is no a priori reason to assume a balanced N cycle, which may be achieved only over longer time scales (e.g., time scales of oceanic mixing) (Codispoti 1989).

Large-Scale Perturbations of the N Cycle

Current evidence reveals large-scale natural (Karl et al. 1995) and human-induced (Vitousek et al. 1997) perturbations in the marine N cycle, paralleling those occurring in the C cycle (Siegenthaler and Sarmiento 1993). River loads of N to the sea have increased dramatically over recent time (Peierls et al. 1991; Caraco and Cole 1999). Resultant coastal eutrophication and increasing hypoxia were mentioned earlier (see "Controls on N Cycle Pathways"). Increasing nutrient loading in coastal areas may account for stimulated uptake and burial of CO_2 on the shelf in these areas (Walsh et al. 1981).

Nitrous oxide has been implicated in atmospheric reactions leading to the destruction of ozone in the stratosphere (Crutzen 1981). It is also a potent greenhouse gas, with a several hundred–fold greater capacity for absorption of radiation per molecule than CO_2 which enables it to contribute to radiative heating in the troposphere (Houghton et al. 1990). Direct manifestation of global perturbation has also been detected in atmospheric N_2O pools. The atmospheric concentration of N_2O, which is about 310 ppb at present, has been increasing at a rate of about 0.2–0.3% per year. (Houghton et al. 1990). The role of the oceans as a source or sink of N_2O to the atmosphere is uncertain (Scranton 1983; Capone 1991). As indicated above, the seas play a central role in N_2O cycling (Capone 1991; Seitzinger and Kroeze 1998) and may act as source (e.g., Law and Owens 1990; Dore et al. 1998) or sink for this gas.

The casual suggestion by Martin et al. (1990) — that fertilization of HNLC areas of the ocean with Fe to stimulate oceanic primary production and carbon export in these areas could help ameliorate increasing atmospheric CO_2 — created a lively discussion during the early 1990s about the merits of such a scheme (Chisholm and Morel 1991). While much of the debate centered on the factors controlling phytoplankton populations in these areas, consideration of the feasibility and merits of large-scale environmental engineering also entered the debate. Suggestions that such a scheme would work were met with skepticism by many oceanographers (Joos et al. 1991; Peng and Broecker 1991). Among possible scenarios proposed, Fuhrman and Capone (1992) suggested that fertilization could stimulate the production of other gases with potential impacts on climate, such as N_2O and dimethyl sulfide (DMS), through the production of suboxic waters with increased surface production and flux. While N_2O was not analyzed during the IRONEX experiment, Turner et al. (1996) did observe increases in DMS.

Paleoecology of the Marine N Cycle

McElroy (1983) and, more recently, Falkowski (1997) have considered the dynamics of oceanic productivity with specific reference to the oceanic N cycle over glacial–interglacial time scales. Considerable evidence exists to document dramatic changes in plankton productivity and N dynamics in major ocean basins over glacial–interglacial periods. Plankton productivity and export production appear to be greater during periods of low sea level (i.e., glacial periods), when shelves are exposed and nutrient flux from runoff and weathering is (presumably) greater (Figure 6). (Farrell et al. 1995; Kumar et al. 1995;

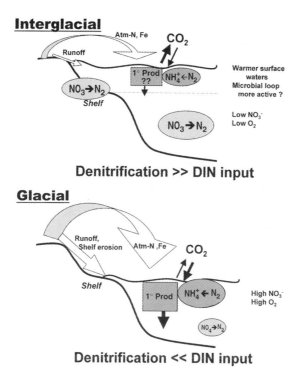

Figure 6. Conceptual model of changes in the N cycle during glacial and interglacial periods. Relative sizes of fluxes between states indicated by sizes of arrows, boxes and spheres.

cf. François and Altabet 1992). This may result in positive feedbacks as CO_2 is drawn down further, with albedo increases with glaciation, and decreased absorption of solar radiation, thereby maintaining the cooling trend. Martin (1990) related the higher productivity during glacial periods to a greater flux of Fe to the oceans during these times and stimulation of phytoplankton in HNLC areas. However, in addition to modification of nutrient delivery during global climate shifts, changes in water column stratification and large-scale circulation need also be considered. Based on an extensive data set of sediment nitrogen isotope analyses, François et al. (1997) have speculated that during the last glacial maximum, increased stratification in the Southern Ocean and plugging of the CO_2 leak from sea to atmosphere may have played a larger role in the lowering of CO_2 than stimulated primary production.

Fe fluxes may also have a direct effect on the balance of denitrification and N_2 fixation (Falkowski 1997). During interglacial periods (such as the present) Fe flux to the oceans is low, and denitrification is hypothesized to be of greater significance than N_2 fixation (which would be Fe-limited), thereby pushing systems to become N limited (Christensen et al. 1987). In contrast, during glacial periods, the oceanic basins may be more oxygenated because of lower sea level, circulation patterns, and lower surface temperatures, thereby restricting oceanic denitrification. Shelf denitrification would presumably be shut off. Downcore distributions of $\delta^{15}N$ in PON in the Pacific Ocean (Ganeshram et al. 1995) and Arabian Sea (Altabet and François 1994) support the contention of a higher level of water column denitrification during interglacial periods (i.e., heavier $\delta^{15}N$ in sediment PON).

Moreover, enhanced Fe flux to the sea during glacial periods (Martin 1990) could also promote N_2 fixation, thereby reversing the effects of interglacial periods on the overall N balance (Falkowski 1997; see Haug et al. 1998). The results of Ganeshram et al. (1995) and Altabet and François (1994) may also be interpreted to indicate increased N_2 fixation during glacial periods (i.e., lighter $\delta^{15}N$ in sediment PON). Further paleoecologic studies are needed to assess the probability and significance of these scenarios.

FUTURE PROSPECTS

Research into the marine N cycle continues apace as we gain ever more detailed knowledge of the relative importance of each pathway within particular ecosystems and the controls on those pathways. Advances in analytical approaches will continue to improve our ability to detect key compounds more sensitively, as well as to resolve and detect new compounds of biological relevance. New pathways and transformations continue to be discovered. Automation of methods for determining pools and processes and remote sensing are making possible analysis of key parameters with unprecedented spatial and temporal resolution. Perhaps one of the most exciting advances to come will be in our developing ability to couple analysis of the spatial and temporal patterns and dynamics of particular N cycle organisms directly with their activities. Further, sensitive means to detect gene expression and regulation in situ using molecular probes (protein, DNA and mRNA) to key enzymes will give us great insights into the real controls on important processes. At the broader scale, we are focusing more on specific functional groups of organism that play key roles in overall ocean biogeochemistry and carbon dynamics, with field research proceeding in an ever more coordinated way with ecosystem modeling. Research over the next decade will likely concentrate more intensively on such groups (e.g., the planktonic N_2 fixers), particularly with respect to their role in the response of the oceans to climate change.

SUMMARY

1. The biogeochemical cycle of nitrogen is relatively complex, with nitrogenous compounds occurring over a range of oxidation states and in a variety of chemical forms and phases.

2. Of the diversity of biological transformations, many are restricted to prokaryotes, and the higher biota depend on them.

3. The bacterial N transformations of N_2 fixation, nitrification, and denitrification largely determine the form and relative availability of key N species, and there is considerable variation among systems with respect to the relative importance of these pathways.

4. The N cycle is an important feature of all marine ecosystems and is a factor affecting the net productivity of many of these systems; it is susceptible to perturbation at various levels.

5. Components of the marine N cycle may provide important feedbacks to marine C cycling and to global climate change.

REFERENCES

Altabet, M. A., and Deuser, W. G. (1985) Seasonal variations in natural abundance of ^{15}N in particles sinking to the deep Sargasso Sea. *Nature* 315:218–219.

Altabet, M. A., and François, R. (1994) Sedimentary nitrogen isotopic ratio as a recorder for surface ocean nitrate utilization. *Global Biogeochem. Cycles* 8:103–116.

Antia, N. J., Harrison, P. J., and Oliveira, L. (1991) The role of dissolved organic nitrogen in phytoplankton nutrition, cell biology and ecology. Phycologia 30:1–89.

Arrigo, K. R., Robinson, D. H., Worthen, D. L., Dunbar, R. B., DiTullio, G. R., VanWoert, M., and Lizotte, M. P. (1999) Phytoplankton community structure and the drawdown of nutrients and CO_2 in the Southern Ocean. *Science* 283:365–367.

Bada, J. L. (1998) Biogeochemistry of organic nitrogen compounds. In: *Nitrogen-containing macromolecules in the bio- and geosphere.* ACS Symposium Series 707:64–73.

Barber, R. T., Sanderson, M. P., Lindley, S. T., Chai, F., Newton, J., Trees, C. C., Foley, D. G., and Chavez, F. P. (1996) Primary productivity and its regulation in the equatorial Pacific during and following the 1991–1992 El Niño. *Deep-Sea Res. II* 43:933–969.

Bedard, C., and Knowles, R. (1989) Physiology, biochemistry, and specific inhibitors of CH_4, NH_4, and CO oxidation by methanotrophs and nitrifiers. *Microbiol. Rev.* 53:68–84.

Berger, W., Fischer, K., Lai, C., and Wu, G. (1987) Ocean productivity and organic carbon flux. *Scripps Institute Oceanography Reference Series* 87–30.

Boynton, W., Garber, J., Summers, R., and Kemp, W. (1995) Inputs transformations, and transport of nitrogen and phosphorus in Chesapeake Bay and selected tributaries. *Estuaries* 18:285–314.

Broecker, W. S., and Peng, T.-H. (1982) *Tracers in the Sea.* Eldigo Press, Palisades, NY.

Bronk, D. A., Glibert, P. M., and Ward, B. (1994) Nitrogen uptake, dissolved organic nitrogen release, and new production. *Science* 265:1843–1846.

Capone, D. G. (1983a) Benthic nitrogen fixation. In E. J. Carpenter and D. G. Capone, eds., *Nitrogen in the Marine Environment.* Academic Press. New York, pp. 105–137.

Capone, D. G. (1983b) N_2 fixation in seagrass communities. *Mar. Tech. Soc. J.* 17:32–37.

Capone, D. G. (1988) Benthic nitrogen fixation. In T. H. Blackburn and J. Sørensen, eds., *Nitrogen Cycling in Coastal Marine Environments.* Wiley, New York, pp. 85–123.

Capone, D. G. (1991) Aspects of the marine nitrogen cycle with relevance to the dynamics of nitrous and nitric oxide. In J. E. Rogers and W. B. Whitman, eds., *Microbial Production and Consumption of Greenhouse Gases: Methane, Nitrogen Oxides, and Halomethanes.* American Society of Microbiology, Washington, DC, pp. 255–275.

Capone, D. G. (1993) Determination of nitrogenase activity in aquatic samples using the acetylene reduction procedure. In P. F. Kemp, B. F. Sherr, E. B. Sherr, and J. J. Cole, eds., *Handbook of Methods in Aquatic Microbial Ecology.* Lewis Publishers, Boca Raton, FL, pp. 621–631.

Capone, D. G. (1996a) A biologically constrained estimate of oceanic N_2O flux. *Mitt. Int. Verein. Limnol.* 25:105–113.

Capone, D. G. (1996b) Microbial nitrogen cycling. In S. Newell and R. Christian, eds., *Manual of Environmental Microbiology, Section IV:Aquatic Environments.* ASM Press, Washington DC, pp. 334–342.

Capone, D. G., and Bautista, M. (1985) Direct evidence for a groundwater source for nitrate in nearshore marine sediments. *Nature* 313:214–216.

Capone, D. G., and Carpenter, E. J. (1982) Nitrogen fixation in the marine environment. *Science* 217:1140–1142.

Capone, D. G., and Carpenter, E. J. (1999) Nitrogen fixation by marine cyanobacteria: Historical and global perspectives. In L. Charpy and A. Larkum, eds., *Marine Cyanobacteria.* l'Institut Ocanographique, Monaco pp. 235–256.

Capone, D. G., Zehr, J., Paerl, H., Bergman, B., and Carpenter, E. J. (1997) *Trichodesmium:* A globally significant marine cyanobacterium. *Science* 276:1221–1229.

Capone, D. G., Subramaniam, A., Montoya, J., Voss, M., Humborg, C., Johansen, A., Siefert, R., and Carpenter, E. J. (1998) An extensive bloom of the N_2-fixing cyanobacterium, *Trichodesmium erythraeum,* in the Central Arabian Sea. *Mar. Ecol. Prog. Ser.* 172:281–292.

Caraco, N. F., and Cole, J. (1999) Human inpact on aquatic nitrogen loads: A regional scale study using large river basins. *Ambio* 28:167–170.

Carpenter, E. J. (1983) Nitrogen fixation by marine *Oscillatoria* (*Trichodesmium*) in the world's oceans. In E. J. Carpenter and D. G. Capone eds., *Nitrogen in the Marine Environment.* Academic Press, New York, pp. 65–103.

Carpenter, E. J., and McCarthy, J. J. (1975) Nitrogen fixation and uptake of combined nitrogenous nutrients by *Oscillatoria* (*Trichodesmium*) *thiebautii* in the western Sargasso Sea. *Limnol. Oceanogr.* 20:389–401.

Carpenter, E. J., Montoya, J. P., Burns, J., Mulholland, M. M., and Capone, D. G. (1999). Extensive bloom of a N_2-fixing symbiotic association in the tropical Atlantic Ocean. *Mar. Ecol. Prog. Ser.* 185:273–283.

Chapman, A. R. O., and Cragie, J. S. (1977) Seasonal growth in *Laminaria longicuris:* Relations with dissolved inorganic nutrients and internal reserves of nitrogen. *Mar. Biol.* 40:197–205.

Chisholm, S., and Morel, F., eds. (1991) *What Controls Phytoplankton Production in Nutrient-Rich Areas of the Open Ocean?* American Society of Limnology and Oceanography, Lawrence, KS.

Christensen, J. P., Murray, J. W., Devol, A. H., and Codispoti, L. A. (1987) Denitrification in continental shelf sediments has a major impact on the oceanic nitrogen budget. *Global Biogeochem. Cycles* 1:97–116.

Christian, R., and Capone, D. G. (1996) Overview of issues. In S. Newell and R. Christian, eds., *Manual of Environmental Microbiology*, Section IV: *Aquatic Environments*. ASM Press, Washington, DC, pp. 245–251.

Cline, J. D., and Ben-Yaakov, S. (1973) Nitrogen/argon ratios by difference thermal conductivity. *Deep-Sea Res.* 20:763–768.

Coale, K. H., et al. (1996) A massive phytoplankon bloom induced by an ecosystem-scale iron fertilization experiment in the equatorial Pacific Ocean. *Nature* 383:495–501.

Codispoti, L. (1989) Phosphorus versus nitrogen limitation of new and export production. In W. H. Berger, V. S. Smetacek, and G. Wefer, eds., *Productivity in the Ocean: Present and Past*. Wiley, New York. pp. 377–394.

Codispoti, L. A. (1995) Is the ocean losing nitrate? *Nature* 376:724.

Codispoti, L. A., and Christensen, J. P. (1985) Nitrification, denitrification and nitrous oxide cycling in the eastern tropical south Pacific Ocean. *Mar. Chem.* 16:277–300.

Cooper, S. R., and Brush, G. S. (1991) Long-term history of Chesapeake Bay anoxia. *Science* 254:992–996.

Cornell, S., Rendell, A., and Jickells, T. (1995) Atmospheric input of dissolved organic nitrogen in the oceans. *Nature* 376:243–246.

Crutzen, P. (1981) Atmospheric chemical processes of the oxides of nitrogen, including nitrous oxide. In C. C. Delwiche, ed., *Denitrification, Nitrification, and Atmospheric Nitrous Oxide*. Wiley, New York, pp. 17–44.

Day Jr, J., Hall, C., Kemp, W., and Yanez-Aranciba, A. (1989) *Estuarine Ecology*. Wiley, New York.

D'Elia, C. (1983) Nitrogen determination in seawater. In E. J. Carpenter and D. G. Capone, eds., *Nitrogen in the Marine Environment*. Academic Press, New York, pp. 731–762.

D'Elia, C. F., and Wiebe, W. J. (1990) Biogeochemical nutrient cycles in coral reef ecosystems. In Z. Dubinsky, ed., *Coral Reefs*. Elsevier Science Publishers, Amsterdam, pp. 49–74.

D'Elia, C. F., Sanders, J. G., and Boynton, W. R. (1986) Nutrient enrichment studies in a coastal plain estuary: Phytoplankton growth in large-scale continuous cultures. *Can. J. Fish. Aquat. Sci.* 43:397–406.

DeMaster, D., and Pope, R. H. (1996) Nutrient dynamics in Amazon shelf waters: Results from AMASSEDS. *Cont. Shelf Res.* 16:263–289.

Devol, A. H. (1991) Direct measurement of nitrogen gas fluxes from continental shelf sediments. *Nature* 349:319–322.

DeVooys, C. G. N. (1979) Primary productivity in aquatic environments. In B. Bolin, ed., *The Global Carbon Cycle*. Wiley, New York, pp. 259–292.

Dore, J. E., Popp, B. N., Karl, D. E., and Sansone, F. J. (1998) A large source of atmospheric nitrous oxide from subtropical North Pacific waters. *Nature* 396:63–66.

Dortch, Q., and Whitledge, T. E. (1992) Does nitrogen or silicon limit phytoplankton production in the Mississippi River plume and nearby regions? *Cont. Shelf Res.* 12:1293–1309.

Duce, R. A. (1986) The impact of atmospheric nitrogen, phosphorus and iron species on marine biological productivity. In P. Buat-Menard, ed., *The Role of Air–Sea Exchange in Geochemical Cycling.* Dordrecht, Reidel, pp. 487–529.

Duce, R. A., and Tindale, N. W. (1991) Atmospheric transport of iron and its deposition in the ocean. *Limnol. Oceanogr.* 36:1715–1726.

Dugdale, R. (1967) Nutrient limitation in the sea: Dynamics, identification and significance. *Limnol. Oceanogr.* 12:685–695.

Dugdale, R. C., and Goering, J. J. (1967) Uptake of new and regenerated forms of nitrogen in primary productivity. *Limnol. Oceanogr.* 12:196–206.

Dugdale, R. C., and Wilkerson, F. P. (1998) Silicate regulation of new production in the equatorial Pacific upwelling. *Nature* 391:270–273.

Eppley, R. W., and Peterson, B. J. (1979) Particulate organic matter flux and planktonic new production in the deep ocean. *Nature* 282:677–680.

Ewel, K., Twilley, R., and Ong, J. (1998) Different kinds of mangrove forests provide different goods and services. *Global Ecol. Biogeogr. Lett.* 7:83–94.

Falkowski, P. (1983) Enzymology of nitrogen assimilation. In E. J. Carpenter and D. G. Capone, eds., *Nitrogen in the Marine Environment.* Academic Press, New York. pp. 839–868.

Falkowski, P. (1997) Evolution of the nitrogen cycle and its influence on biological sequestration of CO_2 in the oceans. *Nature* 387:272–273.

Falkowski, P. G., Ziemann, D., Kobler, Z., and Bienfang, P. K. (1991) Role of eddy pumping in enhancing primary production in the ocean. *Nature* 352:55–58.

Falkowski, P. G., Barber, R. T., and Smetacek, V. (1998) Biogeochemical controls and feedbacks on ocean primary production. *Science* 281:200–206.

Fanning, K. (1992) Nutrient provinces in the sea: concentration ratios, reaction rate ratios, and ideal covariation. *J. Geophys. Res.* 97:5693–5712.

Farrell, J. W., Pedersen, T. F., Calvert, S. E., and Nielsen, B. (1995) Glacial–interglacial changes in nutrient utilization in the equatorial Pacific ocean. *Nature* 377:514–518.

Fenchel, T., King, G. M., and Blackburn, T. H. (1998) *Bacterial Biogeochemistry: The Ecophysiology of Mineral Cycling.* Academic Press, San Diego, CA.

Fisher, T., Peele, E. R., Ammerman, J. W., and Harding, L. W. (1992) Nutrient limitation of phytoplankton in Chesapeake Bay. *Mar. Ecol. Prog. Ser.* 82:51–63.

François, R., and Altabet, M. A. (1992) Glacial to interglacial changes in surface nitrate utilization in the Indian sector of the Southern Ocean as recorded by sediment [15]N. *Paleoceanography* 7:589–606.

François, R., Altabet, M., Yu, E., Sigman, D., Bacon, M., Frank, F., Bohrmann, G., Bareille, G., and Labeyrie, L. (1997) Contribution of Southern Ocean surface-water stratification to low atmospheric CO_2 concentrations during the last glacial period. *Nature* 389:929–935.

Fuhrman, J. A., and Capone, D. G. (1992) Biogeochemical consequences of ocean fertilization. *Limnol. Oceanogr.* 36:1951–1959.

Fuhrman, J. A., and Ferguson, R. L. (1986) Nanomolar concentrations and rapid turnover of dissolved free amino acids in seawater: Agreement between chemical and microbiological measurements. *Mar. Ecol. Prog. Ser.* 33:237–242.

Ganeshram, R. S., Pedersen, T. F., Calvert, S. E., and Murray, J. W. (1995) Large changes in oceanic nutrient inventories from glacial to interglacial periods. *Nature* 376:755–758.

Garside, C., Malone, T. C., Roels, O. A., and Scharfstein, B. A. (1976) An evaluation of sewage derived nutrients and their influence on the Hudson Estuary and New York Bight. *Estuarine Coastal Shelf Sci.* 4:281–289.

Glibert, P. M., and Capone, D. G. (1993) Mineralization and assimilation in aquatic, sediment, and wetland systems. In R. Knowles and H. T. Blackburn, eds., *Nitrogen Isotope Techniques*. Academic Press, San Diego, CA, pp. 243–272.

Gruber, N., and Sarmiento, J. (1997) Global patterns of marine nitrogen fixation and denitrification. *Global Biogeochem. Cycles* 11:235–266.

Hahn, J. (1981) Nitrous oxide in the oceans. In C. C. Delwiche, ed., *Denitrification, Nitrification, and Atmospheric Nitrous Oxide*. Wiley, New York, pp. 191–241.

Hanisak, M. D. (1983) Nitrogen relationships in marine macroalgae. In E J. Carpenter and D. G. Capone, eds., *Nitrogen in the Marine Environment*. Academic Press, New York, pp. 699–730.

Harrison, W. G. (1983) Use of isotopes. In E. J. Carpenter and D. G. Capone, eds., *Nitrogen in the Marine Environment*. Academic Press, New York, pp. 763–807.

Harvey, H. W. (1927) *Biological Chemistry and Physics of Seawater*. Cambridge University Press, Cambridge.

Hattori, A. (1983) Denitrification and dissimilatory nitrogen reduction. In E. J. Carpenter and D. G. Capone, eds., *Nitrogen in the Marine Environment*. Academic Press, New York, pp. 191–232.

Haug, G., Pedersen, T., Sigman, D., Calvert, S., Nielsen, B., and Peterson, L. (1998) Glacial/interglacial variations in production and nitrogen fixation in the Cariaco Basin during the last 580K years. *Paleoceanography* 13:427–3432.

Hedges, J. I., and Keil, R. G. (1995) Sedimentary organic matter preservation: An assessment and speculative synthesis. *Mar. Chem.* 49:81–115.

Henriksen, K., and Kemp, M. (1988) Nitrification in estuarine and coastal marine sediments. In T. H. Blackburn and J. Sørensen, eds., *Nitrogen Cycling in Coastal Marine Environments*. Wiley-Liss, New York, pp. 207–273.

Houghton, J. T., Jenkins, G. J., and Ephraums, J. J., eds. (1990) *Climate Change: The IPCC Scientific Assessment*. Cambridge University Press, Cambridge.

Howarth, R. W. (1988) Nutrient limitation of net primary production in marine ecosystems. *Annu. Rev. Ecol.* 19:89–110.

Howarth, R. W., Marino, R., Lane, J., and Cole, J. J. (1988) Nitrogen fixation in freshwater, estuarine, and marine ecosystems. 1. Rates and importance. *Limnol. Oceanogr.* 33:669.

Hurst, C. J., Knudsen, G. R., McInerney, M. J., Stetzenbach, L. D., and Walter, M. V., eds. (1997) *Manual of Environmental Microbiology*. ASM Press, Washington, DC.

Hutchins, D. A., and Bruland, K. W. (1998) Iron-limited growth and Si:N uptake ratio in a coastal upwelling regime. *Nature* 393:561–564.

Jahnke, R. A. (1996) The global ocean flux of particulate organic carbon: Areal distribution and magnitude. *Global Biogeochem. Cycles* 10:71–88.

Johannes, R. E. (1980) The ecological significance of the submarine discharge of groundwater. *Mar. Ecol. Prog. Ser.* 3:365–373.

Joos, F., Sarmiento, J. L., and Siegenthaler, U. (1991) Estimates of the effect of Southern Ocean iron fertilization on atmospheric CO_2 concentrations. *Nature* 39:772–775.

Justic, D., Rabalais, N., and Turner, R. (1997) Impacts of climate change on net productivity of coastal waters: Implications for carbon budgets and hypoxia. *Climate Res.* 8:225–237.

Kahlil, M. A. K. and Rasmussen, R. A. (1992) The global source of nitrous oxide. *J. Geophys. Res.* 97:14651–14660.

Kaplan, W. A. (1983) Nitrification. In E. J. Carpenter and D. G. Capone, eds., *Nitrogen in the Marine Environment*. Academic Press, New York, pp. 139–190.

Karl, D. M., Letelier, R., Hebel, D. V., Bird, D. F., and Winn, C. D. (1992) *Trichodesmium* blooms and new nitrogen in the north Pacific gyre. In E. J. Carpenter, D. G. Capone, and J. G. Rueter, eds., *Marine Pelagic Cyanobacteria: Trichodesmium and Other Diazotrophs*. Kluwer Academic, Dordrecht, pp. 219–237.

Karl, D. M., Letelier, R., Hebel, D., Tupas, L., Dore, J., Christian, J., and Winn, C. (1995) Ecosystem changes in the North Pacific subtropical gyre attributed to the 1991–92 El Niño. *Nature* 373:230–233.

Karl, D., Letelier, R., Tupas, L., Dore, J., Christian, J., and Hebel, D. (1997) The role of nitrogen fixation in biogeochemical cycling in the subtropical North Pacific ocean. *Nature* 386:533–538.

Kemp, P. F., Sherr, B. F., Sherr, E. B., and Cole, J. J., eds. (1993) *Handbook of Methods in Aquatic Microbial Ecology*. Lewis Publishers, Boca Raton, FL.

Knowles, R. (1982) Denitrification. *Microbiol. Rev.* 46:43–70.

Knowles, R., and Blackburn, T. H., eds. (1993) *Nitrogen Isotope Techniques*. Academic Press, San Diego, CA.

Koike, I., and Sørensen, J. (1988) Nitrate reduction and denitrification in marine sediments. In T. H. Blackburn and J. Sørensen, eds., *Nitrogen Cycling in Coastal Marine Environments*. Wiley, New York, pp. 251–273.

Kumar, N., Anderson, R. F., Mortlock, R. A., Froelich, P. N., Kubic, P., Dittrichhannen, B., and Suter, M. (1995). Increased biological productivity and export production in the glacial southern ocean. *Nature* 378:675–680.

LaPointe, B. (1997) Nutrient thresholds for bottom-up control of macroalgal blooms on coral reefs in Jamaica and southeast Florida. *Limnol. Oceanogr.* 42:1119–1131.

LaPointe, B. E., Littler, M. M., and Littler, D. S. (1992) Nutrient availability to marine macroalgae in siliciclastic versus carbonate-rich coastal waters. *Estuaries* 15:75–82.

Larkum, A. W. D. (1999) Coral reefs. In L. Charpy and A. Larkum, eds., *Marine Cyanobacteria*. l'Institut Oceanographique, Monaco, p. 149–168.

Law, C. S., and Owens, N. J. P. (1990) Significant flux of atmospheric nitrous oxide from the northwest Indian ocean. *Nature* 346:826–828.

Ledwell, J. R., Watson, A. J., and Law, C. S. (1993) Evidence for slow mixing across the pycnocline from an open-ocean tracer-release experiment. *Nature* 364:701–703.

Letelier, R. M., and Karl, D. M. (1996) Role of *Trichodesmium* spp. in the productivity of the subtropical North Pacific ocean. *Mar. Ecol. Prog. Ser.* 133:263–273.

Lewis, M. R., Harrison, W. G., Oakey, N. S., Herbert, D., and Platt, T. (1986) Vertical nitrate fluxes in the oligotrophic ocean. *Science* 234:870–873.

Lipschultz, F., and Owens, N. J. P. (1996) An assessment of nitrogen fixation as a source of nitrogen to the North Atlantic ocean. *Biogeochemistry* 35:261–274.

Liu, K. K. (1979) Geochemistry of inorganic nitrogen compounds in two marine environments: The Santa Barbara Basin and the ocean off Peru. Ph. D. thesis, University of California, Los Angeles.

Longhurst, A. (1998) *Ecological Geography of the Seas.* Academic Press, San Diego, CA.

Lovley, D. R. (1991) Dissimilatory Fe(III) and Mn(IV) reduction. *Microbiol. Rev.* 55:259–287.

Mann, K. (1982) *Ecology of Coastal Waters: A Systems Approach.* University of California Press, Berkeley and Los Angeles.

Martin, J. H. (1990) Glacial–interglacial CO_2 change: The iron hypothesis. *Paleoceanography* 5:1–13.

Martin, J. H. (1992) Iron as a limiting factor in oceanic productivity. In P. Falkowski and A. Woodhead, eds., *Primary Productivity and Biogeochemical Cycles in the Sea.* Plenum Press, New York, pp. 123–137.

Martin, J. H., Gordon, R. M., and Fitzwater, S. E. (1990) Iron in Antartica waters. *Nature* 345:156–158.

Martin, J. H., et al. (1994) Testing the iron hypothesis in ecosystems of the equatorial Pacific ocean. *Nature* 371:123–129.

McCarthy, J. J., and Carpenter, E. J. (1983) Nitrogen cycling in near-surface waters of the open ocean. In E. J. Carpenter and D. G. Capone, eds., *Nitrogen in the Marine Environment.* Academic Press, New York, pp. 487–512.

McCarthy, J. J., Garside, C., Nevins, J. L., and Barber, R. T. (1996) New production along 140°W in the equatorial Pacific during and following the 1992 El Niño event. *Deep-Sea Res. II* 43:1065–1093.

McCarthy, M., Pratum, T., Hedges, J., and Benner, R. (1997) Chemical composition of dissolved organic nitrogen in the ocean. *Nature* 390:150–154.

McCarthy, M., Hedges, J. I., and Benner, R. (1998) Major bacterial contribution to marine organic nitrogen. *Science* 281:231–234.

McElroy, M. B. (1983) Marine biological controls on atmospheric CO_2 climate. *Nature* 302:328–329.

McGillicuddy Jr, D. J., Robinson, A. R., Siegel, D. A., Jannasch, H. W., Johnson, R., Dickey, T. D., McNeil, J., Michaels, A. F., and Knap, A. H. (1998) Influence of mesoscale eddies on new production on the Sargasso Sea. *Nature* 394:263–266.

Meybeck, M. (1993) C, N, P and S in rivers: From sources to global inputs. In R. Wollast, F. T. MacKenzie, and L. Chou, eds., *Interactions of C, N, P and S in Biogeochemical Cycles and Global Change.* Springer-Verlag, Berlin, pp. 163–193.

Michaels, A. F., Siegel, D., Johnson, R. J., Knap, A. H., and Galloway, J. N. (1993) Episodic inputs of atmospheric nitrogen to the Sargasso Sea: Contributions to new production and phytoplankton blooms. *Global Biogeochem. Cycles* 7:339–351.

Michaels, A. F., Bates, N. R., Buesseler, K. O., Carlson, C. A., and Knap, A. H. (1994) Carbon-cycle imbalances in the Sargasso Sea. *Nature* 372:537–540.

Michaels, A. F., Olson, D., Sarmiento, J. L., Ammerman, J. W., Fanning, K., Jahnke, R., Knap, A. H., Lipschultz, F., and Prospero, J. M. (1996) Inputs, losses and transformations of nitrogen and phosphorus in the pelagic North Atlantic ocean. *Biogeochemistry* 35:181–226.

Mills, E. L. (1989) *Biological Oceanography, An Early History 1870–1890*, Cornell University Press, Ithaca, NY, pp. 43–74.

Naqvi, S. W. A., Noronha, R. J., and Reddy, C. V. G. (1982) Denitrification in the Arabian Sea. *Deep-Sea Res.* 29:459–469.

Nelson, D. M., and Treguer, P. (1992) Role of silicon as a limiting nutrient to Antarctic diatoms: Evidence from kinetic studies in the Ross Sea ice-edge zone. *Mar. Ecol. Prog. Ser.* 80:255–264.

Nixon, S., and Pilson, M. E. Q. (1983) Nitrogen in estuaries and coastal marine ecosystems. In E. J. Carpenter and D. G. Capone, eds., *Nitrogen in the Marine Environment*. Academic Press, New York, pp. 565–648.

Olson, R. J. (1981) Differential photoinhibition of marine nitrifying bacteria: A possible mechanism for the formation of the primary nitrite maximum. *J. Mar. Res.* 39:227–238.

Paerl, H. W. (1993) Emerging role of atmospheric nitrogen deposition in coastal eutrophication: Biogeochemical and trophic perspectives. *Can. J. Fish. Aquat. Sci.* 50:2254–2269.

Paerl, H. W., Prufert-Bébout, L. E., and Gou, C. (1994) Iron-stimulated N_2 fixation and growth in natural and cultured populations of the planktonic marine cyanobacteria *Trichodesmium* spp. *Appl. Environ. Microbiol.* 60:1044–1047.

Palenik, B., and Morel, F. M. M. (1990) Amino acid utilization by marine phytoplankton: A novel mechanism. *Limnol. Oceanogr.* 35:260–269.

Parsons, T. R., Maita, Y., and Lalli, C. M. (1984a) *A Manual of Chemical and Biological Methods for Seawater Analysis.* Pergamon Press, Oxford.

Parsons, T. R., Takahashi, M., and Hargraves, B. (1984b) *Biological Oceanographic Processes.* Pergamon Press, Oxford.

Paul, J. H. (1983) Uptake of organic nitrogen. In E. J. Carpenter and D. G. Capone, eds., *Nitrogen in the Marine Environment*. Academic Press, New York, pp. 275–308.

Peierls, B. J., Caraco, N. F., Pace, M. L., and Cole, J. J. (1991) Human influence on river nitrogen. *Nature* 350:386–387.

Peng, T.-H., and Broecker, W. S. (1991) Factors limiting the reduction of atmospheric CO_2 by iron fertilization. *Limnol. Oceanogr.* 36:1919–1927.

Peterson, B. J., and Howarth, R. W. (1987) Sulfur, carbon and nitrogen isotopes used to trace organic matter flow in the salt-marsh estuaries of Sapelo Island, Georgia. *Limnol. Oceanogr.* 32:1195–1213.

Phillips, R. C., and Helfferich, C., eds. (1980) *Handbook of Seagrass Biology: an Ecosystem Perspective.* Garland STPM Press, New York.

Platt, T., Jauhari, P., and Sathyendranath, S. (1992) The importance and measurement of new production. In P. G. Falkowski and A. D. Woodhead, eds., *Primary Productivity and Biogeochemical Cycles in the Sea*. Plenum Press, New York, pp. 273–284.

Redfield, A. C. (1934) On the proportions of organic derivitives in seawater and in their relationships to the composition of other geochemical tracers. In *James Johnstone Memorial Volume*. Liverpool, pp. 176–192.

Redfield, A. C. (1958) The biological control of chemical factors in the environment. *Am. Sci.* 46:205–221.

Redfield, A., Ketchum, B. H., and Richards, F. A. (1963) The influence of organisms on the composition of sea water. In M. N. Hill, ed., *The Sea*, Vol. 2. Academic Press, New York, pp. 26–77

Riley, G. A. (1967) Mathmatical model of nutrient conditions in coastal waters. *Bull. Bingham Oceanogr. Coll.* 19:72–80.

Rueter, J. G., Hutchins, D. A., Smith, R. W., and Unsworth, N. (1992) Iron nutrition of *Trichodesmium*. In E. J. Carpenter, D. G. Capone, and J. G. Rueter, eds., *Marine Pelagic Cyanobacteria: Trichodesmium and Other Diazotrophs*. Kluwer Academic, Dordrecht, pp. 289–306.

Ryther, J. H. (1969) Photosynthesis and fish production in the sea. *Science* 166:72–77.

Ryther, J. H., and Dunstan, W. N. (1971) Nitrogen, phosphorus and eutrophication in the coastal marine environment. *Science* 171:1008–1013.

Sambrotto, R. N., et al. (1993) Elevated consumption of carbon relative to nitrogen in the surface ocean. *Nature* 36:248–250.

Sarmiento, J., Hughes, T., Stouffer, R., and Manabe, S. (1998) Simulated response of the ocean carbon cycle to anthropogenic climate warming. *Nature* 393:245–249.

Schindler, D. E., Carpenter, S. R., Cole, J. J., Kitchell, J. F., and Pace, M. L. (1997) Influence of food web structure on carbon exchange between lakes and the atmosphere. *Science* 277:248–251.

Scranton, M. (1983) Gaseous nitrogen compounds in the marine environment. In E. J. Carpenter and D. G. Capone, eds., *Nitrogen in the Marine Environment*. Academic Press, New York, pp. 37–64.

Seitzinger, S. P. (1988) Denitrification in freshwater and coastal marine ecosystems: Ecological and geochemical significance. *Limnol. Oceanogr.* 33:702–724.

Seitzinger, S. P. (1993) Denitrification and nitrification rates in aquatic sediments. In P. F. Kemp, B. F. Sherr, E. B. Sheer, and J. J. Cole, eds., *Handbook of Methods in Aquatic Microbial Ecology*. Lewis Publishers, Boca Raton, FL, pp. 633–641.

Seitzinger, S., and Kroeze, C. (1998) Global distribution of nitrous oxide production and N inputs in freshwater and coastal marine ecosystems. *Global Biogeochem. Cycles* 12:93–113.

Sharp, J. H. (1983) The distributions of inorganic nitrogen and dissolved and particulate organic nitrogen in the sea. In E. J. Carpenter and D. G. Capone, eds., *Nitrogen in the Marine Environment*. Academic Press, New York, pp. 1–35.

Short, F. T. (1987) Effects of sediment nutrients on seagrasses: Literature review and mesocosm experiment. *Aquat. Bot.* 276:41–57.

Short, F. T., Dennison, W. C., and Capone, D. G. (1990) Phosphorus-limited growth of the tropical seagrass *Syringodium filiforme* in carbonate sediments. *Mar. Ecol. Prog. Ser.* 62:169–174.

Siegenthaler, U., and Sarmiento, J. (1993) Atmospheric carbon dioxide and the oceans. *Nature* 365:119–125.

Smith, S. V. (1984) Phosphorus versus nitrogen limitation in the marine environment. *Limnol. Oceanogr.* 29:1149–1160.

Smith, S. V., Kimmerer, W. J., Laws, E. A., Brock, R. E., and Walsh, T. W. (1981) Kanehoe Bay sewage diversion experiment: Perspectives on ecosystem responses to nutritional perturbations. *Pac. Sci.* 35:279–395.

Soderlund, R., and Svensson, B. H. (1976) The global nitrogen cycle. In B. Svensson and R. Soderlund, eds., *Nitrogen, Phosphorus and Sulphur — Global Cycles*. Scope Report 7 (Stockholm) *Ecol. Bull.* 22:23–73.

Suess, E. (1980) Particulate organic carbon flux in the ocean-surface productivity and oxygen utilization. *Nature* 288:260–262.

Sverdrup, H. (1953) On the conditions for vernal blooming of the phytoplankton. *J. Cons. Perm. Int. Explor. Mer.* 18:287–295.

Thomas, W. H. (1966) Surface nitrogenase nutrients and phytoplankton in the northeastern tropical Pacific ocean. *Limnol. Oceanogr.* 15:393–400.

Turner, R. E., and Rabalais, N. N. (1994) Coastal eutrophication near the Mississippi River delta. *Nature* 368:619.

Turner, S., Nightingale, P., Spokes, L., Liddicoat, M., and Liss, P. (1996) Increased dimethyl sulphide concentrations in sea water from in situ iron enrichment. *Nature* 383:513–517.

Valiela, I. and Teal, J. M. (1979) The nitrogen budget of a salt marsh ecosystem. *Nature* 280:652–656.

Villareal, T., Pilskaln, C., Brzezinski, M., Lipschultz, F., Dennett, M., and Gardner, G. B. (1999) Upwards transport of oceanic nitrate by migrating diatom mats. *Nature* 397:423–425.

Vitousek, P. M., Aber, J., Howarth, R. W. Likens, J. E., Matson, P. A., Schindler, D. W., Schlesinger, W. H., and Tilman, G. D. (1997) Human alteration of the global nitrogen cycle: Causes and consequences. *Ecol. Issues* 1:1–15.

Wada, E., and Hattori, A. (1991) *Nitrogen in the Sea: Forms, Abundances, and Rate Processes*. CRC Press, Boca Raton, FL.

Waksman, S. A., Hotchkiss, M., and Carey, C. L. (1933) Marine bacteria and their role in the cycle of life of the sea. II. Bacteria concerned with the cycle of nitrogen in the sea. *Biol. Bull.* 65:137–167.

Walsh, J. J. (1981) A carbon budget for overfishing off Peru. *Nature* 290:300–304.

Walsh, J. J. (1991) Importance of continental margins in the marine biogeochemical cycling of carbon and nitrogen. *Nature* 350:53–55.

Walsh, J. J., Rowe, G. T., Iverson, R., and McRoy, C. P. (1981) Biological export of shelf carbon is a sink of the global CO_2 cycle. *Nature* 291:196–201.

Ward, B. B. (1995) Functional and taxonomic probes for bacteria in the nitrogen cycle. In I. Joint, ed., *Molecular Ecology of Aquatic Microbes*. NATO AWI, Springer-Verlag, Berlin, pp. 73–86.

Weiss, R. F. (1970) The solubility of nitrogen, oxygen, and argon in water and seawater. *Deep-Sea Res.* 29:459–469.

Woodwell, G., Rich, P., and Hall, C., eds. (1973) *Carbon in Estuaries: Carbon in the Biosphere.* National Technical Information Service, Springfield, VA, CONF-720510.

Yoshida, N. (1988) ^{15}N-depleted N_2O as a product of nitrification. *Nature* 335:528–529.

Zehr, J. P., and Capone, D. G. (1995) Problems and promises of assaying the genetic potential for nitrogen fixation in the marine environment. *Microb. Ecol.* 32:263–281.

Zumft, W. G. (1997) Cell biology and molecular basis of denitrification. *Microbiol. Mol. Biol.* 61:533–616.

16

SYMBIOSIS AND MIXOTROPHY AMONG PELAGIC MICROORGANISMS

David A. Caron

Department of Biological Sciences,
University of Southern California,
Los Angeles, California

INTRODUCTION AND DEFINITIONS

Pelagic microorganisms display an amazing diversity and versatility in their behavior and nutrition (Chapter 2). Physical contact between two microorganisms from different species can initiate an interaction leading to an assortment of outcomes for the individuals involved. Predator–prey interactions (consumption and digestion of one individual by another) are one of the most common types of interchange between species in nature, and indeed they play an important role in structuring pelagic food webs. In addition, however, many other possible outcomes exist when organisms encounter one another. This chapter examines some of the alternatives involving pelagic microorganisms, and the consequences of these behaviors for the nutrition and survival of the participating species.

The terms "symbiosis" and "mixotrophy" have been variously defined in the literature, and their use here warrants some description. "Symbiosis" is broadly defined and refers simply to organisms from different species living together. This may involve the attachment of one microorganism to the surface of another (ectosymbiosis) or one individual living inside the other (endosymbio-

Microbial Ecology of the Oceans, Edited by David L. Kirchman.
ISBN 0-471-29993-6   Copyright © 2000 by Wiley-Liss, Inc.

sis). The possible trophic interactions resulting from these associations are varied. The relationship may be beneficial to one of the individuals but detrimental to the other (parasitism); it may be beneficial to one with no positive or negative effect on the other (commensalism); or it may be beneficial to both individuals (mutualism). These distinct types of symbioses have quite different ramifications for the individuals involved, so it is important to understand the manner in which the term is being applied.

The term "mixotrophy" also has been employed in a variety of ways, but in the broadest sense it refers to a species that obtains its nourishment by combining (or "mixing") different types of nutrition. The most popular use of this term, and the one that is examined in this chapter, refers to the combined use of photosynthetic and heterotrophic nutrition in a single organism. It is a common behavior, and there are a number of different strategies for accomplishing this task (see below).

Nonmicrobial ecologists typically view photosynthesis and heterotrophy as distinct, mutually exclusive modes of nutrition. This simplistic view probably stems from our familiarity with macroscopic organisms of terrestrial ecosystems where the common, colloquial definition of the term "plant" refers to organisms that use light and inorganic substances to produce organic matter, while "animals" use preformed organic compounds (plant and animal matter) for energy and the chemical building blocks for growth. Mixed photosynthetic and heterotrophic nutrition among single-celled, eukaryotic organisms obscures this distinction between "plant nutrition" and "animal nutrition." For this reason, the term "protist" (or "protoctist": see Margulis et al. 1990) is preferred when one is referring to these species, rather than the older terms "protozoa" (which refers to heterotrophic, single-celled eukaryotes) and "microalgae" (which refers to photosynthetic, single-celled eukaryotes). Nevertheless, the older terms are still useful for brevity in describing the dominant (or in some cases, sole) nutritional mode of a protistan species, and therefore they are employed in this chapter. The shortcomings of these historical definitions will become apparent as the nutritional versatility and complexity of these species are made clear.

SYMBIOSIS

Symbiosis (mutualism, commensalism, parasitism) plays a significant role in establishing and maintaining the structure of marine pelagic communities. This situation is in contrast to the notion that symbiosis is not overly important in structuring freshwater communities (Lampert and Sommer 1997). It is not completely clear why this dichotomy exists, although extreme oligotrophy, intense competition for growth-limiting substances, and a long history of coexistence and adaptation in marine environments may contribute to this circumstance. In addition, a large number of protozoan taxa that exist in oceanic ecosystems and form mutualistic associations with photosynthetic microorganisms simply do not inhabit freshwater plankton communities (e.g., planktonic foraminiferans and larger actinopods such as radiolarians and acantharians).

Parasitism

Numerous examples of parasitism involving marine planktonic organisms have been reported (Théodorides 1987), but many of these relationships are poorly characterized, and many others probably exist that have not yet been described. Therefore, drawing firm conclusions regarding the importance of parasitism in establishing and maintaining microbial diversity in the ocean is premature at this time. Nevertheless, considerable speculation now surrounds the role played by microbial parasites and other infectious agents in shaping pelagic food web structure. A better ecological understanding of these interactions will emerge as the details of these relationships become clear.

The distinction between "parasite" and "pathogen" among microorganisms is sometimes a difficult one. Parasites obtain nutrition from their hosts, causing debilitation in the process (e.g., loss of vigor, reduced reproductive output) but not necessarily death of the host. Pathogens produce disease, and their actions often lead to the demise of their hosts. For example, viruses typically are considered pathogens because they infect and kill susceptible hosts. Many "parasites" of microorganisms, however, also cause the death of their hosts, so drawing the line between these behaviors is problematic. Ecologically speaking, both pathogens and parasites are predators because they exploit the living cytoplasm, tissue, or cellular machinery of other species for their own growth and reproduction.

The caveat above notwithstanding, recent synopses of virus activity in marine pelagic ecosystems have concluded that they are potentially important sources of bacterial and cyanobacterial mortality in the plankton (see Chapter 11). Viruses that infect protists are also common in aquatic ecosystems, although they have not received as much attention as cyano- and bacteriophages (Brown 1972; Van Etten et al. 1991). A clear understanding of the overall magnitude of virus-related mortality in energy and carbon flow through microbial plankton communities is just emerging, and over the past decade studies have begun to incorporate these processes into models of pelagic ecosystem function (Bratbak et al. 1992; Thingstad et al. 1993).

Prokaryote mortality as a result of viral infection and lysis removes susceptible host cells from the population. Resistant individuals persist and reproduce. Therefore, in addition to directly reducing host population density, viruses play a role in controlling the diversity and species composition within prokaryote assemblages by selectively eliminating susceptible strains or species (Waterbury and Valois 1993). Furthermore, viruses may be a mechanism for transferring genetic information between host cells if the hosts are not killed in the process of infection and viral release.

Parasitic protozoa from pelagic environments infect other protists as well as metazoa. Parasites of planktonic organisms have been well known for many years (Cachon and Cachon 1987; Théodorides 1987). Like viruses, these species exhibit a fair degree of specificity in host selection and often result in the death of the host. "Parasitoids" of the flagellate genus *Pirsonia* prey on diatoms and are attracted chemotactically to their preferred hosts, to which they attach. They then penetrate the host cell and phagocytize and digest its cytoplasm, producing more infective offspring in the process (Schnepf and Schweikert 1997; Kühn 1998). A similar life cycle exists for the parasitic dinoflagellate *Amoebophrya ceratii*, which infects and kills photosynthetic dinoflagellates of coastal ecosystems (Coats et al. 1996). Infection rates of susceptible host populations at times can reach epidemic levels (20–80% of available hosts parasitized).

The effects of parasitism on a host population can range from relatively benign to devastating. In the former situation, minor losses to the host population (or loss of vigor of individuals within the population) may play only a minor role in affecting overall success of the host species in nature. In other situations, massive infections and high mortality can have catastrophic consequences for the host population. This outcome is more likely in situations where viruses can be effectively transferred between susceptible hosts, such as during nearly monospecific blooms of phytoplankton. Viruses have been implicated in the demise of some phytoplankton blooms in coastal waters that are strongly dominated by a single species (Gastrich et al. 1998).

Commensalism and Mutualism Involving Prokaryotes

A great variety of nonpredatory/nonparasitic relationships exist that involve marine prokaryotes. Many heterotrophic bacteria, for example, participate in nutritional or metabolic "consortia" in which waste product(s) produced by one species serve as substrate(s) for another bacterial species. The metabolic interplay in a number of these relationships from benthic environments has been characterized. The details of substrate use and waste product release are poorly understood in the plankton, but assemblages of heterotrophic bacteria in the water column presumably process organic matter "collectively," leaving largely recalcitrant compounds in the pool of dissolved organic material (see Chapters 5 and 6).

The nutritional–metabolic relationships among bacteria described above are not generally referred to as symbioses, although mutualism is implied in the utilization of available substrates. The bacterial species must live in close (but not necessarily intimate) proximity to one another to facilitate cometabolism. The efficient exchange of metabolites enables them to survive and grow, and species from these consortia typically cannot be cultured separately. Once the metabolic relationships between members are determined, however, individual species can sometimes be grown in pure culture by supplementing the medium with specific metabolites.

At the other end of the spectrum of these collaborative relationships are species whose physical proximity and biochemical integration are so close that independent existence is not possible or is limited to only one of the participants. Extreme examples of this degree of interaction are the chloroplasts and mitochondria of eukaryotic organisms. It is now widely accepted that these organelles have prokaryote origins and have arisen from endosymbiotic events that led to genetic–biochemical incorporation and reduction of the symbiont within the cytoplasm of the host (Margulis 1981).

Between these extremes (loosely interacting autonomous species vs organelle acquisition), there is a huge variety of mutualistic and commensal interactions involving prokaryotes. For example, bacteria colonize the external surfaces (as well as the gut) of virtually every organism that does not actively deter them. Most of these epibiotic–enteric relationships are presumably commensal in that the bacteria may gain some nutritional benefit from the host, or perhaps refuge from predation, with little or no effect on the host. Reviewing the entire spectrum of these relationships is beyond the scope of this chapter, but a few examples provide insight into these physical and metabolic interactions among pelagic marine microorganisms.

Heterotrophic Bacteria

Bacterial–protozoan symbioses are common among parasitic and free-living protozoan species from hypoxic–anoxic ecosystems. In anoxic environments, protozoa can harbor ecto- and/or endosymbiotic bacteria. Much of our knowledge of the existence of these relationships has been obtained from studies of free-living benthic protists, or the microbial flora of such exotic microenvironments as the hindguts of termites and cockroaches. Investigations of cultured species of these protozoa have been useful in establishing the nature of some of these associations (for a review, see Fenchel and Finlay 1995).

Symbioses involving luminescent bacteria are excellent examples of highly developed mutualistic relationships between heterotrophic bacteria and higher organisms in aerobic ecosystems. Luminescence symbioses exist in teleost fish, some cephalopod mollusks, and a few species from diverse phyla (e.g., urochordates and nematodes). Complex morphological adaptations of the hosts exist to accommodate the symbiotic bacteria, and a high degree of specificity exists between host and bacterial symbiont (reviewed in Douglas 1994). The benefits conferred on the host through these relationships include (among other possible factors) species/sex recognition and prey attraction. In turn, benefit to the bacterial symbiont may include refuge from predation and a favorable environment for growth. Studies are just now beginning to unravel the genetic and biochemical bases for establishment and maintenance of these exclusive associations (McFall-Ngai 1998). This work has demonstrated that during the ontogeny of hosts, "competent" bacteria elicit specific developmental changes that are not observed in symbiont-free hosts.

Photosynthetic Prokaryotes

Symbioses between photosynthetic prokaryotes (cyanobacteria) and sponges are common in warm-water benthic ecosystems (Wilkinson 1987). Cyanobacteria also form symbioses with planktonic microorganisms, although these associations are not as common as symbioses involving protistan phototrophs (e.g., photosynthetic dinoflagellates; see below). In some cases, however, symbiotic cyanobacteria may play a unique role in relieving nitrogen limitation. Nitrogen fixation by cyanobacteria is potentially a major source of nitrogen in oligotrophic environments (Karl et al. 1997). Chapters 13 and 15 provide more information on this topic.

Ectosymbiotic associations between cyanobacteria and heterotrophic protists are often observed in surface waters of the open ocean (Figure 1A, B). In particular, a number of genera of heterotrophic dinoflagellates (e.g., *Ornithocercus, Histioneis, Citharistes*) typically possess chroococcoid cyanobacteria attached to the host's surface at specific locations (Gordon et al. 1994). Different hosts often harbor morphologically distinct cyanobacteria, so these relationships have some degree of species specificity. In some cases, invagination of the cell wall of the host creates a special pocket, originally termed the "phaeosome," where the cyanobacteria reside. These relationships are assumed to be mutualistic and are thought to be based on the exchange of specific metabolites.

It has been proposed that the association of cyanobacteria with their hosts enables the establishment of anoxic microenvironments, and these in turn allow nitrogen fixation by the cyanobacteria. Indeed, the abundance of symbiont-bearing heterotrophic dinoflagellates has been correlated with low inorganic nitrogen availability in the plankton (Gordon et al. 1994). Translocation of photosynthate to the host is suspected, and starchlike grains in some host organisms suggest that the host is acquiring photosynthetically derived nutrition from its symbionts. Evidence of feeding (presence of food vacuoles) by some of the hosts has been noted, and these food vacuoles may include the symbiont species as well. This has been proposed as a mechanism for regulating symbiont number, and of course as a source of nutrition for the host.

An example of mutualism between a photosynthetic prokaryote and a photosynthetic eukaryote is the cyanobacterium *Richelia intracellularis* with several diatom genera (Villareal 1989, 1994). This association presumably constitutes a "nutritional collaboration" in which the cyanobacteria provide organic nitrogen to the association via nitrogen fixation. Benefit to the cyanobacterium may be nutritional, but reduced predation due its presence within the frustule of the host also may have survival value (Figure 1C, D).

Commensalism and Mutualism Involving Protists

Numerous protistan species engage in ecto- and endosymbioses with a wide diversity of protists and metazoa (Table 1). Many of the ectosymbioses are

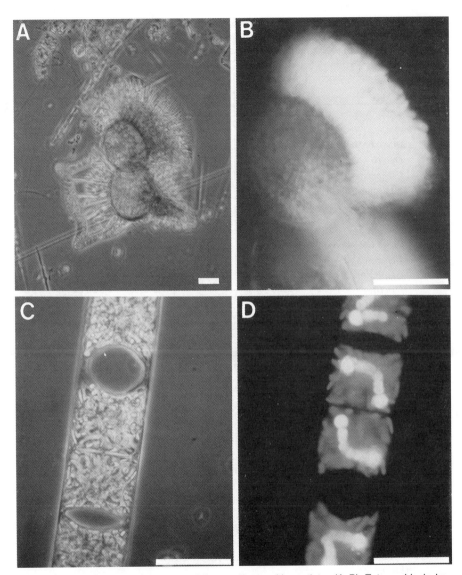

Figure 1. Photosynthetic cyanobacterial mutualisms with protists. (A, B) Ectosymbiosis between the heterotrophic dinoflagellate *Ornithocercus* sp. and a chroococcoid cyanobacterium.
A micrograph of two dinoflagellates (A) shows the ornate structures of this species, while
attached cyanobacteria (brightly-fluorescing cells) are visible in the cingular list region using
epifluorescence micrography (B). (C, D) Endosymbiosis of the cyanobacterium *Richelia intracellularis* with the diatom *Hemiaulus* sp. The symbiont is not apparent within the diatom frustules
under transmitted white light (C) but can be easily demonstrated by epifluorescence microscopy
(D), owing to differences in photosynthetic pigment fluorescence (the brightly fluorescent
chain-forming cyanobacterium fluoresces yellow-orange with blue light excitation, while the
weakly fluorescent diatom fluoresces red). Marker bars 40 μm. (Color versions of this and the
other photomicrographs can be seen at http://www.wiley.com/products/subject/life/kirchman.).

Table 1. Some taxa and representative genera of heterotrophic
protists known to harbor algae in symbiotic (presumably
mutualistic) relationships; references are pertinent summaries
or reviews

Organisms	Ref.
Ectosymbionts	
DINOFLAGELLATES	
Ornithocercus *Citharistes* *Histioneis*	Gordon et al. (1994)
Endosymbionts	
DINOFLAGELLATES	
Peridinium *Glenodinium* *Noctiluca* *Lepidodinium*	Schnepf and Elbrächter (1992)
PLANKTONIC FORAMINIFERA	
Globigerinoides *Orbulina* *Globigerinella* *Globorotalia* *Globoquadrina*	Hemleben et al. (1988), Caron and Swanberg (1990)
POLYCYSTINE RADIOLARIA	
Thalassicolla *Collozoum* *Physematium* *Collosphaera* Numerous other genera	Anderson (1983), Caron and Swanberg (1990)
ACANTHARIA	
Dorataspis *Haliommatidium* *Lithoptera* *Amphilonche* Numerous other genera	Caron et al. (1995)
CILIATES	
Paramecium *Platyophora* *Euplotes* Numerous other genera	Reisser (1986), Dolan (1992)

commensalisms involving the attachment of heterotrophic protists (protozoa) to phytoplankton (Figure 2A) or other heterotrophs, or the association of phytoplankton with the structures of other phytoplankton or heterotrophs (Figure 2B–D). These epizoic (on animals) or epiphytic (on phytoplankton) symbionts presumably gain some benefit from the association with their hosts (e.g., exposure to high nutrient concentrations, or accessibility to high abundances of microbial prey), while hosts appear to be unaffected by their presence. Many of these relationships are probably opportunistic rather than highly specialized, species-specific associations, but there is a tendency for particular host species to be observed with ectocommensals.

Endosymbioses involving phototrophic protists and benthic cnidarians are well-known associations in tropical–subtropical seas, where they contribute to coral reef formation. A number of different species from a variety of algal classes also form endosymbiotic relationships with heterotrophic protists and metazoa in the plankton (Figure 3). Species of chlorophytes, prymnesiophytes, prasinophytes, diatoms, and especially dinoflagellates have been observed in apparently mutualistic associations with a wide diversity of pelagic, heterotrophic species including many protozoa, medusae, siphonophores, and flatworms (Anderson 1983; Hemleben et al. 1988; Stoecker et al. 1989c). Symbiont abundances within these hosts can be quite high, particularly for some protozoa. Solitary radiolarians and planktonic foraminiferans can contain up to several tens of thousands of symbiotic algae per individual protozoan.

In contrast to protistan ectosymbioses, endosymbiotic associations tend to be rather specific and carefully regulated. Most hosts accommodate only a single algal species as endosymbionts, although a single algal species may be an acceptable symbiont to a variety of host species. For example, at least four species of planktonic foraminiferans maintain the same photosynthetic dinoflagellate, *Gymnodinium beii*, as an endosymbiont, while several species of radiolarians maintain a different species of dinoflagellate, *Scrippsiella nutricula* (Gast and Caron 1996). Several dozen freshwater ciliated protozoa harbor *Chlorella*-like symbionts (Dolan 1992). Therefore, some algae appear to be predisposed to symbiotic associations with a variety of host species.

Some notable exceptions to the exclusive relationship between host and symbiont, however, raise questions concerning how these associations are established. For example, turbellarian flatworms have been observed with at least three different types of endosymbiotic alga (Stoecker et al. 1989c), and one species of planktonic foraminiferan forms mutually exclusive relationships with one of two co-occurring algal species (Faber et al. 1988). It is possible that some of these relationships can be explained by the existence of different host species that are not morphologically distinct (so-called cryptic species), but other striking examples exist of cases of multiple symbiont species within a single host. Some benthic foraminiferans can establish symbioses with a variety of diatom species (Lee et al. 1989). Two, and sometimes three endosymbiotic diatom species have been observed in a single protozoan host (Lee et al. 1980). Moreover, "artificial" associations have been induced. Rogerson et al. (1989)

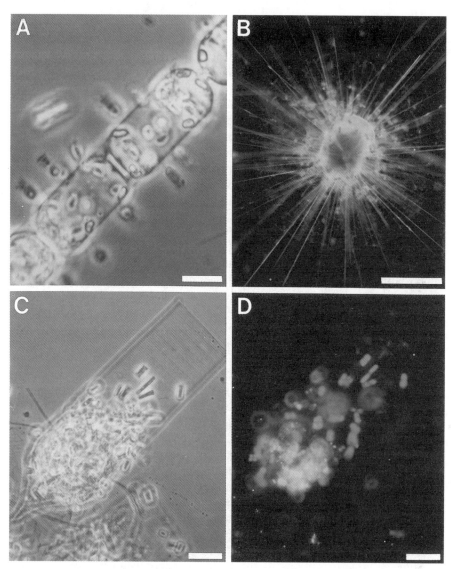

Figure 2. Commensal relationships among protistan species in the plankton. (A) Heterotrophic flagellates attached to the frustule of a chain-forming diatom from Vineyard Sound, Massachusetts. (B) Photosynthetic dinoflagellates of the genus *Pyrocystis* (Elbrächter and Drebes 1978) embedded in the fluid bubble capsule of the oceanic planktonic foraminifer *Hastigerina pelagica*. (C) A tintinnid ciliate from the Ross Sea, Antarctica, with diatoms attached to its lorica. Autofluorescence of chlorophyll (D) indicates the location of the diatoms. Marker bars, 20 μm (A, C, D) and 500 μm (B).

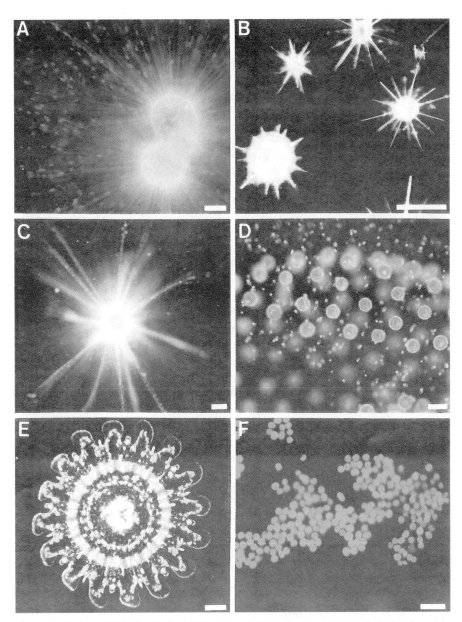

Figure 3. Endosymbiotic "mutualisms" involving photosynthetic protists and planktonic protozoa and metazoa. (A) The planktonic foraminifer *Orbulina universa* and its intracellular dinoflagellate symbionts. (B) Several Acantharia appear yellow-brown because endosymbiotic algae are present. (C) An unidentified radiolarian with its endosymbiotic algae aggregated in the dense pseudopodial network of the host. (D) The colonial spumellarian radiolarian *Collosphaera* sp. with its intracellular dinoflagellate symbionts (small bright dots). (E) A juvenile specimen of the scyphomedusan *Linuche unguiculata* showing dense accumulations of symbiotic algae in its lobes (F) The same specimen as (E) viewed at high magnification by means of epifluorescence microscopy to show the presence of photosynthetic algae (numerous autofluorescent spheres) in a lobe of its umbrella. Marker bars, 50 μm (A, F), 200 μm (B–D), and 1 mm (E).

induced an association between the marine amoeba *Trichosphaerium* and a strain of dinoflagellate (*Symbiodinium*) isolated from an anemone. The association between the dinoflagellate and the amoeba has never been observed in nature.

These examples demonstrate that there is still much to be learned regarding how endosymbioses come about and how they are maintained. The exclusivity of these associations has been attributed to species-specific biochemical and/or genetic interactions between host and symbiont. Research using a ciliate–chlorophyte model (*Paramecium–Chlorella*) indicates that the processes following phagocytosis of symbionts differ from processes for other ingested microorganisms (Reisser 1991). Symbiosis-competent algae somehow alter the phagosomal membrane such that it does not fuse with lysosomes that contain digestive enzymes. Thus, vacuoles with symbionts do not suffer the normal fate of digestion and eventual expulsion from the cell. The exact nature of these interactions is not yet clear.

Endosymbionts may escape digestion and expulsion from suitable hosts, but they do not often remain unchanged. Diatom and dinoflagellate symbionts typically lose their characteristic morphologies in the host cytoplasm. Dinoflagellates do not form thick cell walls in the host, and diatoms do not form silica frustules. This is a reversible process because the algae again form these structures if they are removed from their hosts and grown in their free-living state. Endosymbionts also exhibit physiological changes in their hosts. Experimental studies have demonstrated that algae become "leaky" when they enter into endosymbiotic relationships, often releasing large percentages of photosynthate into the host cytoplasm. This process of translocating photosynthetic products can constitute a substantial portion of the host's nutrition (Caron et al. 1995).

The transmission of endosymbionts between generations of hosts varies with the host taxon. Asexually reproducing species (e.g., some protozoa) can pass symbionts directly to daughter cells resulting from binary fission. Some sexually reproducing species also transmit an "inoculum" to their progeny. Other species depend on reestablishment of the symbiosis each generation. Planktonic foraminiferans and actinopods, for example, reproduce via the release of large numbers of swarmer cells that are smaller than the algae that they harbor as endosymbionts. These organisms must reacquire algae from the environment during their ontogeny.

Ecological Benefits of Protistan Mutualisms

The ecological benefits to the partners of phototroph–heterotroph mutualisms are rather self-evident. The heterotrophic host obtains organic carbon for energy and/or growth from symbiont photosynthesis. Symbiotic algae often have very high rates of primary production and, as mentioned, they can release a large proportion of the total photosynthetically fixed carbon into the cytoplasm of their hosts (Anderson et al. 1983: Reisser 1986; Caron et al. 1995).

Endosymbionts also may contribute directly to the nutrition of their hosts by "cropping" of these algae by the host. Anderson (1983) noted that symbiont abundances in many radiolarian species appear to remain constant despite algal division, implying that symbiont digestion by the host may keep pace with symbiont growth.

Other possible benefits to the host include the removal (utilization) of some of the host's metabolic wastes (which may be used as nutrients for phototrophic growth by the algae) or providing a light-sensing function to the host. Interestingly, it has been reported that the symbiont-bearing planktonic foraminiferan *Globigerinoides sacculifer* is incapable of normal ontogenic development in continuous darkness and attempts to undergo reproduction (gamete formation) within approximately 48 hours (Caron et al. 1982). It is unknown how this light effect is brought about. Further evidence for light-mediated behavior in photosymbioses is the diel movement of symbiotic algae in the cytoplasm of some protozoa and in the tissue of some cnidarians (Bé 1982; Kremer et al. 1990).

The overall benefit of endosymbiosis to a host appears to be rather different for different host–symbiont associations. These differences often are manifested as varying dependences on light and particulate food. The degree to which hosts require the presence of endosymbiotic algae also varies among host species. For example, photosymbioses in some freshwater ciliated protozoa are "facultative." These ciliate species possess algal endosymbionts only during certain periods of the year (Berninger et al. 1986). The photosymbiotic relationships of some marine planktonic foraminiferans and actinopods, however, appear to be obligatory. Normal growth and development does not proceed in their absence (Bé et al. 1982).

Two possible benefits are usually cited for algae involved in endosymbioses. First, the cytoplasm of the host constitutes a microenvironment in which nutrient concentrations are substantially higher than in the surrounding water. This situation is especially true in oligotrophic oceanic ecosystems. Nitrogen and phosphorus wastes (and CO_2) produced by the host thus serve as important sources of major nutrients for endosymbiont primary production. Studies with corals have shown that symbiont-bearing corals placed in the light released less ammonium and phosphate than corals with depleted numbers of algae. These results imply efficient retention of nutrients within these photosymbioses. Prey capture provides part of the diet of the host directly, and remineralized nutrients and carbon resulting from prey digestion are recycled by the symbiotic algae for their own growth, and to augment the diet of the host.

A second possible benefit for the symbiotic algae in endosymbioses is that the host may constitute a refuge from the intense predation pressure on microorganisms in the microbial food web of the water column. This benefit is predicated on growth within the host's tissue/cytoplasm and eventual release to the external environment. This outcome is largely dependent on the specific relationship between host and symbiont, however, and is not always realized.

The abundance of the dinoflagellate symbionts in the planktonic foraminiferan *G. sacculifer* increases steadily during ontogeny of the host (except under extreme starvation conditions), indicating an intracellular environment conducive to symbiont growth and survival. However, the symbionts are digested en masse at the onset of reproduction of this foraminifer (Bé et al. 1983). Obviously, there is no net ecological benefit of this association to the dinoflagellates if they do not escape, so this relationship cannot be strictly considered mutualism.

This particular foraminiferan symbiosis may represent a mutualism in which the host has become exploitative. Mainero and del Rio (1985) noted that "once a mutualistic relationship has arisen, the appearance of cheaters becomes highly probable." Perhaps other host species of this symbiotic dinoflagellate release their symbionts at the onset of host reproduction thereby compensating for losses in the *G. sacculifer* association and resulting in an overall net benefit for endosymbiotic existence of this dinoflagellate.

MIXOTROPHY AMONG PROTISTS

Mixotrophic nutrition, as defined at the beginning of this chapter, is the combined use of phototrophic and heterotrophic nutrition in a single organism. There is a useful distinction here between photosynthetic protists (algae) that are capable of heterotrophic nutrition and heterotrophic protists (protozoa) that acquire photosynthetic ability via the retention of chloroplasts from algal prey.

Phagotrophic Algae

The ability of microscopic phototrophs to use organic material for growth may seem like a remarkable behavior, and indeed the extent to which some phytoplankton are heterotrophic is extraordinary. However, there are some analogies among terrestrial plants. The Venus fly trap is a carnivorous plant that produces plant material by photosynthesis, but also obtains part of its nutrition from the capture and digestion of insects.

There is a long history to the study of microalgal heterotrophy. More than half a century ago it was recognized that a number of marine phytoplankton species could not be cultured axenically (i.e., free of all other living organisms) on completely inorganic media. These phytoplankton species, termed "auxotrophs," were capable of synthesizing most organic substances necessary for their growth but required the presence of other microorganisms (usually bacteria) in their growth medium. The bacteria produced one or more vital "growth factors" (e.g., vitamins, specific lipids) that were taken up by the algae and thus permitted their growth. In return the phytoplankton presumably produced organic compounds that were utilized by the bacteria as substrate.

This loose metabolic mutualism between bacteria and phytoplankton is presumably an important ecological aspect in the coexistence of many of these species (Provasoli and Pintner 1953).

Identification of the organic requirements of marine phytoplankton species allowed many phytoplankton species to be grown axenically in the laboratory by supplementing the culture medium with these substances. Most phytoplankton were able to obtain the organic substances directly from the solution through the process of osmotrophy (direct uptake through the cell membrane). This behavior led to the notion that the organic requirements of most phytoplankton in nature were probably met by osmotrophy, and the significance of phagotrophy as a possible nutritional mode for phytoplankton was not recognized for many years.

It is now clear that a large number of phytoplankton species are capable of meeting part of their nutritional requirements via the ingestion of particulate material (Table 2). This list continues to grow. For example, estimates a decade ago indicated that perhaps 20% of phototrophic dinoflagellates may ingest particulate food (Gaines and Elbrächter 1987). More recent discoveries of dinoflagellate mixotrophic activity by species that were believed to be obligately autotrophic suggest that 20% may have been a low estimate (Bockstahler and Coats 1993; Jacobson and Anderson 1996; Li et al. 1996).

Phagotrophic algae have been noted most commonly in freshwater ecosystems (Bird and Kalff 1986; Bennett et al. 1990; Berninger et al. 1992). These taxa dominate the phytoplankton communities of some lakes at specific times and depths. More recently, studies in pelagic marine communities have begun to demonstrate the presence and occasional high abundances of phagotrophic algae. Abundances of phagotrophic algae ranging up to over 50% of the total abundance of nanoplanktonic phytoplankton have been reported, and rates of bacterivory that can exceed the contribution of heterotrophic flagellates have been observed (Hall et al. 1993; Arenovski et al. 1995; Li et al. 1996; Havskum and Hansen 1997). These reports indicate that a significant portion of the phytoplankton assemblage is capable of particle ingestion. Many of these algae are so small that they are capable of ingesting prey only the size of bacteria and cyanobacteria (Figure 4E, F), but some photosynthetic dinoflagellates also consume phytoplankton and small ciliates ($> 10 \, \mu m$).

Phagotrophic behavior by algae presents several possible benefits for the survival and growth of the alga. These include the acquisition of organic carbon (energy) to supplement poor phototrophic ability and the acquisition of macronutrients (nitrogen, phosphorus) for photosynthetic growth or relief from "micronutrient" limitation. These latter materials might be organic (e.g., vitamins, specific lipids) or inorganic (e.g., iron). Interestingly, there is evidence to support all these possibilities, so it appears that different species of phagotrophic algae conduct heterotrophy to satisfy different nutrition requirements. These differing strategies result in a nearly continuous spectrum of nutrition between the extremes of absolute autotrophy and absolute heterotrophy (Figure 5).

Table 2. Some phytoflagellate genera reported to ingest particles[a]

"Chrysophyceae"[b]	
Catenochrysis	Epipyxis
Chromulina	Ochromonas
Chrysamoeba	Palatinella
Chrysococcus	Pedinella
Chrysosphaerella	Phaeaster
Chrysostephanosphaera	Heterosigma
Cyrtophora	Chlorochromonas
Dinobryon	

Prymnesiophyceae	
Chrysochromulina	Prymnesium
Cocolithus	

Dinophyceae	
Amphidinium	Protoodinium
Ceratium	Dissodinium
Gymnodinium	Gonyaulax
Gyrodinium	Scrippsiella
Massartia	Prorocentrum
Fragilidium	Peridinium
Dinophysis	Alexandrium
Heterocapsa	Blastodinium

Cryptophyceae	
Cryptomonas	Chroomonas

[a]This list does not include heterotrophic protists that ingest algae and retain their chloroplasts in a function state.

[b]Includes Synurophyceae, Dictyochophyceae, Pelagophyceae, Xanthophyceae, and Raphidophyceae, according to Anderson et al. (1998).

Source: Primarily from a summary by Sanders and Porter (1988), with additional information (Schnepf and Elbrächter 1992; Jacobson and Anderson 1996; Skovgaard 1996; Jeong et al. 1997; Tillman 1998).

Poterioochromonas malhamensis is an example of a phagotrophic alga that is predominantly heterotrophic. This alga grows heterotrophically as long as sufficient bacteria are present to support its growth. The alga is even capable of growth in continuous darkness. Photosynthesis is virtually abandoned during active heterotrophic growth (even in the light), and reactivation of the chloroplast does not occur until bacterial abundance drops below a threshold value (Caron et al. 1990; Sanders et al. 1990). Photosynthesis in this species

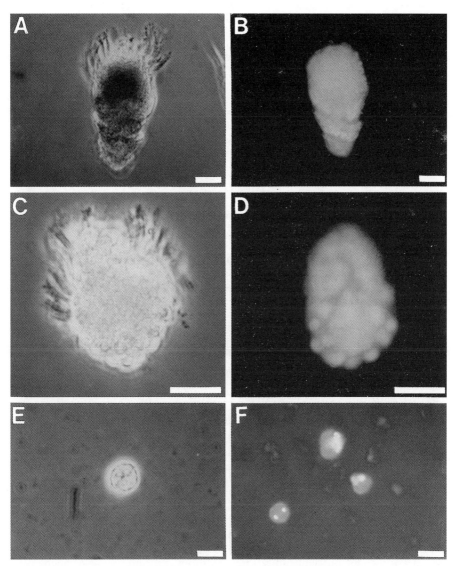

Figure 4. Mixotrophy among protists. Chloroplast retention by protozoa (A–D) and prey ingestion by phagotrophic algae (E, F). The planktonic ciliates *Laboea* sp. (A) and *Mesodinium* sp. (C) exhibit extensive autofluorescence with blue light excitation (B and D, respectively) as a result of presence of large numbers of functional chloroplasts in their cytoplasm. The ability of a pedinellid phytoflagellate (E) to consume particulate food is demonstrated using epifluorescence microscopy by the uptake of fluorescently labeled bacteria (F; brightly fluorescent particles). Marker bars, 25 μm (A, B), 20 μm (C, D) and 10 μm (E, F).

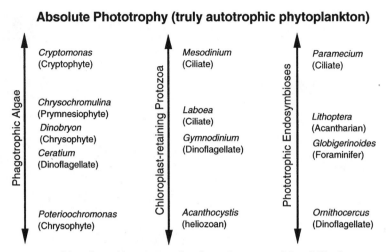

Figure 5. Depiction of various modes of protistan nutrition as a continuum between absolute autotrophy (phototrophy) and absolute heterotrophy. Representative genera of protists (and their taxonomic affiliation) that display these behaviors are indicated on the figure in locations that attempt to approximate their degree of phototrophic/heterotrophic tendency. (Adapted and modified from Jones 1994.)

appears to represent a strategy for survival and perhaps slow growth during periods when prey are unavailable.

At the other end of the spectrum are algae such as some *Cryptomonas* species for which phagotrophic ability has been demonstrated but rarely observed in nature or in the lab (Tranvik et al. 1989). These species appear to depend strongly on photosynthesis for their nutrition, and they are usually incapable of growth in the dark.

Many algal species exist between these extremes, employing mixotrophy to obtain a competitive advantage over purely phototrophic and heterotrophic species. Several phagotrophic algae have been shown to increase their rates of bacterivory under nutrient limitation conditions, presumably as a mechanism for obtaining growth-limiting elements (Nygaard and Tobiesen 1993; Legrand et al. 1998). This explanation is supported by field observations that have demonstrated that phagotrophic algae can occasionally be a significant fraction of the phototrophic protistan assemblages of oligotrophic ocean gyres (Arenovski et al. 1995).

Species of the chrysophyte genus *Dinobryon*, common in both freshwater and coastal marine ecosystems, appear to employ phagotrophy for a variety of reasons. Growth of *Dinobryon divergens* was stimulated under light or iron limitation (Veen 1991). These experimental results and others (Maranger et al. 1998) indicate the relief of limitation (light-limited or trace-metal-limited photosynthesis) via bacterial ingestion. Phosphorus acquisition also has been implicated as the impetus for phagotrophy in field studies of *Dinobryon divergens* (Veen 1991), and heterotrophic supplementation of light-limited photosynthetic growth was indicated in laboratory and field studies of bacterial ingestion by *Dinobryon cylindricum* and *Dinobryon balticum* (Bird and Kalff 1986; Caron et al. 1993; McKenzie et al. 1995). Acquisition of a specific growth factor also appeared to play a role for *D. cylindricum* (Caron et al. 1993). Procurement of specific organic phosphorus compounds has been proposed as an explanation for phagotrophy in another common chrysophyte, the freshwater species *Uroglena americana* (Kimura and Ishida 1986, 1989).

Regardless of the underlying ecological basis for the behavior, phagotrophy by marine phytoplankton constitutes a fundamental departure from the concepts of phototrophy and heterotrophy as nonoverlapping nutrition modes. These species are capable of obtaining nutrition and/or growth factors by means not available to purely autotrophic or heterotrophic protists. The magnitude and ecological significance of this behavior has only recently come to light and is now making its way into theoretical analyses of food web dynamics (Thingstad et al. 1996).

Chloroplast-Retaining Protozoa

Phagotrophic algae are chloroplast-bearing species of protists that have acquired (or retained) the ability to ingest food. Using vernacular terminology, they are "animal-plants," truly photosynthetic protists (algae) that possess the ability to consume particulate food. Based on the endosymbiotic theory of organelle acquisition (see Chapter 2), the existence of chloroplasts in these species represents ancient phagocytic events by heterotrophic protists that have become stable associations between the protists and their symbiotic phototrophs (now the chloroplasts) (Margulis 1981). Many relationships between protozoa and ingested photosynthetic microorganisms exist that have not progressed to the degree that they form stable protist–organelle dependencies. Chloroplast retention by protozoa are such relationships.

Chloroplast retention (used here as synonomous with "chloroplast symbiosis," "plastid symbiosis," and "chloroplast enslavement") is accomplished when a microscopic alga is ingested and digested by a heterotrophic protist and the chloroplasts of the phototroph are retained by the heterotroph in a functional state (Figure 4A–D). In essence, the heterotroph is able to act as a phototroph by virtue of its newly acquired photosynthetic capability. By analogy to the terminology applied above to phagotrophic algae, chloroplast-retaining protozoa are "plant-animals." These relationships are often transient because chloroplast function is not wholly integrated into the new owners' metabolism. Nevertheless, these relationships convey a significant short-term benefit to the heterotroph.

The phenomenon of chloroplast retention was first described in marine mollusks (Trench 1975), but it is now widely recognized as a common occurrence in many planktonic (and benthic) ciliates from marine and freshwater ecosystems, some benthic foraminiferans (Lopez 1979), heliozoa (Patterson and Dürrschmidt 1987), and a number of heterotrophic dinoflagellates (Table 3). The phenomenon is best known among ciliates, where chloroplast retention occurs in a variety of different taxa, and most importantly in several cosmopolitan oligotrichous ciliates (e.g., the genera *Strombidium*, *Laboea*, *Lohmaniella*, *Tontonia*) and the haptorid ciliate *Mesodinium rubrum*. These latter species occasionally constitute a major portion of the total ciliate fauna of plankton communities, and significant fractions of the standing stock of chlorophyll and primary production. *M. rubrum* has been documented at very high abundances in the plankton, where it can cause "red tides" (Crawford 1989). Collectively, chloroplast-retaining ciliates often make up more than 50% of the total ciliate fauna in a variety of estuarine, coastal, and oceanic environments (Stoecker et al. 1987, 1989a, 1996; Putt 1991; Bernard and Rassoulzadegan 1994).

Heterotrophic dinoflagellates probably compose the next most important group of chloroplast-retaining protozoa in the plankton. The term "cleptochloroplasts" (or "kleptochloroplasts"; see Skovgaard 1998) is sometimes applied to these organelles (Schnepf and Elbrächter 1992). The phenomenon of chloroplast retention probably is more common among dinoflagellates than is presently known. Only relatively recently has the extent of heterotrophic nutrition been recognized within this classic "phytoplankton" taxon (Gaines and Elbrächter 1987; Lessard 1991). Erroneous description of heterotrophic dinoflagellates with cleptochloroplasts as "photosynthetic" dinoflagellates

Table 3. Some taxa and genera of heterotrophic protists known to consume algae and retain the chloroplasts of their prey in a functional state

Protists	Ref.
HELIOZOA	
Acanthocystis	Patterson and Dürrschmidt (1987)
Raphidocystis	
Chlamydaster	
DINOFLAGELLATES	
Gymnodinium	Schnepf and Elbrächter (1992)
Amphidinium	
Dinophysis	
CILIATES	
Strombidium	Stoecker et al. (1987, 1991)
Laboea	Stoecker et al. (1988, 1989b)
Lohmaniella	Crawford (1989)
Tontonia	
Mesodinium	

seems plausible. Variable numbers of chloroplasts reported for some dino-flagellate species has been taken as an indication that some of these instances may represent as yet undocumented cases of chloroplast retention by hetero-trophic dinoflagellates (reviewed in Schnepf and Elbrächter 1992).

The general rule for chloroplast-retaining protozoa is that these rela-tionships are not as stable as "true" chloroplasts are in algae. That is, photosynthetic ability of the retained chloroplasts is usually a transient situation. Chloroplast function is gradually lost, and their number slowly decreases in most species that are not provided with prey. Retention times vary. Chloroplasts in the dinoflagellate *Gymnodinium "gracilentum"* re-mained photosynthetically active for a few days and were then lost; but longer retention and activity times have been reported for other dino-flagellate species (Fields and Rhodes 1991) and for other chloroplast-retaining taxa. Chloroplasts acquired from prey by the benthic foraminifer *Elphidium crispum* were shown to be photosynthetically active, but chloroplast number gradually decreased over a period of several weeks when the host was starved (Lee and Lanners 1988). Retention times for chloroplasts in most ciliated protozoa tend to be on the order of a few to several days (Stoecker et al. 1989b). However, chloroplasts appear to be quite stable in the ciliate *M. rubrum.*

Although short-lived, sequestered chloroplasts can provide a significant source of energy for protozoan survival and growth. In possibly the most extreme case, it has been speculated that the ciliate *M. rubrum* is essentially a photosynthetic organism (Crawford 1989). In other cases, the contribution of acquired chloroplasts to the "host" nutrition is less dramatic but significant. Starved individuals of chloroplast-retaining dinoflagellates and ciliates often survive longer in the light than in the dark (Stoecker et al. 1988; Fields and Rhodes 1991). Stoecker et al. (1988) estimated that photosynthesis by chloro-plasts in the ciliate *Laboea strobila* might contribute a third of the ciliate's daily carbon budget. Such contributions might be extremely important for species in competition with purely heterotrophic protists that depend solely on the capture of prey for their nutrition.

COMPLEX NUTRITIONAL MODES IN SMALL CELLS

Symbiosis and mixotrophy within the ocean plankton are largely conducted by prokaryotic and protistan organisms. An amazing feature of these relationships and behaviors is that they often combine very different metabolic processes (e.g., chemosynthesis or photosynthesis with phagocytosis and digestion). While there may be no a priori reason to segregate these processes, one can envision potential complications in maintaining both nutritional modes oper-ating in close proximity to one another. Indeed, many symbiotic relationships and mixotrophic protists tend to spatially or temporally separate these two processes.

In the case of symbioses between independently functioning individuals of different species, the integrity of both partners may be maintained by establishing extracellular associations (e.g., ectocommensalisms). Intracellular symbionts, however, must evade the digestion machinery of their hosts. Prey phagocytosis and digestion in food vacuoles must take place without affecting vacuoles containing the symbionts. In some of these species, this task is accomplished by physically segregating the processes. For example, enzymatic activities associated with catabolism and anabolism differ in the extracapsular and intracapsular regions of the spumellarian radiolarian *Thalassicolla nucleata* (Anderson and Botfield 1983). These differences indicate some degree of physical separation between prey capture/digestion and symbiont photosynthesis.

In other species, both processes take place in close physical proximity, and other adaptations are employed to prevent symbionts from being digested. As described previously, a key to *Paramecium–Chlorella* symbioses (and presumably others) appears to lie in the nature of the phagosomal membrane that encloses the symbiotic algae at the time of ingestion. This membrane is somehow altered in such a way that vacuoles containing symbionts do not fuse with lysosomes, structures that contain digestive enzymes and lead to the formation of food vacuoles.

The situation with chloroplast-retaining protozoa seems even more difficult to comprehend. Ingestion of appropriate prey by these heterotrophs is followed by *partial* digestion but retention of the chloroplasts in a functional state. Exactly how these structures are salvaged from the digestive process is not yet known.

Compartmentalization of photosynthetic and heterotrophic processes may be quite problematic in phagotrophic algae because these species are often smaller than symbiont-bearing protozoa. Plastid digestion presumably is not an issue because their function has been intimately integrated into the metabolism of the protist. However, phagotrophic algae carry out the catabolic processes of prey digestion in close proximity to photosynthesis. There is some indication that melding these processes in a small cell may be difficult because at least some of these species tend to switch between phagotrophy and phototrophy with relatively little overlap of the two processes. Chloroplast activation and photosynthesis in these species are maximized when little phagotrophy is taking place and vice versa (Monroy and Schwartzbach 1984; Brandt and Winter 1987; Sanders et al. 1990).

Presumably all phototrophic protists arose via the ingestion of a phototrophic prey and eventual reduction to an organelle, the chloroplast (Margulis 1981). If this is true, then one question that immediately arises is, Why would such a protist lose the ability to phagocytize prey? One possible explanation has been noted by Raven (1997), who speculated that there should be an energetic cost for a cell to maintain the cellular machinery for both photosynthesis and phagotrophy. This might explain why all algae do not still possess the ability to consume prey, but it would not explain, for example, why all

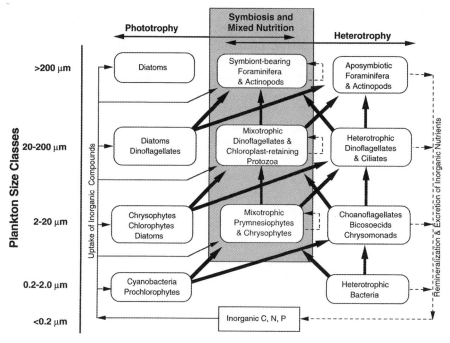

Figure 6. Conceptual model of microbial food webs with photosymbiosis and mixed photo-trophic/heterotrophic nutrition juxtaposed between the traditional roles of primary producers (phototrophy) and consumers (heterotrophy). The diagram is representative, not exhaustive of all taxa involved. Uptake of inorganic nutrients is depicted by thin solid lines, predation by thick solid lines, and recycling of remineralized nutrients by dashed lines. Suspected recycling of nutrients within symbiotic/mixotrophic species is shown by dashed lines leaving and returning to the same box. (Modified from Caron and Finlay 1994.)

protozoa do not possess the ability to retain chloroplasts from ingested algae or why all heterotrophs do not form symbioses with photosynthetic organisms. Clearly, it will be necessary to gain a better knowledge of the unique genetic and biochemical characteristics of symbiotic and mixotrophic species to fully understand these limitations.

This brief overview of symbiosis and mixotrophy by pelagic microorganisms demonstrates the enormous nutritional versatility of prokaryote and protistan species with respect to the processes of heterotrophy and phototrophy. Whether it is mixotrophy within a single organism or a mutualistic association between a phototrophic and heterotrophic species, these behaviors do not fit our conventional definitions of "plant" and "animal." Those biases must give way to the acceptance of a continuum of nutritional modes ranging from absolute phototrophy to absolute heterotrophy (Figure 5), and models of microbial food web structure and dynamics must begin to incorporate these processes (Figure 6).

SUMMARY

1. Nutrition among pelagic unicellular organisms often transcends traditional notions of autotrophy (photosynthesis and chemosynthesis) and heterotrophy; these trophic modes are not mutually exclusive.

2. Many microbes enter into symbiotic relationships whereby two species possessing very different physiological abilities live in close physical proximity (often one inside the other). These associations support a variety of complex (and, in many cases, still poorly understood) biochemical interactions that benefit one or both of the partners.

3. Many protistan taxa possess the ability to conduct both autotrophic (photosynthetic) and heterotrophic processes. Such "mixotrophic" nutrition ranges from prey ingestion and digestion by photosynthetic protists (phagotrophic algae) to the retention of functional chloroplasts by some heterotrophic protists (chloroplast-retaining protozoa).

4. The "nontraditional" trophic interactions of symbiosis and mixotrophy act in concert with predator–prey relationships to structure marine pelagic communities.

REFERENCES

Andersen, R. A., Brett, R. W., Potter, D., and Sexton, J. P. (1998) Phylogeny of the Eustigmatophyceae based upon 18S rDNA, with emphasis on *Nannochloropsis*. *Protist*. 149:61–74.

Anderson, O. R. (1983) *Radiolaria*. Springer-Verlag, New York.

Anderson, O. R., and Botfield, M. (1983) Biochemical and fine structure evidence for cellular specialization in a large spumellarian radiolarian *Thalassicolla nucleata*. *Mar. Biol.* 72:235–241.

Anderson, O. R., Swanberg, N. R., and Bennett, P. (1983) Assimilation of symbiont-derived photosynthesis in some solitary and colonial Radiolaria. *Mar. Biol.* 77:265–269.

Arenovski, A. L., Lim, E. L., and Caron, D. A. (1995) Mixotrophic nanoplankton in oligotrophic surface waters of the Sargasso Sea may employ phagotrophy to obtain major nutrients. *J. Plankton Res.* 17:801–820.

Bé, A. W. H. (1982) Biology of planktonic foraminifera. In T. W. Broadhead, ed., *Foraminifera: Notes for a Short Course*, Vol. 6. University of Tennessee, Knoxville, pp. 51–92.

Bé, A. W. H., Spero, H. J., and Anderson, O. R. (1982) Effects of symbiont elimination and reinfection on the life processes of the planktonic foraminifer *Globigerinoides sacculifer*. *Mar. Biol.* 70:73–86.

Bé, A. W. H., Anderson, O. R., Faber, W. W., Jr., and Caron, D. A. (1983) Sequence of morphological and cytoplasmic changes during gametogenesis in the planktonic foraminifer *Globigerinoides sacculifer* (Brady) *Micropaleontology* 29:310–325.

Bennett, S. J., Sanders, R. W., and Porter, K. G. (1990) Heterotrophic, autotrophic and mixotrophic nanoflagellates: Seasonal abundances and bacterivory in a eutrophic lake. *Limnol. Oceanogr.* 35:1821–1832.

Bernard, C., and Rassoulzadegan, F. (1994) Seasonal variations of mixotrophic ciliates in the northwest Mediterranean Sea. *Mar. Ecol. Prog. Ser.* 108:295–301.

Berninger, U.-G., Finlay, B. J., and Canter, H. M. (1986) The spatial distribution and ecology of zoochlorellae-bearing ciliates in a productive pond. *J. Protozool.* 33:557–563.

Berninger, U.-G., Caron, D. A., and Sanders, R. W. (1992) Mixotrophic algae in three ice-covered lakes of the Pocono Mountains, USA. *Freshwater Biol.* 28:263–272.

Bird, D. F., and Kalff, J. (1986) Bacterial grazing by planktonic lake algae. *Science.* 231:493–495.

Bockstahler, K. R., and Coats, D. W. (1993) Spatial and temporal aspects of mixotrophy in Chesapeake Bay dinoflagellates. *J. Eukaryotic Microbiol.* 40:49–60.

Brandt, P., and Winter, J. (1987) The influence of permanent light and of intermittent light on the reconstitution of the light-harvesting system in regreening *Euglena gracilis. Protoplasma* 136:56–62.

Bratbak, G., Heldal, M., Thingstad, T. F., Rieman, B., and Haslund, O. H. (1992) Incorporation of viruses into the budget of microbial C-transfer. A first approach. *Mar. Ecol. Prog. Ser.* 83:273–280.

Brown, R. M., Jr. (1972) Algal viruses. *Adv. Virus Res.* 17:243–277.

Cachon, J., and Cachon, M. (1987) Parasitic dinoflagellates. In F. J. R. Taylor, ed., *The Biology of Dinoflagellates.* Blackwell Scientific Publications, Oxford, pp. 571–610.

Caron, D. A., and Finlay, B. J. (1994) Protozoan links in food webs. In K. H. Hausmann and N. Ismann, eds., *Progress in Protozoology, Proceedings of the Ninth International Congress of Protozoology, Berlin 1993.* Gustav Fischer Verlag, Stuttgart, pp. 125–130.

Caron, D. A., Bé, A. W. H., and Anderson, O. R. (1982) Effects of variations in light intensity on life processes of the planktonic foraminifer *Globigerinoides sacculifer* in laboratory culture. *J. Mar. Biol. Assoc. UK* 62:435–451.

Caron, D. A., Porter, K. G., and Sanders, R. W. (1990) Carbon, nitrogen and phosphorus budgets for the mixotrophic phytoflagellate *Poterioochromonas malhamensis* (Chrysophyseae) during bacterial ingestion. *Limnol. Oceanogr.* 35:433–443.

Caron, D. A., Sanders, R. W., Lim, E. L., Marrasé, C., Amaral, L. A., Whitney, S., Aoki, R. B., and Porter, K. G. (1993) Light-dependent phagotrophy in the freshwater mixotrophic chrysophyte *Dinobryon cylindricum. Microb. Ecol.* 25:93–111.

Caron, D. A., Michaels, A. F., Swanberg, N. R., and Howse, F. A. (1995) Primary productivity by symbiont-bearing planktonic sarcodines (Acantharia, Radiolaria, Foraminifera) in surface waters near Bermuda. *J. Plankton Res.* 17:103–129.

Coats, D. W., Adam, E. J., Gallegos, C. L., and Hedrick, S. (1996) Parasitism of photosynthetic dinoflagellates in a shallow subestuary of Chesapeake Bay, USA. *Aquat. Microb. Ecol.* 11:1–9.

Crawford, D. W. (1989) *Mesodinium rubrum*: The phytoplankter that wasn't. *Mar. Ecol. Prog. Ser.* 58:161–174.

Dolan, J. R. (1992) Mixotrophy in ciliates: A review of *Chlorella* symbiosis and chloroplast retention. *Mar. Microb. Food Webs* 6:115–132.

Douglas, A. E. (1994) *Symbiotic Interactions.* Oxford University Press, Oxford.

Elbrächter, M., and Drebes, G. (1978) Life cycles, phylogeny and taxonomy of *Dissodinium* and *Pyrocystis* (Dinophyta) *Helgolander Wiss. Meeresunters.* 31:347–366.

Faber, W. W., Jr., Anderson, O. R., Lindsey, J. L., and Caron, D. A. (1988) Algal–foraminiferal symbiosis in the planktonic foraminifer *Globigerinella aequilateralis*.I. Occurrence and stability of two mutually exclusive chrysophyte endosymbionts and their ultrastructure. *J. Foramferan Res.* 18:334–343.

Fenchel, T., and Finlay, B. J. (1995) *Ecology and Evolution in Anoxic Worlds.* Oxford University Press, Oxford.

Fields, S. D., and Rhodes, R. G. (1991) Ingestion and retention of *Chroomonas* spp. (Cryptophyceae) by *Gymnodinium acidotum* (Dinophyceae) *J. Phycol.* 27:525–529.

Gaines, G., and Elbrächter, M. (1987) Heterotrophic nutrition. In F. J. R. Taylor, ed., *The Biology of Dinoflagellates*, Vol. 21. Blackwell Scientific Publications, Oxford, pp. 224–268.

Gast, R. J., and Caron, D. A. (1996) Molecular phylogeny of symbiotic dinoflagellates from Foraminifera and Radiolaria. *Mol. Biol. Evol.* 13:1192–1197.

Gastrich, M. D., Anderson, O. R., Benmayor, S. S., and Cosper, E. M. (1998) Ultrastructural analysis of viral infection in the brown-tide alga, *Aureococcus anophagefferens. Phycologia* 37:300–306.

Gordon, N., Angel, D. L., Neori, A., Kress, N., and Kimor, B. (1994) Heterotrophic dinoflagellates with symbiotic cyanobacteria and nitrogen limitation in the Gulf of Aqaba. *Mar. Ecol. Prog. Ser.* 107:83–88.

Hall, J. A., Barrett, D. P., and James, M. R. (1993) The importance of phytoflagellate, heterotrophic flagellate and cilliate grazing on bacteria and picophytoplankton sized prey in a coastal marine environment. *J. Plankton Res.* 15:1075–1086.

Havskum, H., and Hansen, A. S. (1997) Importance of pigmented and colourless nano-sized protists as grazers on nanoplankton in a phosphate-depleted Norwegian fjord and in enclosures. *Mar. Ecol. Prog. Ser.* 12:139–151.

Hemleben, C., Spindler, M., and Anderson, O. R. (1988) *Modern Planktonic Foraminifera.* Springer-Verlag, New York.

Jacobson, D. M., and Anderson, D. M. (1996) Widespread phagocytosis of ciliates and other protists by marine mixotrophic and heterotrophic thecate dinoflagellates. *J. Phycol.* 32:279–285.

Jeong, H. J., Lee, C. W., Yih, W. H., and Kim, J. S. (1997) *Fragilidium* cf. *mexicanum*, a thecate mixotrophic dinoflagellate which is prey for and a predator on co-occurring thecate heterotrophic dinoflagellate *Protoperidinium* cf. *divergens. Mar. Ecol. Prog. Ser.* 151:299–305.

Jones, R. I. (1994) Mixotrophy in planktonic protists as a spectrum of nutritional strategies. *Mar. Microb. Food Webs* 8:87–96.

Karl, D., Letelier, R., Tupes, L., Dore, J., Christian, J., and Nebel, D. (1997) The role of nitrogen fixation in biogeochemical cycling in the subtropical North Pacific Ocean. *Nature* 388:533–538.

Kimura, B., and Ishida, Y. (1986) Possible phagotrophic feeding of bacteria in a freshwater red tide Chrysophyceae *Uroglena americana*. *Bull. Jpn. Soc. Sci. Fish.* 52:697–701.

Kimura, B., and Ishida, Y. (1989) Phospholipid as a growth factor of *Uroglena americana*, a red tide Chrysophyceae in Lake Biwa. *Nippon Suisan Gakkaishi* 55:799–804.

Kremer, P., Costello, J., Kremer, J., and Canino, M. (1990) Significance of photosynthetic endosymbionts to the carbon budget of the scyphomedusa *Linuche unguiculata*. *Limnol. Oceanogr.* 35:609–624.

Kühn, S. F. (1998) Infection of *Coscinodiscus* spp. by the parasitoid nanoflagellate *Pirsonia diadema*. II. Selective infection behaviour for host species and individual host cells. *J. Plankton Res.* 20:443–454.

Lampert, W., and Sommer, U. (1997) *Limnoecology: The Ecology of Lakes and Streams.* Oxford University Press, New York.

Lee, J. J., and Lanners, E. (1988) The retention of chloroplasts by the foraminifer *Elphidium crispum. Symbiosis* 5:45–60.

Lee, J. J., Reimer, C. W., and McEnery, M. E. (1980) The identification of diatoms isolated as endosymbionts from larger foraminifera from the Gulf of Eilat (Red Sea) and the description of 2 new species, *Fragilaria shiloi* sp. nov. and *Navicula reissii* sp. nov. *Bot. Mar.* 23:41–48.

Lee, J. J., McEnery, M. E., Ter Kuile, B., Erez, J., Röttger, R., Rockwell, R. F., Faber, W. W., Jr., and Lagziel, A. (1989) Identification and distribution of endosymbiotic diatoms in larger Foraminifera. *Micropaleontology* 35:353–366.

Legrand, C., Granéli, E., and Carlsson, P. (1998) Induced phagotrophy in the photosynthetic dinoflagellate *Heterocapsa triquetra. Aquat. Microb. Ecol.* 15:65–75.

Lessard, E. J. (1991) The trophic role of heterotrophic dinoflagellates in diverse marine environments. *Mar. Microb. Food Webs* 5:49–58.

Li, A., Stoecker, D. K., Coats, D. W., and Adam, E. J. (1996) Ingestion of fluorescently labeled and phycoerythrin-containing prey by mixotrophic dinoflagellates. *Aquat. Microb. Ecol.* 10:139–147.

Lopez, E. (1979) Algal chloroplasts in the protoplasm of three species of benthic Foraminifera: Taxonomic affinity, viability and persistence. *Mar. Biol.* 53:201–211.

Mainero, J. S., and del Rio, C. M. (1985) Cheating and taking advantage in mutualistic associations. In D. H. Boucher, ed., *The Biology of Mutualism.* Oxford University Press, New York, pp. 192–216.

Maranger, R., Bird, D. F., and Price, N. M. (1998) Iron acquisition by photosynthetic marine phytoplankton from ingested bacteria. *Nature.* 396:248–251.

Margulis, L. (1981) *Symbiosis in Cell Evolution.* W. H. Freeman, San Francisco.

Margulis, L., Corliss, J. O., Melkonian, M., Chapman, D. J., and McKhann, H. I. (1990) *Handbook of Protoctista.* Jones and Bartlett, Boston.

McFall-Ngai, M. (1998) The development of cooperative associations between animals and bacteria: Establishing détente among domains. *Am. Zool.* 38:3–18.

McKenzie, C. H., Deibel, D., Paranjape, M. A., and Thompson, R. J. (1995) The marine mixotroph *Dinobryon balticum* (Chrysophyceae): Phagotrophy and survival in a cold ocean. *J. Phycol.* 31:19–24.

Monroy, A. F., and Schwartzbach, S. D. (1984) Catabolite repression of chloroplast development in *Euglena*. *Proc. Natl. Acad. Sci. USA* 81:2786–2790.

Nygaard, K., and Tobiesen, A. (1993) Bacterivory in algae: A survival strategy during nutrient limitation. *Limnol. Oceanogr.* 38:273–279.

Patterson, D. J., and Dürrschmidt, M. (1987) Selective retention of chloroplasts by algivorous Heliozoa: Fortuitous chloroplast symbiosis? *Eur. J. Protistol.* 23:51–55.

Provasoli, L., and Pintner, I. J. (1953) Ecological implications of in vitro nutritional requirements of algal flagellates. *Ann. NY Acad. Sci.* 56:839–851.

Putt, M. (1991) Abundance, chlorophyll content and photosynthetic rates of ciliates in the Nordic Seas during summer. *Deep-Sea Res.* 37:1713–1731.

Raven, J. A. (1997) Phagotrophy in phototrophs. *Limnol. Oceanogr.* 42:198–205.

Reisser, W. (1986) Endosymbiotic associations of freshwater protozoa and algae. In J. O. Corliss and D. J. Patterson, eds., *Progress in Protistology*, Vol. 1. Biopress, Bristol, pp. 195–214.

Reisser, W. (1991) Ciliophora as microhabitats of different green algae species: Model systems for an ecological concept of symbiosis formation. *Mar. Microb. Food Webs* 5:75–80.

Rogerson, A., Polne-Fuller, M., Trench, R. K., and Gibor, A. (1989) A laboratory-induced association between the marine amoeba *Trichosphaerium* AM-I-7 and the dinoflagellate *Symbiodinium* #8. *Symbiosis* 7:229–241.

Sanders, R. W., and Porter, K. G. (1988) Phagotrophic phytoflagellates. *Adv. Microb. Ecol.* 10:167–192.

Sanders, R. W., Porter, K. G., and Caron, D. A. (1990) Relationship between phototrophy and phagotrophy in the mixotrophic chrysophyte *Poterioochromonas malhamensis*. *Microb. Ecol.* 19:97–109.

Schnepf, E., and Elbrächter, M. (1992) Nutritional strategies in dinoflagellates. *Eur. J. Protistol.* 28:3–24.

Schnepf, E., and Schweikert, M. (1997) *Pirsonia*, phagotrophic nanoflagellates incertae sedis, feeding on marine diatoms: Attachment, fine structure and taxonomy. *Arch. Protistenkd.* 147:361–371.

Skovgaard, A. (1996) Mixotrophy in *Fragilidium subglobosum* (Dinophyceae): Growth and grazing responses as functions of light intensity. *Mar. Ecol. Prog. Ser.* 143:247–253.

Skovgaard, A. (1998) Role chloroplast retention in a marine dinoflagellate. *Aquat. Microb. Ecol.* 15:293–301.

Stoecker, D., Michaels, A. E., and Davis, L. H. (1987) Large proportion of marine planktonic ciliates found to contain functional chloroplasts. *Nature.* 326:790–792.

Stoecker, D. K., Silver, M. W., Michaels, A. E., and Davis, L. H. (1988) Obligate mixotrophy in *Laboea strobila*, a ciliate which retains chloroplasts. *Mar. Biol.* 99:415–423.

Stoecker, D., Taniguchi, A., and Michaels, A. E. (1989a) Abundance of autotrophic, mixotrophic and heterotrophic planktonic ciliates in shelf and slope waters. *Mar. Ecol. Prog. Ser.* 50:241–254.

Stoecker, D. K., Silver, M. W., Michaels, A. E., and Davis, L. H. (1989b) Enslavement of algal chloroplasts by four *Strombidium* spp. (Ciliophora, Oligotrichida) *Mar. Microb. Food Webs.* 3:79–100.

Stoecker, D. K., Swanberg, N., and Tyler, S. (1989c) Oceanic mixotrophic flatworms. *Mar. Ecol. Prog. Ser.* 58:41–51.

Stoecker, D. K., Putt, M., Davis, L. H., and Michaels, A. E. (1991) Photosynthesis in *Mesodinium rubrum*: Species-specific measurements and comparison to community rates. *Mar. Ecol. Prog. Ser.* 73:245–252.

Stoecker, R. K., Gustafson, D. E., and Verity, P. G. (1996) Micro- and mesoprotozoo-plankton at 140°W in the equatorial Pacific: Heterotrophs and mixotrophs. *Aquat. Microb. Ecol.* 10:273–282.

Théodorides, J. (1987) Parasitology of marine zooplankton. *Adv. Mar. Biol.* 25:117–177.

Thingstad, T. F., Heldal, M., Bratbak, G., and Dundas, I. (1993) Are viruses important partners in pelagic food webs? *Trends Ecol. Evol.* 8:209–213.

Thingstad, T. F., Havskum, H., Garde, K., and Riemann, B. (1996) On the strategy of "eating your competitor': A mathematical analysis of algal mixotrophy. *Ecology.* 77:2108–2118.

Tillman, U. (1998) Phagotrophy by a plastidic haptophyte, *Prymnesium patelliferum. Aquat. Microb. Ecol.* 14:155–160.

Tranvik, L. J., Porter, K. G., and Sieburth, J. M. (1989) Occurrence of bacterivory in *Cryptomonas*, a common freshwater phytoplankter. *Oecologia.* 78:473–476.

Trench, R. K. (1975) Of "leaves that crawl": Functional chloroplasts in animal cells. *Soc. Exp. Biol.* 29:229–265.

Van Etten, J. L., Lane, L. C., and Meints, R. H. (1991) Viruses and viruslike particles of eukaryotic algae. *Microbiol. Rev.* 55:586–620.

Veen, A. (1991) Ecophysiological studies on the phagotrophic phytoflagellate *Dinobryon divergens* Imhof. Ph.D. thesis, Universiteit van Amsterdam.

Villareal, T. A. (1989) Division cycles in the nitrogen-fixing *Rhizosolenia* (Bacil-lariophyceae)–*Richelia* (Nostocaceae) symbiosis. *Br. Phycol. J.* 24:357–365.

Villareal, T. A. (1994) Widespread occurrence of the *Hemiaulus*–cyanobacterial symbio-sis in the southwest North Atlantic ocean. *Bull. Mar. Sci.* 54:1–7.

Waterbury, J. B., and Valois, F. W. (1993) Resistance to co-occurring phages enables marine *Synechococcus* communities to coexist with cyanophages abundant in sea-water. *Appl. Environ. Microbiol.* 59:3393–3399.

Wilkinson, C. R. (1987) Significance of microbial symbionts in sponge evolution and ecology. *Symbiosis.* 4:135–146.

INDEX